Encyclopedia
of Animal Behavior

 # Encyclopedia
# of Animal Behavior

Volume 1: A–C

Edited by
Marc Bekoff

Foreword by
Jane Goodall

GREENWOOD PRESS
Westport, Connecticut • London

Library of Congress Cataloging-in-Publication Data

Encyclopedia of animal behavior / edited by Marc Bekoff; foreword by Jane Goodall.
    p. cm
    Includes bibliographical references (p.)
    ISBN 0-313-32745-9 (set : alk. paper)—ISBN 0-313-32746-7 (vol. 1 : alk. paper)—
    ISBN 0-313-32747-5 (vol. 2 : alk. paper)—ISBN 0-313-33294-0 (vol. 3 : alk. paper)
    1. Animal behavior—Encyclopedias. I. Bekoff, Marc.
QL750.3.E53   2004
591.5′03—dc22   2004056073

British Library Cataloguing in Publication Data is available.

Library of Congress Catalog Card Number: 2004056073
ISBN: 0-313-32745-9 (set code)
        0-313-32746-7 (Vol. 1)
        0-313-32747-5 (Vol. 2)
        0-313-33294-0 (Vol. 3)

First published in 2004

Greenwood Press, 88 Post Road West, Westport, CT 06881
An imprint of Greenwood Publishing Group, Inc.
www.greenwood.com

Printed in the United States of America

The paper used in this book complies with the
Permanent Paper Standard issued by the National
Information Standards Organization (Z39.48-1984).

10 9 8 7 6 5 4 3 2

*These volumes are dedicated to the memory
of one of the most amazing people
I have ever known—my mother Beatrice*

■ Contents

■ Foreword

This is an exciting book. I just wish that it had been available for me when I was a student. My family had 12 volumes of a general information encyclopedia when I was a child and we used it all the time. I loved books and I loved reading. We were not allowed to read at the table during meals—the exception was if someone wanted to look something up in one of those wonderful volumes. Encyclopedias have always been very special to me. You look something up, then your eye is captured by an entry before or following the one you were after. You learn more than you intended. All that information, all that inspiration, all there for you to find!

But when I was a child, back in the 1940s, there was nothing much about animal behavior in our encyclopedias—people did not know that much about the subject. I was given a book called *The Miracle of Life*. It was not written for children, but it was my most valued possession—after *Doctor Doolittle*. It had sections on animals of all kinds in many countries, and it was illustrated with black and white photographs and careful line drawings—showing, for example, all kinds of tongues, feet, mouths, and so on that were adapted for different purposes. I still have it—and when I look at it now, I am amazed by how much we have learned since it was written.

This *Encyclopedia of Animal Behavior* presents up-to-date information and numerous resources, compiled by experts who are highly qualified to talk about the different topics covered in its pages. Entries have been drawn from many different disciplines such as biology, psychology, anthropology, entomology, philosophy, veterinary medicine, and more. The essays have been well-edited and written in such a way that even complex subjects are easy to understand so that the information is available to a wide audience.

One of the most important aspects of this encyclopedia is the exhaustive nature of the information it presents. Another is the engaging photographs and illustrations it offers. We can learn about all aspects of animal behavior from social organization, mating strategies, dominance and play, to the latest information on emotions, awareness, consciousness, and empathy. It details the behavior of a wide range of different species across the animal kingdom, from butterflies to chimpanzees, snakes to elephants, wild animals to domestic ones such as dogs and cats. And it discusses animals both as individuals and as species. Readers will be fascinated and filled with wonder as they learn more and more about the amazing animal kingdom. Their imaginations will be stimulated, for even as they discover the answer to one question, they will realize that that leads to other questions. And they will learn how much we still have to learn.

Several essays discuss the various career options for those wanting to study animal behavior, behavioral ecology, or some other field that will enable the student to combine an interest in animals with earning a living. When I was a child, very few of these options were available, especially to girls, but today there are so many choices that it can be confusing. I get many letters from students who wonder whether they should study ecology, ethology, anthropology, biology—or one of the other "ologies"! For these students, the information on careers in the field of animal behavior will be invaluable. It will also be encouraging for those who are advised by teachers or parents to abandon their love of, or interest in, animals because there is no "real" money to be made that way. In fact, as most of us have found, doing something that we really and truly enjoy and believe in is worth much more than a high salary. There is so much more to discover about animals and the environments they live in.

Fun to read, informative and inspiring, the *Encyclopedia of Animal Behavior* is a book that should be on the shelves of every school and university library, and in every home.

Jane Goodall
Bournemouth, U.K.

■ Preface

When Kevin Downing called to ask if I would undertake the job of editing this encyclopedia, my first response was to say, "No thanks, I'm not that insane!" But his phone call clearly triggered some neurons in my brain, and I realized that I just couldn't say no. I had already edited an *Encyclopedia of Animal Rights and Animal Welfare* for Greenwood (1998) and promised myself that I would never again undertake such a project. But that night I came to realize that editing this encyclopedia was a dream come true, for I love to learn about the behavior of other animals and wanted to share this information with the many people who also yearn to know what there is to know about animals—"What is it like to be a dog or a wolf or a spider?" So, in a moment of weakness, a few days later, while sipping a cup of coffee with Kevin at the Boulder Bookstore, I said, "Sure, I'll do it." And I've not regretted this decision at all.

What You Will Find in This Encyclopedia

The encyclopedia is the most complete and comprehensive collection of original essays on the topic of animal behavior. There are no rivals in its breadth, depth, or scope. Thus, as you might expect, there is an amazing amount of information in the essays in this encyclopedia that will appeal to students of animal behavior, animal lovers, and animal advocates alike. Of course, these groups are not mutually exclusive. Essays range in length from about 300–7,000 words, and there are many illustrations.

As you read the excellent, "user friendly," and up-to-date original essays in this volume, written by an international group of well-respected animal behavior scholars and up-and-coming researchers from many different disciplines including biology, psychology, anthropology, sociology, philosophy, veterinary medicine, literature, law, and religious studies, you will discover that there are many people who have been curious and fortunate enough to follow their dreams to pursue a career in the study of animal behavior and who took the time to write their essays because they wanted to share their zeal and knowledge with you. To begin organizing this encyclopedia and collecting essays, I circulated information about this encyclopedia to numerous professional groups. Almost instantaneously I had more than 200 people eager to share their knowledge with you. As the list of entries grew and word about the encyclopedia spread, people contacted me and asked if they could contribute. They were excited to learn that this encyclopedia was in the works. I feel lucky that so many colleagues wanted to write for this volume—and so, too, are you. The contributors study many different species and a broad range of topics, and much information is provided in the numerous references that are provided at the end of each essay. There also is a list of international organizations, Web sites, and suggested resources to help you find your way.

It is essential to learn more and more about the lives of other animals because learning and knowledge lead to an understanding of animals as individuals and members of a given species, and understanding leads in turn to appreciation and respect for the awesome and mysterious animal beings with whom we share Earth. After reading these wide-ranging essays, in which a broad array of species and diverse behavior patterns are discussed, you will know a great deal about how animals spend their time and energy and how and why they make the decisions they make from day to day. You will come to a better understanding of the ways in which various behavior patterns evolve and how ecological variables can influence behavior. Comparative approaches to the study allow you to see how different species and individuals solve the myriad of problems with which they are faced. You will also discover that detailed information on behavior is essential for many other fields of study such as neurobiology, physiology, anatomy, population biology, conservation, genetics, and evolution. In the absence of behavioral data we often cannot make sense of why individuals hunt or mate they way they do or how to reintroduce species back into the areas in which they once thrived or move them to areas where they will not be bothered by humans. The rapidly evolving field of conservation behavior is an indicator of how important it is to know about animal behavior, behavioral ecology, individual and species differences in behavior, and the cultural traditions of animal societies when trying, for example, to restore ecosystems or to help imperiled or endangered species. This information allows us to be proactive—preventing problems before they start—rather than reactive in our attempts to preserve biodiversity.

The essays in this encyclopedia also consider different levels of analysis of a wide variety of behavioral phenomena, ranging from molecular approaches to the study of behavior to studies of individuals, populations, species, ecosystems, and culture. Topics range from stressing theoretical issues and models to hands-on and practical and applied aspects of animal behavior, including advice on how to live with our companion dogs when behavioral problems arise. There also is much information on careers for those who choose to study animal behavior as their profession.

Who were the people responsible for generating new and important concepts and theories, and how did these ideas develop? It is important to understand and to appreciate the history of animal behavior, so there also are a number of excellent essays on the history of this broad and interdisciplinary field, including the original lectures presented by Konrad Lorenz, Karl von Frisch, and Niko Tinbergen when they shared the Nobel Prize in 1973. There is also an essay on the contributions of Charles Darwin to the study of animal behavior and an essay about William D. Hamilton III's classic research on the genetics and evolution of behavior via kin selection. Often, it is very difficult to assign an exact date to the origin of many seminal ideas because it can be very subjective to say who did what first. However, for those interested in more details about the history of animal behavior, much information is contained in the references that follow each essay and the books listed in "Suggested Resources in the Study of Animal Behavior."

The exciting and remarkable essays in this *Encyclopedia* will provide you with a rich survey of current animal behavior study. Although the vast majority of essays do not provide a point of view and are neutral, a small number of more personal and provocative essays whose authors feel strongly about their subjects have been included—whether Jane Goodall on the necessity to provide sanctuaries for chimpanzees who have been injured both physically and emotionally by humans, Camilla Fox on the behavioral and welfare implications of trapping wildlife, Rupert Sheldrake on his belief in telepathic communication between pets and their human guardians, Jean Greek on her concern that people still do not believe that animals have feelings, and a few others.

How to Use This Encyclopedia

The essays in this encyclopedia are organized as follows. There are major sections arranged alphabetically in which long essays are arranged by topic (and also alphabetically), and there also are long essays that stand alone and short sidebars that are more focused on a specific topic or species. A number of essays could easily have been placed in more than one category, so at the end of each essay there are detailed cross-references to other related essays. These cross-references will be very helpful to you for finding your way around the numerous essays in this encyclopedia.

Acknowledgments: Giving Thanks

A work of such enormous scope requires a lot of cooperation among different people. Anne Thompson provided essential editorial support and encouragement throughout the process of soliciting, receiving, editing, and getting authors' responses to their essays. She also helped to organize the authors and topics into a variety of spreadsheets, a task that would have driven me crazy. Anne was always there to chat, and it's safe to say that without her unwavering support this encyclopedia would never have seen the light of day. Jan Braeunig was a wonderful and cooperative copyeditor, and Elizabeth Kincaid did a fine job of finding many of the photographs. Jan Nystrom, who loves animals as much as I do, also helped with editing and cross-referencing, and suggested many topics that filled out this encyclopedia to make it more comprehensive and complete. For weeks on end it seemed as if Jan always had something to add when we chatted about this project. When I heard her say something like, "Have you thought about entries on this topic?" I mumbled, "uh oh," but her suggestions were always excellent. Jan's enthusiasm and commitment to excellence for this project cannot be overestimated. Jan also selflessly provided much personal support, and I am grateful for her sparkling voice, warm smile, and lightness, along with her compassionate and morale-boosting words of encouragement. Emily Birch also helped me along, but luckily for her she had far less to do with this project and my sanity than she did for the *Encyclopedia of Animal Rights and Animal Welfare* for which she was responsible.

Last, but surely not least, I want to thank all my colleagues who took time out of their busy lives to write exemplary essays. These people share my enthusiasm for learning as much as we can about the lives of other animals, and we all should be grateful for their efforts.

Marc Bekoff
Boulder, Colorado

■ Introduction

What Is It *Like* to Be a Fox? What Does It *Feel Like* to Be a Fox?

My parents love to recall many stories about my life-long interest in animals. My father remembers, with a wide smile, that on a ski trip when I was 6 years old, I asked him what a red fox was feeling as he merrily crossed our path as we traversed a frozen lake. On a recent trip to visit my parents in Florida, my father told me that I was in awe of the magnificence of the fox's red coat and white-tipped tail and lost track of where I was skiing! And, he well remembers that, when I was 4 years old, I yelled at a man for yelling at his dog, and the man chased my father! These two events etched an indelible impression in my heart and in my head. I wanted to study animals when I grew up. My parents have told me that I always "minded animals," that I always wanted to know what they were thinking and feeling— "What is it *like* to be a dog or a cat or a mouse or an ant?"—"What does it *feel like* to be a dog or a cat or a mouse or an ant?" And to this day, learning about the behavior of animals— all animals—has been my passion. When I study coyotes, "I am coyote," and when I study steller's jays, "I am jay." When I study dogs, "I am dog," but I choose not to experience first-hand the odors, the olfactory symphonies, that make up what Paul Auster refers to as their "nasal paradise" in his book *Timbuktu*.

I am lucky to have been able to have studied many different animals in the western United States and in Antarctica. I am also fortunate to live in the mountains outside of Boulder, Colorado, and count among my best friends such animals as the many dogs with whom I have shared my home, red foxes (wily canids who often sit outside my office and watch me write about them), cougars (majestic carnivores whom I have almost tripped over), black bears (magnificent individuals who visit my house), chipmunks, squirrels, robins, jays, various lizards, and a whole host of fascinating spiders and insects. I love to see them, smell them, listen to the cacophony of sounds they produce, and take them into my heart. The loss of any of these symbols of their presence would be a marked absence in my daily life.

I have always been curious about questions such as why dogs play the way they do and whether we can learn about fairness, forgiveness, trust, and morality by studying the details of how individuals "converse" during play and negotiate cooperative playful interactions; why dogs and other animals spend a good deal of time sniffing various parts of others' bodies and odors that make me cringe; how or if animals know who they are; how animals communicate using sounds that I cannot hear; what are the relative contributions of genes (nature) and the environment (nurture) to various behavioral phenomena; why some animals give up the opportunity to mate so they can help raise the young of other group members; what animals know about the states of mind and emotional states of others; if they empathize with one another; why Adélie penguins steal rocks from each other's nests and why these incredible birds wait for another penguin to jump into water in which predatory leopard seals lurk before they themselves jump in; how animals build complex structures such as nests and dams that might require generations to complete; the truth about mating praying mantids—do females really eat their mates; why fish can save energy when they migrate upstream; how and why animals experience uncertainty or self-doubt and refuse to perform tasks when they know they will fail them; how a fruit fly uses its tiny brain to decide if a stimulus in its environment is important and whether or not it should respond to it; how large groups of animals self-organize into a coordinated unit; why a female elephant would leave her favorite toy, a tire, as a tribute to her recently deceased elephant friend; and how animals make complex and rapid choices on the run or on the fly in the amazingly diverse situations in which they find themselves.

The Growth of the Science of Animal Behavior

The study of animal behavior has burgeoned over the past four decades. People worldwide are interested in the behavior of animals because knowledge about animals enriches their lives. There are many more professional journals in animal behavior and behavioral ecology now than 30–40 years ago, and many universities offer undergraduate and advanced degrees in the behavioral sciences. Videos and movies about animals abound. Many people want to remain connected or to reconnect with animals. Our brains are not all that different from those of our ancestors who were more connected to the animals with whom they shared their habitats. Thus, our "old brains" seem to drive us to keep in touch with animals and with nature in general. It isn't natural to be alienated from other beings, and it feels good to interact with them and to know that they are "out there" doing what comes naturally to them.

In 1973, a most exciting and thoroughly unexpected event occurred when Konrad Lorenz, Niko Tinbergen, and Karl von Frisch won the Nobel Prize for Physiology or Medicine for their pioneering work in animal behavior. (Their Nobel Lectures are reprinted in this encyclopedia.) Lorenz, Tinbergen, and von Frisch are called "ethologists," a word that often is reserved for those researchers who are concerned with the evolution or ecology of behavior and who also conduct fieldwork. Lorenz stressed that behavior is something that an animal "has" as well as what he or she "does," and is a phenotype (trait) on which natural selection can act. Nowadays, ethological research is also conducted on captive animals (as was most of Lorenz's research), and for many people the terms "ethology" and "animal behavior" have become synonymous. Some of the essays in this volume show that there are differences in the approaches of biologists and psychologists concerning the types of questions they ask and how they go about studying them [**Sociobiology**, **Levels of Analysis in Animal Behavior**, **Comparative Psychology**, **History of Animal Behavior Studies**].

Winning the Nobel Prize was a most amazing feat for researchers who studied such phenomena as imprinting in geese, homing in wasps, hunting by foxes, and dancing in bees, and some scientists who conducted biomedical research were miffed that such frivolous pursuits merited the most prestigious award, what is called "the prize," for scientific research. And, these three men were also having fun doing their ground-breaking research, and in many scientific circles this was not acceptable. Lorenz has been filmed donning a fox coat and hopping along the ground to see how geese would respond to him! I remember meeting Lorenz at an ethological conference held in Parma, Italy, and his passion and enthusiasm were incredibly contagious. For hours he never repeated a story of the animals with whom he had shared his home. He clearly loved what he did and loved his animal friends who brought so much to his life.

Behavioral Variation between and within Species

Not only are there differences in behavior *between* species (called *inter*specific variation), but also there are marked individual differences *within* species (called *intra*specific variation). These differences make for exciting and informative research concerning, for example, why wolves and dogs differ and why even littermates and siblings may differ from one another. Many of the coyotes whom I studied in the Grand Teton National Park in Wyoming lived in packs, but just down the road coyotes lived either alone or as mated pairs. Thus, making general statements that "the coyote behaves this way or that" is very misleading because "the coyote" does not really exist. The same is true for tool use in chimpanzees and orangutans. Not all of these great apes use tools, and it is challenging to discover why this is so. Intraspecific variation in behavior has been observed in many animals including insects. Lumping all members of a species into one category can be very misleading. A bee is not a bee is not a bee, just as a person is not a person is not a person. Humans and other animals are individuals.

Charles Darwin and Evolutionary Continuity

Many people are also interested in how human animals differ from nonhuman animals. Charles Darwin, considered by many people to the most influential biologist to have ever lived, argued that in many situations the differences among animals are differences in *degree* rather than differences in *kind*. He maintained that there is evolutionary continuity so that it is unlikely, for example, that only humans use tools or have culture. We now know that individuals of many species use tools, have culture, are conscious and have a sense of self, can reason, can draw, can self-medicate, and show very complex patterns of communication that rival what we call "language." So, Darwin was correct. We can learn much about humans by studying the roots of human behavior in nonhumans, and often the differences are not as stark as we think they are. This is not to say that humans are not unique, but rather to say that all animals are unique and that we can learn a lot by using what is called the "comparative" method in which different species are studied with an eye (or nose or ear) toward learning about why they do the things they do in their own particular ways.

Human—Anthropogenic—Effects on Animal Behavior

It is a fact that human influences, also called *anthropogenic* effects, are rampant. We are literally here, there, and everywhere. Humans make a difference in the lives of just about all animals on Earth. We are a powerful and dominating force in Nature, and we are an integral part of the innumerable webs of Nature. And along with this ubiquitous presence come deep responsibilities to step lightly into the lives of other animals. In his book *Prey,* the novelist Michael Crichton writes of "the obstinate egotism that is a hallmark of human interaction with the environment."

Not only do we influence the lives of other animals in an immediate sense, but we also can effect long-lasting and enduring changes in their behavior. A number of essays make this point [**Human (Anthropogenic) Effects on Animal Behavior**, **Pollution and Behavior**, **Edge Effects and Behavior**, **The Effects of Roads and Trails on Animal Movement**], and daily I receive information about yet a new way in which we influence the behavior of other animals. Global warming is influencing the distribution and behavior of animals and resources such as food, water, and resting spots. It has been predicted that between 15% and 37% of species could go extinct between now and 2050 due to global warming. Traditional human processional rituals in Italy can delay the reproductive behavior of local snakes. Trophy hunting is reducing the average size of horns among male bighorn sheep because large rams with big horns are selectively picked off. Selective hunting influences mating behavior such that there is less head-butting among males for access to rams, and there might even be an influence on population genetics among these mountain monarchs. Humans can rapidly change the feeding habits of bears who live around dumpsters so that they become active during the night rather than during the day to avoid humans, and these bears become obese and lazy. Fast food makes them fat. They also enter dens later in the fall and remain in them for shorter periods of time than do bears who do not forage at dumps. Hormones from cattle feedlots can demasculanize males and defeminize wild fish. Fishing can induce sex changes in fish and animals such as cougars, coyotes, foxes; and deer can become so habituated to humans that, rather than flee from us, they become bold and curious and intrude into our neighborhoods. Often researchers inadvertently harm the very animals they want to study. Knowledge of how we affect the behavior of animals can help us make more informed and intelligent choices about how we interfere in their lives.

Understanding and Appreciating the Worlds of Other Animals: The Importance of Curiosity, Open-Mindedness, Patience, and Perseverance

What we learn about other animals can improve their well-being and also ours. The information that we gather about their cognitive skills, their levels of intelligence, and their emotional lives—their passionate nature—informs us that individuals of many species are not robots or automatons, but rather thinking and feeling beings. If they were merely robot-like, why would their behavior fascinate us, why would we bond with them the way we do? This information about their behavior has extremely important practical applications because we can use it when we try to improve individuals' lives after we may have made their lives miserable and have made them dependent on our good will for their very survival. The essay by Jane Goodall on **Sanctuaries** is a wonderful example of what we can do to aid helpless animals, in this case, chimpanzees, recover from severe human-induced trauma.

What is so exciting about this encyclopedia is that numerous "hot topics" are covered that are currently being studied in many different research programs. Much new information is accumulating that shows just how fascinating and complex animal behavior can be. Fish show complex patterns of culture and social cognition, and most likely experience pain. Recent research has shown that fish respond to the pain reliever morphine and that pain-related behaviors are not simple reflexes. Domestic fowl can control how much sperm they produce depending on the promiscuity of a female. Chickens can recognize and remember more than 100 other chickens in their social pecking order. Many individuals show distinct personalities and idiosyncratic quirks, just as humans do. There are extroverts, introverts, agreeable individuals, and neurotic animals. And shy laboratory rats might not live as long as more adventurous rats. It is thought that stress might cause premature aging. Chimpanzees can remember how to count 3 years after they last performed a task that required them to count, and a seal showed that he could remember the concept of "sameness" after a 12-year period. Two elephants, Shirley and Jenni, remembered one another when they were inadvertently reunited after being apart for 20 years. And laboratory rats can survive in the wild and retain some of their wild instincts after many years of laboratory life for their ancestors! They find food, water, hiding holes and establish dominance hierarchies as do their wild relatives. It is difficult to take "the wild" out of an animal.

On the lighter side of things, fish and snakes appear to communicate by flatulating. What a good and economical use of a natural bodily function! Even Aristotle took a break from serious philosophizing and was concerned with animal flatulence. In his *History of*

Animals, a veritable gold mine of natural history about a wide variety of animals, he noted that the "wind" that lions discharge is very pungent. However, he did not postulate that it was used to communicate with other lions! And animals are not immune from rare natural events. Captive hamadryas baboons have been observed to show a reduction of rates of locomotion and threat behavior when there was a solar eclipse. And howler monkeys showed a 42% decrease in population size and major social disorganization after hurricane Iris destroyed the forest in which they lived in southern Belize in October 2001.

The list of new and fascinating discoveries is endless. Solid scientific data, stories, anecdotes, and myths and lore are all needed as we attempt to learn as much as we can about animal behavior. Information from dog parks, field sites, and facilities where animals are kept can all be used to learn about animals. Detailed descriptions of behavior patterns, careful observations, and ethically justified experiments that do not harm the animals in whom we are interested are all important components of a comprehensive approach to animal behavior. Often, when we perform research that is invasive [**Human (Anthropogenic) Effects on Animal Behavior**] we are unable to answer the very questions in which we are interested. Often animals are stressed by our mere presence so we cannot truly study their more natural patterns of behaviors. My colleagues and I believe that this is a major problem that needs to be studied and understood so that the data we collect are as reliable as possible, and the questions in which we are interested are answered with as little ambiguity as possible.

Animals can do amazing things and accomplish incredible feats, but sometimes they do not do what we ask them to do. They have their own points of view, and on occasion they express them freely. An individual might not be motivated to do something because she is tired, not hungry or thirsty, or perhaps she just wants to be left alone. It is also possible that we are not sensitive to the sensory worlds of the animals and that we are asking them to respond to a stimulus to which they are not sensitive—a sound that is outside of their range of hearing, a color that they cannot see, or an odor that they cannot perceive. The sensory world of many animals is quite different among different species and also varies from our own.

One important lesson that I emphasize in my classes is that "*does not* does not mean *cannot*." Just because an animal does not do something does not mean that he or she cannot do a particular task. A wolf might choose not to chase an elk, but this does not mean that he cannot do this. A robin might not learn to discriminate friend from foe, but this does not mean that she cannot do this. We need to discover why this is so and why often individuals make the choices that they do—and among these choices is the choice to not do anything. Not to do something is to do something. Not to decide is to decide.

Humans—researchers and nonresearchers alike—often try to package nature and to sanitize and to simplify the behavior of other animals. Sometimes simple answers to complex questions suffice, and at other times they do not. Experts can disagree, and this is good for science in general and for the study of animal behavior in particular. Disagreements fuel future research for curious minds. Just when we think we know all there is to know, we learn that this is not so. Saying, "I don't know," is one of the best phrases that a researcher can utter because admitting that there are mysteries still to be uncovered and acknowledging disagreements can also fuel future inquiries. The award-winning poet, Mary Oliver, captured it best in her lines from "The Grave": "A dog can never tell you what she knows from the smells of the world, but you know, watching her, that you know almost nothing."

While there are many behavioral phenomena about which we know quite a lot—we can make very accurate and reliable predictions about what an individual is likely to do in a given situation—there are some areas in which we know next to nothing. The minds of other animals are private (as are human minds), and their sensory capacities often are so very different from our own and each others. So, even though we might know much "academically" about the physiology and anatomy of a dog's nose or of a bat's ears, we still do not know with certainty, experientially, what it is like to be a dog or a bat. Wouldn't it be nice to be a dog or a bat or a termite for a while? And, when we study the concept of self-knowledge in animals using mirrors, it is possible that even if we collect data that suggest that dogs do not have as high a degree of self-awareness as do chimpanzees because dogs do not respond with self-directed movements as do chimpanzees when they look at their reflection in a mirror, it remains possible that dogs do have a high degree of self-awareness, but that the use of a mirror does not tap into this ability. Perhaps assessing a dog's response to different odors, including their own, would yield different results. My own study of my dog Jethro's response to his own and to other dogs' "yellow snow" showed that this might be the case (**Dog Scents and "Yellow Snow"**). We need to take into account how animals sense their worlds using different sensory modalities—which are more or less important to them.

Along with unbridled curiosity, cleverness, and creativity, patience is a virtue when it comes to the study of animal behavior. I well remember many hours spent sitting cold and alone among 250,000 Adélie penguins at the Cape Crozier rookery in Antarctica just waiting for them to do so something—anything—other than stealing rocks from each other's nests or sleeping or staring at me trying to figure out who I was—a curious observer or a new land predator! And I also recall falling asleep while waiting for a coyote to wake up and join other pack members who had decided to move to another area in which to hunt and frolic.

Patience is also needed in data analysis. Watching videos over and over again and doing the appropriate statistical analyses can try anyone's patience, but these activities are just as important and exciting as collecting reliable data. Well, maybe they are not all that much fun, but they are essential. And do not give up on some idea just because others think you are wrong. Sometimes you might be heading in the wrong direction, and sometimes you might not. Be patient and analyze the arguments of supporters and critics alike. If the late William Hamilton III had not been persistent in pursuing his revolutionary ideas about the evolution of social behavior via kin selection, the field of animal behavior would have suffered an enormous loss. Had Jane Goodall not insisted on naming the chimpanzees whom she studied at Gombe stream in Tanzania, there would have been a delay in our coming to recognize that individuals had distinct personalities. Goodall also was the first researcher to observe chimpanzees use a blade of grass as a tool to extract a termite meal from a hole, but many other researchers did not believe her until she showed them a video of the activity. Had I given up the study of social play, as some of my colleagues suggested I do when I was a graduate student, I would never have discovered over the next 30 years the important connections between social play and the evolution of fairness, trust, and morality. Years of detailed video analysis (that drive some students crazy), discussions with colleagues from different disciplines, and a belief that I was onto something big kept me going. Imagine if Charles Darwin had given in to his critics when he wrote about his theory of natural selection!

As Donna Haraway notes in her book *The Companion Species Manifesto*: "To do biology with any kind of fidelity, the practitioner *must* tell a story, *must* get the facts, and *must* have the heart to stay hungry for the truth and to abandon a favorite story, a favorite fact, shown to be somehow off the mark. The practitioner must also have the heart to stay with a story through thick and thin, to inherit its discordant resonances, to live its contradictions, when that story gets at a truth about life that matters." I could not agree more with her sentiments.

I, and my esteemed colleagues who contributed to this encyclopedia, know just how difficult, tedious, frustrating, and challenging it can be to answer the simple question, "What is it like to be a dog or a cat or a chimpanzee or a robin or an ant?" It is the challenge of peeling away the layers of complexity and the mysteries that await us that keep us going. And so it will be for you. Most of the essays in this encyclopedia raise more questions than they answer. This state of the art collection sets a high standard for future research and provides an excellent introduction to the study of animal behavior. There can never be too much interest in animal behavior. I bet you cannot prove me wrong! So, what *is* it like to be a dog or a bat or a praying mantis or an octopus? Please read on.

■ Alphabetical List of Entries

(S) indicates a side bar. It is listed under the entry within which it appears.

◼ Guide to Related Topics

ANIMALS

Following is a list of entries that focus primarily on one type of animal. Most of the other entries in the *Encyclopedia of Animal Behavior* also discuss various species. Please see the index for additional references to specific animals.

Ants

Animal Architecture—*Subterranean Ant Nests*

Bats

Bats—*The Behavior of a Mysterious Mammal*
Communication—Auditory—*Bat Sonar*

Bears

Feeding Behavior—*Grizzly Foraging*
Human (Anthropogenic) Effects—*Bears: Understanding, Respecting, and Being Safe around Them*

Birds

Antipredatory Behavior—*Sentinel Behavior*
Behavioral Physiology—*Plumage Color and Vision in Birds*
Caregiving—*Brood Parasitism among Birds*
Cognition—*Grey Parrot Cognition and Communication*
Corvids—*The Crow Family*
Communication—Vocal—*Social System and Acoustic Communication in Spectacled Parrotlets*
Communication—Vocal—*Singing Birds: From Communication to Speciation*
Mimicry—*Magpies*
Play—*Birds at Play*
Reproductive Behavior—*Bowerbirds and Sexual Displays*
Tools—*Tool Use and Manufacture by Birds*
Welfare, Well-Being, and Pain—*Enrichment for Chickens*
Welfare, Well-Being, and Pain—*Feather Pecking in Birds*
Welfare, Well-Being, and Pain—*Rehabilitation of Raptors*

Cats

Cats—*Domestic Cats*
    Wild versus Domestic Behaviors: When Normal
    Behaviors Lead to Problems (S)

Fish

Frogs

Horses

Hyenas

Insects

Lemurs. *See also* Primates

Macaques

Behavioral Physiology

Color Vision in Animals
    Colors—How Do Flowers and Bees Match? (S)
    Do Squid Make a Language on Their Skin? (S)
Insect Vision and Behavior
Plumage Color and Vision in Birds
Thermoregulation
Thermoregulatory Behavior
Turtle Behavior and Physiology
Visual Perception Mechanisms
    Vision, Skull Shape, and Behavior in Dogs (S)

Behaviorism

Burrowing Behavior

Caregiving

Attachment Behaviors
Brood Parasitism among Birds
Brood Parasitism in Freshwater Fish
Fostering Behavior
How Animals Care for their Young
    Helpers in Common Marmosets (S)
Incubation
Mother–Infant Relations in Chimpanzees
Non-Offspring Nursing
Parental Care
    Parental Behavior in Marsupials (S)
Parental Care and Helping Behavior
Parental Care by Insects
Parental Care in Fish
Parental Investment
    Parents Desert Newborns after Break-in! Mother Goes
    for Food as Siblings Battle to the Death! Father Eats Babies! (S)

Cats

Domestic Cats
    Wild versus Domestic Behaviors: When Normal Behaviors
    Lead to Problems (S)

Cephalopods

Octopuses, Squid, and other Mollusks
    Roving Octupuses (S)

BEHAVIOR OF ANIMALS (continued)

Parasite-Induced Behaviors

Personality and Temperament

A Comparative Perspective
Personality in Chimpanzees
Personality, Temperament, and Behavioral Assessment in Animals
Stress, Social Rank and Personality

Play

Birds at Play
Social Play Behavior and Social Morality
    DOG MINDS AND DOG PLAY (S)

Predatory Behavior

Ghost Predators and their Prey
Orb-Web Spiders
Praying Mantids
    "SEXUAL CANNIBALISM": IS IT REALLY "SEXUAL" OR IS IT JUST
    PREDATORY BEHAVIOR? (S)

Recognition

Individual Recognition
Kin Recognition

Reproductive Behavior

Adaptations and Exaptations in the Study of Behavior
Alternative Male Reproductive Tactics in Fish
Assortative Mating
Bowerbirds and Sexual Displays
Captive Breeding
Female Multiple Mating
Marine Turtle Mating and Evolution
Mate Choice
Mate Desertion
Mating Strategies in Marine Isopods
Mollusk Mating Behaviors
Monogamy
Morning Sickness
Sex Change in Fish
Sex, Gender, and Sex Roles
    FEMALE–FEMALE SEXUAL SELECTION (S)
Sexual Behavior in Fruit Flies—Drosophila
Sexual Cannibalism
Sexual Selection

BEHAVIOR OF ANIMALS (continued)

Unisexual Vertebrates

Welfare, Well-Being, and Pain

Animal Welfare and Reward
Back Scratching and Enrichment in Sea Turtles
Behavior and Animal Suffering
Behavioral Assessment of Animal Pain
Behavioral Correlates of Animal Pain and Distress
Carnivores in Captivity
Enrichment for Chickens
Enrichment for Monkeys
Enrichment for New World Monkeys
Experiments in Enriching Captive Snakes
Feather Pecking in Birds
Obsessive–Compulsive Behaviors
Primate Rescue Groups
Psychological Well-Being
Rehabilitation of Marine Mammals
Rehabilitation of Raptors
Sanctuaries
Stress in Dolphins
Veterinary Ethics and Behavior
Wildlife Trapping, Behavior, and Welfare

Wildlife Management and Behavior

Wolf Behavior

Learning to Live in Life or Death Situations

CAREERS IN ANIMAL BEHAVIOR

Careers

Animal Behavior and the Law
Animal Tracking and Animal Behavior
Animal-Assisted Psychotherapy
Applied Animal Behavior
Careers in Animal Behavior Science
Mapping their Minds: Animals on the Other Side of the Lens
Recording Animal Behavior Sounds: The Voice of the Natural World
Significance of Animal Behavior Research
    UNCONVENTIONAL USES OF ANIMAL BEHAVIOR (S)
Veterinary Practice Opportunities for Ethologists
Wildlife Filmmaking
Wildlife Photography
Writing about Animal Behavior: The Animals Are Also Watching Us

HUMANS AND ANIMALS *(continued)*

Careers

Animal Behavior and the Law
Animal Tracking and Animal Behavior
Animal-Assisted Psychotherapy
Applied Animal Behavior
Careers in Animal Behavior Science
Mapping their Minds: Animals on the Other Side of the Lens
Recording Animal Behavior Sounds: The Voice of the Natural World
Significance of Animal Behavior Research
Veterinary Practice Opportunities for Ethologists
Wildlife Filmmaking
Wildlife Photography
Writing about Animal Behavior: The Animals Are Also Watching Us

Cognition

Domestic Dogs Use Humans as Tools
Grey Parrot Cognition and Communication

Conservation and Behavior

Species Reintroduction
   Preble's Meadow Jumping Mouse and Culverts (S)
Wildlife Behavior as a Management Tool in United States National Parks

Ecopsychology

Human—Nature Interconnections

Education

Classroom Activities in Behavior
   Classroom Research (S)

Education—Classroom Activities

All About Chimpanzees!
Insects in the Classroom
"Petscope"
Planarians

History

History of Animal Behavior Studies
   Some Leaders in Animal Behavior Studies (S)
Niko Tinbergen and the "Four Questions" of Ethology

Human (Anthropogenic) Effects

Bears: Understanding, Respecting, and Being Safe around Them
Edge Effects and Behavior
The Effect of Roads and Trails on Animal Movement

LITERATURE, ARTS, AND RELIGION AND ANIMAL BEHAVIOR (continued)

Art and Animal Behavior

Careers

Mapping their Minds: Animals on the Other Side of the Lens
Recording Animal Behavior Sounds: The Voice of the Natural World
Wildlife Filmmaking
Wildlife Photography
Writing about Animal Behavior: The Animals Are Also Watching Us

METHODS OF STUDY. (*See* RESEARCH, EXPLANATION, AND METHODS OF STUDY)

PEOPLE IN ANIMAL BEHAVIOR RESEARCH

Craig, Wallace (1876–1954)

Darwin, Charles (1809–1882)

Frisch, Karl von (1886–1982)

Griffin, Donald Redfield (1915–2003)

Hamilton, William D. III (1936–2000)

History

History of Animal Behavior Studies
    Some Leaders in Animal Behavior Studies (S)
Niko Tinbergen and the "Four Questions" of Ethology

Lorenz, Konrad Z. (1903–1989)

Maynard Smith, John (1920–2004)

Nobel Prize

1973 Nobel Prize for Medicine or Physiology

RESEARCH, EXPLANATION, AND METHODS OF STUDY (continued)

Methods (continued)

DNA Fingerprinting
Ethograms
Molecular Techniques
    PSEUDOREPLICATION (S)
Research Methodology
    EXPERIMENTAL DESIGN (S)
    SONIC TRACKING OF ENDANGERED ATLANTIC SALMON (S)
Zen in the Art of Monkey Watching

Neuroethology

Nobel Prize

1973 Nobel Prize for Medicine or Physiology
Sociobiology

Tinbergen, Niko

Ethology and Stress Diseases

Zoos and Aquariums

Animal Behavior Research in Zoos
Giant Pandas in Captivity
Studying Animal Behavior in Zoos and Aquariums

◼️ Aggressive Behavior
Dominance: Female Chums or Male Hired Guns

Primate Social Groups

Why do nearly all primates form groups? Two dominant hypotheses have been advanced to address this problem. Both are ecologically based. The first, known as the predator–defense hypothesis, serves as a general explanation for all multi-female groups, where a number of adult females reside with several adult males, along with their offspring. This hypothesis asserts that females form groups to reduce the risks of predation. Van Schaik, the main proponent of this idea, claims that predation is also responsible for variation in female relationships, both between and within groups. He speculates that where groups of females are heavily preyed upon, rigid dominance hierarchies will form from increased competition. On the contrary, where predation rates are low, female groups will become more cooperative and less competitive.

The second hypothesis of influence is the resource–defense hypothesis. This hypothesis is proposed for female-bonded groups only, since these females are primarily *philopatric*. This is when males generally transfer to other, non-natal groups after maturing, while females remain with their genetic kin. Basically, the hypothesis suggests the following pattern: Food items preferred by female-bonded species are rare, as well as located in economically defensible and well-distributed patches; individuals that join with allies in successfully defending these preferred patches against competitors will mostly benefit; and genetic kin will be favored over non-kin because, when choosing among potential allies, they are usually more reliable.

The worth of either hypothesis for explaining primate social behavior is still unclear. Ecological pressures remain the focus, however. Unfortunately, this hinders "social" explanations from being advanced, which may be equally valid. Social pressures are important because male and female reproductive strategies often conflict. For example, males are often observed to be aggressive and sexually coercive toward females and their infants in many primate species, and reproductive benefits may often follow. Infanticide is the prime example of males benefiting from aggression in a reproductive manner, whereas it is extremely costly to females. These reproductive strategies and counter-strategies between the sexes may ultimately determine the specifics of primate social organization as molded by evolution. The complexity of these intersexual manipulations is probably influenced by advanced cognitive abilities in primates. A similar explanation has been proposed for the existence the social systems of some primates and other mammals.

Primate Sociality as a Defense against Male Sexual Coercion

The literature dealing with ecological pressures and their potential role in the origin of sociality and its function is rather extensive for primates. Therefore, what follows is an eight-stage hypothesis that promotes a "social" explanation for the evolution of primate sociality. The proposal is referred to as the coercion–defense hypothesis. It is based on sexual selection theory, and it is summarized in the figure on page 3. The hypothesis questions the following statement made by Richard Wrangham (1987, p. 293): ". . . in female-bonded species the fact that a particular set of females lives together is generally affected little by male strategies." Contrary to this projection, the actual or real function of female-bonded primate groups may be to counter male reproductive coercion. Coercion defense may also be central to their origin. The function of and the originating force behind non-female-bonded groups may be coercion defense as well. *Non-female-bonded groups* are those that contain females who are genetically unrelated, because females immigrate from their natal groups at maturity rather (or more often) than males. Food and safety are undoubtedly important. Although, both may be less influential than originally proposed. Critiques of the coercion–defense hypothesis are offered in research papers by C. L. Nunn and C. P. Van Schaik, E. H. M. Sterck, and A. Treves.

Stage 1: Female Choice

Sexual selection is a type of natural selection based exclusively on reproductive competition. It includes both inter- and intrasexual selection, and these are commonly referred to as female choice, and male–male competition, respectively. In that females must care for their offspring while males generally do not, Trivers argues that females should emphasize choice, and males intrasexual competition. Moreover, whereas males are limited only by their access to fertile females, females are always limited by biology in the number of offspring they can bear. Therefore, male primates should be mostly promiscuous, while females should be predominantly choosy. Variance in reproductive success among males should result, as males will be chosen differentially by females to mate. As this reproductive strategy imposed on males by females continues, selection should strongly favor male counterstrategies that improve their plight: primarily, sexual coercion to persuade choosy females to mate.

Stage 2: Male Intersexual Coercion

In 1992 (p.3), Smuts defined *sexual coercion* as ". . . male use of force, or its threat, to increase the chances that a female will mate with the aggressor or to decrease the chances that she will mate with a rival, at some cost to the female." Because this form of coercion involves force by a male to manipulate a female's behavior and her reproductive state, which is costly to the female, the definition also includes *infanticide*, or the killing of infants. It does not, however, include coercion used by males against females in nonsexual contexts, such as in feeding competition. According to Smuts and Smuts, male sexual coercion of females is an equally significant aspect of sexual selection theory overlooked by Darwin as well as neo-Darwinists. This oversight occurs because sexual coercion is often mistakenly considered a part of male–male competition; males gain reproductive advantages over competitors by manipulating fertile females aggressively. Similar consequences for males result from female choice, but sexual coercion is seldom considered within the realm of intersexual competition. Therefore, sexual coercion must be considered to be a related, but distinct, form of sexual selection theory comparable to and equal with both inter- and intrasexual selection.

Stage 3: Male Intrasexual Competition

Males also compete with each other for the opportunity to mate with females. This behavior includes fighting with other males, as well as immigrating from natal groups to peripheral or more distant groups in search of mates, both of which entail high risk for

Coercion–Defense Hypothesis

Outline of the selective pressures mainly responsible for primate sociality as predicted by the coercion–defense hypothesis.

Stage 1

Females choose mates
(intersexual mate choice)

Unselected or coercive males

Stage 2

Sexually and aggressively coerce females
(intersexual coercion)

Stage 3

Gain access to females by competing with other males
(intersexual competition)

Females counter this attempted control by males with

Stage 4

Coercion defense

Reinforced by kin selection and reciprocal altruism in

Stage 5

Female-bonded groups
(many primates)

Stage 6

Males of several species counter female bonding by becoming sexual dimorphic
(e.g., gorillas), or by forming "brotherhoods" (e.g., chimpanzees)

Stage 7

Females then abandon same-sex bonding and resort to the "hired-gun" principle to
defend against sexual coercion from unselected or coercive males

Stage 8

Non-female-bonded groups
(some primates, including humans)

injury or possible death. Males often form male–male cooperatives and linear dominance hierarchies for the same reason. Moreover, these last two forms of competition are used for other purposes, like gaining access to food. Mating competition in alternative forms are also used by males. One example is mating opportunistically, or when out of sight of high-ranking males. This behavior often depends on a male's age and rank, as well as the demography of his social group. Because of the potential reproductive gains, most males are willing to take these risks. Male–male competition for mates, however, must be viewed in combination with female choice and sexual coercion because all three are closely entwined. Smuts and Smuts make clear that competition among males will not automatically translate to mating

All primates, including these lowland gorillas, battle for a number of reasons, still hotly debated.
© Michael Nichols/National Geographic Image Collection

success without females copulating with them willingly. This reality is vitally important, and the circumstance where males coerce females is also crucial for the selection of coercion defense. Together, coercion and coercion defense may be ultimately responsible for the general social structure currently witnessed among modern primates.

Stage 4: Coercion Defense

Females are a limited reproductive resource for males, and this is central to sexual selection theory, as previously noted. Evolutionary biologists have vigorously investigated the many facets of male–male competition, but scant progress has been made in revealing the inter-workings of female choice and male sexual coercion (which includes infanticide). Female mate choice and male sexual coercion is followed by *female resistance*, and this has received even less research attention. Females have evolved a number of mechanisms to counter male manipulation, such as fighting back and mating synchronization, but female–female coalitions with both genetic kin and non-kin to defend against male aggression is a significant feature of species across the primate *taxa* (classification categories). The most prominent social system among a majority of primate species is based on female bonding—when group formation occurs by female kin residing together in a matriline along with several unrelated matrilines.

Stage 5: Female-Bonded Groups

There is little firm evidence to support explanations for why females choose to be social rather than solitary, or why female-bonded groups are common across the primate taxa. Rather than to protect themselves specifically from predators or to gain an advantage in disputes with other groups over choice resources, it is proposed here that females bond in groups primarily to form alliances with other females to protect themselves from male sexual coercion. Concerning food intake, the importance of an offspring-bearing female to consume choice food items is often balanced by a male's similar requirement because of his generally larger size. Ecological pressures may serve in placing limits on group structure, but they do not guide it. Several authors have alluded to this approach in the past, but the idea has not been further developed or made more elaborate. Moreover, females accomplish coercion defense by bonding together to form alliances. Kin selection and reciprocal altruism serve as reinforcements, increasing the fitness of the females. Depending upon the ecological limits of a given species, however, female-bonded groups may have optimal sizes. For example, there is some evidence in red-tail monkeys that reproductive rates of females are negatively correlated with an increase in a group's size. Struhsaker and Leland (1988) found that reproductive rates increased as extra-group male aggression toward females decreased in the smaller of two groups after splitting apart. Both groups used the identical habitat before and after fissioning. Resource competition within the group prior to fissioning appears to have been an unlikely influence, as is sometimes inferred. Male aggression may have a similar effect in multi-male, multi-female primate species, as red-tail monkeys are harem breeders.

Stage 6: Sexual Dimorphism or "Brotherhoods"

Even though most primates are female-bonded, a number of species are not, and the structuring of those societies may also be determined by aggression and coercion defense. To control fertile females and their access to extra-group males, as well as to counter female bonding, males of some species may have been selected to be sexually *dimorphic*, or especially cooperative. For example, gorilla females are sometimes subjected to sexual aggression because they are little more than half the size of their male counterparts. Infanticide among gorillas is common as well. Basic group structure usually consists of one "silverback" male and several unrelated females along with their offspring. Both sexes emigrate from their natal group when mature. Male gorillas remain solitary or travel in all-male groups. They do so until they are able to form their own group by "stealing" willing females from groups of other males. Much like gorillas, mature female chimpanzees leave their natal group to join another. In contrast, male chimpanzees remain with their natal groups and associate with half-brothers throughout their lives. This bonding among brothers is sometimes referred to as "brotherhoods." Note that male chimpanzees are notorious for aggression shown to females, particularly within the sexual context. A similar reputation is held by male spider monkeys, in which they operate in a social system comparable to chimpanzees. Moreover, as with male chimpanzees, male spider monkeys have been observed to form aggressive alliances with other males against females. And like chimpanzees, these male coalitions may be between half-brothers.

Stage 7: "Hired-Gun" Principle

In a 1979 article Wrangham argues that females in several species of primates choose group living because of the protection received from male "friends," particularly protection from infanticide by extra-group males. This argument is referred to as the "hired-gun" principle. It suggests that females form bonds with and seek protection from males, rather than from female kin. These inherently risky "protector friendships" are occasionally observed in female-bonded species such as olive baboons. They are, however, far more common in species where females reside with unrelated females. Here again, gorillas and chimpanzees serve as examples. Infant gorillas are almost always killed by extra-group males upon the death of a silverback protector, even when defended by their mothers. Moreover, adult female chimpanzees and their infants are highly susceptible to severe injury and perhaps even death from extra-group male coalitions when not protected by familiar male, half-brother alliances. Cooperative male aggression against females is an ordinary occurrence, as well, in spider, squirrel and red howler monkeys. Concerning other mammals, male bottlenose dolphins form alliances against females, whereas lion infants are susceptible to infanticide by extra-group brother cooperatives. Coercion defense may also determine whether groups include single or multiple males in female-bonded primate societies, since synchronized female fertility may be a defense against male infanticide. It therefore seems that male aggression and coercion defense are major contributors to primate sociality, including the prominent system that follows.

Stage 8: Non-Female-Bonded Groups

It is said that females in some primate species choose to mate with high-ranking males because dominant males, overall, are genetically superior. In other words, females may be competing for high-quality sperm that dominant males supposedly possess. What females may actually be competing for, however, is male protection from the threat of sexual coercion from other intra-group or extra-group males, and this includes protection from infanticide. Although, it is unlikely that males are choosing either to protect or to coerce females. Males should practice both strategies depending on the circumstances and the resulting genetic costs and benefits. Instead, females may have little choice other than to select single males or "brotherhoods" to protect them (and their infants) from unwelcome male sexual advances and coercion. Given the right circumstances, females choosing the protection of female kin may be simply overwhelmed by male reproductive strategies. Ultimately, this would force females to seek protection offered by male "hired guns." Such social pressures may be critical for the evolution of non-female-bonded groups, while ecological forces may have minor influence. Moreover, male aggression and coercion defense may also be responsible for monogamy in gibbons and siamangs. Furthermore, as male orangutans are observed to force copulation on females, the solitary habits of females may make them especially vulnerable. Female orangutans are without protection from female kin and consistent protection from males. Human females experience an extraordinary degree of male abuse as well. As *Homo sapiens* share a close ancestry and similar dispersal pattern (i.e. female emigration and male philopatry) with common chimpanzees, the abuse that males show females in both species may have a single origin.

See also Aggressive Behavior—*Ritualized Fighting*
Dominance—*Development of Dominance Hierarchies*

Further Resources

Brereton, A. R. 1995. *Coercion-defense hypothesis: The evolution of primate sociality*. Folia Primmatol, 64, 207–214.

Clutton-Brock, T. H. 1995. *Sexual coercion in animal societies*. Animal Behaviour, 49, 1345–1365.

Nunn, C. L. & Van Schaik, C. P. 2000. *Social evolution in primates: The relative roles of ecology and intersexual conflict*. In: *Infanticide by Males and its Implications* (Ed. by C. P. Van Schaik & C. H. Janson), pp. 388–419. Cambridge: Cambridge University Press.

Smuts, B. B. & Smuts, R. W. 1993. *Male aggression and sexual coercion of females in nonhuman primates and other mammals: Evidence and theoretical implications*. In: *Advances in the Study of Behavior* (Ed. by P. J. B. Slater, M. Milinski, J. S. Rosenblatt, & C. T. Snowdon), pp. 1–6. New York: Academic Press, Vol. 22.

Smuts, B. B. 1992. *Male aggression against women: An evolutionary perspective*. Human Nature, 3, 217–249.

Sterck, E. H. M., Watts, D. P. & Van Schaik, C. P. 1997. *The evolution of female social relationships in nonhuman primates*. Behavioral Ecology and Sociobiology, 41, 291–309.

Struhsaker, T. T. & Leland, L. 1988. *Group fission in red-tail monkeys (Cercopithecus ascanius) in the Kibale Forest, Uganda*. In: *A Primate Radiation: Evolutionary Biology of the African Guenons* (Ed. by A. Gautier-Hion, F. Bouliere, J. P. Gautier & J. Kingdon), pp. 364–388. New York: Cambridge University Press.

Treves, A. 1996. *Primate social systems: Conspecific threat and coercion-defense hypotheses*. Folia Primmatol, 69, 81–88.

Trivers, R. L. 1972. *Parental investment and sexual selection*. In: *Sexual Selection and the Descent of Man, 1871–1971* (Ed. by B. Campbell), pp. 136–179. Chicago: Aldine.

Van Schaik, C. P. 1983. *Why are diurnal primates living in groups?* Behaviour, 87, 120–143.

Van Schaik, C. P. 1989. *The ecology of social relationships amongst female primates*. In: *Comparative Socioecology: The Behavioural Ecology of Humans and other Mammals* (Ed. by V. Standen & R. A. Foley), pp. 195–218. Oxford: Blackwell.

Wrangham, R. W. 1987. *Evolution of social structure*. In: *Primate Societies* (Ed. by B. B. Smuts, D. L. Cheney, R. M. Seyfarth, R. W. Wrangham & T. T. Struhsaker), pp. 282–296. Chicago: University of Chicago Press.

Wrangham, R. W. 1980. *An ecological model of female-bonded primate groups*. Behaviour, 75, 262–300.

Wrangham, R. W. 1979. *On the evolution of ape social systems*. Social Science Information, 18, 335–368.

Alyn R. Brereton

■ Aggressive Behavior
Ritualized Fighting

In most species, individuals occasionally come into conflict over access to resources important for their survival and reproduction. Resources such as food, shelter and territories are sources of competition when they are difficult to obtain. The time and material investments made by either sex in producing and raising offspring also represent important reproductive resources for the opposite sex. The availability of mates to invest or co-invest in offspring often is limited and males very often fight over access to females, and sometimes vice versa.

To resolve conflicts over resources, animals use various aggressive behaviors ranging from brief and subtle threats to dramatic, intensive, and potentially injurious fighting. In addition to regular fighting, many species utilize so-called ritualized fighting behaviors to settle conflicts: species-typical display patterns that communicate fighting ability, but that entail less energy use or risk of physical harm than all-out fighting.

The key to understanding the evolution of ritualized fighting is to recognize that all-out fighting will not be in the best interest of either rival if both contestants can reliably predict which one of them would win a real fight using mutual assessment behaviors that are less costly than all-out fighting.

The ability to predict the outcome of an all-out fight without actually engaging in one comes from the fact that individuals in conflict over a resource often are not particularly closely matched in ability or motivation. Most animals' sensory systems are so sensitive that they are able to detect which one of them is the underdog. Ritualized fighting behavior has evolved in such a way that, during a fight, the underdog will withdraw as soon as his probability of winning a full fight is revealed to be low enough that searching for another opportunity to gain the contested resource elsewhere will increase his lifetime fitness (i.e. number of offspring).

Animals routinely avoid full combat because using ritualized assessments of fighting ability provide a more effective and efficient way to estimate each other's *resource-holding power* (RHP) cooperatively and settle conflicts without suffering injury or death. RHP refers to the ability of an individual to procure, defend, and compete for resources. RHP depends upon suites of individual traits and circumstances that include more than conventional notions of fighting ability. For example, in some species the RHP of an individual may be increased by having more information concerning the quality of the resource and how to make best use of it (e.g., by knowing the content and structure of a territory) and the logistics of fighting there (e.g., the locations in a complex fighting arena, such as a spider web or a talus slope, where one can get the best footholds).

RHP can also be increased if an individual has social allies who will help win the fight. For example, male baboons may form alliances to help one another gain access to females for mating. In lions, females help the current males of her pride fend off other male groups that attempt to take over in order to avoid infanticide by new pride males. In many territorial species, territory holders who want to avoid the costs of working out boundaries with a new neighbor may help their current neighbor fight off his challengers, the so-called "dear enemy" phenomenon.

Even in species in which RHP is mainly determined by fighting ability per se, the fact that body size, quality of weaponry, experience, motivation, and health all influence fighting ability helps assure that perfect matches in RHP between opponents will be rare during animal contests. This creates the asymmetries in the probability of winning an all-out fight that has caused the evolution of behavior that allows individuals to assess each other in ritualized contests. Ritualized fighting allows opponents to estimate the cost of full combat so it can be weighed against the benefit of obtaining the resource that they are fighting over.

Elk bulls will spar and fight one another to secure mating access to elk does. Their formidable antlers are dangerous weapons.
© Anthony F. Chiffolo.

Ritualized behaviors are considered signals which have evolved for the purpose of communicating information. In the context of fighting, ritualized behaviors are signals that encode information about RHP, the sender's fighting ability and motivation. The receiver of the signal is the opponent, who must be able to properly assess the signal information and respond accordingly. Through these cooperative behaviors, contestants gain information about each other's full combat capability and their willingness to escalate the fight. In all species in which there has been an evolutionary history of direct competition within species for some resource, thorough investigation of contest behaviors should reveal some form of ritualized fighting.

Ritualized fighting signals are less costly than all-out fighting, but they must be somewhat costly for them to be honest signals (i.e., signals that are not misleading) that rivals have reason to pay attention to. Their cost to produce should reflect the quality of information needed about the opponent, which may vary greatly with circumstances. In the most demanding circumstances the cost of ritualized fighting may approach that of real fighting, perhaps not in terms of the risk of permanent injury or death, but certainly in how much they tax each contestant's metabolic, developmental, and neurological competence. Ritualized fighting behavior may also be important for self-assessment—an accurate up-to-date estimate of one's own fighting ability—which is just as important as assessing one's rival in obtaining a useful measure of relative RHP.

Ritualized fighting behaviors include everything from subtle or distant signaling via postures, vocalizations, and chemical communication to dramatic behaviors entailing obvious demonstrations of strength, coordination, and determination. Posturing is a commonly used no-contact threat display employed by a variety of animals. Aggressive body posture can visually convey important information about an opponent's competitive ability, such as size. For example, many animals posture by presenting the side of their body as they strut or stand, called a broadside display, in order to appear as large as possible to their opponent. Additionally, animals may also increase the size of their profile through body markings that give the impression of greater than actual length or by manipulating physical features, such as fluffing feathers and raising fur. Iguanid lizards have a particularly impressive number of behaviors and structures to exaggerate size, including the extension of skin located on the throat (dewlap), crest raising, back arching and lateral body compression.

Birds, amphibians and some mammals exchange loud, repeated vocalizations during contests. Vocal threat displays can provide contestants with information regarding size, condition or age. In cricket frogs, males in conflict over a mate will decide whether to attack or retreat from an opponent based on acoustic frequency of calls. Fundamental frequency production (the lowest frequency in a sound, perceived as pitch) is constrained by body size. Consequently, vocal displays can reliably indicate body size because small animals cannot produce a call as low in pitch as a larger animal. Vocal display bouts may also depend on strength and stamina, providing information about body condition through relative sound intensity or the length of the display. Baboon males use energetically demanding "wahoo" calls during aggressive interactions; high-ranking and older males call at higher rates and are able to call for a longer period of time.

In some mammal species, olfactory cues used during an encounter with an opponent can provide information about dominance status, stress levels, and health by revealing hormone levels, such as testosterone and corticosteroids. Chemicals that are purposely used to communicate such information between members of the same species are referred to as *pheromones*; if the transmission is accidental but the information is used by the recipient, the chemical is called a *kairomone*; only pheromones qualify as proper signals, and thus as potential components of ritualized chemical fighting displays. One impressive example of pheromone use is the ring-tailed lemur "stink fight," which males engage in over access to females. Males rub secretions from a wrist gland onto their boldly marked tails, which they hold high overhead and wave in their opponent's direction in order to dispense scent and provide an agonistic visual display.

Ritualized fighting behavior may involve body contact or sparring with specialized structures, such as horns, antlers and tusks. Physical contact during ritualized fighting provides a tactile way to assess relative RHP while limiting potential injury. Many animal contests involve a form of ritualized "wrestling"—a cooperative form of fighting that allows opponents to gauge each other's strength and endurance. In sierra dome spiders (*Neriene litigiosa*), males meet each other on a dome-shaped web and lock *pedipalps* (mating structures found on the front of the body) in order to grapple while hanging under the web. During the wrestling match, spiders attempt to jam one another up through the dome of the web by pushing forwards and upwards, using the strength of their legs and pedipalps. Sparring contests with weapons like horns or antlers, can be a relatively safe test of strength and stamina. In caribou, full combat with antlers can cause injury or death; however, almost all fights are resolved with a form of ritualized fighting. Caribou begin antler sparring by slowly and carefully adjusting their antlers before they begin to push each other. Some insects have horns, which they use in a similar fashion to caribou antlers, by locking horns and pushing each other until one opponent gives up or gets dislodged from the fighting platform. Rocky Mountain Bighorn sheep butt heads in incredibly high impact frontal collisions by using their horns in conjunction with injury reducing physical adaptations in the head and neck. Such violent ritualized behaviors can only be understood as less costly than full fighting when one knows the context and style in which full combat would occur—running full speed on very steep rocky slopes trying to inflict blows to the sides of each other's bodies and mate with fleeing females at the same time!

In addition to various techniques, species also differ greatly in the number of ritualized behaviors used in fighting. House crickets employ an extreme number of displays and diverse tactical maneuvers. Male house crickets display by flaring mandibles and *stridulating* (chirps produced by rubbing wings together). They use fighting tactics that include head butts, mandible sparring and foreleg punches. They also engage in harder physical contact, such as head charging, kicking and wrestling, although these behaviors rarely result in significant injury. On the other end of the spectrum, some species only use one threat display. Weakly electric fish use pulses of electricity to communicate aggressive intent to each other, often in territory defense. Unlike most fish that are able to incorporate visual displays into their repertoire, weakly electric fish are nocturnal or live in murky water and are limited to the kind of sensory signal that they can use to communicate fighting ability. Rivals threaten each other by trying to jam the opponent's signal. They compete by altering their own pulse rate to overlap with their opponent's. In response, the loser will either shift its pulse rate to avoid the winner's pulse rate or cease electric signaling for some time after the fight.

Most animal conflicts begin with the least physically costly and injurious tactics in order to assess RHP. Male sierra dome spiders may escalate sequentially through two ritualized and one unritualized fighting behaviors, in which each successive behavior is clearly more risky than the former, and whose energetic costs, on average, are 3.4, 7.3, and 11.3 times higher than resting metabolic rates, respectively. Male sierra domes who differ more than 20% in weight seldom escalate beyond the least expensive fighting behavior. In many species, opponents who are not closely matched in RHP, may not even get close to one another. Qualities of vocalizations, which in some species carry for many miles (e.g., elephants, whales) or long distance assessments of body size and condition (e.g., via pheromones or antler size) may be possible when there is considerable disparity in relative fighting ability. Where RHP differences are large, very efficient assessment methods are effective at communicating competitive ability; decisions that are likely to improve survival and reproduction can be based on easily gathered comparative information on self versus other.

In many species, ritualized fighting behaviors may escalate quantitatively (e.g., intensity and duration) or in quality (e.g., where entirely different behaviors are employed) if both contestants need better information on their rival's ability relative to their own in order to win the match. As relative ability becomes similar, contestants are more likely to escalate, and contest behaviors become increasingly costly in terms of time, metabolic energy, or injury. Cichlid fish of the species *Nannacara anonala*, for example, proceed through a number of stages depending on how close the contestants are matched in size and strength. Cichlids begin their fights by swimming side-to-side, which is a type of broadside display used to assess size. If both fish are not greatly different in size, they proceed to "tail-beat" each other. Tail-beating behavior causes a stream of water to push against the rival, which allows the receiver of the water stream to gauge the opponent's strength more accurately. The individual performing the tail-beating display may assess the ability of the receiver to withstand the stream's force. Cichlid fights may subsequently escalate further with frontal confrontation and mouth wrestling, where the contestants push and pull each other by grasping jaws. Finally, if fish are closely

Ring-tailed lemur (Lemur catta) stink fight, Berenty, Southern Madagascar.
Courtesy of Adrian Warren.

matched in RHP and cannot resolve the conflict with ritualized behaviors, mouth wrestling ceases and full combat ensues; the opponents circle around and bite each other.

The relationship between the effectiveness of a ritualized behavior in demonstrating true ability and the efficiency of performing that behavior (in terms of energy cost and injury risk) is dependent on the degree of difference in RHP between contestants. As the RHP of contestants becomes more similar, more effort is required to demonstrate superiority effectively and reliably. Each contestant will have to perform closer to its maximum potential. In cichlid fish, contestants closely matched in RHP escalate to mouth wrestling, which requires more metabolic energy and physical contact (potentially causing mouth injury) than previous stages of ritualized behavior. Mouth wrestling is less energy efficient than other ritualized fighting behaviors, but provides contestants with better information about maximum strength and stamina and so is more effective at providing contestants with knowledge of relative RHP. A smaller, weaker fish may be able to exaggerate size by spreading its fins and bluff a larger fish into believing that it has similar or better RHP; however, if the contest escalates to mouth wrestling, it could not win a wrestling bout against an opponent that has greater true ability. In any contest, it is in each individual's interest both to deceive an opponent by exaggerating RHP and simultaneously to avoid deception by accurately gauging the RHP of a rival relative to itself during a contest; thus more expensive behaviors make cheating more difficult for both contestants while making true RHP assessment accuracy more possible.

In every species with ritualized fighting behaviors, there are situations likely to trigger escalation to all-out fighting. Opponents that find themselves closely matched may resort to full combat if the information gained from ritualized behaviors is insufficient for making a confident prediction about the outcome of an unritualized fight. Conversely, there may be circumstances where opponents who are grossly mismatched will proceed to all-out fighting. If success in all-out fighting depends more on luck than success in a ritualized fight, an obviously inferior opponent may escalate the fight beyond ritualization. This is more likely to happen when the resource is exceptionally valuable or if the underdog has little to lose in terms of expected future reproduction. In sierra dome spiders, unritualized fighting consists of frenzied biting and grappling where the first contestant to deliver a bite to the body or even the foreleg is very likely to win. As the end of the breeding season nears and the window of the future closes on male sierra domes, escalations to this kind of fighting become more frequent.

One might also expect underdogs to escalate to full combat more often if they have only enough energy to fight effectively in a short and intense all-out fight where luck can play a major role in the outcome. Alternatively, a superior fighter also may escalate to all-out fighting during a contest even though he would be almost certain to win using ritualized behaviors. This may happen in cannibalistic species where the RHP dominant gains a major meal by killing his rival, as in sierra dome spiders. Reducing the costs of display may also cause the RHP dominant to escalate, for example, if ritualized displays take a long time, if they are energetically expensive, or if they run the risk of attracting and increasing vulnerability to predators.

The general principle underlying ritualized fighting is illuminated by considering situations, like those above, in which it escalates to all-out fighting: Ritualized fights are good for both contestants under circumstances that make low-cost assessments reliable and economic, but not otherwise. Many factors potentially can tip the cost-benefit balance toward dangerous unritualized fighting, and most species are fully capable of it. This is one of many ways that even organisms with incredibly small nervous systems exhibit impressive responsiveness to the environment.

See also Aggressive Behavior—*Dominance: Female Chums*
or Male Hired Guns
Applied Animal Behavior—*Social Dynamics*
and Aggression in Dogs
Dominance—*Development of Dominance Hierarchies*

Further Resources

Archer, J. 1988. *The Behavioural Biology of Aggression*. Cambridge: Cambridge University Press.

Bradbury, J. W. & Vehrencamp, S. L. 1998. *Principles of Animal Communication*. Sunderland: Sinauer Associates, Inc.

Huntingford, F. & Turner, A. 1987. *Animal Conflict*. New York: Chapman and Hall Ltd.

Ord, T. J., Blumstein, D. T. & Evans, C. S. 2001. Intra-sexual selection predicts the evolution of signal complexity in lizards. Proceedings of the Royal Society of London, Series B, 268, 737–744.

Suter, R. B. & Keiley. 1983. *Agonistic interactions between male Frontinells pyramitels (Araneae, Linyphiidae)*. Behavioral Ecology and Sociobiology, 15, 1–7.

Paul Watson & Tagide deCarvalho

■ Animal Abuse
Animal and Human Abuse

Studies show that abuse of animals other than humans and human abuse are related. Psychologists and other social scientists develop and apply methods of identifying and treating animal abusers. Since a link exists between the different forms of violence, treatment programs and policies that prevent animal abuse also can lessen instances of violence to humans.

Animal abuse is socially unacceptable, intentional, and unnecessary suffering directed at nonhuman animals. What is socially unacceptable varies in different societies and in different times in the same society. At one time most people did not consider it wrong to kill animals for fur; now in some subcultures, it is unacceptable to wear a fur coat.

Recent studies show that animal abuse is very common. Clifford Flynn in his 1999 survey found that 34.5% of male college students reported abusing animals as children and 43.4% reported witnessing it. The results for women were less than half these rates, but still at high levels. Several decades ago, similar findings about the high incidence of spousal abuse and child abuse led to greater public attention and the development of methods of identifying and treating these forms of abuse. Similar efforts are now underway with regard to animal abuse.

One way to identify animal abusers is to ask. Increasingly, counselors and teachers include questions about companion animals (pets) when they talk to clients and students who are having problems. They ask if they have or ever had companion animals, if they ever lost one, and if they ever harmed or saw someone else harm one. Assessment instruments (carefully worded questionnaires that distinguish between people who do and do not abuse animals) are available to help make this inquiry and to assure that information gathered is accurate and objective. Such information adds to our understanding of animal abuse and its relation to other problems people have.

People who abuse animals often do have other problems. They are more likely to be violent to humans as well. They often try to deal with situations that they find frustrating through violence. Animal abuse is a symptom of *Conduct Disorder of Childhood*, a condition in which children repeatedly commit various antisocial acts, such as violence, property damage, and theft. Other serious disorders of both childhood and adulthood also often include a history of animal abuse.

If a person has abused animals or even witnessed animal abuse, some kind of help is necessary. People with a history of severe abuse of animals and other emotional or behavioral problems may have to live in a special residence where they can receive constant treatment and supervision for a year or longer. Green Chimneys in New York is a residential treatment center for children with severe problems, often including animal abuse. The children live on the active farm, learning to relate to animals in more caring and responsible ways.

Counselors treat children and adults with less severe problems in weekly sessions. *AniCare Child* is a treatment approach that emphasizes teaching the child to empathize with other beings, including animals. The child also learns to recognize and manage his or her emotions and to solve problems in constructive ways. *AniCare* is a related approach for adults. It emphasizes helping the person to be accountable for his or her behavior. Often animal abusers do not admit to themselves or others that what they did is wrong. They are not willing to accept that their behavior is a problem. They develop "stories" that deny the presence or importance of the abuse, or that distort their role in the abuse, or that somehow justify it. The counselor must work to help the client see that there is a problem and that he or she must accept responsibility for it.

Both residential and out-patient treatment approaches such as these must be evaluated. Psychologists and other social scientists are undertaking outcome studies to see if and how the therapy works.

Abusing animals is against the law. All fifty states in the United States have anticruelty laws. In the past decade many states have made these laws stronger. In 41 states at least some forms of serious animal abuse are a felony, a more serious form of crime. Also, in 24 states, the judge can require people convicted of animal abuse to undergo counseling in addition to serving time in prison. It is important that counselors receive training in how to work effectively with these people. Psychologists and other mental health providers also can serve as expert witnesses in the trials of suspected animal abusers to inform the court of the seriousness of this behavior and of its relation to human violence.

The Federal Bureau of Investigation keeps track of the numbers of different kinds of crimes committed in the United States. Some animal advocates are working to require the FBI to add animal cruelty to the list of crimes reported in its annual crime report. This would provide needed data about trends in animal abuse and also help persuade the criminal justice system and the public generally that animal abuse is a serious crime.

In addition to identifying, prosecuting, and treating people who already have abused animals, it is important to prevent abuse in the first place. Secondary prevention programs identify people "at risk"—those more likely to become abusers and to have other emotional and behavioral problems. Children at risk might be those who do not have adequate supervision in their home. The Forget Me Not Farm in California takes children like these and provides them with gardening and animal husbandry experiences on a farm.

Primary prevention involves efforts to educate everyone about the seriousness and importance of animal abuse before they have this problem. In the long term, this is the most important way to eliminate animal abuse. Humane education is part of the curriculum in many schools, usually beginning at the elementary level. In addition to instruction in the care and welfare of companion animals, it increasingly includes information about animal abuse and its link to spousal and child abuse.

Some social scientists and animal advocates are working to make the entire curriculum more sensitive to issues of animal abuse and exploitation. A group of educators in California are developing charter schools that have as an organizing theme the issue of violence, including animal abuse.

Even more broadly, primary prevention deals with reducing violence in the popular culture. George Gerbner in his 1995 study found that children's television shows present animals predominantly in violent ways—as villains, threats, or victims. Efforts to persuade the television industry to reduce the level of violence and the way that animals are portrayed would be helpful.

Because of the relation between animal abuse and human abuse, it is important for personnel in agencies involved in both areas of abuse to work together. Animal control and humane officers, people who help injured and abused animals, can be trained to identify and report spousal and child abuse. Human service, domestic violence, child protective, and probation and parole personnel similarly can be trained to identify and report animal abuse. Increasingly, various jurisdictions are instituting these cross-training and cross-reporting policies. In San Diego County, a policy requires humane and animal control officers and child and adult protective services personnel to report both human and animal abuse to the appropriate authorities.

These and other related agency personnel can form networks to combine their resources and to increase awareness of the connection between animal abuse and other forms of violence. In Massachusetts, the Link Up Education Network includes groups of veterinarians, humane officers, police, domestic violence child protection counselors, attorneys, social workers, physicians, and animal advocates.

Shelters for abused women are facilities that provide help for the many women who are abused by their spouses. Frank Ascione found in a series of studies that 85% of women in shelters report instances of animal abuse in their households. Often threat of abuse of the companion animal is a reason that women delay leaving their homes when they are abused. Increasingly, women's shelters are working with animal shelters, veterinary offices, and animal fostering programs to provide safe havens, temporary shelter, for the companion animals of abused women.

Many laws and statutes mandate physicians, therapists, and other professionals to report spousal and child abuse, even if it involves breaking the confidential relation between themselves and their clients. Link networks are working to extend these measures so that these professionals at least are permitted to and can not be sued for reporting animal abuse. Veterinarians are an important group of professionals to involve in this identification and reporting of animal abuse.

In summary, recently gained understanding of violent behavior and its relation to animal abuse have led to various programs, treatments, and policies all directed to the early identification, treatment, and reduction of violence in our society.

See also Animal Abuse—*Violence to Human and Nonhuman*
Animals

Further Resources

Arluke, A. & Lockwood, R. (Eds.) 1997. Special theme issue: Animal cruelty. Society and Animals, 5, 3 (whole issue).

Ascione, F. & Arkow, P. (Eds.) 1999. *Child Abuse, Domestic Violence, and Animal Abuse.* West Lafayette, Indiana: Purdue University Press.

Flynn, C. P. 1999. *Animal abuse in childhood and later support for interpersonal violence in families.* Society & Animals, 7, 161–172.

Gerbner, George, 1995. *Animal Issues in the Media.* Encino, CA: Ark Trust.

Jory, B. & Randour, M. L. 1999. *The AniCare Model of Treatment for Animal Abuse.* Washington Grove, MA: Psychologists for the Ethical Treatment of Animals.

Lockwood, R. & Ascione, F. (Eds.) 1998. *Cruelty to Animals and Interpersonal Violence.* West Lafayette, IN: Purdue University Press.

Randour, M. L., Krinsk, S. & Wolf, J. 2002. *AniCare Child: An Assessment and Treatment Approach for Childhood Animal Abuse.* Washington Grove, MA: Psychologists for the Ethical Treatment of Animals.

Kenneth Shapiro

■ Animal Abuse
Violence to Human and Nonhuman Animals

> *"The time will come when men such as I will look upon the murder of animals as they now look upon the murder of men."*
>
> LEONARDO DA VINCI, ITALIAN PAINTER, SCULPTOR,
> ARCHITECT, MUSICIAN, ENGINEER, AND SCIENTIST

> *"The greatness of a nation and its moral progress can be judged by the way its animals are treated."*
>
> MAHATMA GANDHI

Historically, people have believed there was a connection between how humans treat each other and how they treat animals. Now there is a growing awareness and documentation of the link that exists between human violence toward other humans and human abuse of nonhuman animals. Social scientists from many disciplines including psychology, criminology, and sociology are studying this association. Some experts refer to this connection as "The Link." Some people believe we should not be concerned about animal abuse simply because it has implications for human welfare, but because animals are entitled to live lives free from mistreatment.

The first issue to consider is a definition of animal abuse. Frank R. Ascione (2001, p. 2), who is a leading researcher in the field, uses the following definition: "socially unacceptable behavior that causes unnecessary pain, suffering or distress to and/or death of an animal." This definition does not include the killing involved in meat production and hunting.

Animal abuse can take many forms. This includes physical torture and death, sexual abuse, psychological abuse, and neglect, which is a failure to provide adequately for basic needs such as food, water, shelter, and medical care. These types of animal abuse are similar to patterns of child abuse. In many societies pets are often viewed as family members. It is not surprising to find that pets and other nonhuman animals can be victims just like human family members.

Who abuses animals? Animal abuse is committed by people of all ages from young children to the elderly. Males are more likely to abuse animals than females. Some of the many possible reasons that a person may abuse an animal are:

- to control or train an animal;
- to make the animal itself become aggressive (as in a fighting or attack dog);

- to harm or threaten to harm an animal such as a pet in an attempt to control the pet's human;
- to punish the pet's human;
- to entertain others by shocking them with violence toward an animal;
- to express dislike of a species of animal; or
- to harm an animal in anger instead of the person they are really angry at, because the animal cannot retaliate.

Regardless of the reason for the animal abuse, research studies show there is a strong probability that a person who harms an animal will also engage in acts of violence toward other humans. This does not mean that harming an animal causes a person to harm humans. What this means is that animal abusers are more likely to hurt people than those who do not abuse animals.

News reports provide many examples of this problem. The FBI states that a childhood pattern of animal abuse (including carving up stuffed animals or photographs of animals) is a risk factor for future violence. Adolescents who were responsible for seven different school shootings between 1997 and 2001 had histories of animal abuse (Barnard & Hogan, n. d.). Many infamous serial killers had histories of animal abuse including Ted Bundy, Albert DeSalvo (The Boston Strangler), and Jeffrey Dahmer, who killed at least 17 people.

Children who have a mental health diagnosis of Conduct Disorder are often cruel to animals. Conduct Disorder is a serious childhood mental illness, which is associated with a child becoming a violent juvenile offender. In fact, animal cruelty may be one of the earliest diagnostic symptoms to appear in these children. Cruelty to animals is often an aspect of behavior that distinguishes children with mental health problems from children who do not have recognized mental health problems.

In one study of over 5,000 children from ages 4 to 16 (conducted by Achenbach and colleagues and reported by Ascione 2001, p.2) children's caregivers were asked how often their children injured animals. For every age level it was found that both boys and girls who had been referred to mental health services had higher rates of animal cruelty than children who had not been referred. These cruelty rates ranged from the highest of 34% for the 4–5 year old boys with mental health referrals to the lowest rates of near zero percent for older girls who did not have a mental health referral. At all age levels the boys with mental health referrals had the highest rates of cruelty, followed by girls with mental health problems, then boys with no mental health referrals and finally the lowest rates were the girls with no mental health referrals.

Of course there are problems with this type of study. Caregivers may be reluctant to tell the truth about a child's bad behavior. They may not know about a child's cruelty, since children would be likely to hide animal cruelty and other unacceptable behavior from adults. For these reasons the real frequency of cruelty is likely to be higher than what the study found. Other research confirms that parents do underestimate how often children engage in animal abuse. When parent reports of abuse are compared with the child's self-report of animal abuse, the children report more abuse than the parents. This problem aside, a study with such a large sample should be taken seriously. It clearly shows there is a relationship between the frequency of a child engaging in animal cruelty and the child having mental health problems.

Ascione also reported a study by Youssef and colleagues of over 2,000 adolescents who were not being treated for mental health problems (2001, p.3). This study found that children who reported being violent toward other people were more than 4 times as likely to also report being violent toward animals than children who did not report violence toward humans.

This connection between animal abuse committed by children and continued risk for violent antisocial behavior is also found in studies of male and female adult jail inmates. Several studies show that criminals jailed for violent crimes have much higher rates of past animal abuse than criminals jailed for nonviolent crimes. Individuals with a past history of animal abuse are much more likely to commit a variety of different crimes including drug and violent crimes than people with no history of animal abuse.

Animal abuse is not limited to children. Animal abuse occurs in families that are experiencing different kinds of problems. Parents may threaten to harm, injure, or dispose of a child's pet to punish a child for misbehavior. One veterinarian told of a husband who brought a young healthy dog into her office to be put to sleep. The veterinarian was suspicious and called the wife before proceeding. The wife said the couple was divorcing. The man was going to have the dog killed to punish his wife for the divorce. The dog in this case was returned unharmed to the wife.

Pets living in families in which there is domestic violence are at risk to be abused. Studies show that people who abuse a spouse may also threaten, injure, or kill a family pet in order to control or punish family members. The frequency of this pet abuse problem ranges from 46.5% to 71% of the pet-owning women who seek safety in shelters. These episodes of violence toward the family pet often occur in front of the children. It is typical that the children will copy these abusive behaviors. (Ascione 1998; Flynn 2000)

It is common that women who want to escape domestic violence by fleeing to a shelter delay going because they are afraid of the harm that might be done to a pet who must be left at home. These fears are justified. For example, a shelter worker reported a case of a husband cutting off a dog's ears with garden shears and sending the photo and the ears to the wife who had escaped to the shelter. It is important to know in these situations that not only is there risk to the pet but the pet is often a source of support for the abused person. So a woman fleeing to a shelter must both risk injury to her pet and leave behind the pet who is probably an important source of emotional support.

The Humane Society of the United States has recently reported on the link between forms of human and animal violence appearing in the elderly. Senior citizens living alone who may be unable to care for themselves may have pets whom they are also not able to care for. In these cases a neglected pet can be a warning that the person needs help in self-care.

There is another other type of animal abuse associated with the elderly. A person who is providing care for an elderly person may neglect or harm the pet as a way of controlling the elderly person. The caregiver may also take funds, which were meant for the pet's care for their personal use. Family members are most often the ones who abuse elders. The states of Illinois and California have realized this link. They now have programs that require cross reporting of elder and animal abuse.

It should be clear from this brief report that animal abuse is a common occurrence and it is not limited to any segment of the population. Humans of all ages commit animal abuse. Animal abuse is a symptom of a serious mental health problem. People who abuse animals are more likely to harm other people than people who do not abuse animals.

What can be done to end animal abuse? People can become aware that animal abuse is a serious problem not only for the animal victim but for people too. Workers in helping and health professions can be trained to ask about the well being of pets in a family as part of their normal routines. These professionals can be trained to know risks associated with animal abuse and how to help when someone reports animal abuse. Agencies that deal with human and animal victims can cooperate in identifying both humans and animals who may be in harm's way. Schools can adopt policies that require teaching kindness and respect for all living creatures when teaching citizenship.

See also Animal Abuse—*Animal and Human Abuse*

Further Resources

Ascione, Frank R. 1998. *Battered women's reports of their partners' and their children's cruelty to animals.* In: *Cruelty to Animals and Interpersonal Violence: Readings in Research and Application* (Ed. by R. Lockwood & F. R. Ascione) pp. 290–304. West Lafayette, IN: Purdue University Press.

Ascione, Frank R. 2001. *Animal abuse and youth violence.* OJJDP Juvenile Justice Bulletin, September, 2001. Retrieved June 5, 2003, from http://www.ncjrs.org/pdffiles1/ojjdp/188677.pdf

Ascione, F. R. & Arkow, P. (Eds.). 1999. *Child Abuse, Domestic Violence and Animal Abuse: Linking the Circles of Compassion for Prevention and Intervention.* West Lafayette, IN: Purdue University Press.

Barnard, N. D. & Hogan, A. R. (n. d.) Animal abuse: The start of some things bad. Retrieved 7/6/2003 http://www.pcrm.org/issues/Commentary/commentary0105.html.

Elder abuse and animal cruelty. (n. d.). Humane Society of the United States. Retrieved 5/30/2003, from http://www.hsus.org/ace/19053.

Flynn, C. P. 2000. *Battered women and their animal companions: Symbolic interaction between human and nonhuman animals.* Society & Animals, 8(2). Retrieved June 22, 2003, from http://www.psyeta.org/sa/sa8.2/flynn.shtml.

Flynn, C. P. 2001. *Acknowledging the "zoological connection": A sociological analysis of animal cruelty.* Society & Animals, 9(1). Retrieved June 22, 2003, from http://www.psyeta.org/sa/sa9.1/flynn.shtml.

Lockwood, R. & Ascione, F. R. (Eds.). 1998. *Cruelty to Animals and Interpersonal Violence: Readings in Research and Application.* West Lafayette, IN: Purdue University Press.

Kathleen C. Gerbasi

■ Animal Architecture
Subterranean Ant Nests

The animal kingdom presents us with a huge variety of animals that are consummate builders and architects. Alongside the familiar examples of bird nests and beaver dams, we find wasps that build nests of paper, mud or saliva; beetles that drill elaborate galleries in wood; caddis fly larvae that construct movable houses from tiny pebbles or pieces of cut leaves; others that build seine nets to trap floating creatures for food; termites that build huge castles of clay; and rodents that build large mounds of earth or plant debris (including cactus pieces) as shelter. The list is very long.

Insect builders and the structures they build have received considerable attention, but insects that excavate nests in the soil have not. I have been interested in the kind of architecture produced by ants when they excavate nests in soil—that is, architecture produced by removal, rather than construction. Throughout most of the world, ground-nesting ants are a common feature of the landscape, but aside from dumps of excavated soil around their nest entrance, there is little hint of what lies underground. There are two basic ways in which this underground structure can be studied—we may excavate the nest carefully, exposing and mapping each chamber and tunnel. This produces a reasonable record of the architecture, but also provides information on how the various members of the ant colony are distributed within the nest. Alternately, we may make a cast of the hollow space by filling it with a hardening liquid of some type. Once hardened, the cast is excavated and mounted for study or display. Information on the distribution of colony members is mostly lost. The most successful casting materials are dental plaster and molten aluminum or zinc. Plaster casts are fragile, and usually come out of the ground in a number of pieces that must be reconstructed with support. Molten metals produce strong casts that need little support, but the equipment needed to melt metal in the field is considerable. All methods of casting work best in very sandy soil.

Casting produces beautiful, often spectacular structures that bring home the abilities of ants as architects. The nest of the Florida harvester ant (seen as a plaster cast in the photo) was almost 3 meters deep, contains 135 chambers and 12 meters of vertical tunnels. Building it required the removal of about 20 kg (44 lb) of sand by about 5,000 worker ants weighing a total of 20 g (.7 oz). The colony moves to a new nest once or twice a year, and when they do, they excavate the new nest in a remarkable 5 days.

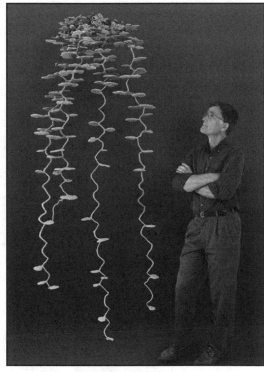

A plaster cast of a Florida harvester ant nest.
Courtesy of Walter R. Tschinkel

The nest's architecture is intimately related to the colony's social structure and life cycle. The ant brood (immature ants) are located mostly in the bottom third of the nest, as are the youngest workers that emerge from them. These young workers take care of the brood, but as they age, they gradually drift upward in the nest, taking on more general tasks such as transporting, storing and preparing food (mostly seeds in this species), or repair and maintenance work. Chambers filled with stored seed are located in the upper third of the nest, but never closer to the surface than about 50 cm (20 in). Still older workers are located in the chambers near the surface. Here they receive food brought in by foragers, take out the trash and discard it, and defend the nest against intruders. Only the very oldest workers leave the nest to forage for insects and seeds, and live only a few more weeks once they have begun doing so. This vertical arrangement by age is not passive—young ants displaced to upper chambers return to lower ones, and old workers displaced to lower chambers return to upper ones. Ants somehow "know" where they are. Moreover, when a nest moves, the workers, brood and seeds come to reside in similar relative locations in the new nest.

All harvester ant nests show similar elements arranged in typical patterns. Chambers are larger, more complex in shape and closer together near the surface, becoming smaller, simpler and farther apart with depth. The tunnels connecting them are helical, like spiral staircases, and chambers open off the outside of the helix.

Of the 15 species of ant nests I have cast, most consist of a more-or-less vertical tunnel(s) connecting horizontal chambers, but there is great variation in the expression of these basic elements—chamber size, shape, density and number vary, as do the number, depth and size of the tunnels. Nests of some species consist of a single tunnel/chamber unit, whereas others may contain dozens or even hundreds of them packed close together.

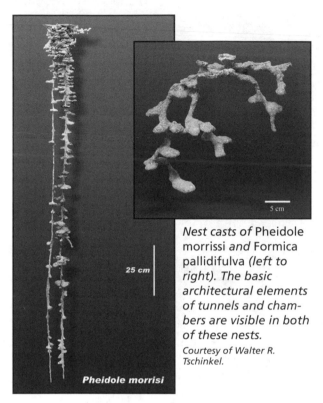

Nest casts of Pheidole morrissi *and* Formica pallidifulva *(left to right). The basic architectural elements of tunnels and chambers are visible in both of these nests.*
Courtesy of Walter R. Tschinkel.

It is primarily the variation of these basic elements that gives each species of ant nest its species-typical appearance. The next two figures are examples of the nests of two species of ants. Until proven otherwise, we can assume that nest architecture is somehow linked to social function, and that the nest helps to organize the colony, much like human buildings serve to organize the work of those within.

Ants build these intricate nests without a blueprint, without a boss and in total darkness. How they do this is only dimly understood. By "understood," we mean primarily that we have a set of working hypotheses that await testing. Generally, it is likely that all the necessary behavioral programs reside within the nervous system of each individual worker. This program also provides ways in which each worker interacts with the nest under construction as well as with other workers, and is probably subject to both positive and negative feedback that stimulate an increase in some actions and a reduction in others. For example, it is possible that when a worker detects other workers digging, it is motivated to dig in the same area. This leads to organized chamber enlargement or tunnel extension. The details of these interactions lead to a particular chamber or tunnel size and shape. At the opposite extreme, digging may cease when the density of digging workers at a growing chamber margin falls below a certain threshold, making chamber size proportional to the available work force. Through such interactions, the architecture of the excavated nest emerges from the collective action of many workers. The ability to create these architectures is distributed across many workers, rather than being embodied in one or a few. Most such hypotheses have yet to be tested.

Further Resources

Collias, N. E. & E. C. Collias (Eds.) 1976. *External constructions of animals*, Halstead Press (professional level, contributed papers).

Conway, J. R. 1983. *Nest architecture and population of the honey ant*, Myrmecocystus mexicanus *Wesmael (Formicidae), in Colorado*. Southwestern Naturalist 28, 21–31.

Flegg, J. & Hosking, D. 1990. *Animal Builders*. London: Belitha Press.

Frisch K. von. 1974. *Animal architecture*. New York: Harcourt Brace Jovanovich.

Hancocks D. 1973. *Master Builders of the Animal World*, New York: Harper & Row.

Hansell M. H. 1984. *Animal Architecture and Building Behaviour*. London, New York: Longman.

Hölldobler, B. & Wilson, E. O. 1990. *The Ants*. Cambridge, MA: Belknap/Harvard Press.

McClintock, J. 2003. *The secret life of ants*. Discover 24(11), 77–81.

Tschinkel, W. R. 1987. *Seasonal life history and nest architecture of a winter-active ant* Prenolepis imparis. Insectes Sociaux, 34, 143–164.

Walter R. Tschinkel

■ Animal Models of Human Psychology
An Assessment of Their Effectiveness

About 8–10% of psychological research involves using nonhuman animals in laboratory settings (APA 1984). The purpose of most of this research is to use animals as models of human behavior. By comparison, other research by comparative psychologists and ethologists studies other animals primarily to understand them, not humans; it is often conducted in their natural habitat rather than the laboratory and typically involves less suffering and harm to the animals.

The use of nonhuman animals as models of human behavior is one of several available research strategies—others include clinical studies of patients and epidemiological studies comparing differences, such as life-style, between patients and normal populations.

The animal model strategy involves a trade-off. By keeping nonhuman animals in laboratories, psychologists gain experimental control—they can manage the conditions to which animals in the experiment are subjected, including the conditions in which animals are raised. This often increases reliability, the confidence in the consistency and replicability (repeatability) of the results. However, the investigators give up direct observation of the human behavior that is the actual object of study. For no matter how carefully constructed, the model is never identical with the modeled (LaFollette & Shanks 1996). A model is always only an analogy, consisting of similarities and differences. The indirect study involved in the strategy of using models decreases validity, the measure of the match between the object studied in the research and the actual object of study. For this reason, animal model research always requires a test of "extrapolability." The investigator must test the results of studying the animal model to see if they also obtain in the human setting. This also means that no model ever becomes a fully trustworthy stand-in for the modeled. Validation is an ongoing process.

Psychologists have attempted to develop models of most human problems with a psychological component: anxiety, depression, schizophrenia, ulcers, obesity, child abuse, aging, headache, cigarette smoking (Shapiro 1998). Psychological research involves behavioral, cognitive, and emotional variables, particular measurable aspects of the model. However, increasingly, it also involves physiological variables and is often not distinguishable from biomedical research. For example, to produce an animal model of anxiety, investigators can: breed animals selectively to obtain a more timid and anxious disposition; perform surgery or electrical stimulation on the nervous system to produce the experience of anxiety or susceptibility to it; and create environmental conditions that induce anxiety (deprivation of care, introduction of feared objects).

A recent review of animal models of anxiety disorders (Zinbarg & Mineka 2001) describes studies in which investigators bred rats for timidity and found them more susceptible to the development of intense fears and phobias (fear of specific objects). They induced anxiety through conditioning, linking formerly emotionally neutral stimuli to negative consequences (electric shock). Other studies tested whether simply observing another animal's conditioning can produce intense fears. Finally, studies investigated the relation of perceived control over the environment to frightening or traumatic experiences. Monkeys reared in the first year of life in a condition in which they have no control over the receipt of food or water are then tested in frightening situations. They are more susceptible to the emotional consequences of fear-provoking situations than are monkeys reared in conditions where they can control their environment.

Contemporary debate over the ethics of our use of nonhuman animals in biomedical and behavioral research keys on both issues of rights and of usefulness. If the strategy of using animals as models is effective, then, according to this latter utilitarian ethic, it is not only justified, but a positive good.

Most philosophers agree that nonhuman animals should be given some ethical consideration, although they differ on whether or not and to what degree these ethics limit the use of nonhuman animals. However, scientists agree less on the effectiveness of animal research and, to date, have devoted less effort examining it.

The question of effectiveness is complex. An animal model may pass a test of validation but still not provide new information or improve treatment of the modeled disorder. As important and necessary as is the test of validation, gains in understanding and treatment success are the bottom line considerations—not the degree of similarity between the model and the modeled. An animal model could be very similar to the modeled and still not be informative; contrarily, a relatively dissimilar or low fidelity model could provide a new insight. In fact, investigators can learn from differences as well as from similarities.

Sorting out the contribution of animal-based research from insights originating in clinical observations is difficult as often these two enterprises are competing for the same research dollars (Greek & Greek 2000) and are not aware of or responsive to each other's research (Shapiro 1998). Following a line of research suggested by animal models can be misleading (Kaufman & Hahner 1991) or can shape research in directions that are not the most productive available (Drewett & Kani 1981). An animal model can be effective but not necessary because of the existence of alternative strategies equally or more effective. An animal model effective in the past may no longer be necessary as new and sophisticated technology is developed that allows the direct observation of clinical populations of humans (e.g., imaging techniques that allow observation of brain functioning in conscious subjects).

Finally, any gains in understanding and treatment success must be offset by the costs to the animals in suffering. As measured by various instruments (Scale of Invasiveness, Shapiro & Field 1987), the degree of pain, distress, and harm involved in animal research is considerable.

The evaluation of animal research is a scientific question: We can devise studies to answer these questions empirically. It is also a historical question: We need to examine the past record of scientific work to evaluate the sources of gains in understanding and treatment success.

The animal model research does not itself test whether its results are providing new information or gains in treatment success. Recently Shapiro (1998) and Dagg (1999) have developed and applied methods, borrowed from the social sciences, which directly test the effectiveness of selected animal models. Shapiro applied a battery of tests to evaluate costs and benefits of a set of animal models of eating disorders (anorexia, bulimia). The battery includes *outcome studies* (studies of the effects of treatment), *citation analysis* (a technique of evaluating the impact of published research by counting the number of times it is cited in relevant scientific literature), survey, and a measure of suffering borne by the animals. Results show that animal models of eating disorders have not contributed to more effective treatment, which remains relatively unsuccessful. The frequency of citation of animal model studies is relatively low compared to the average published scientific study. A survey of clinicians specializing in the treatment of eating disorders found that they were largely unaware of and uninfluenced by animal model studies. Application of the Scale of Invasiveness found that these animal studies typically involve considerable pain, distress, and harm.

Using analysis of a set of animal model studies by Canadian researchers, Dagg found that a large number of them were cited rarely. In addition to measuring invasiveness, she included data about the numbers of animals involved in the studies as a further measure of costs to the animals. Both she and Shapiro recommend that animal care committees (committees that oversee the use of animals in research) use these various measures to evaluate proposed ongoing research.

In addition to increased scrutiny of the effectiveness of animal models in the study of human psychology, future developments that will shape the science-based component of the debate include dramatic innovations in genetic engineering and technologies that allow a return to direct observation of humans. The impact of the former is difficult to predict as it involves work on both human and nonhuman genomes. In addition to these developments, advances in moral philosophy and the pressure of animal rights organizations will impact public attitudes and policies toward animal research.

In conclusion, the strategy of using animals as models of human psychology and psychological disorders raises ethical questions. It is also a trade-off in terms of science. Evaluation of this science is difficult and has not been adequately addressed to date. It is not simply a matter of comparing similarities and differences between the model and the modeled, a difficult task in itself. It involves the evaluation of the gains in understanding and treatment success contributed by animal models compared to other available and evolving approaches.

Further Resources

American Psychological Association. 1984. *The use of animals in psychological research. Backgrounder*. Washington, DC: Author.

Dagg, A. 1999. *Responsible animal research: Three "flags" to consider*. Journal of Applied Animal Welfare Science, 2, 337–346.

Drewett, R. & Kani, W. 1981. *Animal experiments in the behavioural sciences*. In *Animals in Research* (Ed. by D. Sperlinger), pp. 175–201. New York: Wiley.

Greek, C. R., & Greek, J. S. 2000. *Sacred Cows and Golden Geese: The Human Cost of Experiments on Animals*. New York: Continuum.

Kaufman, S. & Hahner, K. (Eds.). 1991. *Perspectives on Medical Research*. Volume 1. NY: Medical Research Modernization Committee.

LaFollette, H., & Shanks, N. 1996. *Brute science: Dilemmas of Animal Experimentation*. London: Routledge.

Overmeir, J. B. & Burke, P. D. (Eds.). 1992. *Animal Models of Human Pathology: A Bibliography of a Quarter Century of Behavioral Research, 1967–1992*. Washington, DC: Glossary.

Shapiro, K. J. 1998. *Animal Models of Human Psychology: Critique of Science, Ethics, and Policy*. Seattle: Hogrefe and Huber.

Shapiro, K. J., & Field, P. 1987. *A new scale of invasiveness in animal experimentation*. PSYeta Bulletin, 7, 5–8.

Zinbarg, R. E. & Mineka, S. 2001. *Understanding, treating, and preventing anxiety, phobias, and anxiety disorders*. In: *Animal Research and Human Health* (Ed. by M. E. Carroll & J. B. Overmier), pp. 19–29. Washington, DC: American Psychological Association.

Kenneth Shapiro

■ Animals in Myth and Lore
Animals in Native American Lore

As in cultures all over the world, animals play roles in the myths and legends of Native America. In North America, myths featuring Coyote are especially common; other prominent animals are Wolf, Bear, Raven, and Eagle. In many cases, the behavior of the mythic

An ancient Native American petroglyph of a coyote in New Mexico.
Courtesy of the National Park Service

animal can be seen to reflect that of the "real," biological animal. Coyote, in particular, plays the combined role of clown, trickster, and culture hero for many North American tribes, and this reminds us of the raffish behavior often observed in *Canis latrans*.

In fact, in Native California and some adjacent areas of North America, the mythic animals are much more than characters in stories; they represent a race of beings who preceded human beings on earth, who laid down the laws and customs that humans should follow. They are sometimes called "The Spirit People" or "The First People." Indeed, they maintain an important presence to this day. The person called Bear still exists in the spirit world; humans who want to hunt bears pray to him, requesting permission to kill him. But he also exists in the form of every individual bear in this material world, who may then be killed by the hunter who has prayed effectively. For many tribes, Coyote is the most important of the Spirit People, and Native Americans have sometimes called him "the Indian God." To be sure, he did not actually create the world, but he *organized* it in the way we know today; he *ordained* the way people should live. For many tribes, it is Coyote who brings fire to humankind, who gives them salmon and acorns, and who in fact invents death, not only for the ecological motive of preventing evolution, but also for a psychological motive—"so that people will be sorry for each other."

A simple story that illustrates this is from the Karuk tribe, who live on the Klamath River in northwestern California; it shows how Coyote "laid down the law" for the respective roles of men and women—"so that they wouldn't get lazy." (Some Native American stories are shortened here; the originals could last all night.)

> The First People said, "Let the river flow *down*stream on one side, and *up*stream on the other side; let it be that way." So all right, when they traveled *down*stream by boat, they drifted down, downstream. But they'd travel back up on the other side of the river; they'd drift *up*stream too, as it flowed upstream, that water. And then Coyote said, "No way! Let it not be like that, let it all flow *down*stream. Let the young husbands have to *push* their way up there, when they row upstream."
>
> And then again the First People said, "Women carry their packbaskets Uphill, up there they put firewood in them, they make basket-loads. Then the women leave for home; and they just leave them there, those basket-loads." And they said, "They'll just *walk* home, those basketloads." And then Coyote said, "That's not right! No way! Let the young wives just *carry* the loads." So that's how it is; now they don't walk any more, those basket-loads don't.

This story may not show much relationship to the character of the biological coyote, but it illustrates the powerful role of Coyote in myth. Other texts remind us more of *Canis latrans* in his omnivorous, scavenging aspect. Corresponding to the white man's saying that "A coyote will eat anything that doesn't eat it first," we have the following narrative from the

Cupeño Indians of southern California. (Readers should be warned that these stories are not intended as cute animal tales for children; they can be grim and bloody, and they were intended for all ages, like the Ancient Greek myths or Shakespeare's *Macbeth*.)

> So Coyote was living there. And then Coyote said, "We have no food put away." And Coyote went hunting, he fetched his bow. And he went, he took his daughter. So he kept circling around in a wash; his tail went switch-switch-switch. And his daughter kept coming slowly behind him. So he kept jumping, with his daughter, and he would catch an ant. And he kept coming, coming, coming; they arrived there. And Coyote's wife was there at home; she was gathering all the grass, to cook something there.
>
> And then it was evening, and Coyote didn't see anything to kill. So he grabbed his daughter; he hit her with a stick. And he killed that daughter of his. And from there he carried her home on his back. So he arrived there; his wife was by the door. So then he took all the hair off his daughter. And there was a big place for cooking; he cut up his daughter there, Coyote did. And he cooked her; he boil-boil-boiled her; her meat, it got real green.
>
> And then she came from there, his daughter. And she spoke as she came, "Pine tree, stick, pine tree, grass stick, jingle-ingle-ingle." So the mother looked around, "Ah! it must be her voice I hear," she said. "She's turned into a ghost," she said. "Well," said the wife, "this guy must have killed her." So right then she got up, the old woman Coyote. And she gathered all the grass while Coyote was cooking his food, and she burned-burned-burned everything. So then their house went up in smoke. And Coyote said, "Oh, old woman, don't burn me!" So all that house of theirs burned, and Coyote burned up, he died.

Observers of coyotes report on their round-the-clock activeness, which seems to include a strong element of playfulness. As one writer says, "Coyotes *do* have a sense of humor. How else to explain, for instance, the well-known propensity of experienced coyotes to dig up traps, turn them over, and urinate or defecate on them?" The Nez Perce tribe of Idaho tell this story:

> Two coyotes went up the river. From there they saw people living down by the river. And each friend said to the other, "Go ahead." Then each said, "No, you go." Then the first said, "So you go first; they'll see you, and soon they'll say, "That coyote's going on the trail." The second said, "*I'm* not a coyote." The first said, "But you're the same as me, we're just alike, and we're both coyotes." The second said, "No, I'm really Anutherwun." So they argued. Then the second said, "Now you go first." Then people said, "That coyote's going upriver." Then his friend said, "See? That's what they said about you; you're a coyote." And the first said, "Then you go too; they'll say the same about you."
>
> The second said, "All right, now I'll go." And he too slowly started walking on the trail from there. And people said, "Ah, yet another one." Then the second came and said, "See? I'm not a coyote, I'm Anutherwun. See, the people said I'm Anutherwun." That's all.

Among the most salient characteristics of the biological coyote is its opportunism and adaptability. Sheepmen and trappers are quoted as saying, "The coyote will be here long after we are all gone!" It is estimated that there are now more coyotes within the Los Angeles city limits than there were in aboriginal times; they thrive on garbage, and on people's perception of them as cute. As white people have populated the continent, the original range of *Canis latrans*, in the western United States, has expanded eastward; the animal is now found from Alaska to Central America, and from the Pacific to the Atlantic.

This aspect of the coyote turns up in Native American myths repeatedly: Coyote "drowns," but immediately afterward he is seducing teenage girls. He "dies," but a few minutes later—like Wyle E. Coyote in the cartoons, when he falls off a cliff or gets blown up with dynamite—Old Man Coyote is back, and he's planning the life of the world to come. The end of a Karuk myth gives one version:

> When his corpse was thrown in the river, he floated ashore at Requa, and he lay there on the beach. After a while the yellowjackets ate him, and he lay there like that. Soon the ants ate him, too, and he lay there like that. By now there were just bones lying there, and still he lay there like that. But there was meat there in his testicles. And the yellowjackets thought, "Let's eat this!" And so they ate that meat. And when they bit it, then Coyote jumped up, and he said "Atutututututu!" And he picked up a stick, and he hit them. Coyote did that; he almost clubbed through them. That's why they all have small waists.

It turns out, of course, that many characteristics of the biological coyote are also typical of another animal, *Homo sapiens*. Both species are tricky, gluttonous, humorous, pragmatic, opportunistic omnivores, skilled at survival. In fact, Old Man Coyote, who was, after all, created in the human mind, often seems more like human beings than like *Canis latrans*. For instance, the mythic Coyote is insatiably lecherous, whereas biological coyotes are rather faithful as mates and parents, at least for a year at a time. Some writers have gone so far as to say, "We have met Coyote, and he is us." But it may be closer to the truth to say that he is a contradiction and at the same time a mediator; he teaches us what *not* to do, he bridges the gap between *Canis* and *Homo*, and he brings us humans closer to our fellow animals.

The role of trickster and culture hero in Native North American myths seems to be assigned especially often to creatures that violate the distinction between herbivores and carnivores—animals that are omnivorous, and indeed carrion-eating. In some tribes of the Pacific Northwest, this role is filled by the raven. Another omnivorous animal which plays important mythic roles, though generally not that of trickster, is the bear—which, as some Indians say, "is almost a person." Indeed, the bear eats many of the same wild foods as humans, and the two species are serious competitors for such items as berries, gophers, and salmon; it is perhaps natural to think that they might interbreed. The story of "the girl who married the bear" is widespread in the Pacific Northwest, the Subarctic, and even in Siberia. This version is from the Tagish tribe of the Yukon:

Bears play a prominent role in many Native American myths.
© Anthony F. Chiffolo.

> Once there was a girl who picked berries in the summer. She went with her family, and they picked berries and dried them. When she went with her women folk, they would

see bear shit on the trail. Girls have to be careful about bear shit; they shouldn't walk over it. But this girl always jumped over it and kicked it.

When she was quite big, they went out picking berries. She saw some bear shit. She said all kinds of words to it and kicked it and jumped over it. When they were all coming home, the girl saw some nice berries and stopped to get them. The others went ahead. When she had picked the berries and was starting to get up, her berries all spilled out of her basket. She leaned down and was picking them up.

Suddenly she saw a young man. He was very good looking. He had red paint on his face. He stopped and talked to her. He said, "Those berries you are picking are no good. Let's go up a little ways and fill your baskets up. There are some good berries growing up there. I'll walk home with you. You needn't be afraid."

When they had picked them all, he said, "It's time to eat." He made a fire. They cooked gopher, quite a lot of it, and they ate some. Then the man said, "It's too late to go home now. We'll go home tomorrow. It's summer, and there's no need to fix a big camp."

So they stayed there. When they went to bed, he said, "Don't lift your head in the morning and look at me, even if you wake up before I do." So they went to bed. Next morning they woke up. The man said, "Don't be afraid; I'm going home with you." Then he slapped her right on top of her head, and he put a circle around the girl's head the way the sun goes. He did this so she would forget.

Then after this she forgot all about going home. She just went around with him picking berries. Every time they camped, it seemed like a month to her, but it was really only a day. They started in May. They kept traveling and going. At last she knew. They were traveling again, and it was getting late. And she came to her senses and knew it.

It was cold. He said, "It's time to make a camp. We must make a home." He started digging a den. She knew then that he was a bear. She knew her brothers used to go there to hunt and to eat bear. In the spring they took the dogs there, and they hunted bears in April.

Just when the man was digging, he looked like a bear. The rest of the time he seemed like a human being. The girl didn't know how else to stay alive, so she stayed with him as long as he was good to her.

It was really late in the fall. He said, "Well, I guess we'll go home now. We have enough food and berries." Really they went into the den. They stayed there and slept. They woke up once a month and got up to eat. Soon the girl found that she was carrying a baby. She had two little babies—one was a girl, and one was a boy. She had them in February in the den. This is when bears have their cubs.

[One night the bear said:] "You're my wife, and I'm going to leave soon. It looks like your brothers are going to come up here soon, before the snow is gone. I want you to know that I am going to do something bad. I'm going to fight back."

"Don't do it!" she said, "They're my brothers. Don't kill them; let them kill you! You have treated me good; why did you live with me if you are going to kill them?"

"Well, all right," he said, "I won't fight, but I want you to know what will happen!"

Next morning he said, "Well, it's close, it's close. Wake up!" Just when they were getting up, they heard a noise. "The dogs are barking," he said. "Well, I'll leave." When he went, he said, "You are not going to see me again!"

For a long time there was no noise. She went out of the den. She heard her brothers. They had already killed the bear, and she felt bad.

At this point the girl is reunited with her family. But her children are half bear; when they suck at her breasts, their claws and teeth make the blood stream down. Ultimately she and her children turn completely into bears, and vanish into the forest.

Stories like these are still told and memorized among American Indians—and new stories, especially about Coyote, are still being made up. With the recent renewal of interest in

Native American traditions and in Native spirituality, among both Indians and whites, these narratives remain richly alive. We are learning how to learn from them, and they will continue to teach us.

See also Animals in Myth and Lore—*Fairy Tales and Myths*
of Animal Behavior

Further Resources

Bekoff, M., Ed.1978. *Coyotes: Biology, Behavior, and Management*. New York: Academic Press.
Bright, W. 1993. *A Coyote Reader*. Berkeley: University of California Press.
Leydet, F. 1977. *The Coyote: Defiant Songdog of the West*. San Francisco: Chronicle Books.
McClellan, C. 1970. *The Girl Who Married the Bear*. (National Museum of Man, Publications in ethnology, 2.) Ottawa: National Museums of Canada.
Rockwell, D. 1991. *Giving Voice to Bear: North American Indian Myths, Rituals, and Images of the Bear*. Niwot, CO: Roberts Rinehart.

William Bright

■ Animals in Myth and Lore
Fairy Tales and Myths of Animal Behavior

The many myths of animal behavior would take a lifetime to explore. Some are unfair: Wolves are bad. Some are complimentary: Owls are wise. Our human inclination to characterize animals with human traits is so strong, it has even penetrated spiritualism, giving rise to totems, shamanism, and heroic chieftains adorned with the talons or fangs of slain animals to embody their strength and courage. Many understated myths are hilariously contradictory: Animals live as "mom, dad, and the kids" but every animal is "he." Others are not funny: Animals lack emotions so feel nothing when their children are lost but are otherwise aggressive and vengeful. Mythology spills onto some animal behaviorists as well: Naming animals makes it harder to study them objectively; scientists do not get attached to their study animals if they do not name them; and scientists feel nothing for the animals they spend their lives trying to understand.

This essay considers myths about animal behavior perpetuated in fairy tales. Fairy tales lie quietly between the two greatest influences on our thinking about animal behavior, René Descartes and Charles Darwin. Descartes, a 17th century philosopher—famous for his *cogito ergo sum*, "I think, therefore I am"—used logic and reason to conclude that animals had neither mind nor feelings. The unfortunate Cartesian consequence was that animals were reduced to furry, feathery, or scaly machines who were mentally and morally inferior to people. Descartes created a huge unassailable chasm between humans and animals. Darwin, a 19th century biologist, introduced evolution with a way animals might have appeared on Earth without passing through the Garden of Eden. Species arose from natural selection, a blind process in which traits that worked well were passed down across generations (they were "selected" for), whereas those that did not and were eliminated (were "selected" against). Natural selection was like artificial selection, used by breeders of everything from roses to racehorses, except that breeders selected traits purposefully, but natural selection selected them blindly. Natural selection implied that all animals ultimately arose from the same primeval soup. Darwin thus made it impossible for us to ignore our own animal status and forged a resolute link between humans and the other animals. Donald R. Griffin transformed the study of animal behavior by suggesting publicly that animals are intelligent, aware, and can reason.

As if in grand counterpoint to these conflicting views, we have all been raised on fairy tales. In *faërie*, the realm where fairies exist and magic is never questioned, animals are unabashedly human. The main myth about animal behavior in fairy tales is probably that animals act like people and authors have therefore committed *anthropomorphism*, the sin of inappropriately giving animals human characteristics. But have they? Does blatant anthropomorphism in fairy tales teach us that animals *must* be very different from us—*never* summon courage, feel a commitment, or get sentimental—because they don't act in real life the way they act in fairy tales?

Or, do fairy tales teach us the opposite? Do they verify the commonality between humans and animals we all sense? Maybe this is why we easily accept moral lessons from talking animals who exemplify honor over dishonor, industry over sloth, and humility over pride. The consummate storyteller Hans Christian Andersen said in "The Ice Maiden" that young children can talk to the animals: Rudy was a quiet little Swiss boy who listened carefully to his grandfather's lessons but learned more

> . . . from other sources, particularly from the domestic animals who belonged to the house. One was a large dog, called Ajola, which had belonged to his father; and the other was a tom-cat. This cat stood very high in Rudy's favor, for he had taught him to climb. When the cat said, "Come out on the roof with me," Rudy quite understood him, for the language of fowls, ducks, cats, and dogs, is as easily understood by a young child as his own native tongue.

Andersen contends that "some children retain these ideas later than others, and they are considered backwards and childish for their age. People say so; but is it so?"

Hans Christian Andersen first published "Tales Told for Children" in 1835, a set of four tales that included "The Princess and the Pea." The immediate—and enduring—popularity of Andersen's tales is ironic because, as an aspiring actor and playwright, he thought the tales a waste of time. Many, like "The Ugly Duckling," are biographic, revealing Andersen's personal anxieties while touching on universal themes. Others are satiric, like "The Emperor's New Suit." He said he always told a story to the little ones but remembered to give mother and father something to think about as well. Memorable tales include "The Little Mermaid," "The Snow Queen," and "Thumbelina."

In "Thumbelina," animals are unabashedly human and readers may not see them as animals, nor indeed, Andersen's many slurs on society, since the tale has the preoccupying theme of the female role in betrothal. Thumbelina was a beautiful inch-high maiden born from a barleycorn and cradled elegantly in a polished walnut shell with a fragrant rose leaf for a quilt. Kidnapped, Thumbelina was aghast when an ". . . old toad bowed low to her in the water, and said, "Here is my son, he will be your husband, and you will live happily in the marsh by the stream" and all her son could say for himself was "croak, croak, croak." The toad swam away with the elegant little bed, leaving Thumbelina to sit alone on a green leaf to weep at the thought of living with the old toad and having her ugly son for a husband. The little fishes who swam about in the water beneath had seen the old toad and heard what she said, so they lifted their heads above the water to look at the little maiden. As soon as they caught sight of her, they saw she was very pretty, and it made them very sorry to think that she must go and live with the ugly toads. "No, it must never be!" So they assembled together in the water, round the green stalk which held the leaf on which the little maiden stood, and gnawed it away at the root with their teeth. Then the leaf floated down the stream, carrying Thumbelina far away out of reach of land.

Struggling through the stubble of a corn field—very hard work for a tiny maiden—Thumbelina encounters a field mouse standing ". . . in warmth and comfort, with a whole roomful of corn, a kitchen, and a beautiful dining room. Poor little Thumbelina stood before the door just like a little beggar-girl, and begged for a small piece of barleycorn, for she had been without a morsel to eat for two days. The field mouse, who was really a good old field mouse, said, "You poor little creature, come into my warm room and dine with me." She was very pleased with Thumbelina, and bid her ". . . quite welcome to stay with me all the winter, if you like; but you must keep my rooms clean and neat, and tell me stories, for I shall like to hear them very much." And Thumbelina did all the field mouse asked her, and found herself very comfortable.

By and by, the field mouse announced that they shall soon have a visitor, a neighbor who was better off than the kindly field mouse and who wore a beautiful black velvet coat. "If you could only have him for a husband," the field mouse lamented, "you would be well provided for indeed. But he is blind, so you must tell him some of your prettiest stories." But Thumbelina did not feel at all interested about this neighbor, for he was a mole. He was indeed very rich and learned, but he always spoke slightingly of the sun and the pretty flowers, because he had never seen them. Thumbelina was obliged to sing to him, but sang so sweetly that the mole fell in love with her.

Forced to await her wedding to the mole, Thumbelina rescued a bird whom the mole dragged in. They became good friends, and the bird persuaded Thumbelina to tuck herself under his cozy feathers that Fall and fly with him to a warm place where there were flowers all the time. When they arrived, the bird realized she could not live in the trees with him so he dropped her into the middle of a large white flower where she was stunned to see ". . . a tiny little man, as white and transparent as if he had been made of crystal! He had a gold crown on his head, and delicate wings at his shoulders, and was not much larger than Thumbelina herself. He was the angel of the flower; for a tiny man and a tiny woman dwell in every flower; and this was the king of them all." The little prince was frightened at the bird who dropped Thumbelina off because it was like a giant. "But when he saw Thumbelina, he was delighted, and thought her the prettiest little maiden he had ever seen. He took the gold crown from his head, and placed it on hers, and asked her name, and if she would be his wife, and queen over all the flowers. This certainly was a very different sort of husband than a toad or a mole in black velvet so she said 'Yes' to the handsome prince."

In many tales, the animals have human qualities but the emphasis is on the human hero or heroine, making it easy for people to develop unconscious misconceptions about animals. Fairy tale animals often appear when they are needed, which suggests they have human understanding. Having vanquished the wicked witch but thwarted by a large body of water in the original version, Hansel and Gretel are carried across by a white duck who suddenly appears. In the original tale by the Brothers Grimm, Cinderella is aided by white birds who live in the hazel tree she planted at her mother's grave. When the king announces a three-day festival for all the maids of the land from whom his son, the prince, will pick a bride, Cinderella is forbidden from attending the festival until she picks all the lentils out of the cinders at the hearth. Alone, she calls out to the "tame pigeons, turtledoves, and all the birds beneath the sky," which appear and help her quickly sort the good from bad lentils. She then hastens to her mother's grave beneath the hazel tree and pleas, "Shake and quiver, little tree, throw gold and silver down to me." A white bird throws down a gold and silver dress with slippers embroidered with silk and silver, which she dons and goes to the festival. Each of the three days of the royal festival, the birds help Cinderella pick lentils from the ashes, the white bird in the hazel tree gives her a more beautiful gown and slippers than before, and she attends the festival. Each day, the prince refuses to dance with anyone but her. Each night, she runs away from him. On the last day, the prince covers the stairs with pitch and traps one of her pure-gold slippers. When the prince searches for the girl who fits the slipper, both wicked stepsisters cut off part of their feet to mislead the prince into thinking the coveted shoe fit. Pigeons uncover their ruse by calling out from the hazel tree as the royal coach goes by: "Rook di goo, rook di goo! There's blood in the shoe. The shoe is too tight, this bride is not right!" When Cinderella is finally identified as the right bride, the white doves peck out the eyes of the evil stepsisters during her wedding, punishing their wickedness and falsehood with blindness as long as they lived.

The most glaring modern myth about animal behavior in fairy tales, affectionately perpetrated by Walt Disney and his army of animators, is that animals are omnipresent in fairy tales. Commercial productions of many of the most popular fairy tales, however, led by Disney, feature animals far more abundantly than in the original tales—which undoubtedly helped the enduring success of these classic versions. In the original "Snow White" by the prolific Brothers' Grimm, animals play two brief roles. Snow White is a little princess whose father marries a wicked queen and, in the way of fairy tales, disappears. The Queen has a magic mirror which feeds her unquenchable vanity by assuring her that she is the fairest in the land. Snow White grows into a beautiful young maiden and, because the Queen is no longer the fairest, sustains three attacks on her life. The last pushes her into a deep sleep, a glass coffin, a cadre of devoted dwarfs, and the eventual appearance of a handsome lad who becomes smitten with her still form and awakens her, well refreshed, with a kiss. One role of animals occurs during the first attack on Snow White's life. The Queen bids her best huntsman to take Snow White into the woods, kill her, and return with her liver and lung as proof. The huntsman returns instead with a pig's liver and lung (which the Queen consumes in a voracious version of burning an effigy). The other role is a brief acknowledgment that Snow White is not accosted by the woodland animals as she flees the kingdom; the implication is that the wild animals should have torn her to pieces and devoured her.

In colorful contrast, Disney's version of "Snow White and the Seven Dwarfs" is filled with animals at every turn who are spontaneously loyal and keenly aware of Snow White's situation. At the castle, Snow White does her chores surrounded by white doves. Banished to the woods after the huntsman's attempt on her life, she talks to a little lost bluebird, which leads to a veritable band of bluebirds, bunnies, deer, chipmunks, squirrels, and even a turtle. They understand that Snow White needs a place to stay, and lead her to a small, very messy cottage with seven of everything. Together, they tidy up. (In the original tale, the dwarfs' cottage is spotless but Snow White can stay there only if she does all their chores.). Birds attack the Queen when she appears as a little old lady who persuades Snow White to try a poison apple. When Snow White does so and falls into an enchanted swoon, the animals race to the gold mines to fetch the dwarfs, who tumble over themselves in haste. Alas! They are too late. The animals are as sorrowful as the dwarfs during Snow White's swoon and as joyous at her recovery when the suitor shows up with the ultimate rescue. The wicked Queen is aided, of course, by nasty black vultures and cackling ravens.

Black birds play a different role than white birds, and essays have undoubtedly been written about the use of color in fairy tales symbolizing good and evil. In the dark tale "Anne Lisbeth," a great black raven darts at Anne Lisbeth as she travels along a high road near the sea where her abandoned son drowned, and causes Anne to say "Ah, what bird of ill omen art thou?" "The Seven Ravens" is a tale where people were turned into animals for being bad but saved by a good person and turned back into people again. In a moment of anger, a father curses his sons and turns them into ravens for failing to return with water for their baby sister's emergency baptism. She survives and grows into a delicate beauty, never realizing she has seven brothers. The townspeople's whispered gossip that her life cost the lives of her seven brothers overwhelms her with guilt so she journeys far and wide in search of them. Finally, she donates one of her own fingers to use as a key to open the glass mountain that imprisons her brothers, which releases them from the curse. They are happily reunited.

It was not easy to find a fairy tale where the animals do what animals do. In "The Wolf and the Seven Young Kids," a tale that can be found in various versions from Europe to Africa and Asia, we are immediately assured that: "Once upon a time there was an old goat. She had seven little kids, and loved them all, just as a mother loves her children." Before leaving her children to feed (in real life, goat kids are followers rather than tuckers, so the kids would have followed their mom; fawns and gazelle kids would have stayed behind), the mother goat tells her children to guard against the wolf. You will know him by his rough voice and black paws. He will surely eat you if you let him in. The wolf is very bad, as most are in fairy tales, but this one is also very clever. The kids recognize his rough voice so he chews on chalk to soothe it so it is smooth like the mother goat's voice. The ruse doesn't work and the kids will not let him in. He has the baker make him a fake foot out of dough and has the miller cover the dough with flour to make it look white like the mother's foot. This ruse works, the kids let him in, and he eats all but one kid in haste, then understandably falls into a deep sleep. The goat mother and remaining kid slit open his stomach, release the gobbled kids, and—in the way of fairy tales—the wolf remains asleep as they fill him up with stones. He wakes up thirsty, tries to drink, and drowns (undoubtedly punished for his wanton gluttony).

As the tale shows, most fairy tale animals are people dressed up in animal suits, a ready impression from real life when one encounters the clear gaze of apes, dolphins, and many other species. In "The Frog King" or "Iron Heinrich," a frog recognizes human suffering and reacts with compassion. He hears a lovely princess crying at the well into which her golden ball has tumbled, and promises to retrieve it for her if she will let him sit at her side, eat from her plate, and sleep in her bed. She promises, but rushes off when the ball is safely back in her hand and forgets. Luckily for the frog, the King remonstrates his daughter for forgetting her promises to someone who has helped her, so she is forced to spend time with the frog. This breaks the (inevitable) spell and restores the frog to his rightful place as a handsome prince. Heinrich, of whom we only hear about at the end of this little tale, is his faithful hand-servant. As the prince is restored to his human form, we hear three metal "pops!" Each is the sound of a steal band breaking loose from around Heinrich's heart, bands that had bound him while his master was a frog.

Fairy tales include an extraordinary range of marriages between humans and animals. Slimy suitors are immensely popular. Stories like "The Frog King" emanate from Britain to China. For some reason, beautiful European fairy tale maidens were often forced to marry a hog or a hedgehog. Not all human–animal marriages involve frogs. Fairy tale people also marry monkeys, dogs, cats, mice, and tortoises with regularity. Then, of course, there are also many supernatural sweethearts in the land of *faërie* where magic is taken seriously.

See also Animals in Myth and Lore—*Animals in Native American Lore*

Further Resources

Griffin, D. R. 1981. *The Question of Animal Awareness*, Kaufman Inc: Los Altos, CA.

Ann Weaver

■ Animals in Myth and Lore
Mythology and Animal Behavior

Mythology is one means by which humans come to terms with their world and their place in it. Nonhuman animals play a fundamental role in mythologies across cultures. Mythmakers use nonhuman animals in myths to define the relationship between animals and humans, to characterize attributes of the animals themselves and to embody forces and attributes of the natural and supernatural world.

Nonhuman animal behavior is one of the central components of myths. Mythology may attempt to explain behavior and/or define the spiritual function of the animal in the myth. The process of incorporating animal behavior into myth generally reveals the specific perceptions of the mythmakers and the philosophy of their culture with respect to the animal. Myths then aid to further propagate these perceptions across a culture and over time. In this way, mythology operates to determine the perception and the actions of the recipients toward the animal itself.

Often, when a myth is based in a culture willing to coexist with the animal subject, the myth can reveal attributes of the animal's behavior. For example, trickster mythologies of Native Americans involving species such as ravens and coyotes emphasize the behavioral flexibility of these animals. Cultures more interested in exploiting and eradicating a particular species may actually obscure its behavior—such as promoting the idea that domestic pigs are by choice dirty animals. These two approaches to animal behavior in mythmaking tend to illustrate pre-twentieth century perceptions when many humans still had important relationships with diverse species. More recently, as people are less often in direct contact with nonhuman animals, the phenomenon of romanticizing these species (especially in much "New Age" mythology) has resulted in the perpetuation of myths obscuring the actual behavior of the animal and instead has highlighted the narcissism of the mythmaker.

Different species have fared differently in the process of mythmaking. Dove species tend to be subject to positive mythologies such as their identification with the spirit, or the spiritual part of the Trinity, in Christian symbolism and art. Snake species, on the other hand, have had a mixed history of relatively positive associations such as the Aztec's "Master of Life" plumed serpent Quetzalcoatl and negative associations such as the Judeo–Christian's satanic serpent in the Garden of Eden. The changing role of the domestic cat in mythology, especially western, over time is discussed in depth below as an example of how perception and cultural milieu affect the role of one species in mythology.

Mythology and the Behavior of the Domestic Cat
Felis sylvestris catus

"She sights a Bird—she chuckles—
She flattens—then she crawls—
She runs without the look of feet—
Her eyes increase to Balls—"

EMILY DICKINSON

Since their earliest known associations with humans, domestic cats have been a focus of mythology. Domestic cats have been worshipped as goddesses and persecuted as demons. Rarely have they been viewed with neutrality. From the start of the feline–human relationship, the foraging behavior of the cat has played a central role in the characterization of its attributes and in the portrayal of the cat–human relationship. In situations where the cat has been elevated to a deity status, its hunting prowess is often a focal myth. The domestic cat's social and reproductive behavior has tended to receive much less focus and attention.

Throughout history, feline mythology regarding the cat has revealed biases of the myth-makers. Their earliest domestication likely occurred in Egypt, and ancient Egyptian mythology contributed a great deal to building a foundation by which other cultures viewed the cat (either positively or negatively).

At present, humans appear to possess a mixed bag of mythology regarding the cat. It is only recently that researchers have attempted to document actual domestic feline behavior in their natural contexts—including the home and the feral colony. These researchers have identified a variety of unique and interesting behaviors hitherto obscured by cultural biases—such as complex social interactions, flexible social behaviors and cooperative breeding behavior. However, despite these recent observations, many of the beliefs and mythologies regarding cats formulated in their early relationships with humans, remain.

The Beginning—Domestication

Cats were probably first domesticated from the African wildcat (*Felis sylvestris lybica*) between 7000 and 4000 BCE. Fossil remains from this time have been found in archeological sites in Jericho (6000–7000 BCE) where wildcats naturally occurred and in Cyprus (8000 BCE) where they did not naturally occur. The clearest evidence for an association between cats and humans comes from Egypt (6000–7000 BCE).

The most commonly accepted theory of domestication is that African wildcats were attracted to agricultural stores and garbage because of the rodents. Once their utility as hunters was noted, their continued presence was encouraged. The further spread of domestic cats may have been facilitated through the movement of humans and associated rodents around the globe. Since the origins of the feline–human relationship, the foraging behavior of the cat, specifically its reliance on mice and rats for primary sustenance, has retained a central place in portrayals of this animal. This aspect of feline mythology does reveal qualities of the cat's natural behavior since rodents appear to form the core of the feral cat's diet.

Ancient Egypt and the Classical World—the Cat as Hunter and Mother and Goddess of the Moon

Domestic cats were apparently widespread in Egypt by 4000 BCE and are a common image in mosaics and paintings from this time. They were associated with a variety of deities and religious beliefs. For example, the *Egyptian Book of the Dead* includes the legend of a cat, Atoum, allowing the solar boat of Ra, the sun god, to bring light to the day by killing Apopis, a serpent that blocked the way. As this important component of ancient Egyptian religious belief demonstrates, the foraging behavior of the cat was a central component in its deification.

Bastet was the deity most consistently associated with domestic cats in ancient Egypt. She was initially represented as a lioness, but by 1000 BCE the goddess tended to be represented as a cat or as a woman with the head of a cat. Bastet was the daughter of Ra and sister of Sekmet and was associated with joy, love and fertility. She was also the goddess of mothering and midwifery. This latter association resonates with recent discoveries by animal behaviorists that feral domestic cats may aid one another in birthing kittens.

Both the popularity of Bastet and the rodent-catching abilities of domestic cats probably contributed to the high level of esteem and protection afforded this species in Egypt during the "New Kingdom" (from 1567 BCE). According to Herodotes, writing his *Histories* in 5th century BCE, Bastet's primary temple in Bubastis was the home of thousands of well-fed felines. During this time, killing a cat was punishable by death, and one legend suggests that the destruction of a cat by a Roman army led the Egyptians to retreat for fear that other cats would be harmed. It is also alleged that Egypt outlawed the export of cats outside its borders, thus effectively preventing the movement of cats across the world until later.

It was during this time that mummification and cat cemeteries were common across Egypt. Although most of the mummies removed from Egypt during the 19th century were turned into fertilizer, a few survived to be examined by scientists. It is interesting to note that these cats did not appear to die natural deaths but rather were strangled at a young age and may have been raised to be mummified and sold as votives.

Because cats were probably first domesticated in Egypt and subsequently spread across the world, the Egyptian worship of the cat probably spread along with the animal. Certainly Egyptian thought affected the Greco–Roman and Gallic views of the cat. It is likely that ancient Egyptian mythology served to set up the cat's central status in the mythology and superstition of other cultures across the world.

Until the rise of the Roman Empire, domestic cats were rare in parts of the world other than Egypt. However, they do play an important role in Greek mythology, where Artemis, the moon goddess (Phoebe), was associated with Bastet. One legend regarding Artemis, included in Ovid's *Metamorphoses*, involves her transforming into a cat and taking refuge in Egypt to escape a serpent. Apart from the association of the cat with the moon goddess, lunar forces were also thought by the ancient Greeks to control the changing of cats' pupils.

An ancient Egyptian tomb painting showing a man using a cat to hunt birds.
Courtesy of the Library of Congress.

The hunting drive of the cat finds its way in to the Greek Aesop's fables in the story of a cat in love with a man who asked Venus to turn her into a woman. She retained the urge to hunt, however, and could not resist chasing a mouse across the room.

According to some theories, the spread of the Roman Empire across Europe was also responsible for the spread of the cat, since the appearance of domestic cat remains tends to coincide with the Roman occupation in places such as Great Britain. The Romans appeared to value the cat for its hunting behavior and maintained the same feline association with their moon goddess, Diana, as the Greeks had with Artemis.

Asia—the Cat as Companion and Storm Demon

Many Asian religions, especially Hinduism and Buddhism, differ from Greco–Roman and Judeo–Christian traditions in that that they do not maintain that nonhuman animals are entirely distinct spiritually from humans. This religious tolerance and incorporation of nonhuman animals into spiritual ideas may have contributed to the fact that in Asia, especially prior to the Middle Ages, cats were viewed in a positive light for both their hunting prowess and their companionship.

In the Buddhist faith, cats were thought to intercede between Buddha and faithful followers because of their purity. Buddhist monks raised cats and, according to legend, every temple in Japan owned two cats to catch mice. It is said that emperor Ichijo ordered that cats should be pampered. Although some cats were viewed as potential witches in Japan during the Middle Ages, tortoiseshell cats retained an aura of good luck.

The Birman breed (different from the Burmese breed) is subject of a legend itself. It is said that in Burma (now Myanmar) centuries ago white cats guarded the temple of Lao-Tsun, the home of the goddess Tsun-Kyan-Kse. When a head priest lay dying after being attacked by bandits, his beloved companion, the temple cat, Sinh, placed his paws on the priest and faced the goddess Tsun-Kyan-Kse. The cat was turned from pure white to white with color points around the face, legs and tail and pure white feet. His eyes were turned a bright blue.

Less positively, many premedieval and medieval (as well as some European) cultures viewed cats as associated with storms. One such association is the Japanese demon of lightning, Raiju, who sometimes appeared as a cat. When a tree showed the marks of lightning it was thought to have been scratched by Raiju. This interpretation is based upon actual feline marking behavior. These storm associations in general may in part be a result of the tendency of cats to sense storms before humans do and often to display fearful behavior before and during storms.

Middle East—Honored Companions

In the Middle East, prior to the rise of Islam, some groups of Arabs worshipped a Golden Cat. Muhammed himself is said to have cut off his sleeve so as not to wake his cat Muezza who was sleeping in his arms. In the 13th century the Sultan El-Daher-Beybars bequeathed a park to house stray cats.

Medieval Europe—The Cat as Demon

In Europe in the early Middle Ages cats were still considered valuable as hunters. The earliest English legend involving cats is that of Dick Whittington and his cat. Whittingham was a poor boy in London whose hunting cat allegedly helped garner the wealth that positioned him later to become Lord Mayor of London three times.

The increasing power of the Christian church in Europe resulted in the proliferation of negative mythology and concomitant mistreatment of the domestic cat during the mid- and late-Middle Ages. The negative mythology spread by the church overcame many of the positive associations people had based upon natural cat hunting and companionship behavior. One unfortunate outcome of this overriding negative mythology was that, in obscuring the effectiveness of the cat as a hunter, it also obscured the potential for the cat to reduce the rat population that was, at that time, still regularly spreading bubonic plague.

The negative reaction to the cat during this time was likely part of the generalized culture of the Middle Ages that increasingly viewed humans as entirely separate types of entities from nonhuman animals. This idea, made clear in Judeo–Christian belief in Genesis also had its roots in Greco–Roman philosophy. By viewing the cat as creature entirely distinct and "soul-less," people during the Middle Ages divorced themselves from actual observation of the cat's behavior. This enabled the spread of superstitions regarding the cat. These superstitions as promulgated by the Christian Church were probably reactions to the cult of Bastet and Artemis/Diana both of whom were associated with cats. Furthermore, Hecate, Goddess of Magic in classical Greece, was a particular target of Church mythology, and she was often conflated with Diana/Artemis.

The most damaging myth regarding cats of this time was that they were often witches' familiars or even the devil himself. For example, the aversion to black cats can be traced at least as far back as to Pope Gregory IX's (1227–1241) papal bull declaring that that devil was like a black cat. Similarly, Baldwin's *Beware the Cat* suggested that a witch could take her cat's body nine times. This reinforced the idea of the cat as familiar and probably introduced the idea of cats having nine lives. These sorts of writings, combined with the fact that the sixteenth and seventeenth centuries were the most extensive period for trials for witchcraft in Europe, resulted in the domestic cat being subject to tortures and execution on mass scales across Europe.

It was in the seventeenth century, when the French Cardinal Richeleau began to keep cats at court, that the fortune of cats improved in Europe, although many of the superstitions and myths persist to this day.

Post-medieval Philosophy and Storytelling

The centuries following the Middle Ages included the mechanistic musings of Descartes as well as less mechanistic discussions of Enlightenment philosophers and the symbolic presentations of the Romantics. Although the view of nature and nonhuman animals became progressively less negative during these shifts in western thought, there was a continued tendency to view nonhuman animals through the lens of human thought rather than through any attempt at direct observation.

This type of human-perception-based take on the cat acted at times to obscure actual behavior. One example is Rudyard Kipling's "The Cat Who Walked by Himself." In this story, the myth of the independence of the cat is underscored. The myth of the independent cat is probably a result of the difference between the cat's and the dog's social behaviors. Cats are capable of existing and hunting solitarily. However, the level of affiliation shown by a cat to its caretaker is an indication of the capacity of the domestic cat for a complex social system. Indeed, researchers are only beginning to collect data on the complexity of cat social groups. One thing that is clear at this point is that domestic cats are well-equipped with signals for social life, and that cats in social groups maintain complicated interactions with variety of other individuals ranging from agonistic to affiliative to apparently cooperative.

Pantomine poster for Puss in Boots, 1874.
© *The Art Archive / Private Collection / Eileen Tweedy.*

Despite being primarily about humans, some stories based on mythologies collected during the post-medieval period contain grains of true reflection of the cat's behavior. The tale of Dick Whittingham and the cat that helped make his fortune shares a theme with many fairy tales such as the North African Berber Story "The Clever Cat" and the German Brothers Grimm's "Puss'n Boots." All of these stories focus on the impressive foraging behavior of the cat as well its ability to be a companion. The cat's intelligence and trickiness are also heralded in many of these stories. In many ways, these stories illuminate the flexibility of foraging and social behavior observed in feral and pet domestic cats now being more clearly recognized by researchers.

It was with the naturalist explorers and Darwin's revolution that nonhuman animals, including the cat, started to be viewed as entities outside of human thought, worth study on their own. Modern public perception of cat behavior is starting to be somewhat influenced by observations by cat behavior researchers. However, because of the history of strong feelings and extreme superstitions regarding the cat, the image of the cat and its behavior tends to remain distinct from actual observations. This is because, in general, we as humans, especially in Western society, retain our legacy of medieval and post-medieval thought, with the idea of nonhuman animals as mechanistic projections of our own hopes and fears. For a species such as the cat, so central to many mythologies, the perception of the actual animal's behavior is seen through a mixed veil of myth and superstition. We do retain some of the revealing myths of the hunter and the companion but also the obscuring myths of the aloof and wicked creature. It is clear that only by lifting this veil on the cat and other species can we begin to view the animal as itself.

See also Animals in Myth and Lore—*Animals in Native American Lore*

Further Resources

Clutton-Brock, J. 1993. *Cats Ancient and Modern.* Cambridge, MA: Harvard University Press.

Dickinson, E. 1999. *The Poems of Emily Dickinson* (Ed. by R. W. Franklin), p. 159. Cambridge, MA: The Belknap Press.

Lindemans, M. 2004. *Encyclopedia Mythica.* http://www.pantheon.org/

Sunquist, M. & Sunquist, F. 2002. *Wild Cats of the World.* Chicago: University of Chicago Press.

Turner, D. C. & Bateson, P. 2000. *The Domestic Cat: The Biology of its Behaviour.* Cambridge, UK: Cambridge University Press.

Jennifer Calkins

The Wolf in Fairy Tales

Douglas W. Smith

Few animals provoke the human imagination and have more tales and stories about them than does the wolf. Expansive in its range (most of the northern hemisphere) and historically abundant, the wolf appears in the stories of many cultures. Add to this a creature that is active primarily at night, communicates by howling (most who have heard wolves howl say it sends chills down their spines), and kills for a living (including feeding on humans who died and were left during the bubonic plague outbreak in Europe), and you have the recipe for some of the most vivid, emotional, and long lasting stories for any animal known to humans. "Little Red Riding Hood," "Peter and the Wolf," "Three Little Pigs," "Romulus and Remus," "Boy Who Cried Wolf," "Wolf at the Door," "Wolf in Sheep's Clothing" and other stories, including many by aboriginal cultures, are known to most people. Embedded in our psyche is what we *think* we know about wolves, rather than true stories about wolves given to us by modern day science.

Walter Crane's Illustration to "Little Red Riding Hood."
© Art Resource, NY.

This is so true that what is portrayed in stories about the wolf is often construed as fact by the general population. Without any exposure to or education about wolves, many people believe that they are blood thirsty, ruthless killers, who kill for the fun of it. They don't realize that wolves most often lose the battle with prey and in fact could die themselves.

Historically, these legends and myths tilted human attitudes toward the negative. The mythology of the wolf was so great and pervasive that most people had an opinion about wolves, whereas for many other species people had no opinion or had not thought about it. More often than not this opinion about wolves was negative. This attitude has persisted for a few thousand years. Beginning with when humans domesticated goats and sheep, about 10,000 years ago, wolves became our competitor. Only in the last couple of decades has this view changed. I have met, and still know people whose hatred of wolves is so deep and vitriolic that they curse the day when widespread killing of wolves by any means was banned.

Today, because many of the stories have been exposed as false, the public's view of the wolf is changing. Many people view wolves positively now. With still strong opposition as well as growing support, wolf issues are extremely polarizing.

(continued)

The Wolf in Fairy Tales (continued)

Wolf management is among the most intense controversies in wildlife management today. Most wolf managers must be sociologists as well.

Wolf managers and researchers still struggle to advance an accurate and factual story of the wolf. Telling the good with the bad is part of it, but a new view is emerging—that wolves are vital components of ecosystems, shaping much of the biological diversity that we see in natural systems. This new image of the wolf has been slow to catch on because the old, negative view is so entrenched. Nonetheless, inroads on a more ecologically aware public are occurring. I am hopeful that this enlightened view will spread more quickly for if it does not, ecosystems may be what suffer, because after all "the wolf is at the door" for many species today.

■|Anthropomorphism

One of the earliest, and still common, methods of attempting to understand the behavior of other animals is termed *anthropomorphism*. Anthropomorphism is the attribution of human characteristics to nonhuman entities—usually, but not always, other animals. Since the end of the nineteenth century, scientists have traditionally regarded anthropomorphism as a serious error that must be avoided. This was a consequence of the too ready interpretation of the behavior of other animals, using labels that were difficult to define or recognize even in other people, let alone other species. Anthropomorphism was considered particularly troublesome when referring to the emotions, feelings, or mental abilities of animals. What is love in a dolphin, anger in a chicken, or intentional spite in a spider? Do animals really ponder decisions, and think about, infer, or evaluate the consequences of their actions? It seemed more objective and parsimonious to attribute the behavior of animals, even monkeys and apes, to elementary processes of reflexes, instincts, and conditioning. Ironically, it was not considered an error by most scientists to see similarities between humans and other species in behaviors such as eating, fighting, mating, playing, or caring for offspring and even in the mechanisms underlying the similarities.

The distrust of anthropomorphic descriptions of animal behavior was elevated due to its close association with another early feature of the study of animal behavior: reliance on anecdotes. *Anecdotes* are typically stories about the unusual actions of specific animals rather than careful descriptions of what animals typically do and experiments on the mechanisms underlying their behavior on topics such as sensory stimuli, hormones, and prior history. Thus, in one famous example, a dog, Blackie, that seemed consciously and purposefully to have figured out how to open a gate latch, was shown to have acquired the behavior gradually by steps that seemed more accidental than consciously intelligent. Such observations were very important in developing a rigorous science of animal behavior based on careful studies that attempt to unravel the causes of behavior, rather than merely collecting stories from untrained pet owners and farmers about the accomplishments of animals or observations by travelers of little known exotic animals. Unfortunately, early students of animal behavior, including Charles Darwin (not to say Aristotle), were forced to rely on such anecdotes in trying to compare human and nonhuman animals in behavior and mentality, as well as in anatomy and physiology.

Nevertheless, the lure of anthropomorphic language in describing animal behavior is great and is still found today in journal articles and books written by eminent scientists. Why? Some scholars decry this continued use of what they consider unscientific and unnecessary language. Others claim that while anthropomorphism is highly engrained in us as a species, we should be constantly on guard and dismiss any attempts to make claims

A trained grizzly bear and his trainer strike an identical pose for the camera.
© *Joel Sartore/National Geographic Image Collection.*

about the mental and emotional lives of other animals as sloppy and sentimental. Still other scientists have recognized that anthopomorphism is more than just naive thinking and recognize that it exists and persists because it was, and is, adaptive. This view suggests that anthropomorphism is evolutionarily beneficial. How can this be if, objectively, anthropomorphism often seems uncritical and wildly speculative?

First, it is important to realize that many anecdotes are often based on real, but rare or hard to observe events. Careful recording and documentation of unusual observations are often the basis for developing a natural history for a species, especially those for which little is known. Thus, careful documentation of unusual behavior can provide an impetus to systematic large-scale research—research that would not have been even contemplated if a report of some unusual problem-solving ability or complex skill had not previously been reported. It is in these early descriptions that anthropomorphic vocabulary is often evident, because the author has no referent for what they see other than their own experience. This leads to the second reason anthropomorphism persists: It works! Farmers, animal handlers, herders, and others who seem most successful at training, breeding, and raising animals are apt to be anthropomorphic about them. They find that being gentle with animals, not stressing them physically or using loud commands or punishment, are more likely to have animals that produce the most milk, successfully rear offspring, win races (e.g., Seabiscuit) or stay healthy.

In addition, there is a third reason, the most important of all, for why anthropomorphism is still found among scientists. There is an increasing amount of solid scientific work showing that animals are much more than reflexive or simply conditioned beings. While people have reflexes and are easily conditioned, as in learning to avoid a hot burner on a stove, they also solve problems, have feelings, and mentally rehearse the consequences of their actions: They can plan. We now know that many nonhuman animals convey sophisticated information to one another about food resources and enemies, solve problems not just by trial and error but by observing what others do, and react to positive and aversive events in their lives with emotional responses very similar to those in people. It is interesting that, as mentioned above, to view animals as eating (which appears due to hunger) or drinking (which appears due to thirst) is noncontroversial. However, hearing a lost animal emit a distress call or an injured animal whimper is not considered presumptive evidence that they are anxious or in pain.

However, this modern recognition of the value of anthropomorphism is still controversial and often rejected by those who are, in effect, unwittingly anthropomorphic in many respects. When Donald Griffin's writings in the 1970s, *as a scientist*, seemed to be encouraging unfettered and untestable speculation about animal behavior (see Griffin 1978), the critical reaction was swift and reached its zenith in the book by Kennedy in 1992. Now a more nuanced approach is needed.

In order to understand the cognitive accomplishments of a bee or beetle, squid or chimpanzee, we need to evaluate how they perceive *their* world. In doing so, technology can assist us, but we need to remind ourselves constantly that we are using our human senses and human-based technology, and are processing the information with a human brain. It is also important to realize that infants may perceive the world differently than children, males differently than females, and teenagers differently than the elderly. Sensory capacities change with age and experience. Furthermore, within each of the above groups there can be great differences among people in how they perceive the external world and their internal states, in how they process information and perform, and in how they experience and express emotions. Thus, it becomes clear that the tools we use to describe the behavior of other species is just an extension of how we describe, infer, and predict the behavior of other people that we deal with in our own daily lives: parents, teachers, children, salespeople, and politicians.

This leads to the conclusion that, both in everyday life and in studying animals, we need to use anthropomorphism critically—not attempt to abolish all anthropomorphism and associated concepts. The choice is not, then, between anthropomorphism and a sterile alien vocabulary, but in how to harness our tendency to relate to the behavior of other beings in terms of our own modes of perceiving, thinking, feeling, and behaving. What we need to apply is a *critical anthropomorphism*, an anthropomorphism that it is a legitimate, and perhaps particularly creative, way to do science if it is used to develop hypotheses that can be tested in a rigorous manner. Critical anthropomorphism is essentially a means of using various sources of information including natural history, careful descriptions of behavior, and scientific knowledge of how brains and senses work together with our perceptions, intuitions, and feelings, identifying with the animal, mathematical modeling, and so forth in order to generate ideas that may prove useful in understanding and predicting how an animal responds in planned (experimental) and unplanned interventions. For example, studies in which brain activity is imaged in alert animals can allow us to compare the neurological processes underlying not only how different people, but other species, respond to visual stimuli, odors, pain, and problems, as well as the changes in the brain that occur during learning. One of the most remarkable discoveries was the finding that many mammals, including people, have a "face recognition module" in the brain that finally disproves the view that people learn the meaning of facial expressions from scratch.

Critical anthropomorphism also helps us guard against the most flagrant errors of anthropomorphic thinking. For example, we need to be on guard, as scientists, against using animals as allegorical or moral examples for human behavior (e.g., see how the ant is always working, so get back to the books). We need to avoid personification of animals, such as considering camels to be arrogant because they keep their noses up in the air. We need to avoid elevating superficial comparisons (that cat is acting exactly as a toddler) into explanations that keep us from careful analysis.

Recently, an edited book (Mitchell, Thompson, & Miles 1997) and a monograph (Crist 2000) have been devoted to exploring the problems and perils of anthropomorphism and anecdotes in modern studies of animal behavior. Anthropomorphism actually informs many of the most creative scientific accomplishments in science. In science, anthropomorphism is harmful primarily when it is unacknowledged, or unrecognized, and used as the basis for accepting conclusions through circumventing the need to actually test them.

The Story of IM

Are two heads better than one? Double-headed animals are occasionally born and are not likely to survive. We know that bicephalic ("Siamese") human twins are not happy with their lot, and in one well-publicized recent case, embarked on a highly risky surgery rather than stay together that led to the death of both. Other surgeries have been more successful.

Snakes are one of the most common vertebrate animals in which two-headed individuals are produced, presumably by a developmental accident. In fact, a book published many years ago delightfully recounts the information on these animals from ancient times. I have been fortunate in having been able to study three two-headed snakes: a plains gartersnake, a black rat snake, and a ringneck snake. All these were captured in the wild as newly born or hatched animals and this is true of most recorded examples. The snake I studied most was a two-headed black rat snake named IM. We observed its behavior in the laboratory for almost 20 years. The left head was named Instinct and the right head was named Mind, so I called the snake IM. The reason both letters are capitalized should be clear.

What we found can be summarized simply. Like almost all snakes, IM ate relatively large prey, and the two heads often fought over food when one could not swallow it fast enough. Both heads would attempt to swallow the prey in competitions that would take over an hour. Clearly, this would not be very adaptive in the wild because snakes are highly vulnerable to predators when they are preoccupied with swallowing prey. The two heads of IM never seemed to learn to cooperate, but over his first five years of life, the two heads ate almost exactly the same mass of food (usually mice). The left head ate more, but smaller, prey than did the right head! Initially we thought that this suggested cooperation since preliminary x-rays showed only one stomach, so the food was going to the same place. We later discovered that the snake had two complete digestive systems down to the colon. Thus, each head could have had his own hunger system!

Less easily resolved was the observation that the heads often disagreed on where to go. When hunting prey in an obstacle course (a pegboard with pegs sticking up), the two heads would often get hung up on a peg as it approached a mouse and thus the prey would have time to escape. It is tempting to view the behavior of IM as an overt example of what goes on in the heads of people and other animals when faced with conflicting demands and choices. Like IM, some people become confused, paralyzed, or in other ways unable to make rational decisions about how to behave. Perhaps the most productive people are those who can make decisions quickly when faced with such conflicts. This is speculation. The answer to why IM did not learn to cooperate may be that being two-headed is clearly abnormal and something the species could not have adapted to through evolution. Animals often seem to adapt to losing a limb or sense organ more easily than to an extra appendage, especially if it has a brain (or a mind of its own).

Notice how difficult it is to think and talk about IM and his dilemmas in life without being anthropomorphic, even about a snake. However, we can be critically anthropomorphic and readily change our interpretation when new data come to light, as in the x-ray of IM's two stomachs.

This leads to the phenomenon of *anthropomorphism by omission*, as described by Rivas and Burghardt (2002)—the failure to consider that other animals have a different world than ours. We can, without realizing it, attribute human traits to nonhuman animals by failing to consider that many species perceive the world in a different manner than we do. Many examples, from communication to conservation ecology, can be listed. If we do not acknowledge that different species have different perspectives and priorities than we do, we may draw conclusions that are erroneous about, for example, the song of the nightingale. The only way to avoid it is to consider the internal and external world of the animal and try to evaluate *what it is like to be the animal*. The idea of studying the private worlds of other animals was pioneered almost 100 years ago by Jacob von Uexküll, who attempted to bring the latest neural, physiological, and perceptual findings to bear in understanding the behavior of animals by considering both their inner world (*Innenwelt*) and how they perceived and responded to their environment (*Umwelt*).

More recently, Bekoff has extended this view to advocate a *biocentric anthropomorphism*. Although it is true that we will never get to experience the world fully like another animal would, by doing our best to accomplish this we will get closer to understanding the life of the animal. Certainly we will never obtain access to the full inner life and private experiences of another human being, but some creative people get much closer to accomplishing this than others, and are thus considered particularly insightful, empathic, or privy to human nature writ large.

Anthropomorphism by omission can be present and be a detrimental influence on work in a variety of disciplines including high profile behavioral ecology, state of the art cognitive ethology, language studies, theoretical ecology (warning colors and mimicry), zoos, and decision making in conservation. Finally, it is not enough to avoid anthropomorphic vocabulary and claim to be strictly objective. Anthropomorphism comes in many guises and can catch you unawares. The most easily recognized ones are not the problem, but, more often, the illusion that one is immune is the problem. If anthropomorphism is a natural tendency of human beings, scientists are not immune; lurking unseen, it can compromise efforts in many areas. By using critical anthropomorphism and trying to wear the animal's "shoes" we can overcome part of our natural bias and accomplish a more legitimate understanding of the life of the animals and better understanding of nature. Students and researchers should put themselves in the position of their study animals, not only as a novel, additional approach, but also as a required step in conducting good science. Actually consulting prior work on a species is important in research designs and interpreting behavior.

See also Cognition—*Cognitive Ethology: The Comparative*
       Study of Animal Minds
    Cognition—*Animal Consciousness*
    Cognition—*Fairness in Monkeys*
    Emotions—*Emotions and Affective Experiences*
    Empathy
    Friendship in Animals
    Play—*Social Play Behavior and Social Morality*

Further Resources

Bekoff, M. 2000. *Animal emotions: exploring passionate natures*. Bioscience, 50, 861–870.
Burghardt, G. M. 1985. *Animal awareness: current perceptions and historical perspective*. American Psychologist, 40, 905–919.

Burghardt, G. M. 1991. *Cognitive ethology and critical anthropomorphism: A snake with two heads and hognose snakes that play dead*. In *Cognitive ethology: the minds of other animals*. (Ed. by C. A. Ristau), pp. 53–90. San Francisco: Erlbaum.

Burghardt, G. M. 1997. *Amending Tinbergen: A fifth aim for ethology*. In *Anthropomorphism, Anecdotes, and Animals*. (Ed. by R. W. Mitchell, N. S. Thompson, & H. L. Miles), pp. 254–276. Albany: State University of New York Press.

Crist, E. 2000. *Images of Animals: Anthropomorphism and Animal Mind*. Philadelphia: Temple University Press.

Fisher, J. A. 1990. *The myth of anthropomorphism*. In *Interpretation and explanation in the study of animal behavior. Vol 1. Interpretation, intentionality, and communication* (Ed. by M. Bekoff & D. Jamieson), pp. 96–116. Boulder, CO: Westview Press.

Griffin, D. R. 1978. *Prospects for a cognitive ethology*. Behavioral and Brain Sciences, 1, 527–538.

Kennedy, J. S. 1992. *The New Anthropomorphism*. New York: Cambridge University Press.

Lockwood, R. 1989. *Anthropomorphism is not a four-letter word*. In *Perceptions of Animals in American Culture* (Ed. by R. J. Hoage), pp. 41–56. Washington, DC: Smithsonian Institution Press.

Mitchell, R. W., Thompson, N. S., and H. L. Miles (Eds.). 1997. *Anthropomorphism, Anecdotes, and Animals*. Albany: State University of New York Press.

Rivas, J. A. & G. M. Burghardt. 2002. *Crotalomorphism: A metaphor for understanding anthropomorphism by omission*. In *The Cognitive Animal: Empirical and Theoretical Perspectives on Animal Cognition* (Ed. by M. Bekoff, C. Allen, & G. M. Burghardt, pp. 9–17. Cambridge, MA: MIT Press

Gordon M. Burghardt

■ Antipredatory Behavior
A Brief Overview

Virtually every animal at some point in its life has to avoid predation and there are a remarkable variety of ways to do so. Many of them work by making it unprofitable for a predator to find, pursue, kill, or eat a prey species.

Studies on great tits, a small European bird, foraging on worms have demonstrated that even slight increases in the time it takes to find prey can have a large effect on their foraging decisions. Thus, from the prey species' perspective, even slight camouflage (or crypsis) might save its life.

We commonly see morphological and behavioral adaptations to increase the searching time for predators. For instance, decorator crabs in the waters off New England camouflage themselves by placing sea grass on their carapace. Species without the ability to change their appearance are often quite selective about the background they settle on. Such "background selection" helps camouflage an individual and therefore makes it more difficult for a predator to detect them. Ptarmigan, a bird that lives in alpine and tundra habitats, annually molts its feathers into white plumage for the winter and a mottled dark plumage for the summer. Snowshoe hares have similar morphological adaptations. However, these animals are faced with a quandary once the snow melts.

This pufferfish blows itself up so as to appear larger and more formidable to any potential predators.
© Anthony F. Chiffolo.

Male ptarmigan (who retain their white feathers for a few more weeks than females) actively soil their winter plumage after the snow has melted. By doing so, they reduce their conspicuousness before molting into their more cryptic summer plumage.

What can prey do once detected? Many species simply escape. Studies of noctuid moths being hunted by bats are a wonderful example of a system where the neurobiology of antipredator behavior is understood. Bats that hunt noctuid moths produce two types of ultrasonic cries: a slow-paced sound that is used when the bats are generally orienting and looking for food, and a faster-paced cry that is used in the final stages of attack to help them track moth movement in detail. In response to this, moths have evolved two sorts of neurons. The first (called the A1) fires at moderate intensity to the slow-paced bat calls. Moths use binaural cues to locate the bat (i.e., because they have ears on both sides of their body, differences in the right and left A1 cell firing rate allows left–right localization). However, if a bat gets too close and engages in the fast-paced echolocation cries, the moth's second neuron (called the A2) begins firing. This second neuron is connected to the flight muscles and when it fires, it results in erratic flight which, if the moth is lucky, results in escape. Another example which illustrates how selection acts on prey to avoid predators is seen in crickets who must avoid predation by bats and parasitoid flies. Studies of cricket hearing have demonstrated that they are most sensitive to the sounds produced by other crickets and to high-frequency sounds produced by bats. Parasitoid flies that lay eggs in crickets have exploited the cricket's calling behavior. Female fly hearing is most sensitive to the frequencies contained in the cricket calls. On islands where the parasitoids have recently arrived, crickets are more vulnerable to the flies than on the mainland; over time they begin to modify their calling behavior. These examples illustrate how selection to avoid predators has resulted in receptors exquisitely tuned to be responsive to the sounds of predators, and the opportunity for coevolutionary arms races to evolve between predators and their prey.

Many moths display false eyes, known as eye spots, to confuse birds.
© *Getty Images/Digital Vision.*

Some species misdirect attack, and in doing so, increase the chance that they will survive an encounter. For instance, those Antarctic krill that have a great likelihood of being killed, will engage in a tail flip and leave their exoskeleton floating in the water. If the predator attacks the exoskeleton, the krill has saved its life at the cost of having to re-grow an exoskeleton. Some lizards will attract a predator's attention to their tail by waving it around. Upon being attacked, their tail breaks off and the lizard flees. Tails may be important in social communication, but the lizard has survived the attack to engage in a future social encounter.

Other species engage in startling displays. Moths may uncover their hind wings which contain large "eye-spots." Vertebrates are particularly sensitive to looming objects and eye-spots which generate startle responses. Thus, when a foraging bird suddenly has the eye-spots of a moth flash beneath it, the bird may hesitate for a moment. If the moth is lucky, this moment may be sufficient to allow escape. Octopuses are well-known masters of disguise, but most avoid predation by being nocturnal. Recently, a diurnal Indo-Malayan octopus has been discovered that dynamically mimics poisonous species of snakes and fish. In doing so, the octopus escapes predators, which avoid it.

A broad class of adaptations helps species avoid predation by discouraging pursuit. Such pursuit-deterrent signals include the conspicuous *stotting* (a stiff gaited bounce) of ungulates (hoofed mammals) when chased by carnivores. On the African savannah, Thompson's gazelles *stot* when they are being pursued by cheetah. Stotting is likely to work

by signaling to the predator that it is not profitable to continue pursuit; after all, a prey that
can jump up and down is in sufficiently good
condition to avoid capture. From the cheetah's
perspective, they, like other *optimal foragers* are
likely to be interested in maximizing their energy
input over time. Thus, pursuing a healthy and
stotting gazelle is likely not to be profitable.

Many species *aggregate* (gather in dense
groups) to decrease predation risk. If such ag-
gregations do not attract more predators, and if
predators will kill the same number of prey regardless of group size, then by aggregating, the
per-capita risk of predation decreases. In response
to this increased safety in numbers, individuals in

*This behavior, called stotting, discourages
predators from pursuing these gazelle.*
© Getty Images/Digital Vision.

a group are able to allocate more time to foraging and less to antipredator vigilance. These
"group size effects" are widely reported in birds and mammals and illustrate dynamic risk
assessment.

See also Antipredatory Behavior—*Sentinel Behavior*
Antipredatory Behavior—*Predator–Prey
Communication*
Antipredatory Behavior—*Vigilance*
Communication—Vocal—*Jump-yips of Black-tailed
Prairie Dogs*
Social Organization—*Socioecology of Marmots*

Further Resources

Caro, T. M. 1986. *The functions of stotting in Thompson's gazelles: Some tests of the predictions.* Animal
Behaviour, 34, 663–684.
Gerhardt, H. C. & Huber, F. 2002. *Acoustic Communication in Insects and Anurans.* Chicago: University
of Chicago Press.
Lima, S. L. 1995. *Back to the basics of anti-predatory vigilance: The group size effect.* Animal Behaviour,
49, 11–20.
Lima, S. L. & Dill, L. M. 1990. *Behavioral decisions made under the risk of predation: A review and
prospectus.* Canadian Journal of Zoology, 68, 619–640.
Owen, D. 1980. *Camouflage and Mimicry.* Chicago: University of Chicago Press.

Daniel T. Blumstein

■ Antipredatory Behavior
Finding Food While Avoiding Predators

In every backyard and schoolyard animals are balancing success at eating with their risk of
being eaten by predators. Familiar little birds such as sparrows and chickadees could be
attacked by predators such as hawks, owls, and cats. Bird feeders are rich sources of food,

Chickadee at feeder.
Courtesy of Peter A. Bednekoff

Alert squirrel holding food.
© Anthony F. Chiffolo.

but their positions give avenues for attack by and escape from predators. Birds prefer to feed in positions that are difficult to attack and easy to escape from. When several feeders are available, the most popular will likely be high enough off the ground to thwart attack by cats and close enough to trees and bushes to allow escape should a hawk or owl appear.

Squirrels also feed while avoiding predators. Because squirrels flee up trees when attacked, they are more at risk when farther from trees. When farther from a tree, squirrels eat less, choose foods that require less concentration while eating, and are more likely to pick the food up and carry it to safety before consuming it. Also, when farther from a tree, they do not let potential threats get as close before running to safety. Thus potential danger alters many of the actions of squirrels.

The behavior of sparrows, chickadees, and squirrels illustrates a common balancing act performed by animals. In order to survive and grow, animals must eat, but they also must avoid being eaten by other animals. Sometimes animals can find rich food patches without encountering predators, but more often animals often can only get more food by placing themselves in greater danger. This important balancing act takes on many forms involving how much to move, when to feed, and where to feed.

How Much to Move

One way to avoid predators is to hide. When predators are nearby, many animals hide or freeze until the predators move away. Hiding and freezing prevent an animal from getting food. In general, a forager can find more food if it searches through a larger area. While covering a larger area, it is also more likely to cross paths with predators. Besides simply crossing paths with more predators, moving foragers may be easier for predators to detect. In experiments, tadpoles move less when their main predators, dragonfly larvae, are nearby. If tadpoles are anesthetized to keep them from moving, they are less likely to be killed. Movement is needed to find food but movement often leads to danger.

When to Feed

Even if animals have to move sometimes in order to feed, they do not need to move all of the time. They can balance food and safety most effectively if they feed during the safest times. If animals need more food than they can get during the safe times, they must also feed during more dangerous times. For small birds like dark-eyed juncos, feeding normally starts early on a winter day and lasts until late afternoon. To feed more, juncos extend their feeding into the moments of low light and deep shadows around dawn and dusk. During these times, owls, cats, and other predators are more likely to sneak up on the juncos.

For other animals, night is safe and brightly lit times are risky. Bats often begin feeding a bit before nightfall. Bats that feed on insects are safe at night from predatory birds, but insects are often most abundant during the day. Bats can get more food if they start feeding before dark, but this also brings greater risk. Bats generally emerge in the fading light of the evening when they can find plenty of food without too much danger. Feeding at night is also safer for fish such as minnows and small salmon. These little fish can grow quickly only by feeding during dangerous daylight hours when they are easily seen by predators such as big fish and herons.

Animals that feed under the cover of darkness cannot feed as safely every night because some nights are darker than others. Nocturnal voles, hamster-like rodents that live in meadows and woodlands, reduce their nighttime feeding around the time of the full moon because the increased light makes them more vulnerable to their main predators, owls.

Where to Feed

During the times they are foraging, animals often choose between places that offer different amounts of food and danger. For example, snails in streams find more of their food, algae, on the sunny side of rocks, but the tops of rocks are also more exposed to fish predators. Snails often compromise and feed on the sides of rocks where algae grows but the snails are less exposed to predators. Food and danger can both be related to exposure. Exposure to sunlight allows more photosynthesis, but exposure often leaves foragers more vulnerable to predators. Similarly, sunfish can find more zooplankton, small floating animals, to eat in the open water portions of lakes because these areas are more productive for phytoplankton, tiny floating plants, which support the zooplankton. The open areas, however, provide no refuge

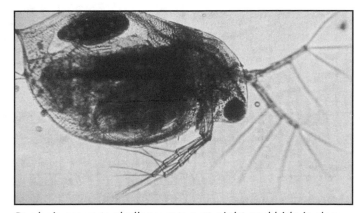

Daphnia move to shallow waters at night and hide in deep water during the day.
© *Lester V. Bergman/CORBIS.*

from attack, while the weedy portions provide refuge but less food. Sunfish often live in the weedy areas when small and move into the open areas once they are large enough to escape many attacks.

In the open water of lakes and oceans, there are few places to hide. Many organisms such as cladocerans in the genus *Daphnia*, midge larvae, and small fish hide in the deep, dark water during the day and venture into the shallow water to feed during the night. They cannot find as much food and grow as quickly in the deep water, but at least their predators cannot easily find them. Fishermen often fish at night when more fish are near the surface. In lakes that lack predators, *Daphnia* often do not make this daily commute and instead stay at all times in the shallow water where there is more food.

In other cases, the attack behavior of the predator determines which areas are safe and which are dangerous. In the forests of northern Europe, pygmy owls swoop down on outer, lower branches of trees. Small birds feed on these branches only when they are very hungry or when stronger birds keep them out of other parts of the trees.

When a group of animals are feeding together, individuals on the edge of the group will first come into contact with new sources of both food and danger. Whirligig beetles move to edge positions when hungry and minnows move to central positions when alarmed. Animals may choose positions that balance threats from different sorts of predators. For example, grasshoppers are safest from birds when low in grass, safest from lizards and small mammals when high on grass, and are found at mid-height when both sorts of predators are around. For small animals, safety and death may be just inches apart.

Scientists have gone a step further and measured how much you need to "pay" small animals to venture into the open. Gerbils are equally likely to visit a patch 1 m (3.3 ft) into the open as a patch under bushes only if the patch in the open provides 4 to 8 times as much food. Blue tits, European relatives of chickadees, require a much higher feeding rate in order to move 1.5 m (5 ft) from the edge of the trees than to feed under the trees. It does not require much extra energy to move these short distances, so such dramatic differences in habitat use are best explained by changes in danger.

The trade-off between foraging and danger can become larger due to the actions and interactions of foraging animals. Marmots and ground squirrels flee from danger by running back to their burrows and prefer to feed near their burrows. Because they first eat much of the nearby food, they then need to go further from their burrow to feed rapidly. Many lizards also feed from a safe, central place. Such lizards can find more prey farther out, but at a cost. The actions of lizards also produce a gradient of food and danger for their potential prey. For the grasshoppers they eat, food is less depleted close to the lizard's central spot, but obviously there is greater risk of attack by the lizard. Thus foraging and danger have cascading effects across several trophic levels.

Animals may also crowd into a safe area. Then the food found there will have to be shared among many mouths. By congregating, animals reduce each other's feeding rates through competition and also decrease each other's danger through safety-in-numbers advantages. When avoiding predatory perch, 92% of small crucian carp concentrate in the safer shallows, making the shallows both safer and more depleted of food.

Potential predation effects the behavior of many animals in many different ways. It is easy to observe the effects of potential predation risk on the behavior of foraging animals. On the other hand, it is difficult to observe acts of predation. In most situations, acts of predation are rare. How can predation be important if it is so rare? The answer is that ever-present precautions prevent most potential predation from happening. When such precautions are absent, the situation is dramatically different. Animals on some islands have evolved for long periods without predators. Introduced predators can easily kill large numbers of these animals that lack defenses. In Australia, foxes can kill large numbers of some native mammals that apparently lack effective antipredator defenses. In New Zealand, a single dog is estimated to have killed 500 kiwis during a period of perhaps 6 weeks. Antipredator behavior by potential prey is the reason we do not regularly witness such slaughter at our doorsteps. The potential for predation is important because it makes cautious behavior common. Realized predation is rare because cautious behavior is common. When no predation occurs, this is likely due to the successful efforts of potential prey in avoiding their predators.

See also Antipredatory Behavior—*Sentinel Behavior*
Antipredatory Behavior—*Vigilance*
Mimicry

Further Resources

Alcock, J. 2001. *Animal Behavior: An Evolutionary Approach*, 7th ed. Sunderland, MA: Sinauer. (particularly chapters 7 & 8)

Attenborough, D. 1990. *The Trials of Life: A Natural History of Animal Behavior*. Boston: Little, Brown.

Blumstein, D. T. 2002. *Moving to suburbia: ontogenetic and evolutionary consequences of life on predator-free islands*. Journal of Biogeography, 29, 685–692.

Krause, J. & Ruxton, G. D. 2002. *Living in Groups*. Oxford: Oxford University Press.

Lima S. L. 1998a. *Stress and decision making under the risk of predation: recent developments from behavioral, reproductive, and ecological perspectives*. Advances in the Study of Behavior, 27, 215–290.

Lima S. L. 1998b. *Nonlethal effects in the ecology of predator-prey interactions*. BioScience, 48, 25–34.

Lima S. L. & Dill L. M. 1990. *Behavioral decisions made under the risk of predation: A review and prospectus*. Canadian Journal of Zoology, 68, 619–640.

Peter A. Bednekoff & Rebekka L. Darner

■ Antipredatory Behavior
Hiding Behavior of Young Ungulates

On the grasslands of East Africa a day-old wildebeest calf stands in its mother's shadow. Since the hour of its birth this little antelope has been able to walk and even to run and he has followed his mother's every move. It may be difficult to see the calf as it hovers just beside or beneath the mother. This is part of the wildebeest protection package, designed to improve the calf's chances of surviving its youth. Protection is vital because there are many dangers, including jackals, hyenas, leopards, cheetahs, wild dogs, and lions. The young wildebeest is an easy, safe kill for any of these predators, so death always hovers nearby. Consequently, the mother is very alert. She sacrifices time that could be spent feeding or resting to time that she spends standing and scanning for predators. If she spots a predator she moves away, calf following at heel, before the predator can get close enough for a successful attack. If a predator pursues, the mother positions herself between it and the calf as they run away. The calf is remarkably fast for such a youngster but it does not have either the speed or, especially, the endurance of an adult wildebeest. Thus, at some point in the run from the predator, the calf abruptly drops to the ground while the mother continues to run. Because the mother has been partly concealing the calf during the run, because the calf drops rapidly and freezes, and because the mother keeps running, the predator has poor information about the calf's location. With only a vague idea of where to search, finding the calf will be difficult—the calf is lying motionless, jaw pressed to the ground, ears back—and probably not worth the time.

For a predator, the time and energy cost of searching for and pursuing prey must be outweighed by the energy gain of eating that prey. Given that a wildebeest calf is fairly small, the energy gain that it represents is not worth an extensive search.

Finally, if the predator stops pursuing the mother and begins to search, the mother comes back to stand and watch. If the predator gets close to the calf the mother may attack using her horns and hooves. The mother will certainly attack a smaller predator such as a jackal, but she will not attack a lion.

The wildebeest protection package has the following key elements:

1. The young is relatively large at birth and is able to walk and run shortly after birth.
2. The young follows the mother and stays close to or under her.

3. The mother is very vigilant and moves herself and the calf away from predators that she detects at a distance.

4. If a predator pursues, the mother and young run and the mother positions herself between the calf and the predator. At some point during the flight, the calf drops and freezes while the mother continues to run.

5. If the predator searches for the calf, the mother will intervene if she can.

This effective, impressive protection package is not just a specialization of wildebeest. The package is called the *following strategy* and it is used by many species of hoofed mammals. Wildebeest are a bit unusual because most antelope species (there are about 137 species in Africa and Asia) do not use it. *Following* is the strategy used by all of the odd-toed hoofed mammals—the horses, zebras, tapirs and relatives, and by even-toed non-cud-chewing mammals such as hippos, pigs and peccaries. Superimposing this information on the family tree of hoofed mammals shows that *following* was the original, or ancestral protection package.

Nearly all of the cud-chewing or ruminant mammals use another strategy, called *hiding*. Hiding is a very specialized set of behavioral tasks performed by the mother and young. As will be explained later, *hiding* is really an extremely modified version of *following*, but is so specialized or derived that it looks like something entirely different. This is the way that evolution works. When aspects of organisms (such as bones, hair, lungs, behavior) become modified over time as natural selection operates, quite often the current modified version is so different from the ancestral version that seeing the connection requires some detective work.

The hiding strategy is illustrated by the pronghorn, another open country grassland species. Pronghorn evolved in the North American grasslands which, until just 10,000 years ago, were full of many species of hoofed mammals and an equally diverse set of predators, including lions, hyenas, wolves, and cheetahs.

When pronghorn fawns are born (mothers always have twins), the mother licks the young and eats the birth membranes, including the placenta. Then, after resting for about an hour she leads the fawns away from the birth site. The fawns can walk but they are quite shaky. The 400–500 yard trip from the birth site may take as much as 20 minutes, the mother moving slowly and patiently as the fawns struggle to keep up.

While she walks, the mother is very alert. After she has reached what she deems an acceptable distance from the birth site (which despite the mother's cleaning probably still has an odor that would attract predators), the mother stops and gives the fawns a signal. They walk away from the mother, looking for a place to recline. The mother retreats, but keeps an eye on the fawns. As each fawn starts to recline, the mother swings her head toward it and watches as the fawn sinks out of sight. In those two seconds while she focuses on the disappearing fawn, the mother memorizes the location—usually in flat grassland with few landmarks—then moves briskly away or even runs.

Now, if a coyote on the prowl comes near the mother, it sees only a solitary female. There is no evidence of young and the mother's location does not provide any useful information on where to search. The fawns, which do not yet have the hoof glands that leave odor, have walked away from the mother, so there is no scent trail to them.

The *hiding strategy* is thus about information control. The mother and young are acting in a coordinated way to deny information to a predator. First, they are denying information about the existence of the fawn. A coyote that encounters a pronghorn mother has no idea whether a fawn is near. The mother puts on a remarkably good version of a "poker face," and will continue to pretend uncon-cern unless the coyote, by random casting about, gets too close to the fawn. Even at 100 meters (330 ft) or more, the mother knows exactly where the fawn is. She observes the coyote's path, is able to make a men-tal projection of that path, and is able to judge the perpendicular dis-tance between the projected line and her fawn. If that distance is less than 15–20 meters (50–65 ft), she runs toward the coyote and either attacks it, striking with her forelegs, or prances in front of it with her big white rump patch flared, trying to entice the coyote into chasing her.

An alert pronghorn calf lies on the ground. Its mother is almost certainly nearby.
© Michael S. Quinton / National Geographic Image Collection.

This spatial acuity means that a mother may stand, apparently unconcerned, as a coyote passes between her and her fawn, as long as the anticipated trajectory of the coyote is ac-ceptable. The fawn's job during this bluff is to remain motionless, jaw pressed flat to the ground, ears laid back.

Clearly, the prime goal of the hiding strategy is to hide the very existence of the young. In this way, hiding seems superior to following; a hyena that pursues a wildebeest mother and young can search for the dropped young secure in the knowledge that there is some-thing to be found, but a coyote that searches in the vicinity of a pronghorn mother has no such assurance.

To perfectly conceal the existence of the fawn, a mother would ideally stay away until the fawn is fully grown and the fawn would ideally remain immobile until that time. Of course that is impossible—because the fawn needs to eat and to develop normally, it needs to move. So the mother visits the fawn, but infrequently. When fawns are 0–2 weeks old these visits occur at about $3\frac{1}{2}$ hour intervals. At a reunion, the mother stands while the fawn suckles and she also licks the fawn's anus and urinary orifice. As a part of the hiding strat-egy, the fawn will only defecate or urinate when stimulated by the mother. The mother eats the fawn's urine and feces. This helps to conceal the fawn and it also allows the mother to produce antibodies that she will put into milk that are designed to fight whatever pathogens are in the fawn's urine and feces. Because the fawn will not get another meal for 3–4 hours, because it is growing rapidly, and because it must make enough heat to keep warm while lying motionless in the cold wet spring grass, it takes on a big load of milk. This ability to gorge on milk is another part of the hiding strategy. If young of a following species are given comparably sized milk meals they die, because the milk backs up into the rumen and stops the normal fermentation.

After the fawn has taken on its big milk load and offloaded its excrement with im-munological information to the mother, it typically has a short bout of play, in which it ca-vorts and runs in loops away from and back to the mother. Then the mother leads the fawn to a new location and gives it a signal to find a new hiding place. The fawn walks away, nose down, while the mother begins to walk away at an angle, giving every impression that she is alone. But as before, as the fawn begins to recline, the mother turns to stare at it to memorize the spot. If the fawn survives, this routine will be repeated at 3–4 hour intervals for the next 2 weeks. Then the fawn will begin a gradual transition out of hiding in which the reunion periods become longer and longer until the fawn's activity cycle essentially matches the mother's.

During the pure hiding phase the mother's behavior shows other specializations. First, the mother adjusts her distance from the hidden fawn so that her own position offers little useful information to the coyote. To do this the mother controls both the mean (average) distance from the fawn and the variance (degree of spread about the mean) in this distance. The ideal mean distance is an example of what ecologists call a trade-off. A trade-off emerges when an actor must balance conflicting goals. For pronghorn mothers, one goal is to actively defend the fawn by attack or luring if a coyote gets too close. This goal alone would be served best if the mother stood next to the fawn. But hovering over the fawn is in conflict with another goal, which is to deny the coyote any useful information on where to search. To serve the latter goal mother would ideally remain at some huge distance from the fawn—a mile or more—so that a coyote would have no chance of finding the fawn if it started a systematic search starting at the mother's position. But the mother, at a mile from her fawn, cannot monitor coyote trajectories in the vicinity of her fawn, and she certainly cannot get to the fawn to defend it in a few seconds. So the ideal distance is one that allows mothers to get to their fawns quickly, but still dissuades a coyote from searching with the mother's position as a starting point. How would coyote be "dissuaded?" Because coyotes eat many kinds of prey, the key for pronghorn mothers is to stay at an average distance that presents the coyote with a time and energy cost that is greater than the time and energy cost for pursuing alternate prey. The average distance that pronghorn mothers stay away from fawns is 70 meters (230 ft), and at this distance a coyote could expect a slightly higher rate of energy gain by hunting ground squirrels than by searching systematically for the fawn. But if a coyote "knew" that mothers were always 70 meters from fawns it could profit by searching a narrow doughnut 70 meters from the mother. For this reason mothers also control the variance in distance from the fawn. They do not simply remain at 70 meters. Sometimes they are 1/2 mile or more away and sometimes they are within 30 meters (98 ft). The range in distances is huge and the consequence for coyotes is that there is no adaptive search doughnut.

The second general tactic that mothers use is active searching for coyotes in the 30–45 minutes preceding a visit to the fawn. During this time the mother runs to a number of vantage points and stands motionless for a few minutes at each. Although it is difficult to prove that the mother is searching for predators, her behavior certainly gives this impression.

The third general tactic that mothers use is open deceit. The four activity states of pronghorn are recline, stand, move, and feed. Mothers with hidden fawns modify their schedule of activity states so that activity gives no clues about the amount of time remaining until reunion with the fawn. Obviously, mother must be moving to return to her fawn, so a coyote might profit by stopping to observe a moving female. But mothers in the final half-hour before reunion, in addition to searching for predators, also engage in sham feeding and reclining. When a pronghorn without a hidden fawn to protect reclines, it typically stays down for 1–2 hours. Mothers on their way to fawns often recline, giving every impression that they are going down for that length of time, but then get up and resume the approach 10 minutes later. The purpose of the sham reclining bouts (mothers also do sham feeding) is to broadcast false information about time remaining to the visit to the fawn.

Nearly all the ruminant mammals practice the hiding strategy but no other species has been studied at the same level of detail as pronghorn. Thus we do not know if mothers in the several hundred other species use all the tricks that pronghorn mothers use. But it is clear that all of the species use the basic hiding routine: Young move away from the mother and recline, mothers stay away for several hours and return to young for short feeding/activity bouts before the young recline again. In most ecological settings, hiding must be the superior strategy because so many species that live in habitats from dense forests to grasslands use it. As noted earlier, an overlay of the distribution of the hiding and following strategies on the family tree of hoofed mammals shows that the following was the earlier or ancestral strategy, and that hiding first appeared later, at the point in the tree where the ruminants appeared.

The family tree also shows that in a few instances species of ruminants (such as wildebeest) have more recently switched back to following. The selective pressures that favor the secondary return are not known, but the ability of some species to make the switch illustrates another aspect of evolutionary change. Evolutionary modifications are sometimes made by switching off developmental pathways, rather than by dismantling them. Thus at some point in the future, a changing environment that makes the ancestral pattern advantageous may prompt a rapid evolutionary response, when intact ancestral developmental pathways are switched back on.

Hiding originally evolved from following as a modification of the collapse and freeze response of followers. Recall the wildebeest calf running beside its mother, a hyena in pursuit. At some point the calf drops and freezes while the mother continues to run. To evolve hiding out of this we simply require that the calf drops and freezes instantly when a predator is sighted. Subsequently, natural selection can work to make the drop and freeze response occur even when a predator is absent. Then natural selection can work on the duration of freezing, on the ability of the stomach to accept huge loads of milk, and on the mother's behavior to make her uninformative about the existence and location of the calf.

The hiding and following strategies of hoofed mammals offer a clear example of how behavior evolves. The hiding behavior of pronghorn offers an equally clear example of behavioral adaptation—the fine-tuning of behavior by natural selection to serve particular goals. Hiders also illustrate the principle that evolutionary change in anatomy and physiology (in this case the ability of hider young to gorge on milk) often is precipitated by an evolutionary change in behavior. Ultimately, the message here is that predators exert powerful and far-reaching effects on the biology of prey.

See also Caregiving—*How Animals Care for their Young*
        Caregiving—*Incubation*
        Caregiving—*Parental Care*
        Caregiving—*Parental Investment*

Further Resources

Byers, J. A. 1997. *American Pronghorn. Social Adaptations and the Ghosts of Predators Past*. Chicago: University of Chicago Press.

Byers, J. A. 2003. *Built for Speed. A Year in the Life of Pronghorn*. Cambridge, MA: Harvard University Press.

Byers, J. A. 2003. *Pronghorn*. In *Wild Mammals of North America*, 2nd Edn. (Ed. By G. Feldhamer, B. Thompson, & J. Chapman). Baltimore: Johns Hopkins University Press.

Byers, J. A. & Byers, K. Z. 1983. *Do pronghorn mothers reveal the locations of their hidden fawns?* Behavioral Ecology and Sociobiology, 13, 147–156.

John A. Byers

■ Antipredatory Behavior
Predator–Prey Communication

A discussion of communication between predator and prey might seem absurd to many readers. After all, we normally think of ourselves as eager to communicate with our friends or family, but equally keen to minimize correspondence with our victims or enemies. From this perspective, predators should be stealthy, not "talkative" with their prey. And prey should stay out of the way of their predators, not keep in touch with them. But, it turns out that this way of thinking about predator–prey relationships and communication oversimplifies both. The fact that the lives of predators and prey are entangled often means that demands for communication are forced on both parties, even though their interests are in conflict. To understand why this is so, we must understand how communication systems originate and function.

Predators and prey impose changes on one another both evolutionarily and developmentally. These changes reflect a kind of "race" between predator and prey, because innovations in antipredator behavior put predators "behind" in their capacity to capture prey, and innovations in predatory behavior put prey "behind" in their capacity to avoid capture. British animal behaviorists Richard Dawkins and John Krebs have proposed that such races are like arms races between countries. A significant component of an arms race involves spying to anticipate what the opponent country will develop next. But spying is a two-edged sword because the spied-upon country may be able to "turn" a spy. Turned spies can be potent tools, used not only to identify other agents but also as a channel for fake intelligence. This example illustrates how a connection, once established, can be turned against the connector, and also provides a useful model for thinking about the origin and functioning of communication systems.

How can predators get an edge in their "arms race" with their prey? Predators can become more effective in part by "spying," that is, by acquiring the information needed to make at least two distinctions: (1) between potential prey individuals that are and are not feasible as food sources, and (2) between ways of attacking that are and are not workable for a particular prey item. The capacity to make these distinctions early in an encounter assists predators in avoiding failed or costly predation efforts. (Predation efforts can be costly when prey are very effective at defending themselves, either by escaping or fighting back, and so run up the expense of the predatory effort.) These assessment processes are illustrated by the predatory behavior of jumping spiders of the genus *Portia*, as revealed by the research of animal behaviorists Robert Jackson and Stimson Wilcox. *Portia* employs a surprising variety of predatory methods, including luring its intended victim by strumming the victim's web in a pattern that mimics the struggles of an insect caught there. *Portia* uses its exceptionally acute vision and other senses to assess the effect of strumming, in a trial-and-error approach to predation. If the victim responds to strumming, these spiders persist with that strumming pattern, but they switch to another pattern when the target is unresponsive. However, if the intended prey approaches too aggressively, *Portia* may try an alternative to strumming, leaving the web, circling behind it, and beginning a long approach from off the web that ends with a surprise aerial attack in which *Portia* swings in on a length of its own silk.

The efficiency of prey can be impaired by unnecessary antipredator behavior that interferes with other important activities such as feeding, reproducing, and socializing. Prey can limit such inefficiency by spying on potential predators and responding only to real predatory threats, not those that only appear dangerous. However, such distinctions are often subtle, and demand some of the same kinds of probing used by *Portia* to assess its prey. California ground squirrels use such assessment probes with Northern Pacific Rattlesnakes. These snakes prey primarily on young squirrels. My colleagues Richard Coss, Matthew Rowe, and I have discovered that adult squirrels are less vulnerable than young because they have the ability to dodge strikes and to neutralize rattlesnake venom when a strike does land. Consequently adults can be very assertive in defending their young, sometimes confronting, harassing and even attacking rattlesnakes. But adults *do* need to assess the danger posed by rattlesnakes not only for their pups but also for themselves, because a nonlethal bite can still injure an adult squirrel. Danger from these rattlesnakes depends heavily on two factors: (1) *Body size*—larger snakes are more dangerous, and (2) *Body temperature*—a snake's body temperature tends to follow its surrounding temperature, and warmer snakes are more dangerous. Like *Portia*, these squirrels use interactions for assessment, aggressively confronting rattlesnakes and provoking them to rattle, a defensive signal normally intended for predators of rattlesnakes, not their prey. These rattling sounds contain cues about both the body size and temperature of the snake, acoustic cues that are especially valuable for risk assessment in the darkness of squirrel burrows. Squirrels proceed with greater caution when their probes reveal acoustic evidence of a larger, warmer snake.

Once predator or prey establishes a connection for assessing the other, that connection may be turned on the assessor (i.e., the "spying nation"). It follows that assessment processes such as those used by *Portia* and ground squirrels can be exploited by their adversaries. Many researchers in animal communication are converging on the conclusion that signals often originate through a process of turning an assessment connection back on the assessor. For example, both larger and warmer rattlesnakes produce louder rattling sounds. If California ground squirrels reduce their harassment of snakes when they detect a loud rattling sound, this assessment process favors the ability to produce louder rattling. In fact, rattlesnakes do appear to have responded to such assessment by rattling louder, in apparent exaggeration of their size. More generally, assessment connections become "handles" that other parties can "grab" and use to manage the behavior of the assessing individual. The "tools" used to grab those handles are what we call signals. In other words, signals acquire their power for communication through exploitation of the assessment systems of others, and this exploitation often involves modification of the cues already in use by others for assessment. When a signal and a corresponding assessment system are in place, we have a communication system. Such a model of communication can help us understand the many forms of predator–prey communication known to exist, from signaling by prey as a means of discouraging predation, to signals emitted by predators to assess the vulnerability of their prey.

See also Antipredatory Behavior—*A Brief Overview*
　　　　　Antipredatory Behavior—*Sentinel Behavior*
　　　　　Antipredatory Behavior—*Vigilance*

Further Resources

Caro, T. M. 1986. *The functions of stotting: A review of the hypotheses.* Animal Behaviour, 34, 649–662.
Gil-da-Costa, R., Palleroni, A., Hauser, M. D., Touchton, J. & Kelley, J. P. 2002. *Rapid acquisition of an alarm response by a neotropical primate to a newly introduced avian predator.* Proceedings of the Royal Society, Series B, 270, 605–610.

Guilford, T. & Dawkins, M. S. 1991. *Receiver psychology and the evolution of animal signals.* Animal Behaviour, 42, 1–14.

Owings, D. H. & Morton, E. S. 1998. *Animal Vocal Communication: A New Approach.* Cambridge: Cambridge University Press.

Owings, D. H., Rowe, M. P. & Rundus, A. S. 2002. *The rattling sound of rattlesnakes* (Crotalus viridis) *as a communicative resource for ground squirrels* (Spermophilus beecheyi) *and burrowing owls* (Athene cunicularia). Journal of Comparative Psychology, 116, 197–205.

Wilcox, S. & Jackson, R. R. 2002. *Jumping spider tricksters: deceit, predation, and cognition.* In: *The Cognitive Animal* (Ed. by M. Bekoff, C. Allen, & G. Burghardt), pp. 27–33. Cambridge, MA: MIT Press.

Donald H. Owings

■ Antipredatory Behavior
Sentinel Behavior

One Florida scrub jay sits on top of a scrub oak while other members of its group search for food in the thick vegetation below. This single sentinel sits in a relaxed posture yet scans the horizon continually. After several minutes, another scrub jay takes up a sentinel position on another oak. The first sentinel drops quickly to the sandy ground and begins to forage. Several moments later, the new sentinel emits a harsh alarm call when it spots a merlin flying fast across the tops of the oaks. All the jays dive into dense vegetation and the merlin flies on in search of easier prey.

Sentinel behavior has been defined as coordinated watchfulness, usually from exposed positions. In this definition, "coordinated" means that the number of sentinels is more consistent than expected by chance. For example, Florida scrub jays consistently have one sentinel with a few occasions of zero or multiple sentinels. Here coordination describes an outcome, not a process that leads to that outcome. Much of this essay will be devoted to discussing processing that could lead to this striking outcome.

Sentinel behavior has a puzzling taxonomic distribution among birds and mammals: It is found in many groups but has been reported only in a fraction of each group. Besides the crow and jay family, sentinel behavior has been reported for several kinds of parrots, tropical tanagers from Brazil, birds known as babblers from Israel and India, a cuckoo relative from Puerto Rico, a weaver bird of southern Africa, rock haunting possums from Australia, dwarf mongooses and suricates from Africa, various monkeys, hyraxes from southern Africa, and klipspringers—a kind of dwarf antelope from Africa. Because sentinel behavior is spread across multiple taxonomic groups but fairly rare within each taxonomic group, it is most likely

Many animals are watchful of potential predators when they are exposed. This photo shows three antelope exhibiting sentinel behavior, or coordinated watchfulness.
© *Getty Images/PhotoDisc.*

due to a particular combination of common ecological factors. The conditions need to be general so that they apply to many animals in many places, but the combination needs to be particular so that it is not met more often. These animals differ quite a bit in size, shape, and diet. What characteristics might they share with each other that they do not share with their close relatives?

A big clue comes from the locations in which sentinel behavior has been reported, such as portions of Africa, Australia, southern Brazil, and the Middle East. The locations fall around the tropics in dry areas where small trees mix with other vegetation. In such environments, the vegetation may be dense near the ground but forms a single, fairly short layer. Florida scrub jays, for example, forage among a tangle of palmettos, twisted oaks, lichens, and other vegetation. Species with sentinels all forage among thick vegetation where they can see little, but they can get a clear view around their environment if they move the short distance to the top of the vegetation. This good view is probably the reason individuals watch from high positions when not foraging.

Sentinels watch from high positions. What are they watching for? The most common answer is birds of prey such as falcons, hawks, and eagles. In dwarf mongooses, sentinel behavior is common only in populations where the mongooses are frequently attacked by pale chanting goshawks. In suricates (also called meerkats), sentinel behavior is much more common in natural areas with lots of hawks than in ranch areas with fewer hawks. In Florida scrub jays, sentinel behavior is most common when large numbers of hawks are present in the area. Birds of prey rely on surprise for their attacks. If warned of a possible attack, animals can take cover in burrows or in thick vegetation. From atop the vegetation, sentinels can detect a falcon, hawk, and eagle when it is still far away.

Thus resting on a high perch may be a safe thing to do in savanna and scrub habitats when under attack from birds of prey. Another big question remains: Why is sentinel behavior coordinated? All members of a group might rest in high positions at the same time, or at random times. This would not meet the definition for sentinel behavior. In an analysis of the costs and benefits of sentinel behavior, each individual gets a benefit from taking up a high position if it can see approaching danger well enough to avoid it. The key condition for coordination is whether this individual shares its knowledge of danger with others. If it shares its information, other individuals benefit from its safety. As long as information spreads from sentinels to foragers, the safest thing for an individual to do is forage when someone else is a sentinel and be a sentinel when no one else is doing so. This yields coordination among sentinels.

How does information spread? Many times the sentinel gives an alarm call. This is not special since many species without sentinels give alarm calls. In other situations, the information may spread because any sudden departure from a high position is obvious to foragers on the ground. Here the rush to save itself would lead the sentinel to warn others. In some species, sentinels give the opposite of alarm calls—soft calls when there is no danger. These calls seem to tell foragers that a sentinel is present and that it is safe to forage intensely. Such calls may help coordinate sentinel behavior but probably are not essential to its origins.

How can sentinel behavior be explained? When we analyze the value of behavior, we think about costs and benefits. Thus far this article has focused on the possibility that sentinels are relatively safe. The rest of this article will compare this explanation to three alternative hypotheses. Scientists compare hypotheses to explain observed information and use hypotheses to predict new observations. Scientific hypotheses can be tested by testing their assumptions and by testing their predictions. As either new information is obtained or new hypotheses are proposed, the favored hypothesis may change. Ideally a hypothesis will explain what is known, predict new observations, and make as few assumptions as possible.

The four hypotheses about sentinel behavior make different assumptions and lead to different predictions. Here are the four hypotheses:

1. *Sentinel safety*: Sentinels receive a direct benefit from being able to detect predators. Solitary animals are predicted to rest in high positions like sentinels. Sentinel coordination comes about when information about attacks spreads from sentinels to foragers. Animals are predicted to spend more time as sentinels if given extra food.

2. *Kin selection*: Sentinels endanger themselves, but help protect related individuals. Animals are predicted to be sentinels more often when they have more relatives in a group.

3. *Reciprocity*: Sentinels endanger themselves in the short-term, but the cost is more than paid back when other individuals are sentinels at other times. Within a group, individuals are predicted to take predictable "shifts" as sentinels and, if others skip their shift, to react by reducing their time as sentinels or by punishing the slacker.

4. *Social prestige*: Sentinels endanger themselves in order to signal their quality and compete for social prestige. Animals are predicted to spend more time as sentinels if given extra food, to be sentinels more when others are sentinels more, and to competitively prevent others from being sentinels.

The strongest test among hypotheses is to find an assumption or prediction that is made by one hypothesis but forbidden by others. The hypotheses differ in their assumptions about whether sentinels are safe or at risk. Are sentinels able to detect attacks? In both dwarf mongooses and Florida scrub jays, sentinels were roughly 10 times as likely to give alarm calls as were foragers. This is indirect evidence that sentinels detect predators well enough to be safe. More direct evidence comes from a large study of suricates in South Africa. Here sentinels were safer from attack than were other animals.

These Canada geese parents are very protective of their young and will post a sentry to watch out for predators.
© Anthony F. Chiffolo

The hypotheses also differ in that sentinel safety predicts that solitary animals will take up high positions like sentinels. Solitary animals are predicted not to be sentinels under the other three hypotheses because the benefits can only be achieved indirectly through other animals. Tim Clutton-Brock and his colleagues have observed sentinel-like behavior by solitary suricates.

Thus the sentinel safety hypothesis performs better than the other three hypotheses in accounting for both observations of sentinel safety and the behavior of solitary animals. It is still worth testing how well the hypotheses explain other observations. Sometimes no existing hypothesis is adequate and scientists search for new hypotheses.

In other tests, a hypothesis makes special predictions, but other hypotheses do not forbid these predictions. Such tests are weaker because they cannot show us that any hypothesis is false. Scientists still favor the hypothesis that makes the special predictions.

Two hypotheses, sentinel safety and social prestige, predict that animals will be sentinels more when they are given extra food. The same response is not specifically predicted by kin selection or reciprocity, but is not forbidden these hypotheses. The predicted response has been supported for three species: More sentinels were observed after researchers fed suricates with hard-boiled eggs, Arabian babblers with mealworms, and Florida scrub jays with peanuts.

The kin selection, reciprocity, and social prestige hypotheses each make novel predictions. Are any of these supported by existing evidence?

Kin selection predicts that animals are more likely to be sentinels when they are more related to the rest of their groups. The sorts of groups that have sentinels generally are extended family groups. In Florida scrub jays, for example, groups form from young hatched in previous summers that continue to live with their parents. Thus there might be a broad link between sentinel behavior and kin selection. The specific prediction, however, is not supported. Sentinels may have no relatives in the group and within large groups individuals with many relatives are not more likely to be sentinels than are individuals with few relatives. Among suricates, some of the most active sentinels are individuals who have recently joined the group and at that point have no relatives within it. Currently kin selection does not help us explain sentinel behavior.

Reciprocity is familiar to humans because humans often trade favors with each taking turns at performing some task, such as washing the dishes or taking out the trash. In such cases, duties tend to follow a predictable rotation and each tends to check that others do not skip their turns. Could sentinel behavior be explained as similar behavior? So far, studies have found no evidence that sentinel behavior involves taking turns. Sentinel bouts do not follow a simple rotation. When individuals are fed extra food, they become sentinels more often then other individuals who work their sentinel bouts around these extra bouts. Thus the predictions based on reciprocity are not supported for sentinel behavior.

Amotz Zahavi has suggested that Arabian babblers compete for social status when becoming sentinels. Sentinel behavior is linked to social status in that adults are sentinels more than juveniles and males are sentinels more than females. This correlation, however, is not evidence that sentinel behavior causes social status. It seems more likely that adults and males are sentinels more because they are also best at finding food. When Jonathan Wright and colleagues (2001) fed Arabian babblers, they became sentinels more often. The top two males in the group did not demonstrably compete when they each could be sentinels more. Thus the hypothesis about social status does less to predict social behavior than the simpler hypothesis that feeding and sentinel behavior are linked.

See also Antipredatory Behavior—*Finding Food While Avoiding Predators*
Antipredatory Behavior—*Vigilance*

Further Resources

Alcock, J. 2001. *Animal Behavior: An Evolutionary Approach.* 7th ed. Sunderland, MA: Sinauer.

Bednekoff, P. A. 1997. *Mutualism among safe, selfish sentinels: A dynamic game.* American Naturalist, 150, 373–392.

Bednekoff, P. A. & Woolfenden, G. W. 2003. *Florida scrub-jays* (Aphelocoma coerulescens) *are sentinels more when well-fed (even with no kin nearby).* Ethology, 110, 895–904.

Blumstein, D.T. 1999. *Selfish sentinels.* Science, 284, 1633–1634.

Clutton-Brock, T. H., O'Riain, M. J., Brotherton, P. N. M., Gaynor, D., Kansky, R., Griffin, A. S., & Manser, M. 1999. *Selfish sentinels in cooperative mammals.* Science, 284, 1640–1644.

Goodenough, J., McGuire, B., Wallace, R. A. 2001. *Perspectives on Animal Behavior*. 2nd ed. New York: J. Wiley.

Wright, J., Maklakov, A. A., & Khazin, V. 2001: *State-dependent sentinels: an experimental study in the Arabian babbler*. Proceeding of the Royal Society of London B, 268, 821–826.

Zahavi, A. 1990. *Arabian babblers: The quest for social status in a cooperative breeder*. In: *Cooperative Breeding in Birds: Long-term Studies of Ecology and Behavior* (Ed. by P. B Stacy & W. D. Koenig), pp. 103–130. Cambridge: Cambridge University Press.

Peter A. Bednekoff

■ Antipredatory Behavior
Vigilance

Animals use their eyes, ears, noses and other senses to detect danger in the world around them. Scientists that study animal behavior call this *vigilance*. Almost all the research on vigilance behavior concerns visual scanning of the environment and most studies have been conducted on birds. But increasing numbers of studies examine mammals and this has produced some surprising discoveries.

As one might expect, animals are more vigilant when danger seems close at hand. At such times, they spend more time with eyes open, nostrils sniffing the air, and ears pricked up. Their senses are gathering information about the world around them. This may give them early warning to escape attack, warn their fellows, or counterattack. Humans behave this way too. Next time you see people waiting to cross a street at a busy intersection, watch their faces.

This baboon is very cautious while venturing to the ground to get a drink.
© Anthony F. Chiffolo.

Look how their eyes scan far and wide. Compare their wariness or level of alertness to other people who are having a conversation or eating a meal.

Scientists who study animal behavior observe the faces of animals and measure the amount of time those animals spend with eyes open, heads up, or otherwise alert. That's how one can compare the vigilance of animals in risky situations to the vigilance of those in safer conditions. For example, a common risky situation for monkeys that spend their lives in the trees are the occasional times they are forced to come down low in the trees. Every monkey ever studied is more vigilant when it is near the ground, probably because it is worried about predators that hunt near the ground, such as leopards, constrictor snakes, or even humans. Another common risky situation occurs when animals move into an area where predators often spend time, like waterholes. African antelopes approaching a waterhole are very wary. As they drop their heads to drink they are particularly vulnerable so they often interrupt drinking to look around. One can tell that they are nervous and alert because they will start at unexpected noises or movements.

By measuring the time animals spend vigilant, scientists have learned many interesting things about concealment and the challenges animals face every day. A vigilant animal can avoid dangers posed by predators and members of the same species (*conspecifics*) better than an animal that is unaware of its surroundings. In addition to predators, some conspecifics also need to be watched closely. For example, adult male red colobus monkeys in Uganda spent more time watching their rivals when an adult female in their group was sexually receptive. Infanticide is another threat that is widespread in mammals and stems from conspecifics. Unrelated adults may attack infants to remove competitors or hasten ovulation in the infants' mother. This danger is often countered by maternal vigilance. For example, all members of a black howler monkey group raise their vigilance when there is a newborn in the group. Mothers and fathers of newborns are particularly vigilant toward strangers in many species.

One of the earliest comparisons scientists made was to study the vigilance behavior of birds in small and large flocks to learn more about safety in numbers. They noticed that birds in big flocks were less alert than birds in small flocks, thereby saving time for other important activities. Birds that spend less time vigilant can spend more time collecting food, because their eyes can focus on food (usually down on the ground) rather than danger (usually above and to the sides). As a consequence, birds in big flocks can feed more efficiently than birds in small flocks. This discovery made quite a stir in the scientific community and started a research program that has lasted 30 years and continues to this day. Why did scientists become so interested in the finding that birds in large flocks were less vigilant than birds in small flocks?

It was thought that animals like these elk lived in groups in order to reduce time being vigilant.
© Anthony F. Chiffolo.

An animal that spends less time scanning its surroundings may have more time for other important activities (feeding, playing, socializing, resting, etc.). These other activities help a bird to survive and achieve success in its group. For example, a well-fed, relaxed sparrow in a big flock might survive a harsh winter more easily than an undernourished, wary sparrow in a smaller group. The sparrow in a big group is taking advantage of safety in numbers to improve its survival chances and maximize its chances of reproducing and contributing to the next generation of sparrows.

If the scientists are correct, vigilance behavior provides an important clue to understanding why animals join groups in the first place. In groups, animals must share food, resting spots, and mates. Why join a group if you have to share all the things that make you fat and happy? The answer—as common sense may suggest—is that animals form groups for safety. It makes sense to share food, shelter, and mates when you are also sharing protection. So, many scientists suggested that birds in big flocks reduced the time they spent vigilant because they felt safer. It wasn't only feeding birds that enjoyed safety in numbers. One study showed that mallard ducks in big groups slept more soundly than those in small groups. Animal behavior researchers had found something truly important—a daily, minute-by-minute advantage of living in a group that could even make it worthwhile to share with members of the group.

For about 15 years, scientists studying vigilance were satisfied with the explanation that sharing with your associates pays off because you can save time otherwise spent vigilant—and scientists then spent a lot of time measuring exactly how much time could be saved by living in a group and in reporting the fantastic ways animals used their vigilance behavior and their fellow group members for safety. But some scientists began to study animals in more exotic situations than birds at a feeder. Some animals catch their food in the air (like bats and swallows), others crop the tops of trees (like giraffes), and yet others hang upside down from tree branches (like sloths and some monkeys). These animals seemed to be able to watch their surroundings while eating. Do they gain from safety in numbers? As more and more questions without clear answers began to accumulate, some scientists became less satisfied with the idea that all animals will reduce their vigilance in groups. Other scientists thought this was a very convincing theory that required no improvement, but then they stumbled on some observations that didn't fit the theory. Both types of questions about theories are needed. Science progresses when many curious people ask tough questions and devise new tests and methods to answer the questions.

Among the most interesting challenges to the theory came from studies that found groups in which animals did not reduce their vigilance—or actually increased their vigilance—in larger groups. It took a while for these findings to sink in but the new findings became more accepted when scientists remembered that not all groups are the same. Everyone knows groups of animals come in different sizes, but groups of animals also come in different social organizations—the size, shape, and structure of a group. For example, social organization varies when the species, sex, age, and status of animals in groups changes over time or across different groups. Regarding the shape of groups, Marc Bekoff found that western evening grosbeaks feeding in a group shaped like a ring were less vigilant than grosbeaks that formed a group shaped like a line—even if the same number of birds were in either group. Vigilance also changes with the composition of the group as described above for black howler monkeys in groups with newborn infants.

When one starts to look at a group as more than a simple accumulation of animals, fascinating questions jump to mind. What if the members of a group are not friendly to each other—will they still share in the chores of vigilance? The vigilance of scavengers—such as bald eagles feeding on dead salmon which had washed up on shore or Tasmanian devils gorging on carrion—helped answer this question. When scavengers come from far and wide to feed on dead animals they form a temporary group of competitors (individuals seeking the same limited resource). Neither bald eagles nor Tasmanian devils reduce their vigilance when they scavenge in larger groups. They may even increase vigilance! This prompted a new question: What happens when danger comes from within your group as well as outside of it? Several scientists studied competitive monkeys and apes to answer this question. Monkeys and apes (the closest living relatives of humans) form groups that vary markedly in social organization. Some groups are just composed of family members and they did share the duties of vigilance as expected. Other monkeys and apes find themselves born into groups with family members but also unrelated animals that they don't always trust. Such groups contain bullies and they contain future friends that one hasn't yet learned to trust. A number of studies showed that monkeys and apes in competitive groups spent surprising amounts of time watching each other! When one travels with enemies and strangers, one must watch them to keep a competitive edge. This proved to be another case in which animals in larger groups did not reduce vigilance. As a result, scientists interested in complex social organization are still seeking the significant advantages of living in a group that can offset the minute-by-minute costs of sharing with associates.

We can sum up the history of the scientific study of vigilance rather simply. Armed with a first, good theory that worked in many cases, scientists began to ask more in-depth questions about how animals spend their time and protect themselves. Along the way, they built a second, better theory that worked in more cases. It was better because it explained more of the fascinating diversity of animal life around us. Anyone with a good grasp of the intricacies of vigilance behavior and the ability to observe animals in safe and risky situations can make a contribution to this exciting and rapidly changing field of animal behavior.

Our understanding of animal vigilance behavior is still growing but many questions remain to be answered about animal vigilance. Does a long-lasting glance mean the animal feels more risk, that it is looking at something interesting, or some other reason? Can vigilance toward conspecifics tell us who is important in a complex social group and who can be ignored? We now understand that vigilance is a way of gathering information—not just detecting predators—so anything interesting in an animals' environment deserves a glance and maybe further investigation. Sometimes associates deserve more attention than the trees, high grass, and other potential hiding places for predators. For instance, some animals face little risk from predators because they are very well defended, very good at escaping, or they live in habitats with few predators. Nevertheless, they remain vigilant, often to observe their associates and to learn about the changing world around them. These and other questions continue to intrigue scientists who study animal behavior. The answers are providing insights into the evolution of the brain as an information-processing organ and the origins of grouping as a behavior in animals, among other things.

See also Antipredatory Behavior—*Finding Food while Avoiding Predators*
Antipredatory Behavior—*Sentinel Behavior*

Further Resources

Alcock, J. 1999. *Animal Behavior: An Evolutionary Approach.* Sunderland, MA: Sinauer Associates.

Bekoff, M. 1995. *Vigilance, flock size, and flock geometry: Information gathering by western evening grosbeaks* (Aves, fringillidae). Ethology, 99, 150–161.

Edmunds, M. 1974. *Defence in Animals: A Survey of Anti-Predator Defences.* London: Longman.

Elgar, M. A. 1989. *Predator vigilance and group size in mammals and birds: A critical review of the empirical evidence.* Biological Review, 64, 13–33.

Lima, S. L. 1990. *The influence of models on the interpretation of vigilance.* In: *Interpretation and Explanation in the Study of Animal Behavior.* (Ed. by M. Bekoff & D. Jamieson), pp. 246–267. Boulder, CO: Westview Press.

Lima, S. L. & Bednekoff, P. A. 1999. *Back to the basics of anti-predatory vigilance: Can non-vigilant animals detect attack?* Animal Behaviour, 58, 537–543.

Logan, K. A. & Sweanor, L. L. 2001. *Details of a Desert Carnivore.* Washington, D.C.: Island Press.

Treves, A. 2000. *Theory and method in studies of vigilance and aggregation.* Animal Behaviour, 60, 711–722.

Treves, A. & Pizzagalli, D. 2002. *Vigilance and perception of social stimuli: Views from ethology and social neuroscience.* In: *The Cognitive Animal: Empirical and Theoretical Perspectives on Animal Cognition.* (Ed. by M. Bekoff, C. Allen & G. Burghardt), pp. 463–469. Cambridge, MA: MIT Press.

Adrian Treves

■ Applied Animal Behavior
Mine-Sniffing Dogs

Wars do not end when the peace treaty is signed and everybody goes home. Land mines and other unexploded remnants of war can continue to kill and maim people for years to come, as well as blocking access to towns, homes and farms. Such access creates food, an economy, a community, and a life. Just the possibility of one mine can prevent the return of entire communities. The mines must therefore be found and removed.

Removing mines is necessarily a hazardous occupation because mines are secretive by nature and design—they are probably the best sit-and-wait predator on the planet. A mine that cost $5 (U.S.) to buy, 10 seconds to arm and 2 minutes to place, can sit quietly in the soil for 20 years before exploding instantly when disturbed. That mine takes a 12-person team a full day to locate and destroy, at a cost of about $1000 (U.S.). Clearly, any method for reducing that cost must be explored and exploited.

One solution is that most mysterious of mammalian senses, olfaction. The skills of dogs as odor detection devices are well known. Their ability is exploited in many different roles, ranging from drugs and bomb detection, to search and rescue. So why not mine detection?

Why not, indeed! Today, about 750 dogs are used globally to search for mines, and with impressive success. Mine detection dogs are trained to move slowly, nose to ground, in a careful search pattern across the landscape. They maintain a high level of concentration at all times, and take frequent breaks in order to avoid fatigue. The job of the handler is to ensure that the dog's nose covers all of the ground being searched. Usually, the minefield is subdivided in a structured way, such as into boxes marked out as a 10 × 25 meter grid. The dog will be trained to search lines 10 m wide on a long lead.

Alternatively, the dog can be trained to work on a short lead. Here, the handler has more direct control over the dog's search and can watch every square centimeter that the dog covers. As well, the handler walks the ground beside the dog—an impressive demonstration to returning refugees that their land has been properly cleared.

Clearly, the training needs to be effective enough to ensure that the dog can work safely in the minefield. In order to develop their skills, these dogs go to school for between 6 and 12 months before being licensed to work in minefields. Their schooling continues for life, as their detection and searching skills are tuned, refined and tested on a daily basis. The training seems to work, as accidents are impressively rare (e.g., no accidents involving dogs have occurred in Afghanistan since 1997).

Staff Sgt. Thomas Ellis, 33, of Sarasota, Florida, and his dog Arras on a mine-sniffing exercise on the compound of the Tuzla Air Base, Friday, Feb. 23, 1996. Arras, a 70-pound Belgian Malinois is one of six dogs trained on the base. Arras previously worked sniffing for bombs with the military police and was trained in a 20-week program to detect mines.
© AP / Wide World Photo.

In Afghanistan and Bosnia-Herzegovina, about 150 (Afghanistan) and 30 (Bosnia) dogs work 5–6 days a week on the daunting problem of detecting the undetectable. How mine detection dogs achieve that task is the subject of the first study ever to link searches by dogs to the availability of odor signals given off by mines.

Detecting the Mine

Listen to a person speaking a foreign language. The sounds are easily detected, but without an assigned meaning, the signals are just noise. Learning the language involves a two-stage recognition process: First, the sounds must be broken down into defined units; second, each unit must be assigned a meaning. The process is one of plucking sense from nonsense, or of linking signal detection to meaning. Once the two are joined, recognition has occurred and the language is understood.

For a dog, learning to detect a mine involves the same process. Its nose is constantly bombarded with chemical signals, called odors. Most of those odors are noise—they have no meaning and are of no interest. Any dog has a simple odor language, consisting of concepts such as "rabbit," or "the female dog who lives next door," and effortlessly separates those recognizable odors from background noise.

The two problems in mine detection are, first, for the dog to assign meaning to the odor signals given off by a mine, and second, to detect that very weak signal consistently.

Unfortunately, mines have no interest in communicating with dogs. Compounding the problem is that the main explosive substance used in mines, TNT, has very low volatility (and thus detectability). The problem can be compared to listening in on a whispered conversation at a cocktail party. Not only are the signals hard to distinguish, they are swamped by a noisy background, and are not intended to be received or interpreted by you anyway.

Dogs *can* detect mines. But asking dogs to find mines pushes their detection skills to the limit. A significant unknown in the detection process is the availability of signals to detect. Just what odor signals does a mine provide, and are there conditions when signal availability falls below the recognition threshold for the dog? Clearly, such conditions might cause the dog to miss the mine, and must be avoided.

A Challenge for the Scientist

The study in Afghanistan and Bosnia is addressing these questions. The dogs are tasked with searching for mines in test mine fields (the mines are real, but triggers have been removed) under carefully controlled conditions. Immediately after the dog has found (or missed) a mine, soil samples are taken and weather conditions are recorded. The behavior of the dog is continuously filmed for later analysis, and the handler is interviewed.

Why soil samples? TNT and related molecules that leak out of the mine migrate slowly to the surface, assisted by soil moisture and electrostatic processes. Once at the surface, they remain bound to dust particles, available to the dog's nose as it passes by.

The availability of molecules of explosive in each sample is being assessed by chemists. The dog searches at the ground/air interface; thus the chemists should similarly measure availability in the surface layer of soil.

In Afghanistan, the dogs search for mines in hot and dry conditions, with strong winds and temperatures ranging from 10° to 40° C (50° to 104° F). Above 40° C (104° F) it is too hot for the dogs and their handlers to work. In Bosnia, the dogs search in cool and damp

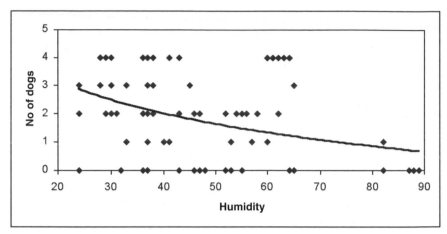

Weather and humidity affect the dogs' abilities to detect mines.
Courtesy of Ian G. McLean.

conditions, with temperatures from 5° to 30° C (41° to 86° F), and frequent rain. The chemists must therefore also search for mines under those conditions. Thus the trials are run several times to encompass all the weather conditions faced by the dogs.

For the scientist, establishing and running such a complex experiment in a post-conflict area is a significant challenge.

What We Found

How does weather affect the ability of the dog to find mines? Results show that cool weather, wet weather, and heavy soils all make the job more difficult. It turns out that the hot, dry conditions and sandy soils of Afghanistan make mine detection easier for dogs than the more temperate conditions and heavier wet soils of Bosnia.

How close to a mine does a dog need to get before detecting it? The dogs are trained to "point" the mine—to go right up to it. So it is not possible determine the maximum distance at which they can detect mines. However, a technique developed recently in South Africa allows this question to be asked in a different way. Usually mine detection dogs are taken to the minefield. But an entirely different approach is to bring the minefield to the dog. This is achieved by vacuuming odor of minefield onto a filter, and then taking the filter to a specially trained dog who gives the answer "Yes!" (mine present), or "No" (no mines here).

Now we can ask the dog the maximum distance at which mines are detectable. The answer is in the following graph. Dogs detect the mines with decreasing probability up to distances of 10 m (30 ft). Considering the weak odor signal available from the mine, this result shows an amazing detection capability.

Positive Outcomes

Worldwide, 2001 was the first of the last 50 years in which it is believed that fewer mines were laid than were cleared. Fortunately, clearance rates are improving, due in large part to the increasing use of dogs for mine detection. Improved understanding of both how dogs detect mines, and of the limits to that detection skill, will serve to improve the quality and safety of clearance operations globally.

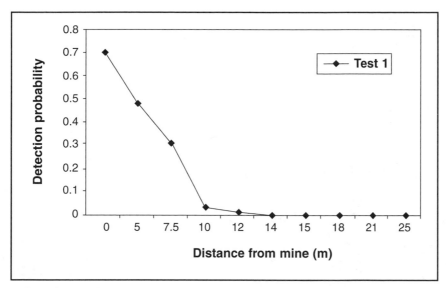

Linking detection and distance: Mine detection dogs can detect a mine at distances of up to 10 m (30 ft) from the mine.
Courtesy of Ian G. McLean.

Currently, about 20,000 people work on mine-clearance programs in up to 60 countries around the world. At great personal risk, they return thousands of acres of mine-contaminated land to economic productivity every year. In each country, they work towards a vision of a safe and productive environment, a functioning economy, and a stable socio-political landscape.

That vision is ambitious enough. But beyond even that, by supporting research such as the dog's nose project, the deminers of Afghanistan and Bosnia allow others to benefit from the devastation wrought in their countries.

See also Applied Animal Behavior—*Scat-Sniffing Dogs*

Further Resources

The following four web sites provide a good entry to the literature on de-mining, although they go far beyond the use of dogs.

Geneva International Centre for Humanitarian Demining
    www.gichd.ch
        The Geneva International Centre for Humanitarian Demining (GICHD) supports humanitarian mine action through operational assistance, research, and implementation of the Anti-Personnel Mine Ban Convention. It is an independent foundation supported by 18 governments. The Centre has a Mine Dog Detection project and has also found that African Giant Pouched Rats show promise for use in mine detection.
Journal of Mine Action.
        Available as an official publication of the Mine Action Information Center at James Madison University, Harrisonburg, Virginia 22807. 3 volumes per year. See http://www.hdic.jmu.edu
Journal of Mines and Unexploded Ordnance (a recently established electronic journal) http://www.demine.org
McLean, I. G. 2003. *Mine Detection Dogs: Training, Operations and Odour Detection.* Geneva: GICHD
        (Note: This is an edited book that will only be available through the GICHD as publisher, but it gives by far the best overview of issues on dogs.) See http://www.gichd.ch

Ian G. McLean

■ Applied Animal Behavior
Scat-Sniffing Dogs

Domesticated dogs, like their wild canid ancestors, have highly sensitive noses—their sense of smell is estimated to be more than 1,000 times keener than that of humans. Conservation biologists have learned how to harness this sensitivity to find wildlife feces, or *scat*, in forests and other natural settings. Specially trained "scat-sniffing dogs" are being used to detect scat from numerous species, including bears, foxes, fisher, marten, cougars, bobcats, lynx, black-footed ferrets, and wolves.

Scat is a valuable resource for biologists, indicating which species has been there long after the animal is gone. Furthermore, a scat imparts a wealth of knowledge about the health and history of the individual that deposited it: A single sample's DNA and other contents can reveal the sex, individual identification, fertility status, diet, parasites, and pathogens of its former host. In some cases, stress hormones are also analyzed to determine the stress level of the animal. This information can provide important insights into the status and habitat needs of wildlife populations.

Systematically locating scat in the wild is difficult—some animals avoid trails and/or hide their feces, males and females may differ in how visibly they deposit their droppings, and small or degraded scats are virtually imperceptible on the forest floor. The idea of using dogs to help detect wildlife scat was developed in the late 1990s by University of Washington researcher Dr. Samuel Wasser and professional dog trainer Barbara Davenport. Using a protocol similar to that used for training drug and bomb detector dogs, scat-sniffing dogs are taught to associate an enticing reward—a tennis ball on a string—with finding scat from a particular species.

Domesticated dogs' sense of smell is estimated to be more than 1,000 times keener than that of humans. Conservation biologists have learned how to harness this sensitivity to find wildlife feces, or scat, in natural settings.
Courtesy of Paula MacKay.

In practice, scat-sniffing dogs are a cross between narcotics dogs and search and rescue dogs, as they are used to conduct searches across very large areas. Given the physical demands of this occupation, the best candidates are large, agile working breeds that have ample drive and energy. They must also be very object-focused, as the tennis ball serves as an ongoing incentive for seeking out scat.

Communication and mutual trust between a scat-sniffing dog and its handler are key to the success of this research method. The handler must be highly attuned to the dog's behavior and should be able to accurately read when the dog is on a scent. A handler must also help guide the dog in situations when tricky air currents or

Using a protocol similar to that used for training drug and bomb detector dogs, scat-sniffing dogs are taught to associate an enticing reward—a tennis ball on a string—with finding scat from a particular species.
Courtesy of Paula MacKay.

other environmental conditions make it difficult for the dog to locate the scat. Reciprocally, the handler depends on the dog to distinguish between "target" and "non-target" scats, and to be consistent in indicating when and where a target scat has been detected. Once a "find" has been properly communicated, a tennis ball seals the deal.

Scat-sniffing dogs enable wildlife biologists to survey remote areas at minimal cost. They also allow for the scientific study of wild animals without handling them or influencing their natural behavior. Such noninvasive research is becoming increasingly urgent as more and more species are threatened by habitat loss and other effects of a growing human population. Using their innate sense of scent to lead the way, scat-sniffing dogs are an important new asset to wildlife conservation.

See also Applied Animal Behavior—*Mine-Sniffing Dogs*

Further Resources

MacKay, P. 2002. *For scientists, a canine poop patrol.* E Magazine, 13 (4), 14.

MacKay, P. 2003. *Dogs join wildlife researchers in Vermont.* Northern Woodlands, 10 (2), 14.

MacKay, P. 2003. *Scent of the wild.* The Bark, 23, 70–73.

Meadows, R. 2002. *Scat-sniffing dogs.* Zoogoer, 31 (5), 22–27.

Smith, D. A., K. Ralls, B. Davenport, B. Adams, & J. E. Maldonado. 2001. *Canine assistants for conservationists.* Science, 291, 435.

Paula MacKay

■ Applied Animal Behavior
Social Dynamics and Aggression in Dogs

Konrad Lorenz realized that aggression is a normal behavior, ubiquitous across the animal species, including humans. Whatever sociologists and religionists purport, aggression plays an important and functional role in the maintenance of stable societies. It serves both cohesive and dispersive functions, maintaining the integrity and harmony of a group at minimal energy cost and protecting against attack from interlopers. Both cohesive and dispersive types of aggression are affective—that is, they involve significant mood change.

On the whole, the social interactions between members of wild dog packs are relatively peaceful, with little if any, serious aggression evident between pack members. The pack is in harmony, governed by a set of unwritten rules communicated by posturing and body language. Each individual knows exactly where he (or she) stands within this dynamic but evolving society. Some basic rules are: age and seniority before youth and beauty, size and strength do matter, and possession is nine tenths of the law. Every individual knows and respects these unwritten rules and, in the event of transgression, will be reminded of them by others. The initial response from higher rankers is usually one of warning—vocal (a growl) or symbolic (a lunge, lip lift, or snap). The warning is usually heeded so no serious aggression, with its biologically profligate consequences, occurs. This type of interaction exemplifies the *cohesive* function of *aggression*, which serves an internal policing function analogous to law enforcement in our own society.

Dispersive aggression is designed to protect an individual or group from outside threats. It is defensive in nature and its orientation is toward intruders. Dispersive aggression is associated with exaggerated behavioral displays. The idea is not for the dog to engage in a fight but to avert one while achieving the requisite behavioral goal. Growling and barking are often fierce and protracted with this type of aggression. Piloerection and posturing cause the dog to look larger, and its movements are deliberately intimidating. Dispersive aggression may be engaged in by an individual or by pack members acting in concert: Either way the goal is the same. Reasons for dispersive aggression include defense of territory and protection of the pack.

Another type of aggression, *predatory aggression*, though destructive, is performed unemotionally in the course of the vital but mundane business of procuring food. We have all seen film of wild dogs pursuing, gang tacking, and killing a prey species; such images are firmly emblazoned on our minds. In viewing such deadly strikes we may wince and then shrug saying, that's nature—and we'd be right. No change in affect is involved in this type of aggression. The cognitive change is one of intense focus on the task in hand, coupled with excitement.

Canine aggression can thus be classified as affective or predatory, with affective aggression being subdivided into cohesive (offensive) and dispersive (defensive) types. This typology works well when discussing the physiology or psychopharmacology of aggression; however, it does little to explain the precise circumstances of the aggression. For example, aggression between males to establish rank may be considered "cohesive," but the term "cohesive" does not describe the cause of the aggression. In 1968, Moyer coined a descriptive terminology for the various types of aggression. His classification has stood the test of time and is the most widely used classification in discussions about animal aggression. To acknowledge Moyer's classification is to bow down to a behavioral Mecca.

Moyer classified aggression as instrumental (learned aggression), fear-based, territorial, intermale, irritable, maternal, or predatory. Modifications of this classification system are commonly used to describe the various types of canine aggression. Instrumental aggression, since it is learned and serves as an instrument of a dog's will, has been interpreted as dominance-related aggression, though the role of true social dominance in propagating this type of aggression in a domestic setting has recently been challenged.

Dominance Aggression

Dominance is perhaps the most widely used term in canine training and behaviorism and also the most misconstrued one. True dominance of a dog with respect to its owners is relatively rare. When encountered, dominant dogs tend to be more aloof than their less confident counterparts and act like the "James Deans" of the dog world. Dominant dogs will take issue with another dog or person when pressed over a valued resource, if disturbed while resting, or in response to admonishments. When true dominants respond to a challenge, their reaction is often symbolic and restrained, like their wild pack equivalents. They know how to handle themselves and seem fully aware of their standing in the group. They can afford to mutter disapproval because they are confident in the effect that such warning will have. They know they can have their own way if they so desire, and so will often defer, almost gratuitously, to minor flauntings of the rules. So, for example, with regard to possessiveness, it is as if they are signaling, "This is mine, but if you can borrow it if you want." It is said, and seems to be true, that the hallmark of true dominance is deference, not conflict.

Dominant dogs also have territorial responsibilities. In nature, it is their job to warn the pack of approach by outsiders and to initiate defense against unwelcome visitors. The same instinct drives socially dominant dogs to bark at approaching strangers and repel borders if necessary. However, if the approaching person or other dog is welcome, the warning message quickly transforms to one of greeting.

Truly dominant dogs are infrequently aggressive and are usually not a problem for owners who transmit clear signals of overall leadership. Problems occur in dogs that have a modicum of dominance coupled with anxiety or frank fearfulness and whose owners do not display (canine) leadership qualities. These dogs may be better classified as subdominants or betas. They do not have the confidence to defer in the face of insignificant challenges. They protect things they do not need to protect, resist minor physical interferences, and respond disproportionately to any perceived challenge. Such dogs do not appear aloof or independent; instead they appear hyper-vigilant, moody, or insecure, and their attitude seems to reflect "What's mine is mine, and what's yours is mine." Their aggressive responses are often fast and without prior warning, and following an incident they appear contrite, remorseful, or seemingly unaware of what has occurred. Their response following an aggressive bout has been interpreted as a beacon of their insecurity.

Regarding classical owner-directed, dominance-related aggression:

- Dogs that manifest it are anxious and insecure.
- It is facilitated by unhealthy social dynamics.
- Genetic factors are involved.
- It is dyadic (strictly between two individuals).
- There is a learned component.

All behaviorists are familiar with the classical features of dominance-related aggression by dogs toward their owners. It occurs in the contexts of protection of what the individual dog deems to be a valued resource (e.g., food, bones, rawhides, tissues, toys or other objects), space-guarding (e.g., crate, bed, resting place), personal space guarding (e.g., in response to being petted on the head, kissed, hugged, leaned over, etc.), and self-protection (if stared at, admonished, gesticulated toward, threatened, or physically punished). Owners acknowledge that their dogs are not aggressive all the time. Most admit that their dog is a good companion for 95 to 98% of the time and only acts out occasionally. The aggression is predictable within certain contexts: Dominance-related aggression varies according to the time of day (being more likely at night), the place (more likely if the dog is on the same level as the person), the circumstance, and the person involved. People whom the dog does not "respect" are more likely to be subjected to this type of aggression.

A diagnosis of dominance-related aggression can be confirmed using customized questionnaires such as that shown in the following chart. Aggression toward familiar people/family members in five of the categories is confirmatory of a problem. Scores are sometimes assigned to the aggressive response in each category to calculate a total aggression score. The latter can be used for quantitative assessment of the problem and monitoring of its course.

Behavioral programs designed to reduce instrumental aggression are designed to elevate the status of the person concerned relative to the dog. A concerted family effort is required. Most dominance-control programs or "leadership programs" involve two basic components:

1. avoiding further aggression by strategic methods; and
2. requiring the dog to earn privileges and valued resources by obeying a command and displaying a submissive posture.

Canine dominance aggression assessment chart

Check appropriate box if your dog exhibits any of the behaviors listed on the chart toward a family member:

	GROWL (Score 1)	LIP LIFT (Score 2)	SNAP (Score 4)	BITE (Score 8)	No aggressive response	Not tried
Touch dog's food or add food while eating						
Walk past dog while eating						
Take away real bone, rawhide, or delicious food						
Walk by dog when s/he has a real bone/rawhide						
Touch delicious food when dog is eating						
Take away a stolen object						
Physically wake dog up or disturb resting dog						
Restrain dog when it wants to go someplace						
Lift dog						
Pet dog						
Medicate dog						
Handle dog's face/mouth						
Handle dog's feet						
Trim the dog's foenails						
Groom dog						
Bathe or towel off						
Take off or put on collar						
Pull dog back by the collar or scruff						
Reach for or grab dog by the collar						
Hold dog by the muzzle						
Stare at the dog						
Reprimand dog in loud voice						
Visually threaten dog: newspaper or hand						
Hit the dog						
Walk by dog in crate						
Walk by/talk to dog on furniture						
Remove dog from furniture: physically or verbally						
Make dog respond to command						

Courtesy of Nicholas Dodman.

The success of such programs in addressing the problem of owner-directed aggression has been established. Most owners (70–90%) seeking help with this problem report major improvements in their dogs' aggressive behavior in a period of 2 months. Some of these owners regard their dog as being "cured." Of course, the program only works for those people who engage in it. Those who cannot participate (e.g., young children under 6 years old, old persons, people who are ill, and outsiders) remain at risk if they inadvertently challenge the dog in any of the ways described above.

Not all dominance-related aggression problems are between dogs and their owners. Some are between dogs living in the same household. Household "packs," when stable, are as peaceful as wild dog packs. There is often a little rough signaling between dogs, even occasional noisy scraps, but on the whole, dogs in most household packs learn to coexist and even enjoy each other's company. Occasionally, however, owners choose to add a new canine family member to their pack, and that's when the problems can start. Pure dominance struggles between dogs between say, a youngster coming up through the ranks and a senior dog, are usually short lived and come to a spontaneous conclusion in a matter of weeks. This is not always the case, though, and some dogs just will not tolerate each other. The best solution in such cases is to find another home for one or other member of the feuding parties. Another type of interdog aggression is termed "sibling rivalry." The feuding parties are not necessarily siblings but do live under the same roof. In this case, the underdog forms a tacit alliance with the owner and uses the alliance to springboard into aggression toward the other. It is as if the owner's presence gives the dog the confidence to attack (perhaps because fights that occur are inevitably broken up by the owner). The dogs usually do not fight when the owner is absent. The goal of treatment is to allow the "natural" pack order to prevail. Masterly lack of intervention in early struggles is helpful in the initial stages, but the problem can become so severe as to threaten one or other dog's life. At this stage, separation and chaperoned reintroduction may be necessary. Both dogs should be on lead and under good control at these times. If a fight begins, the two dogs should be separated and held apart by a couple of feet until they have settled down. The true alpha should then be petted and praised while the beta is tied up and forced to watch. This reestablishes the alpha in its true position and, after some weeks of such treatment, both dogs get the picture of who is boss. The correct order of leadership should be owners over both dogs, and the elder incumbent dog over a younger new resident.

Another more global way of addressing dominance-related aggression is with medication. Drugs that increase serotonin levels in the brain are most effective since serotonin is instrumental in aggression and the maintenance of dominant status. Fluoxetine (Prozac®) is employed most commonly for this purpose, and the results of treatment can be dramatic. Other selective serotonin reuptake inhibitors (SSRIs) like paroxetine, sertraline, and fluvoxamine, are probably equally effective. The relationship between brain serotonin levels and aggression is a reciprocal one and, fortunately, paradoxical reactions are uncommon with serotonin modulating drugs.

Fear Aggression

On face value, fear aggression might appear to be a relatively uncomplicated entity, easy to define, explain, and eminently understandable. An affective defensive type of aggression, fear aggression centers on self-preservation by means of threat, intimidation and ultimately violence. There can be no doubt that fear aggression's function is dispersive or that it can be initiated, augmented or attenuated by learning. There is also little doubt that genetic factors play a role in its development in an individual.

In the dog pack or family home, fear aggression by dogs is usually directed toward unwelcome visitors, as opposed to familiar individuals. It may occur in any location where the dog feels threatened, but is often more convincingly displayed when the aggressor is on its own territory. One of the hallmarks of fear aggression is that it appears as a "class action suit" directed toward whole groups of people or other dogs. Human targets of fear aggression may be people of a particular age, race, sex, or appearance. Some fear aggressive dogs express their ire toward extremely narrow groups of people, such as men with white beards or people who smoke. For fear aggressive dogs, men and children usually pose the greatest threat, perhaps because of their generally more invasive and unpredictable behavior. People who present an unusual appearance, perhaps because of garb or infirmity, are also frequently singled out. Other dogs may evoke fear aggressive responses. Again, the class action concept applies with whole groups of dogs, say, all large dogs, all black dogs, or all small dogs, being the usual victims. Sometimes fear aggression toward people or other dogs can generalize so that all unfamiliar people or all unfamiliar dogs are potential substrates for aggression.

Lack of proper socialization plays a key role in the development of fear aggression. Though it is impossible to socialize a dog toward every social challenge it will face in later life, some generalization of this learning should occur. It is also important to protect young dogs from adverse experiences, especially during the sensitive period of learning (birth to 12–14 weeks).

One concept of fear aggression is that it is simply a fearful response by an insecure dog, but that is not the whole story. In a behaviorist's clinic it is often noted that fear aggressive dogs display "dominance-related" aggression, too. Insecure they may be, but pushovers they are not. They will often hold their ground over issues with familiar people where fear is clearly not the driving force. A purely fearful dog will not challenge a fear-inducing stimulus unless its back is almost literally against the wall. Instead, they run and hide or display submissive (deferent) and diminutive postures. Their policies, when facing a challenge, appear to be those of escape or appeasement. But when fearfulness is fortified by modicum of naturally-imbued dominance, the fearful dog is likely to take affirmative action to eliminate the source of the fear. A good offense is often the best defense, seems to be their policy. When this approach is successful at driving away the feared person, say, the mail carrier, the result is rewarding (negative reinforcement), ensuring that the behavior is repeated and intensifies.

Dog trainers have a most appropriate term to describe dogs of this persuasion: They call them "shy and sharp." It may be these dogs' dual nature that causes confusion regarding the true etiology of dominance aggression. Because dogs exhibiting dominance aggression often have comorbid fearful behaviors, the concept has arisen that dominant–aggressive dogs are motivated by fear. The reverse (only) may be a more appropriate explanation. Dominance might be necessary for the development of fear aggression, but fear may not be a necessary component of dominance.

Some fear aggressive dogs appear more dominant than fearful, causing them to develop a highly proactive fear response. Others are more fearful than dominant, leading to some reticence in coming forward. The latter dogs tend to warn rather than lunge and bite. In addition, they tend to be more confident about aggressive displays when accompanied by their owner, when the "victim" shows signs of fear, or when on they are on "home turf" (including the safe haven their owner's car).

The clinical appearance of the fear aggressive dog tends to be ambivalent, as might be expected if the condition is a composite of two conditions. When confronted by their nemesis, dogs of this persuasion are often in conflict, caught between two drives. Initially, they may come forward in an intimidating way only to withdraw, barking threats, as their fear overwhelms them: The cycle is then repeated. So-called approach–avoidance behavior of this type is a classical conflict behavior. It is as if the dog is oscillating between the more dominant and more fearful aspects of its personality. Simultaneously the dog's body posture displays ambivalence. Squinting eyes, flattened ears, furtive sideways glances, raised hackles, hunkered stance, and tucked tail indicate fearfulness. Growling, lip lifts, open-mouthed threats, snapping, lunging and biting indicate aggression. A wagging tail, often displayed under these circumstances, is no reason to trust a fear aggressive dog. The wagging tail indicates excitement but not friendliness.

When attempting to treat such a dog behaviorally, it is important to realize that cure is virtually impossible. Unfortunately, learning is never "unlearned" and mistrust will always exist in some circumstances. The goals of therapy are to keep people or other dogs safe from attack, avoid unnecessary exposure to challenges, and to alter the dog's threshold for this type of aggression and its intensity. Proper leadership and control, physical restraint, counterconditioning, and desensitization are all helpful measures. It is also helpful to ensure that the dog gets plenty of aerobic exercise, eats optimal rations, and responds consistently to one word commands. The best tool for retraining a fear aggressive dog is a head halter. Simply fitting one of these devices often dramatically reduces a fearful dog's aggressive response. While the physical control afforded by head halters is instantaneous, training dogs to respond differently when confronted by a fearful stimulus takes more time. With respect to training, as Konrad Lorenz correctly opined: *Art and science aren't enough, patience is the basic stuff* (1994, p. 42). That is most certainly so when it comes to retraining a fearful dog.

Drugs employed to reduce dogs' fear are useful adjuncts to retraining. SSRIs, certain anxiolytics, such as buspirone, and beta adrenergic blockers are all valid treatments. *SSRIs* stabilize mood and reduce social phobias in that way. *Anxiolytics* eliminate anxiety more directly. *Beta blockers* effectively nix the "fight or flight" branch of the autonomic nervous system (that division of the nervous system that deals with subconscious reactions).

Territorial Aggression

Along with the many benefits that accrue from being the dominant dog in a pack, there are also certain duties and responsibilities. One of these responsibilities involves patrolling and marking the periphery of the territory to ward off intruders. Alarm barking by dominant dogs serves to alert other pack members to a possible breach in security so that concerted efforts may be initiated to repel borders. Territorial guarding is really an expanded version of the local or personal space-guarding penchant of the dominant dog. The dominant dog is the initiator and prime instigator of territorial defense, which involves an affective and dispersive type of aggression. In the home, more confident dominant dogs assume this erstwhile pack function, alerting house members to any encroaching visitors. Once visitors have been acknowledged as welcome, the dominant dog's warning soon turns to greeting as it settles down to enjoy the company.

Territorial aggression by a truly dominant dog rarely poses a physical threat to human visitors, especially if family members display reasonable leadership qualities. However, depending on their behavior and body language, other dogs encroaching on the territory may pose a far more serious territorial threat to an incumbent dominant dog. Serious fighting can ensue if an overly confident usurper ignores warnings and takes liberties by crossing into a dominant dog's territory. It should be borne in mind that territory as perceived by the dog often extends well beyond the legally defined plot plan. Rather, the dog views its territory as extending well beyond the owner's home and yard into the surrounding streets. In fact, it embraces any area that the dog normally patrols and marks with urine. The owner's car is another area perceived as territory and which will be aggressively protected by a territorial dog.

A variation on the theme of territorial aggression driven by dominance is that of apparent territorial aggression spawned by fear. It seems that some fearful dogs are not confident enough to become aggressive unless they are on their own territory. On neutral ground, they may simply appear furtive and restless, tending to stay close to their owners and avoiding any advances by strangers. On their own turf it is a different matter. With home field (territorial) advantage they will bark and growl aggressively, and perhaps lunge at or bite unwelcome visitors. As with more confident fear aggressive dogs, the targets of their aggression are often men or children, people of unusual appearance or mannerism, and people who are tentative around them. Uniformed visitors come in for special attention and may be at serious risk of being bitten. One reason why this is so is that uniformed visitors inevitably leave, sometimes in short order, powerfully reinforcing the aggressive behavior.

The clinical manifestation of this type of aggression is dramatic with much barking, posturing, and lunging, as its primary function is to drive the infiltrator away. Such dogs do not tolerate well the presence of strangers who are allowed into the house by the owners. Instead of the "hail fellow, well met" approach of their more dominant counterparts, they maintain a certain shiftiness and hypervigilance. If the visitor starts to speak loudly or gesticulate, if they stand up and move around, or go to leave, another aggressive display may be stirred up. As with fearful dogs, the attacks of fear/territorial dogs are of the bite-and-run variety, tending to be from the side or from behind. Clothing may be torn and visitors' calves or thighs may be punctured. The type of attack is aptly referred to as a "cheap shot."

Treatment of territorial aggression, from whatever cause, entails creating a healthy lifestyle for the dog with appropriate outlets and interests, and maintaining control of the dog and the safety of visitors at all times. A head halter is an ideal tool with which a dog can be taught how to behave. While it is sometimes necessary to put the dog in a crate or in another room, it will learn nothing when sequestered in this way, so such an approach should not be the sole method of dealing with the problem. Instead, introducing the dog on lead to visitors, controlling and rewarding it for good behavior, is a more educational approach. Counter-conditioning is an invaluable adjunct to such spontaneous or prearranged retraining sessions. Finally, feeding territorial/fear aggressive dogs low protein rations has been shown to significantly impact the level of aggression, causing a dramatic reduction in some cases. Adjunctive pharmacological treatment for territorial aggression is the same as for fear aggression.

Intermale Aggression

Intermale aggression is a specialized form of dominance aggression that occurs between male dogs confronting each other over territory, resources, or mates. When a female in heat is the prize, the aggression is sometimes designated as sexual. One of the main internal motivators for this type of aggression in intact (uncastrated) male dogs is the sex hormone, testosterone. Testosterone depresses brain serotonin and increases impulsivity and aggression. Neutering substantially reduces or eliminates intermale aggression in about 60% of dogs, though some dogs continue to display aggression to other male dogs, presumably on account of a learned component. Alternatively, male brain mechanisms may continue to be activated in some castrates despite an absence of testosterone. A neutered male is, after all, a neutered male and not an "it."

Second to neutering, treatment of intermale aggression is largely by management and owner control, though some dietary measures, such as feeding a low protein diet or adding tryptophan to the rations, may be helpful. Alternatively, in severe cases, an SSRI, like fluoxetine (Prozac®) can be extremely helpful.

Irritable Aggression

Whereas dominance aggression is a relatively predictable trait, and fear aggression is a response to a real or perceived threat, irritable aggression is a somewhat nonspecific response to hyper-arousal or over stimulation. If a dog is approached or interfered with in this irritable state, it may lash out unpredictably at the nearest living creature or object. Heightened irritability accounts for redirected aggression, pain-induced aggression, and postictal aggression, and certain other subtypes of aggression. There is no treatment for irritable aggression other than to appreciate a dog's irritable state and avoid interferences with it at that time. Alternatively, if the irritability is predictable (e.g., at a veterinary office visit), a dog can be preemptively muzzled.

Maternal Aggression

Maternal aggression is yet another variation on the theme of dominance aggression, though it is highly specialized in that it refers to the specific instance of a dam protecting her young, in this case, a bitch protecting her puppies. Maternal aggression, like other maternal behaviors, appears abruptly after parturition (giving birth) as if a switch has been thrown. Falling progesterone, rising estrogen, and oxytocin release all contribute, though prolactin appears most instrumental. If the level of maternal aggression is plotted against the serum prolactin level, the two curves can be directly superimposed on each other. Maternal aggression is a functional form of aggression that has obvious survival benefits. It is rarely a problem in the clinical setting, but occasionally humans may fall foul of a bitch's overprotectiveness of her pups. If treatment is indicated, medication with synthetic progesterone and neutering to prevent recurrence are logical measures.

Predatory Aggression

Although predatory aggression looks aggressive and is destructive to prey, it is in many ways quite distinct from other forms of aggression. Its functions are neither cohesive nor dispersive, and it is performed with relatively little change in affect. This cold-blooded type of aggression is performed by dogs in a natural setting simply in the course of procurement of food. Domestic dogs exhibit various levels of predatory aggression with some breeds more highly endowed than others, according to their heritage.

Typically, predatory aggression expresses itself as an intense interest in any small, rapidly moving varmints or birds, and sometimes larger animals like sheep and goats. While a solitary predatory attack is eminently possible, this type of aggression also manifests itself as pack or group aggression when more than one dog is involved. The pack hunting instinct, so prevalent in the wild, sometimes rekindles itself in a backyard or neighborhood situation, reducing the possibility of the victim's escape and its chances of survival.

Predatory aggression directed appropriately toward genuine prey animals, while objectionable to us, is nevertheless natural behavior for the dog. Problems arise when dogs' predatory instincts become redirected onto inappropriate improper substrates such as bicycles, cars, skateboards, joggers, running children, or small dogs. An essential ingredient in this type of aggression is movement, which triggers the predatory attack. When more than one dog is present, the odds of triggering are increased and the others will "pack" along for the ride. The most alarming and dangerous of predatory aggression by dogs occurs when it is directed toward human neonates, babies fresh back from the hospital, who are not yet recognized as fellow pack members. Fortunately, this version of predatory aggression is extremely rare.

There is no treatment for predatory aggression. It is just a force to be recognized and to be reckoned with. The chief strategy when dealing with predatory aggression is to avoid allowing the dog to get into a situation in which its predatory instincts may be awakened. In this context measures such as not allowing the dog outside unless it is supervised and under control, the judicious use of crates and gates, and the installation of a solid perimeter fence, are all recommended. Some drug treatments, particularly those that increase brain serotonin, may be helpful but should not be relied upon.

How to Deal with Aggression

All dogs are capable of some level of aggressive behavior depending upon the circumstance. Many different types of aggression can be displayed, often more than one type in any one individual. Some individuals are, by nature or nurture, more aggressive than others. For such dogs, it is imperative to provide them enough exercise, an appropriate diet, and a gainful existence. It is also important for owners to establish proper psychological and physical control of their dogs by elevating their own leadership status and using the correct training tools. Head halters and even muzzles have their place in the management of aggression and in retraining. Counterconditioning, desensitization, and medical treatment (as indicated) may also be helpful.

Occasionally, dysfunctional pathological types of aggression are encountered. To reduce aggression arising from these causes, first the problem has to be diagnosed. The first clue is an unusual expression of aggression in terms of its lack of appropriateness, high intensity, or intermittent nature. Once the underlying cause has been ascertained it can often be rectified. While pathological aggression may arise for any one of a number of reasons ranging from brain tumors to toxins, two of the most common causes are hypothyroidism and partial seizures. Borderline low or frankly low thyroid levels increase anxiety, helping to fuel aggression. When thyroid levels are restored to an optimal range by treatment with synthetic thyroid hormone supplements, aggression arising from this cause can be greatly reduced or eliminated. The hallmarks of partial seizure-based aggression in dogs are pre-ictal mood changes heralding aggressive encounters, exaggerated aggressive responses to trivial cues, autonomic nervous system activation, postictal mood change/depression, sporadic and/or prolonged bouts of aggression and, clinical response to treatment with anticonvulsants. Another possible cause of dysfunctional aggression is Attention Deficit Hyperactivity Disorder (ADHD). This is diagnosed by observing a positive (calming) response to stimulant medications. If the diagnosis is confirmed, treatment is effected using the same class of medication. While medical causes of aggression may be relatively uncommon, they should weeded out and treated before attempting to fit aggression into an ethological paradigm.

One final thought. While aggression in dogs is the most common behavioral problem reported by dog owners, most of the aggressive attacks are considerably restrained and designed more to signify the dog's strong displeasure with some intervention. Dogs rarely set out to physically maim or injure an individual. In fact, serious dog bite attacks resulting in lethal injury to humans occur only 10 or 12 times per year in the United States despite the coexistence of over 60 million dogs and around 260 million human beings. When this demographic is viewed in the light of some 7,000 homicides annually in the United States, it becomes amply clear which is the most aggressive species. The media has sensitized us to fear meeting a large dog on a dark street at night, whereas in fact we should be more afraid of bumping into a human stranger.

See also Aggressive Behavior—*Ritualized Fighting*
Careers—*Veterinary Practice Opportunities*
for Ethologists
Communication—Vocal—*Social Communication*
in Dogs: The Subtleties of Silent Language
Dominance—*Development of Dominance*
Hierarchies
Social Organization—*Social Order and*
Communication in Dogs
Welfare, Well-Being, and Pain—*Psychological*
Well-Being

Further Resources

Dodman, N. 1995. *The Dog Who Loved Too Much*. New York: Bantam Books.
Dodman N. 1999. *Dogs Behaving Badly*. New York: Bantam Books.
Dodman N. 2002. *If Only They Could Speak*. New York: WW Norton & Company.
Dodman N. H. & Shuster L. (Eds.). 1998. *Psychopharmacology of Animal Behavior Problems*. Malden, MA: Blackwell Science, Inc.
Lorenz, K. 1994. *Man Meets Dog*. New York: Kodansha America Inc.
Moyer, K. E. 1968. *Kinds of aggression and their physiological basis*. Communications in Behavioral Biology, 2, 65–87.

Nicholas Dodman

■|Art and Animal Behavior

The study of animal behavior has many applications, but one that is perhaps not immediately obvious is its contribution to our understanding of ancient art.

Images of animals permeate the artwork of many cultures, highlighting the fundamental role they have played in our lives for thousands of years. Art historians examine these images for their aesthetic qualities, but also in an attempt to understand more about their makers. The species chosen and their relationship to humans in artworks can reveal much about their practical, symbolic, and religious significance within different cultures. However, art historians often lack experience in natural history. As a consequence, they frequently ignore or misinterpret the behavior of depicted animals. Yet, this too has the potential to tell us much about the minds and knowledge of our predecessors. More importantly, looking at animal images from an ethological perspective can clarify depictions that might otherwise remain ambiguous.

Prehistoric art is full of animal imagery, painted upon and carved into the walls of caves, and etched onto the surface of small portable objects. These depictions include extinct species such as mammoths, woolly rhinoceros, and cave bears, but also many animals which we still see today, such as horses, bison, and deer. Often these images are simply static representations of particular species. However, some animals appear to be engaged in specific activities which are revealed by their body postures. The positioning of head, tail, and legs, orientation of the body in space, and juxtaposition to other depicted animals often provide sufficient detail to identify the behavior represented. It is thus possible to recognize animals walking, running, sleeping, feeding, grooming, urinating, defecating, threatening, fighting, courting, mating, and nursing.

Knowledge of natural behavior can provide new insights. For example, representations of bison found in a cave in Altamira, Spain, were assumed to be dead animals because they appear to be collapsed on the ground, with their heads turned to one side and their legs contracted. However, comparison with living bison revealed that the ancient paintings more likely depict these animals rolling in the dust, an activity in which they frequently engage. Similarly, knowledge of the aggressive behavior of reindeer has revealed that a stag, engraved upon a section of antler from Kesslerloch, Switzerland, is not grazing passively, but adopting a threat posture with its head lowered and horns presented. So, too, the depiction of a bull nuzzling the tail of a cow standing before him from Teyjat, France, undoubtedly represents genital sniffing, the process by which males assess a female's level of fertility.

Images such as these indicate that ancient peoples must have watched animals closely and that they had a sophisticated understanding of their natural behavior. Hunting would have been an initial motivation, as knowing about the habits of their prey would have helped hunters to anticipate what different species might do in particular circumstances. Then, with the rise of agriculture, knowledge of the social and sexual behavior of animals would have aided early attempts at domestication and the refinement of husbandry procedures. These, too, are revealed in ancient art, often changing the apparent meaning of animal depictions. For example, paintings from Egypt and Mesopotamia of cows being milked often include a calf standing nearby. It has been suggested that the presence of the young animal is not simply an artistic flourish, but indicates an awareness of the "milk ejection reflex," whereby the sight and smell of their calf causes cows to release the hormone oxytocin, which allows them to let down their milk. This would explain why Egyptian depictions often show calves physically restrained in front of the cow, to keep them in view throughout the milking.

Behavior also often accounts for the association of certain animals with particular deities and religious concepts. In ancient Egypt, the choice of the bulti fish (*Tilapia nilotica*) as a symbol of fertility and rebirth was likely based upon its mouth-brooding behavior. Bulti fish hold their fertilized eggs in their mouths until they hatch, and after emerging, the fry often return to their parent's mouth when disturbed. The apparent "death" and "resurrection" of the fry thus made the bulti fish a logical symbol of regeneration. In Mycenaean art, the symbolic depiction of animals was based directly upon their behavior. Animals whose behavior was marked by metamorphosis or hibernation (e.g., butterflies) were used exclusively as funerary symbols to ensure the deceased's transformation and survival in the afterlife. In contrast, images of animals associated with aggression and strength (e.g., lions) or dynamic movement (e.g., swimming dolphins) are usually found in contexts associated with war (such as engravings on daggers).

Sometimes, ancient artists drew parallels between human and animal behavior. These associations are lost if they are not viewed from an ethological perspective. For example, Minoan wall frescoes in a room on the island of Thera show two boys on one wall, and pairs of antelope on adjoining walls. The male antelope stand side-by-side, tails raised, staring intensely at one another, while the boys, bedecked with jewelry and sporting one glove, are boxing. The antelope have been tentatively identified as impala, animals that engage in ritualized displays during sexual and social competition. The juxtaposition of boxers and animals was thus based upon the physical similarity of their posturing and the comparable function of their behavior.

Of course, ancient depictions of animal behavior are not always correct. Sometimes, this is because the draftsman's intention was to show an idealized, rather than a realistic, image. However, mistakes must also have occurred because artists did not have firsthand experience of the animals they depicted. For example, an ancient Egyptian representation of leopards (*Panthera pardus*) shows them mating like domesticated animals, with the female standing upright. Clearly, the artist had not witnessed these cats mating, because they would surely have noticed that females are always prone during copulation. Some apparent mistakes, however, may also be due to our misunderstanding of the artist's method of expression.

Applying the findings of ethology to art enables us to understand what ancient depictions of animals actually represent. This, in turn, tells us much about the knowledge of their makers. Animal behavior gives us a unique ability to see what ancient artists actually saw and is, thus, effectively a window on the past.

Further Resources

Clark, K. 1977. *Animals and Men: Their Relationship as Reflected in Western Art from Prehistory to the Present Day*. New York: William Morrow and Co.

Collins, B. J. (Ed.) 2002. *A History of the Animal World in the Ancient Near East*. Leiden, The Netherlands: Brill Academic Publishers.

Guthrie, R. D. 2000. *Paleolithic art as a resource in artiodactl paleobiology*. In: *Antelopes, Deer, and Relatives: Fossil Record, Behavioral Ecology, Systematics, and Conservation* (Ed. by E. S. Vrba & G. B. Schaller), pp. 96–127. New Haven, CT: Yale University Press.

Ikram, S. 1991. *Animal mating motifs in Egyptian funerary representations*. Gottinger Miszellen, 124, 51–68.

Klingender, F. 1971. *Animals in Art and Thought to the End of the Middle Ages*. London: Routledge and Kegan Paul.

Linda Evans

Bats
The Behavior of a Mysterious Mammal

Bats are one of the most enigmatic of mammals. Most species are nocturnal and because bats are the only true-flying mammals, the study of their behavior has been challenging. With the application of technologies such as radiotelemetry and night-vision devices, the dark world of bats has become more illuminated.

Behaviors in the Roost

Early field studies of bat behavior were conducted by investigators entering caves and observing roosting bats. Studies of the interactions between mothers and their pups of the Brazilian free-tailed bat (*Tadarida brasilinesis*) give clear indication that bat behavior is both complex and sophisticated. Because Brazilian free-tailed bats live in maternity colonies of millions of females and their newborn young, biologists assumed that mothers returning to the roost 6 or 7 hours after foraging, would suckle the first newborn they encountered in the roost because finding their own offspring in the millions of pups would require too much energy. This simplistic model of mother–infant care was shattered when biologists banded the bats and followed the behavior of the mothers when they returned after feeding. Females did indeed search for their own young, fending off other pups hungry for a milk snack, the search sometimes taking up to 9 hours. The mother finds her offspring in the sea of millions of newborns by remembering the voice of its young which calls out for its mother, and when close-in, the mother uses smell as a final discriminator.

But not all bats species are adverse to feeding other colony member's young. The evening bat (*Nycticeius humeralis*) is known to nurse communally. However, such behavior is organized and is not randomly acted out. Young are nursed by their own mother solely until they are at least 8 days old, and the frequency of communal suckling increases when the young begin to forage for themselves. Thus it appears that the energetic investment of communal nursing by females occurs only after a youngster has lived long enough that the likelihood of survival is high.

In other species, females that don't suckle unrelated pups will still care about the safety of youngsters in the roost. When most of the adult females are out foraging, some females in colonies of fringed myotis (*Myotis thysanodes*) and little brown myotis (*M. lucifugus*) stay home and babysit, keeping an eye and ear on the brood and even descending to retrieve infants that have fallen from the roost. Adult bats also help each other. The common vampire bat (*Desmodus rotundus*) is known to food-share with other individuals in the colony who were unable to find a blood meal that night. In an amazing case of caregiving behavior, unrelated females of the Rodriguez fruit bat (*Pteropus rodricensis*) used helper-assisted birth to coach an inexperienced and troubled pregnant female during parturition by showing her the proper birthing posture (hanging from the thumbs) that she then mimicked. They also appeared to comfort her by licking and wrapping their wings around her.

Behaviors while Foraging

Many species of bats are known to take advantage of vocal information generated by the foraging activities of other bats. Behaviors such as eavesdropping on the echolocation calls of other bats to find good feeding areas coupled with following behavior, and imitative and group foraging are common. Recent work has shown that bats may cooperate to drink at small water holes in the West, establishing a distinctive flight path that all individuals follow to avoid collisions while drinking on the wing.

Echolocation

The evolution of echolocation is an event that distinguishes bats from most other mammals, and evidence of sonar ability can be found in the inner ear anatomy of fossil bats dating back 55 million years. High frequency sounds (*ultrasound*: above the human hearing range) emitted from the bat's mouth bounce back to its ears in the form of echoes as the sound waves strike objects in the environment. The bat's brain interprets these echoes into three dimensional depictions of the bat's environment in a similar way the human brain interprets light waves received by the eyes. The volume of echolocatory sounds generated by even the smallest bat is about 110 decibels at a distance of 4 inches (a little louder than a smoke detector's alarm!).

Bats differ the pulse rate of their vocalizations to save energy. Pulses generated at a tempo of about 25 times per second are referred to as the *search phase*. If the search phase echoes reveal an object that sounds like a prey item, the bat will fly toward it and increase its pulse rate to about 50 times per second, referred to as the *approach phase*. When within about 3 feet of a potential prey item, the bat enters the *terminal phase* of echolocation and pulses its voice about 250 times per second. The terminal phase is commonly referred to as a "feeding buzz" because of its buzz- or zipper-like quality when translated through ultrasonic devices that make the sounds audible to humans. The feeding buzz is the most discriminatory of the echolocation phases and supplies the bat with information concerning the size of the object as well as its flight speed and flight direction. The echolocation calls of bats are tonal and may include the use of harmonics that provide greater discriminatory ability.

In addition, some species of bats emit a *constant frequency* call that does not change frequency throughout the call sequence. These are referred to as CF bats. Most species, however, sweep their calls through many frequencies which increases the discriminatory ability of the echo. These bats are known as *frequency modulating*, or FM, bats. The echolocatory ability of bats allows them to avoid obstacles and catch night-flying insects in complete darkness. Although echolocation is the predominant mode of sensory acuity, no species of bats are blind, and, in fact, many of them have excellent vision to augment their echolocation ability.

See also Communication—Auditory—Bat Sonar

Further Resources

Altringham, J. D. 1996. *Bats: Biology and Behavior*. Oxford: Oxford University Press.
Fenton, M. B. 1983. *Just Bats*. Toronto: University of Toronto Press.
Racey, P. A., & S. M. Swift (Eds). 1995. *Ecology, Evolution and Behaviour of Bats*. Oxford: Oxford University Press.
Wilson, D. E. 1997. *Bats in Question*. Washington, London: Smithsonian Institution Press.

Rick A. Adams

■ Behavioral Phylogeny
The Evolutionary Origins of Behavior

Animal behavior is the result of millions of years of evolutionary history, but how do we study the evolutionary origins of behavior? Occasionally behavior leaves a trace in the fossil record. For example, we can tell some dinosaurs took care of their young, because we see fossilized remains of mothers close to their babies and hatching eggs. Fossil footprints also inform us that some ancient animals traveled in herds. But the fossil record tells us little about the details of behavioral evolution—whether those mothers provided food or protection from predators, or whether the herds were led by a single individual or by a group. Instead, behavioral biologists rely on the comparative method and behavioral phylogenies to understand the details of behavioral evolution.

The *comparative method* refers to any study in which differences among existing species are used to infer something about the traits of their ancestors. A *phylogeny* or *phylogenetic tree* is a visual diagram by which we describe the evolutionary relationships among species. By considering species-typical behavior in a phylogenetic context, we can learn a great deal about how behavior evolved. Let's imagine we are interested in the evolution of laughter. Because smiles and sounds do not appear in the fossil record, we use a comparative study, looking for laughter in nonhuman primates. We find that most monkeys and apes smile, but they do so as a sign of social submission (to appease a threatening animal). Other great apes—our close phylogenetic relatives—also laugh when they are tickled during rough-and-tumble play. Recent studies of humans suggest that we too use laughter and smiles to cement social bonds, and only infrequently because we have heard or seen something funny. Putting it all together, we can conclude that our primate ancestors probably smiled to pacify potential aggressors, that laughter was brought in when smiles were used in social play, and that this gave rise finally to a link between laughter and humor.

By mapping behavioral traits onto a phylogeny we can interpret their evolutionary origins more accurately. For example, we are able to determine if different animals share behavior through common ancestry (homologous behavior) or whether similar behavior has evolved via independent evolutionary events in otherwise distantly related animals (convergent behavior). Behavioral phylogenies are also essential for teasing out evolutionary pressures or constraints, such as ecological forces, that act on behavior.

Shared Behavior through Descent

Species that are closely related often share similar traits simply because both inherited those traits from a common ancestor. For example, you probably look more like your sisters and brothers than you do like random strangers simply because you and your siblings share the multitude of genes, cultural traditions, and other traits inherited from your parents. Similarly, species that arose from the same phylogenetic ancestors will often retain the behavioral characteristics of their ancestors, and consequently share those traits with other descendents.

Species-typical behavior is also a product of the environment in which animals are found. Through natural selection, behavior patterns that help individuals to survive and produce offspring in the context of a particular environment will become increasingly common over evolutionary time. Thus, the properties of the physical environment (ecological habitat), predator, prey and parasite species (which are also evolving simultaneously in the same environment), and the complex web of social interactions with other animals of the same species, all shape the evolution of behavior.

Because natural selection is such a potent evolutionary force, only those behavior patterns that are important to survival and reproduction are likely to persist unchanged over evolutionary time. Using behavioral phylogenies we can identify which behavior patterns have persisted and look for explanations in terms of their importance to the animals exhibiting that behavior. Many lizards are territorial, but instead of fighting when disputes arise, they generally resolve conflicts using visual displays consisting of repeated up-and-down head movements or headbobs. Because these displays are energetically costly, animals which perform many displays are showing their opponents they are in excellent physical condition and subsequently more likely to win fights. By using displays to assess opponents before committing to combat, lizards circumvent many of the injuries and other costs associated with fighting behavior, which is especially important when they are unlikely to win. Looking at headbob displays across species, we find that the details of display structure (number, type and timing of headbobs produced) vary considerably across species, but that many, if not most, lizard species produce headbob displays of some form. Placing these behavior patterns in a phylogenetic context, we find that the details of display structure are well explained by the social and physical environments in which each species is found, being clearly the result of natural selection and offering a quick response to a constantly changing environment. But the use of headbob displays evolved very early in lizards and has therefore been retained over very long periods of evolutionary time. We can conclude from this that the risk of physical combat is considerable, and performing vigorous headbob displays is an excellent method for lizards to demonstrate superior fighting ability without having to resort to combat.

A phylogenetic approach can be particularly useful in interpreting behavior that seems unrelated to survival or reproduction. Why should a species exhibit any behavior that does not improve its ability to survive or reproduce? If possessing a behavior presents little or no additional cost to an animal, then a behavior might be passed on between parent and descendant species even though it no longer serves any function. Females of several parasitic bird species seek out nests of other species, where they lay their eggs to be cared for by the nest owner. The most notorious culprits of this brood parasitism are the Old World cuckoos (hence the term "cuckoldry"). Targeted host species often evolve methods to counter the tactics of brood parasites, such as recognizing and ejecting foreign eggs from their nests. But such ejecting behavior is puzzling when it is exhibited by species like the North American loggerhead shrike that does not occur (and never has) in regions frequented by cuckoos or any other brood parasite. How do we explain the presence of such behavior? The story becomes clearer when we look into the phylogenetic history of shrikes and see they are closely related to several Old World bird species that regularly eject cuckoo intrusions. The ejecting behavior seen in loggerhead shrikes is apparently an evolutionary "left-over" from a time when such behavior did in fact serve a valuable function.

Independent Evolution of Similar Behavior

In contrast, we expect animals derived from different ancestors to possess traits that reflect their unique evolutionary history. Yet many unrelated animals exhibit behavior remarkably similar in appearance and function despite differences in their evolutionary background. Behavioral phylogenies are imperative in, first, identifying whether these behavior patterns do indeed result from independent evolutionary events and, second, determining

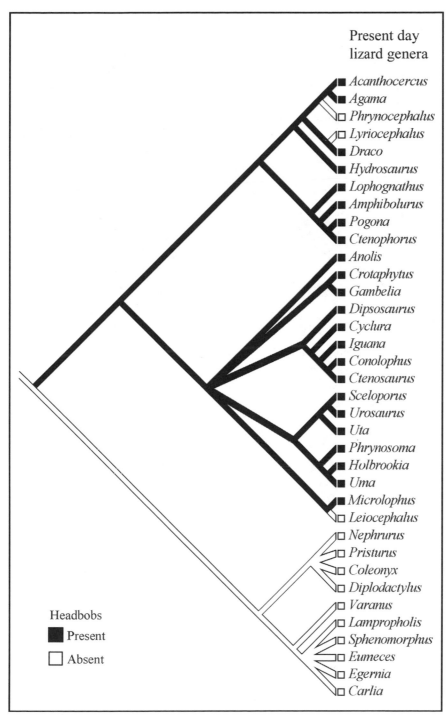

Present day
lizard genera

■ *Acanthocercus*
■ *Agama*
□ *Phrynocephalus*
□ *Lyriocephalus*
■ *Draco*
■ *Hydrosaurus*
■ *Lophognathus*
■ *Amphibolurus*
■ *Pogona*
■ *Ctenophorus*
■ *Anolis*
■ *Crotaphytus*
■ *Gambelia*
■ *Dipsosaurus*
■ *Cyclura*
■ *Iguana*
■ *Conolophus*
■ *Ctenosaurus*
■ *Sceloporus*
■ *Urosaurus*
■ *Uta*
■ *Phrynosoma*
■ *Holbrookia*
■ *Uma*
■ *Microlophus*
□ *Leiocephalus*
□ *Nephrurus*
□ *Pristurus*
□ *Coleonyx*
□ *Diplodactylus*
□ *Varanus*
□ *Lampropholis*
□ *Sphenomorphus*
□ *Eumeces*
□ *Egernia*
□ *Carlia*

Headbobs
■ Present
□ Absent

A behavioral phylogeny illustrating how the comparative method can be used in conjunction with a phylogenetic tree to map the evolutionary history of a behavior, in this case the evolution of headbob displays used in territorial defense by lizards.

Courtesy of Terry Ord.

what shared evolutionary factors have facilitated the independent evolution of similar, convergent behavior.

Often, when the evolutionary relationships between animals expressing similar behavior are clearly remote, it is obvious convergence has occurred. For example, black-headed gulls and California ground squirrels have both evolved antipredatory defenses involving synchronized "mobbing" behavior to confuse and distract predators from entering nesting areas. Considering how distantly related these animals are, it is unlikely mobbing was inherited from the same common ancestor. However, with closer relatives, identifying events of true convergence can be considerably more difficult and requires careful inspection of a detailed phylogeny to ensure that the shared presence of the behavior is not the result of common descent. For example, many European songbirds (e.g., blackbirds and chaffinchs) produce a high-pitched "seet" alarm call to warn other group members of the presence of predatory hawks. Such calls need to alert nearby individuals without attracting the attention of the predator itself, a need that substantially limits the natural pool of sounds with the appropriate acoustic properties. The result are signals remarkable in their similarity across different bird species. Only by mapping these calls onto a phylogeny, can we deduce they have arisen through multiple, independent, evolutionary events (i.e., convergent evolution).

Habitat characteristics also determine the properties of animal signals. For example, denser habitats introduce obstructions between the sender and receiver of signals and reduce the distance over which signals remain effective. Many birds have increased the transmission distance of their songs by tailoring calls to the acoustic environment in which they are typically given. Birds in similar physical habitats may thus give remarkably similar calls, despite being phylogenetically unrelated.

Behavioral Sequences

Phylogenies have also been informative in determining the sequence of evolutionary changes in behavior without having to rely on an incomplete and sparse fossil record. If we believe the threat of predation (or parasitism) has led to the evolution of novel behavioral defenses, such as mobbing (or egg ejecting) behavior, this hypothesis assumes a causal factor predating the evolution of behavior. If, after tracing the evolutionary history of an animal, we find our proposed causal factor occurred *after* the development of behavior, then we would obviously have to rethink our hypothesis.

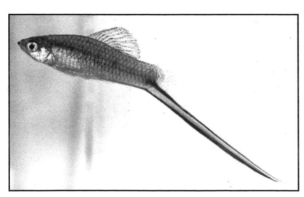

A swordtail fish from Central America.
© *Getty Images/Digital Vision.*

With the development of sophisticated phylogenetic techniques in the early 1990s, biologists have been able to trace the evolutionary history of behavioral traits with greater confidence. One of the first results looking at sequences of behavioral evolution was the discovery of a fascinating phenomenon—behavioral predispositions. Several live-bearing freshwater fish in Central America are called swordtails because they possess a long filament protruding from the caudal (tail) fin. These "swords" are developed by males and are believed to attract females—the longer the sword, the more attractive a male. Males of the closely related platyfish, on the other hand, do not develop swords, yet incredibly females show preferences for males that have been experimentally manipulated to possess artificial swords. Reconstructing historical sequences of behavior onto a phylogeny reveals female preference for swords predate the evolution of the swords themselves. This "preexisting bias" hypothesis created much excitement (and its fair share of controversy) in the animal behavior community. It was some of the first evidence that psychological predispositions in animals—essentially biases resulting from the way the brain is wired—can have a profound affect on the evolution of elaborate behavioral traits.

Behavioral Phylogenies and Biological Sleuthing

There are many examples of how effective phylogenies have been in solving evolutionary puzzles, and how some patient detective work has given us profound insight into the processes that lead to the evolution of complex behavioral traits. It is becoming increasingly important to take phylogenies into account in any comparative study of behavior, otherwise it is difficult (if not impossible) to make reasonable inferences about the current function and evolutionary history of behavioral traits.

Behavioral phylogenies may also be valuable in helping to understand the ecology of rare or endangered species that are difficult to study directly because they either inhabit geographically remote areas or are hard to observe in the wild. By extrapolating from a behavioral phylogeny inclusive of closely related and better-studied species, we can gain an informed estimate of the behavioral ecology of these rarer animals. When time is of the essence, research funding limited, or detailed study impractical, this information may be extremely important in devising appropriate conservation strategies for saving endangered species.

Behavioral phylogenetics—understanding the evolutionary history of behavior—has been central to the study of animal behavior since the inception of the field. In the 1940s, Konrad Lorenz used behavioral phylogenies to show how duck displays evolved from grooming and foraging behavior. Nicholas and Elsie Collias combined behavioral phylogenies with a lifetime of data collected on nest building techniques used by different bird species to uncover the sequence of evolutionary changes leading from simple twig nests to the amazingly elaborate works of African weaver birds.

How different behaviors have evolved over time, what behaviors were likely to have been expressed by extinct ancestors, and from what forms unique behavior in present day species have evolved, are questions that have fascinated biologists for generations. By comparing and contrasting related species we can gain a window into the past and piece together the steps that have led to the behavior present in animal societies today. In revealing the pressures and constraints acting on behavior, phylogenies provide us with a unique picture of how the diversity, and similarity, of behavior has evolved in the animal world.

Further Resources

Alcock, J. 2001. *Animal Behavior: An Evolutionary Approach.* 7th edn. Sunderland, Massachusetts: Sinauer.

Collias, N. E. & E. C. Collias. 1984. *Nest Building and Bird Behavior.* Princeton: Princeton University Press.

Lorenz, K. 1941. *Comparative studies of the motor patterns of Anatinae.* As republished in *Foundations of Animal Behavior* (Ed. by L. D. Houck & L. C. Drickamer), pp. 683–696. Chicago: University of Chicago Press.

Martins, E. P. 1996. *Phylogenies and the Comparative Method in Animal Behavior.* New York: Oxford University Press.

Martins, E. P. 2000. *Adaptation and the comparative method.* Trends in Ecology and Evolution, 15, 295–299.

Provine, R. R. 2000. *Laughter: A Scientific Investigation.* New York: Viking Press.

Terry J. Ord & Emilia P. Martins

■ Behavioral Physiology
Color Vision in Animals

Good color vision is important for communication between animals, but it is only found in primates, birds, fish, reptiles and insects. How they experience color and its range is different for each animal group. Color sending and receiving is always a two-way interaction; any specialty of vision evolves because something needs to be recognized, so each group will have a sender and a receiver specializing in matching colors for some specific reason. Male animals send color signals on the skin that are specialized for females to pick up; flowers send signals to just the right pollinator animals. We have to look for a pairing.

Color is part of light and light is electromagnetic energy in the narrow range of wavelengths that excites receptor molecules. Within this range, a specific wavelength gives us the impression of red or green or blue. But just because an eye receives the light in a particular wavelength doesn't mean that the animal behind the eye can tell color. With one kind of receptor, monochromat animals know light is there but can't discriminate its wavelength. Two kinds of photopigments sensitive to two different wavelengths allow the eye to tell the difference. Animals with these are *dichromats* and can discriminate color poorly; most mammals are dichromats. Three photopigments in *trichromats* give pretty good color vision; primates and insects are trichromats. With four photopigments, *tetrachromat* animals like fish and birds have a wide range of potential colors to see.

Four steps limit what color is seen. All light energy on our planet comes from the sun, so how much is available and in which wavelengths will vary. There's obviously more light out there at noon than at dusk, and in the open versus in shade, at the ocean surface than in the deep. Signal senders reflect light off surfaces. Pigments in skin or feathers absorb a lot of wavelengths and only reflect a few, such as the robin's red. Interference bends light rays before passing them to receivers, so hummingbirds' throat patches give a color seen only at a specific angle. After senders reflect light, it again passes through an environment. There's little light around at night or in deep shade, so nocturnal or forest animals are often dichromats with poor color vision. When color wouldn't be transmitted (e.g., red isn't transmitted deep in the ocean), red deep-water fish aren't visible to predators.

The last step in color transmission is reception. The range and sensitivity are important—we humans are good in the red–green area, whereas insects seem to be almost red blind. In turn, we can't receive ultraviolet as insects and birds can. Also, most pigment reception is not very fine tuned, so trichromats' three pigments are good enough for discrimination. Birds have a wide range with their four pigments and also have pigment-containing oil droplets in the retina that narrow and sharpen color perception, and can see color really well.

Birds are one of the groups using color signal for reproduction. When a male bird wants to signal to a potential mate or to a rival with color, there are problems and opportunities. Males have colorful feathers (e.g., the red-winged blackbird male). If we'd named the species by the female, we would have called it the streaky brown bird. Male hummingbirds

Colors—How Do Flowers and Bees Match?

Jennifer A. Mather

The pairing of bees and flowers is quite complicated. Plants need bees for reproduction—to a plant, a bee is a mobile penis. Plants need bees to visit and then go find another flower of the same species. To attract bees a plant is tuned to bee senses, so flowers are colored violet, blue and yellow. Ultraviolet reception is in the upper part of bee's eyes and is used in navigation by looking at the sky. To attract bees, a flower gives a reward of nectar and pollen. Some flowers don't provide rewards but mimic rewarding flowers, and some insects cheat and chew a hole in the flower to get nectar without picking up or leaving pollen. To get a bee to leave, flowers limit their reward by not having much nectar in each flower. Even better, some flowers only have nectar available at one time of the day. Bees can learn to time their visits easily and specialize in one flower type for a couple of hours before switching to another.

Flowers target their attractiveness just right for bees to see. Flowering plants often have an inflorescence, a large collection of flowers to attract bees from a distance. Up close, bees are more interested in edges and lines, so flowers signal the location of the nectar and pollen with visual guides, lines and spots visible in the ultraviolet that point to the center. They give off scents too—strongest where pollen and nectar are. Some flowers have these visual guides only while nectar is available so bees don't waste their time when there's no reward.

Bees' behavior matches what flowers want them to do. Bees have good short term memory, and learn flower color or smell in only a few trials. They learn how to handle a particular flower as well so it's efficient for a bee to specialize on a particular flower type, and their spatial memory helps them return to a patch of flowers. In the short term, bees use this memory of where to return to, which is what the plant needs them to do to pass pollen to others of its species. In the long term, they are more flexible—conditions change. Pollen transfer is a difficult business; only 1 in 100 visits results in pollination. To make it more likely, plants pattern the reward in an inflorescence, and bees pattern their flower visiting accordingly. In some plants pollen can only be received by older flowers. In these, bees start at the bottom, where the ovary is ready to receive pollen. They spiral upwards getting less and less nectar, getting to younger flowers that have lots of pollen. Soon the flowers have little nectar and the bee leaves, usually to land at the bottom flower of an inflorescence of the same plant species, pass on pollen and get more nectar and pollen to feed the young. This is partnership at its best, finely tuned to reward both members.

See also Behavioral Physiology—*Insect Vision and Behavior*

Further Resources

Barth F. G. 1985. *Insects and Flowers: the Biology of a Partnership*. (Trans. by M. A. Biederman-Thompson). Princeton, NJ: Princeton University Press.

Chittka, L. & Thompson, J. D. (Eds.) 2001. *The Ecology of Pollination*. Cambridge, UK: Cambridge University Press.

are colorful and females less so. Why only one sex? Males of these species compete for the favor of the females, who watch and choose the best one. Females don't need to have the gaudy colors that males have, so they have drab brownish colors that blend with the background instead.

If you're using color to attract potential mates there is a conflict between discriminability and information. A male bird wants a sexual signal fairly simple and easy for a female to discriminate. Take the red-winged blackbird. It has easily discriminated red and yellow wing patches. A female should be able to tell it from a yellow-headed blackbird with its white wing patch or from a Brewer's blackbird with none at all. If not, there won't be living offspring and she's missed her chance to have young. Males want to make themselves obvious; they stand out on cattail stems, spread their wings and flare the feathers so everyone knows they are there. What happens if the signal is lost? Researchers blackened over the colored epaulets on the wings and watched. Males lost their territories and their reproductive opportunities without their badges of color.

A color patch can also vary between individuals, giving information about "quality" of a male for females. They look at what his color patch is and how big and how intense it is; a large bright color patch means a particular male is in good shape. That's what females want for the father of their offspring. This is exactly what happens in yellow warblers and white-crowned sparrows, which have a lot of variation in the intensity and range of color patches in males. Why doesn't natural selection push the males to all have brightly colored patches? Males with dull patches are better provisioners for their young. If you can't be flashy, at least be a good provider.

Fish also have a huge range of colors. *Sexual dimorphism*, different appearances in the members of different sexes, is common. The same rule of flashy for courtship and drab for concealment can work the opposite of our stereotype, though. Male damselfish, which fiercely defend a patch of rock on which females have laid eggs, are much more drab than the females. Tropical reef fish can use colors for species identification (when males and females are alike), thus the butterflyfish species are widely different. Look at pictures of the raccoon, the saddleback, the reticulated and the oval butterflyfish. All these species are from the same genus *Chaetodon* and all are found in Hawaii, but none of them is likely to be confused with any other because they are so obviously different by color. The colors that advertise their species membership are in the blue and yellow wavelengths not yet filtered out in the shallow clear water.

Color is used in sexual communication in the bluehead wrasse in the Caribbean. They have an interesting sex life; they live in loose groups, and a male and a female rise above the group to release sperm and eggs at the same time. Juveniles and females are a fairly uniform yellowish, with or without stripes. The large "terminal" male, the only male member of the group, is colored differently—an indigo blue head, a white bar just behind the gills and a bluegreen body. There is a cost for this display, though; researchers have found that the average length of time that a terminal male bluehead lives is 10 days. This brings up another problem with color signals—eavesdropping. The vertical bands of blue, black and white are very obvious, so females are attracted to him, but predators also find him easy to see. What happens after he is removed from the competition? The biggest female changes sex and color and becomes the terminal male, until he too is seen and eaten.

Color signals are used in sexual selection by guppies, but the trade-off is different within the species. They live in streams in Central America and males have attractive red pigmentation on the sides of their bodies. The red is picked up well by fish, poorly by predators. Pigments that pick up color are not evenly distributed throughout the fishes'

Do Squid Make a Language on Their Skin

Jennifer A. Mather

In the 1980s Moynihan made a startling claim, that squid might make a visual parallel to human spoken language with the patterns on their skins. He had been watching Caribbean reef squid, *Sepioteuthis sepioidea*, and noticed how many different patterns of bars, dot lines and mottles could be made on different areas of the skin. He said these could be controlled separately and might be like nouns, verbs, adverbs and adjectives in our spoken language. But after putting this idea forward, he went back to studying birds.

The problem with this idea is that language is much more than a collection of signal units; it is a complex but creative communication system. Any language that a squid might make has to evolve as a cooperation between senders and receivers, and human speech and bee dances are examples of how this sending–receiving system can be made up of units in an abstract system. A language should be creative, in that new combinations can be made of old units to mean new things, like the whole new human dialect that has sprung up with the common use of computers. It should be learned, and different variations could be learned in different populations, such as the French, Mandarin and Swahili human languages. Most of all, a language system should be used to communicate about the outside world. The male octopus' white papillae signal that he's a male ready to mate; that is not outside him. But the bee's dance, telling that she has found food at a particular direction from the hive, is communication, and so is this paragraph. All of these communications tell us something about the structure of language.

How does the squid skin display system measure up to this high standard? Recent research suggests not too well. Moynihan was right about the huge number of seemingly independent pieces of the skin display system, as R. T. Hanlon and J. B. Messenger also found for cuttlefish, and there are different combinations of these pieces. But adding more pieces of display seems to extend but not change the "message" that a squid is sending. The squid are certainly creative in making new sets with these local pattern combinations, and aggressive and sexual displays are different in subadults than adults. These differences across the lifespan might show learning, or they might show simple maturation. Disappointingly, the patterns squid make to one another seem to be only communicating their internal state—whether they are aggressive or submissive, interested in mating or not interested. There seems to be no cooperation in their groups. They don't even signal to others, for instance, that a predatory barracuda is on the prowl. Moynihan was probably wrong; it's not the visual equivalent of a language. Still, it's a fascinating communication system, just now being better explored, and it reminds us that even wrong guesses can lead to important research.

Further Resources

Hanlon, R. T. & Messenger, J. B. 1996. *Cephalopod Behaviour*. Cambridge: Cambridge University Press.

Mather, J. A. 2004. *Cephalopod skin displays: From concealment to communication*. In: *Evolution of Communication Systems* (Ed. by K. Oller & U. Greibel), in the Theoretical Biology series. Cambridge, MA: MIT press.

Moynihan, M. H. & Radaniche, A. F. 1982. *The behavior and natural history of the Caribbean reef squid Sepioteuthis sepioidea with a consideration of social, signal and defensive patterns for difficult and dangerous environments*. Advances in Ethology, 125, 1–150.

retinas. Pigments sensitive to green wavelengths are in the upper part of the retina for food finding and those sensitive to red are in the lateral area of the eyes so they can see each other. Guppies in open streams with lots of sunshine streaming in through the water are easily visible by bird predators, so these populations have much less bright red. Guppies in shaded areas have much less danger from predation and their red patches are often much larger. There are two balanced pressures, from females to be conspicuously red and be chosen to reproduce and by predators to be dull and live to do so. Which pressure wins depends on where a fish lives.

Color discrimination tuned to feeding opportunities can be two types of match between sender and receiver. It can be a signal by a particular food that it's ready for animals to sample. There must be an advantage to the food species to being eaten, which sounds strange at first. Or color matches can be ones that will directly lead to the prey not being eaten. These can be warning signals of inedibility or camouflage that makes the predator unable to find the animal.

The match that has probably led to human's trichromat vision is that of primates and fruits. Many primates forage in the rain forest for fruits, which are often visible as ripe by their change in color from green to red or yellow (e.g., bananas). This is an advantage to the fruiting trees because the part of the tree's system that will start the next generation isn't the fruit but the seed within it. Monkeys eat the flesh and spit out or swallow the seeds, only to leave them behind. With the monkey leaving the seed, the unmoving tree's problem of spreading seeds is solved. As two of our three photopigments are at the red–green end of the spectrum, humans and other primates evolved to discriminate fruit from leaves and ripe fruits from unripe ones. The fruits reward us with food.

A special case of color vision matching color signals is flower pollinators. Bees get nectar and pollen from flowers and, in turn, take pollen from one to another and fertilize the flower. Both gain; neither would live without the other. Again, there's a sender–receiver match. Insects are trichromats with a pigment sensitive to color in the ultraviolet range. Many flowers have "honey guides," lines in ultraviolet along the petals that show the bees where to probe for nectar. Bees learn specific flowers by color, shape and odor and also where to navigate back and forth from the hive to find them again. There's a gap in the insect's sensitivity, though—they have no photopigment to pick up red. Yet there are lots of red flowers. One way the gap is filled is that hummingbirds, which do see red, are attracted to these flowers. Many of them are also tube-shaped, built to match this group of long-beaked pollinators.

Color patterns are also a way of avoiding being eaten by predators. Wasps and coral snakes advertise their danger with obvious vertical stripes or bands in reds, yellows and blacks. In the opposite match, many edible animals use camouflage colors. The octopus is an expert at this. With yellow, red and brown color sacs and white and bluegreen reflection areas in its skin, it can look like pretty well any ocean background. The paradox here is that octopuses, like other mollusks, can't see color. They make their fabulous color camouflage to avoid being seen by the good eyes of fish, who are the "designers" of octopus skin.

If color vision is so great, why don't all animals have it? There are disadvantages, too. Being able to see color allows an animal to pick out objects and to see distance better. But color vision is only useful in the bright light of daytime, so in dusk or low light it's a disadvantage. If things aren't colored or are colored in wavelengths it can't pick it up, an animal may see poorly with color vision. And species that don't pick up color can discriminate another characteristic of light, its direction of polarization—maybe it's their version of color. Like anything else, the color we appreciate so much is a trade-off in many ways. It's a special balance—between sender and receiver, between one evolutionary pressure and another, between usefulness and handicap.

See also Behavioral Physiology—*Insect Vision and Behavior*
Behavioral Physiology—*Plumage Color and Vision in Birds*

Further Resources

Bradbury, J. W. & Vehrencamp, S. L. 1998. *Principles of Animal Communication*. Sunderland, MA: Sinauer.
Esmark, Y., Amundsen, T. & Rosenquist, G. 2000. *Animal Signals: Signalling and Signal Design in Animal Communication*. Trondheim, Norway: Tapir Academic Press.

Jennifer A. Mather

■ Behavioral Physiology
Insect Vision and Behavior

Like humans, insects use their eyes for gathering information about the light in their environment to make behavioral decisions. Male insects use vision to distinguish females from other males. Females use vision to assess the quality of potential mates. Females use vision to find and recognize potential places to lay their eggs. Both sexes use vision to find food resources and to guide their locomotion.

However, insect eyes are very different from ours. Human eyes, like those of other vertebrates, are camera eyes in which a single lens focuses an image of the world on the retina, an array of light-sensitive cells or photoreceptors. Each photoreceptor converts the light falling on it to produce a nervous signal that is sent to the brain. In general and up to a point, the brighter the light falling on the retina the stronger the signal sent to the brain.

In our eyes there are two classes of photoreceptors, rods and cones. Rods respond best very weak green light and permit us to see in dim light. In contrast, cones are less sensitive to light but play an important role in color vision in bright light. There are three types of cones that respond to red, green, and blue light, respectively. Our brain discriminates colors by comparing the responses of the three types of cones to light falling on the retina. For example, blue light will cause the blue-sensitive cones to send a much stronger signal to the brain than the red- and green-sensitive cones. Red light would produce a very different pattern of response from the cones in the retina.

The retina is densely packed with these various sorts of photoreceptors (several hundreds of thousands per eye) and so our ability to resolve detail in our visual field is quite

Butterfly eye: Empress Leilia butterfly.
Courtesy of Ronald L. Rutowski

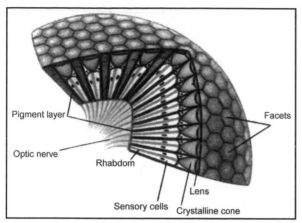

Pigment layer

Optic nerve

Rhabdom

Facets

Lens

Sensory cells   Crystalline cone

Cross section showing ommatidia and the basic elements of their structure.

Courtesy of Ronald L. Rutowski

good. At a distance of 6 meters (20 ft) we can recognize an E that is as little as a 2 cm (3/4 in) tall.

Insect eyes are very different in structure and are called compound eyes. The basic unit, called an *ommatidium*, is a tube with a lens at one end and a set of light-sensitive cells at the other. Light enters this tube through the lens and travels down the tube to the photoreceptors. There are as many as six light-sensitive cells in an ommatidium and they vary in the color of light to which they are sensitive. By comparing the responses of these cells, the brain can extract information from the output of an ommatidium about the color and intensity of the light entering the ommatidium, but that is all. Contrary to many popular representations on the web and elsewhere, a single ommatidium does not provide all the information needed to produce image.

The typical insect eye is called a compound eye because it has many hundreds or even thousands of ommatidia. Their combined output permits the brain to construct a color image of the world. But this image often differs from the one our brain constructs in at least three major ways. First, a typical insect's view of the world is less detailed than ours. Because of the small size of insects, their eyes are small and there are limits to how many ommatidia an eye can have. This limit is well below the number of rods and cones that we have. For example, one eye of a medium-sized butterfly has about 6,000 ommatidia whereas the human eye has over 300,000 rods and cones. The practical consequence of this is that the butterfly cannot see as well as we do. For an insect to be able to resolve an E at 6 m the E would have to be about 40 cm (15.5 in) high!

Second, insect vision is faster than ours. Their photoreceptors respond to and recover quickly from being stimulated with light. Consider the temporal resolution of the human eye as indicated by the design of motion picture films. A motion picture actually consists of a series of still pictures of a moving object that are projected in rapid sequence. As long as the sequence of images is projected at a rate of greater than about 18 per second we do not see them as separate images, but as the object in continuous smooth motion, because of the relatively slow rate at which our photoreceptors recover from stimulation with light. Insect photoreceptors recover much more quickly and so would see pictures projected at 18 per second as separate images. Behavioral studies and electrophysiological studies of insect eyes suggest that the rate at which images are projected would have to be increased to 100 images per second or more to be seen as a continuous image.

Third, the colors of the world for an insect will be different than they are for us. Again, there are several types of photoreceptors in each ommatidium that differ in the colors to which they respond in much the same way that cones differ in the colors of light to which they respond. But in the eye of, for example, a honeybee the cells are sensitive to blue, green, and ultraviolet (UV) light.

The spectrum of visible colors in the animal kingdom extends from red wavelengths (about 700 nanometers) through the yellows (ca. 600 nm), green (ca. 550 nm), and blues (450 nm), and into the UV wavelengths (300–400 nm) (1 nm = 1 billionth of a meter). We can see and recognize colors from 700 down to 400 nm, but we cannot see UV wavelengths.

The sensitivities of the cells in a typical insect ommatidium permit them to see and distinguish wavelengths from 600 down to 300 nm. Hence insects can see wavelengths of light that we cannot (UV) and vice versa.

Do these features of insect vision have consequences for the behavior of these animals? Yes. Consider, for example, the insect's ability to see UV wavelengths. This ability means the door is open to the evolution of color signals in this range of wavelengths. Interestingly, UV vision in insects was only discovered in the early part of the 1900s and it was not until the 1930s that scientists began examining flowers and other insects to see if their coloration and color patterns contained elements in the UV. The results were stunning. Especially striking were the UV color patterns discovered in flowers and butterflies. More recently, the color signals of birds have been found to have pattern elements in the UV.

Appearance to humans and UV pattern: Orange Sulphur butterflies.
Courtesy of Ronald L. Rutowski

But are these pattern elements behaviorally significant? Evidence continues to grow that they are and the students of animal behavior must pay attention to the potential for UV components in the animal signals they observe (although we cannot see them ourselves) and the potential for UV sensitivity in the animals they are studying.

With respect to the low spatial resolution of insect eyes the implications are less clear but here are a couple. They will not be able to see each other a distance of more that a few meters. A common butterfly such as the Painted Lady has a wingspan of about 4 cm (1.5 in) and we can see them when they are 10 m (32 ft) away or more. However, both behavioral experiments and calculations made from the optics and structure of butterfly eyes suggests that they can only see members of their species at distances of 3 m (10 ft) or less. Also, the wings of many butterflies like Painted Ladies display amazingly intricate patterns and it is easy to assume that these patterns are important visual signals in interactions between butterflies. Given the low resolution of the insect eye this seems unlikely except at very close range. In general the size of the world that can be seen clearly is much smaller for insects than it is for us.

Mentioned here just three major ways in which the visual experience of insects might differ from ours. There are others including the insect's ability to detect the polarization of light in signals that space does not permit us to cover here. As our understanding of the vision of insects and other animals improves through studies of eye morphology and physiology and behavioral experiments, we increasingly appreciate how cautious we must be not to assume that we see what they see when we think about and try to explain their behavior toward their environment, each other, and us.

See also Behavioral Physiology—*Color Vision in Animals*
        Behavioral Physiology—*Colors—How do Flowers and Bees Match?*
        Behavioral Physiology—*Plumage Color and Vision in Birds*

Further Resources

Bradbury, J. W. & Vehrencamp, S. L. 1998. *Principles of Animal Communication*. Sunderland, MA: Sinauer Associates, Inc.

Dusenbury, D. B. 1992. *Sensory Ecology: How Organisms Acquire and Respond to Information*. New York: W. H. Freeman and Co.

Hailman, J. P. 1977. *Optical Signals: Animal Communication and Light*. Bloomington, Indiana: University Press.

Land, M. F. & Nilsson, D-E. 2001. *Animal Eyes*. Oxford: Oxford University Press.

Land, M. F. 1997. *Visual acuity in insects*. Annual Review of Entomology, 42, 147–177.

Wald, G. 1959. *Life and light*. Scientific American. October 1959.

Ronald L. Rutowski

■ Behavioral Physiology
Plumage Color and Vision in Birds

Color is an important component in birds' daily lives. They use color to locate food sources such as nectar-producing flowers and ripe fruits. They identify members of their own species with color patterns. For example, the yellow-shafted flicker has a black "moustache" of plumage beside its bill, whereas the glided flicker has a red one. Males compete for territory through colorful displays and attract mates with plumage color and patterns. Thus, it is not surprising that behaviorists have long been interested in determining plumage color's role as a signal and an important means of communication.

Many birds use aggressive signal displays involving crests, long plumes and plumage color to maintain territories. The red fan parrot extends red and blue neck feathers around its head during aggressive encounters. In male chaffinches, colorful breasts are used to establish dominance. The color signal is such a powerful trait that females with breasts dyed pink became dominant over other females and even some males. Red-winged blackbird males have hidden red plumage patches (epaulets) on their wings that they flash at competitors. When males' epaulets were dyed black in one research study, the males were more likely to lose their territory than males with epaulets.

In some species, the male's plumage functions to attract females. The male peacock spreads large iridescent green feathers to get the female's attention. House finch males' plumage color ranges from yellow to red. When presented with males of different colors during experimental testing, female finches were more attracted to the redder males. In a similar experiment, goldfinch females preferred brighter yellow males. Zebra finch bills vary from orange to red in color. Rather than altering bill or plumage color, behaviorists used colored leg bands to enhance male color. Experiments showed that females preferred males with red legs bands as opposed to those with orange or blue.

In the previously mentioned experiments and many others, brighter or more intensely colored males maintained better territories and attracted more mates, but behavior is seldom so predictable across species. Black pied flycatcher males attract more mates than brown males, but when brown birds were dyed black, their mating success did not improve. Dying yellow-headed blackbirds' heads black in an experiment did not reduce their mating success. Likewise, when yellow throats and Bullock's orioles' plumage colors were reduced, the change did not alter their attractiveness to females.

Much of the problem of evaluating color as an important behavioral signal is that behaviorists see color through human eyes. Although humans see color very well, bird vision is quite different. To compensate for this difference, it is important to understand how the eye detects color. When the lens focuses a light image onto the retina, two types of photoreceptor cells transmit that image to the brain: rods and cones. Rods are more sensitive to light, but do not distinguish color; they enable organisms to see at night, but only in black and white. Cones are sensitive to wavelengths or colors of light and allow organisms to see color during the day when light intensity is higher. The human eye has three different cones (trichromatic) that are sensitive at wavelengths of 480 nm (blue), 505 nm (green) and 575 nm (red). Working together, these cones detect colors along the spectrum from 380 nm to 750 nm: violet, blue, green, yellow, orange, and red.

Birds, on the other hand, have at least four cones (tetrachromatic) that are sensitive at 370 nm (ultraviolet), 450 nm (violet-blue), 480 nm (blue) and 570 nm (red). The 370 nm cone allows birds to see ultraviolet, a color that is invisible to the human eye. Some birds also have a cone sensitive at 630 nm that enables them to see more hues of red than humans. In addition to different cones, birds have oil droplets associated with some of their cones that are colored with yellow, orange or red carotenoid pigments. Behaviorists are not sure how the droplets function, but some researchers suggest the pigments may make certain colors more vibrant to birds.

Realizing that human color perception is very different from that of birds, behaviorists now employ new technologies to evaluate plumage color. In the past, plumage color was measured by comparing it to Munsell color chips. The chips were used as a standard to evaluate color from a human perspective, but may not have been a good indicator of color differences seen by birds. Today, with more portable and inexpensive spectrophotometers available, behaviorists are able to measure the *hue* (actual wavelength) and determine its *chroma* (saturation or purity of the wavelength) and *value* (how much white is in the color) of plumage. Behaviorists can also evaluate dyes they use to change plumage colors for experimental tests to be sure the dyes are producing the measurable differences and effective colors. Researchers can also check for ultraviolet reflectance in products. For instance, some nontoxic correction fluids used in the past to turn feathers white reflected ultraviolet whereas others did not.

Behaviorists concerned with the function of colored plumage must also consider ultraviolet as a visible plumage color in order to get an accurate portrayal of a bird's color signal. For example, blue tit males and females look similar, but when plumage is measured with a spectrophotometer, males' crest feathers are ultraviolet. Some white feathers reflect ultraviolet, while others do not. Recent studies have found that ultraviolet colors are just as important as visible colors to birds. Starlings choose mates based on their ultraviolet color. Zebra finch females avoid males with no ultraviolet plumage. Blue throat males have a chestnut colored throat patch that reflects ultraviolet. When the throat patch color was reduced experimentally in males, the males had difficulty attracting mates.

The fact that birds do not sense color in the same way as humans has opened a whole new area of study. To understand avian behavior completely, behaviorists now realize they must first understand the function of both ultraviolet color and color pigments in birds' eyes before they can completely understand plumage color as a means of communication.

See also Behavioral Physiology—*Color Vision in Animals*
Behavioral Physiology—*Insect Vision and Behavior*

Further Resources

Anderson, M. 1994. *Sexual Selection*. Princeton, NJ: Princeton University Press.

Bennett, A. T. D., Cuthill, I. C. & Norris, K. J. 1994. *Sexual selection and the mismeasure of color*. American Naturalist, 144, 848–860.

Burton, R. 1985. *Bird Behavior*. New York: Alfred A. Knopf.

Cuthill, I. C, Bennett, A. T. D., Partridge, J. C. & Maier, E. J. 1999. *Ultraviolet reflectance and the assessment of avian sexual dichromatism*. American Naturalist, 153, 183–200.

Fogden, M. & Fogden, P. 1974. *Animals and their Colors: Camouflage, warning coloration, courtship and territorial display, mimicry*. New York: Crown Publishers, Inc.

Jacobs, G. H. 1981. *Comparative Color Vision*. New York: Academic Press.

Sinclair, S. 1985. *How Animals See*. New York: Facts on File Publications.

Susan U. Linville

Behavioral Physiology
Thermoregulation

Most people would agree that in the winter one is much less likely to see spiders and insects moving about in their yard than birds or small mammals. This just a small example of how animal behavior is linked to how animals control their temperature. An organism's behavior is intimately tied in with its physiological condition. Proteins control the majority of physiological processes in the body, and the ability for a protein to work properly is influenced by temperature. Because proteins control many behavioral processes, animals try to control the temperature of their body (T_b).

In a nutshell, an animal's control of T_b is about the transfer of heat energy between the animal and its environment. There are four major ways in which heat energy is transferred to/from animals and their environment: *radiation (R)*, *conduction (K)*, *convection (C)*, and *evaporation*. Heat can be gained by increasing *metabolic rate, radiation, convection*, or *conduction*. The rate of heat exchange for radiation, conduction and convection depends on the temperature difference between the individual and the environment, the amount of surface area through which heat loss can occur, and the thermal conductivity coefficients for each mode of heat loss. The contributions of each of these three mechanisms (radiation, convection, and conduction) to T_b depend on the external air temperature. The rate of *evaporation* depends on the difference between the partial pressures of sweat and air. Evaporation is the only mechanism for heat loss when air temperature exceeds body temperature.

As stated in above, all animals can manipulate heat exchange via radiation, conduction, convection, and evaporation. However, animals are typically divided into two major groups with respect to how they *obtain heat*. *Endotherms* use metabolism to generate body heat but *ectotherms* don't. *Ectotherms* gain the majority of their heat from external sources. Birds and small mammals are endotherms and are active in the winter because they are able to generate their own heat. Using metabolic heat is an energetically expensive way to live. Endothermy requires an abundant supply of food, but in exchange allows animals to live in a variety of habitats and to be active at all hours of the day and night and throughout the entire year. Insects and spiders are ectotherms, relying on external sources for their heat. Endothermy and ectothermy are not mutually exclusive; some animals use a combination of both mechanisms to control T_b. For example, some pythons use rhythmic contractions of trunk muscles to warm their eggs, whereas roadrunners (birds) are known to spread their wings (increasing the surface area exposed to the sun) to increase their T_b. The division of endotherms and ectotherms does not parallel the presence/absence of a vertebrate. Some vertebrates are ectotherms, some are endotherms, but generally most invertebrates are ectotherms.

Both endotherms and ectotherms use conduction, convection, and evaporative cooling to maintain an optimal temperature for metabolic processes. Endotherms control heat exchange via convection by manipulating the blood vessels at the surface of the skin. When an endotherm needs to cool off, the surface blood vessels dilate. When the surface blood vessels constrict, heat loss is minimized (you get goose bumps when this occurs). Panting and sweating are ways in which endotherms cool off using evaporative cooling. In addition, endotherms may vary their metabolic rate to further control T_b. For example, when the stresses of maintaining a optimal temperature become too difficult, many small mammals enter torpor, a state of adaptive hypothermia. During this physiological state, T_b is reduced and the animal becomes immobile. Some mammals enter torpor for just a few hours (pocket mice), whereas others may enter this state for weeks (bears).

Terrestrial ectotherms often control T_b by using behavior. A lizard moving back and forth between the sun and shade is obviously using behavior to control T_b. Fiddler crabs move into and out of their burrows to control their T_b. A variety of factors affect thermoregulatory behavior including an organism's size, shape, and orientation to the sun. On cool mornings, one will see a butterfly open its wings fully in the sunlight. The butterfly is basking, having the greatest surface area available for warming by the sun. On hot summer days, butterflies will consume nectar from flowers with their wings folded up, decreasing the surface area warmed by the sun.

Conduction is the transfer of heat energy between an animal's body and a substrate. Animals are known to use behavioral thermoregulation to modulate heat exchange via conduction. Snakes are known to choose flat, large rocks to bask upon in the early morning and spend the night under thicker, rounder rocks. The flat rocks warm more quickly in the morning sun than the round rocks. And, as night approaches, the round, thicker rocks cool more slowly than the thin, flat rocks.

Group living can also play a role in temperature regulation. Animals that live in groups often build shelters to control heat exchange. Tent caterpillars build their tents on the west side of trees to capture the afternoon heat. The caterpillars found in these tents have a higher T_b at night than caterpillars that are forced to leave the tent. The increased T_b leads to a faster development time in many invertebrates.

Group members can also use behavior to regulate T_b. The larvae of the common honeybee need to develop at temperatures in the range of 32–36° C. If the temperature in the nest falls, the worker bees will shiver their wing muscles to generate heat. If the nest becomes too warm, the workers can decrease its temperature by fanning cool air into the nest. They can also cool the nest by collecting water and spreading it on the surface of the comb, inducing evaporative cooling. Ants are known to move developing larvae and queens within the nest, migrating up and down during the day to find optimal temperature within the nest.

Overall, the aquatic environment possesses more uniform temperatures than terrestrial environments. Aquatic ectotherms have temperatures that approximate the water temperature in which they are found. However, some aquatic ectotherms (like great white sharks) slow heat loss by using a countercurrent heat exchanger. They line up their veins and capillaries so that the warm blood of the core of the body warms the blood coming in from extremities. Aquatic endotherms usually have layers of insulation (hair, blubber) to slow heat loss. In fact, some marine mammals are so well insulated that they risk overheating when land or water temperatures exceed 15°C.

Further Resources

Heath, D. 1986. *Thermoregulation in Vertebrates*. Annual Review of Physiology, 48, 593–638.

Hertz., P. E, Huey, R. B. & Stevenson, R. D. 1993. *Evaluating temperature regulation by field-active ectotherms: The fallacy of the inappropriate questions*. American Naturalist, 142, 796–818.

Huey, R. B., Peterson, C. R., Arnold, S. J. & Porter, W. P. 1989. *Hot rocks and not so hot rocks: Retreat site selection by garter snakes and its thermal consequences*. Ecology, 70, 931–944.

Mary Crowe

■ Behavioral Physiology
Thermoregulatory Behavior

As you read this, it is likely that you are feeling neither too hot nor too cold. Whether it is winter or summer or whether you are indoors or outdoors, you have probably chosen the right clothing, dialed in the right temperature on your thermostat, perhaps even poured yourself a hot or cold drink, before settling down to read. These behaviors are not typically performed for their own sake. Rather, they are performed to accomplish one of the most vital of all biological goals: the regulation of our internal thermal environment.

Think of those times in your life when you have been too hot or too cold and try to imagine yourself reading a book. It is difficult to imagine because such periods of thermal discomfort tend to obliterate the possibility of engaging and focusing the mind on difficult cognitive tasks. There are a number of reasons for this, not least of which is that thermal discomfort, like severe hunger, thirst, or sleep deprivation, demands immediate remedy. We simply cannot think deep thoughts when we are shivering or sweating.

Thermoregulation falls within the broader domain of homeostasis, a concept introduced by the American physiologist, Walter Cannon (1871–1945). *Homeostasis* refers to the process by which all animals, including humans, regulate physiological parameters within a range conducive to survival. Blood sugar, blood pressure, body water, oxygen levels, and temperature are just a few examples of the many aspects of our internal environment (or *internal milieu*, as the French physiologist, Claude Bernard, famously expressed it) that are regulated through physiological adjustment. Some of these systems are purely physiological; that is, behavior does not play a significant role in the process. For example, when we rise from a sitting position, stretch receptors in the major artery leaving the heart detect (and with experience anticipate) a fall in blood pressure and trigger a reflex that stimulates an increase in heart rate to prevent us from fainting due to a loss of blood flow to the brain.

Many homeostatic processes, however, rely heavily on behavior. For example, when a person suffers a gunshot wound and is losing blood, many physiological responses are activated to minimize the impact of this blood loss on the victim's survival. Ultimately, however, the only way to replenish this fluid loss is through behavior, that is, drinking (leaving aside the modern possibility of having fluid delivered directly into the bloodstream through a catheter). Thus, many homeostatic processes depend on a vital combination of physiological and behavioral responses.

Although the French physiologist Claude Bernard (1813–1878) is widely regarded as the founder of regulatory physiology in its modern form, it was not until much later that the role of behavior was fully appreciated. Curt Richter (1894–1988), working at Johns Hopkins University, was one of the pioneers in the study of the contributions of behavior to physiological regulation. The rationale for Richter's insistence on the importance of regulatory behavior can be found in a brief report published over 60 years ago (Wilkins & Richter 1940). This report documents the case of a 3-year-old boy who had been admitted into a hospital in Baltimore because of abnormal sex organ development. In addition, this boy exhibited an excessive craving for salt that began when he was one year of age. The boy's problems were blamed on deficient functioning of the adrenal cortex, an endocrine organ that produces a hormone, aldosterone, that plays a role in retaining salt. Unfortunately, doctors at the hospital placed him on a standard low-salt hospital diet, whereupon he died within 7 days. As was subsequently appreciated, the boy's craving for salt was a behavioral adjustment to his malfunctioning adrenal cortex. Thus, by preventing the boy from regulating his need for salt behaviorally, the doctors had inadvertently killed him.

Every aspect of biological functioning—from gene expression to enzymatic functioning to neuronal processing to muscular contractions to habitat selection—is intimately connected with temperature. Its centrality to biological function is undoubtedly due to its fundamental role in all things physical, as with gravity. But unlike gravity, temperature varies within our bodies and throughout our environment, thus providing the need and the opportunity to regulate temperature through physiological and behavioral adjustments.

Humans and other mammals, as well as birds, were once referred to as *warm-blooded* because of their ability to stay warm in any environment. In contrast, fish, reptiles, and amphibians were once referred to as *cold-blooded* because their body temperatures were typically identical to the water or air temperature in which they were caught. We now refer to these groups as *endotherms* and *ectotherms*, respectively, to focus on the animal's ability to produce heat internally or its dependence on external sources of heat (*endo* meaning internal, and *ecto* meaning external), rather than the temperature of the blood. But, more importantly, all animals, endotherms and ectotherms alike, use behavior to regulate their body temperatures within narrow ranges. Indeed, the use of behavior to regulate temperature is more evolutionarily ancient than such physiological mechanisms as shivering, non-shivering forms of heat production (as with brown adipose tissue, a heat-producing form of fat that helps to keep many infant mammals warm and makes it possible for hibernating mammals to warm up from near-freezing body temperatures), and sweating.

Lizards and other ectotherms use behavior during the day to warm up when body temperature dips too low and cool down when it rises too high. For much of the day, a lizard is thermally comfortable and can perform many necessary activities—such as eating, drinking, and mating—without having to concern itself with thermoregulation. But when temperatures exceed certain thresholds, behavior must be diverted to address thermal concerns. If too cold, the lizard might shuttle to a hot rock and bask in the sun. If too hot, it might shuttle to a cool, shaded area. Thus, thermoregulatory need places constraints on the behavior that lizards and other ectotherms can express at any given moment. Of course, at night, with the setting of the sun, lizards and other ectotherms enter a state of torpor and must cease all behavioral activity until the next morning when their heat source returns. In this way, the evolutionary invention of endothermy freed behavior from much of its commitment to thermoregulation and made possible a variety of lifestyles, such as nocturnality and the habitation of extreme thermal environments, that are not possible for most ectotherms. This is not to say that endotherms do not use behavior to thermoregulate, only that the commitment of behavioral resources is less intense than it is for ectotherms.

We are all familiar with the physiological changes that accompany the production of a fever: First we feel chilled as body temperature increases; then we attain and maintain a higher body temperature but feel thermally comfortable (despite the aches that accompany sickness); and then the fever breaks as body temperature returns to normal. Each of these fever-episode phases is accompanied by physiological adjustments, such as shivering when we feel chilled and sweating when the fever breaks. But what about ectothermic lizards? Do they get fevers when they are sick? If so, how?

The answer, of course, is by using behavior. When we get a fever, we retreat to bed and get under the covers when we feel chilled, and toss off the covers when the fever breaks. Lizards also use behavior by shuttling to warmer temperatures and staying in such warm regions longer when they are infected. In fact, it was the ability to examine such fever behaviors in lizards that allowed for the first demonstration in any animal that a fever is an effective tool in fighting infection: Lizards that were prevented from shuttling to warmer temperatures were much more likely to die after infection than lizards that were not prevented from doing so. (Higher body temperatures help to fight infections in a number of ways, including boosting the immune system and starving invading bacteria of iron, a vital nutrient on which they rely increasingly at higher temperatures.) As it turns out, fever is an ancient response to infection, and behavior is a vital component of this response.

Many infant mammals face particularly serious thermal challenges. For example, infant rats are born without fur. This fact, combined with their small size, makes it difficult for them to retain endogenously produced heat; for newborn rats, even 70°F (21°C) (room temperature for humans) represents an extreme thermal challenge. The mother rat helps her young withstand the cold by building a well-insulated nest and by hovering over her young when she lactates. In rabbits, however, the mother attends to her young for just a few minutes a day, leaving them alone in their nest to fend for themselves. How do such small, naked, and blind infants survive?

The answer lies first in the fact that species such as rats, rabbits, and dogs that give birth to relatively undeveloped young (that is, *altricial* young, in contrast to *precocial* young such as horses) tend to give birth to litters. Giving birth to multiple young provides each pup with the opportunity to huddle with its littermates. To appreciate the value of huddling, we must first examine a basic physical principle that governs the flow of heat from all physical objects: The Surface Law.

The Surface Law refers to the fact that as an object increases in size (but does not change its shape), its surface area (expressed as the square of some linear dimension) increases more slowly than its volume (expressed as the cube of the same linear dimension). For example, the surface area of a sphere is equal to $2\pi r^2$ (where r is the radius of the sphere) while its volume is equal to $4\pi r^3/3$. Now consider the ratio of surface area to volume, which we can estimate by dividing $2\pi r^2$ by $4\pi r^3/3$ and ignoring the constants in the formulae: The result is $r^2/r^3 = r^{2/3}$. What this relationship means is that as objects or animals increase in size, the amount of surface (for example, skin) through which heat can be lost does not increase as rapidly as the volume. This is analogous to filling a room with more and more people without opening the windows: The result is a very hot room.

Infant rats need to stay warm when air temperatures plummet, and one way that they do this is by using the Surface Law to their advantage: Huddling allows each pup to be part of a larger group, thereby increasing the volume of pups and decreasing the relative amount of surface through which heat can be lost from this now-larger entity. As a consequence, the heat that each pup produces can be retained more effectively within the group. But lest you think that this is a case of a pup behaving altruistically by sharing its heat with its siblings, it is important to note that each pup within the group is simply doing what is best for itself, and by doing so the group of pups benefits. Thus, overheated pups in the center of a huddle will push their way to the outside of the huddle and cooling pups on the outside will dive in. In other words, the huddle is a very active and dynamic entity, with each individual thermoregulating behaviorally to optimize its body temperature.

Huddling is not restricted to infants. On the contrary, many adults utilize the benefits of huddling in extremely cold environments. For example, the breeding and reproductive cycle of male emperor penguins is such that the females leave them alone during the coldest Antarctic winter months. These males have the task of incubating an egg, which sits perched on the penguin's feet with a flap of abdominal skin overlaying it. Exposed on an ice shelf for many months without food and exposed to temperatures that dip as low as −76°F (−60°C), hundreds or thousands of males may gather together in one enormous huddle and, by doing so, survive the winter and successfully incubate their eggs. As with infant rats, the males move in and out of the huddle in order to maintain their thermal comfort.

Thermoregulation is a biological imperative, and nearly all animals—from worms to humans—use behavior to meet their thermal needs of the moment. Although thermoregulatory behavior has been investigated for relatively few years, it is now appreciated that its flexibility, rapid responsiveness to experience, and relatively low energetic costs have contributed to its retention and refinement throughout the animal kingdom. Indeed, thermoregulatory behavior continues to be essential even in those animals—such as mammals and birds—that have evolved exquisite physiological mechanisms for gaining and losing heat. In these animals, physiology and behavior work hand-in-hand in elegant and complex ways, leading many to question the wisdom of continuing to regard physiology and behavior as separate processes.

See also Behavioral Physiology—*Thermoregulation*

Further Resources

Alberts, J. R. 1978. *Huddling by rat pups: Group behavioral mechanisms of temperature regulation and energy conservation.* Journal of Comparative and Physiological Psychology, 92, 231–245.

Blumberg, M. S. 2002. *Body Heat: Temperature and Life on Earth.* Cambridge, MA: Harvard University Press.

Cannon, W. B. 1932. *The Wisdom of the Body.* New York: Norton.

Cooper, K. E. 1995. *Fever and Antipyresis.* Cambridge: Cambridge University Press.

Kluger, M. J. 1979. *Fever: Its Biology, Evolution, and Function.* Princeton: Princeton University Press.

Satinoff, E. 1996. *Behavioral thermoregulation in the cold.* In: *Handbook of Physiology* (Ed. by M. J. Fregly & C. M. Blatteis), pp. 481–505. Oxford: Oxford University Press.

Sokoloff, G., Blumberg, M. S. & Adams, M. M. 2000. *A comparative analysis of huddling in infant Norway rats and Syrian golden hamsters: Does endothermy modulate behavior?* Behavioral Neuroscience, 114, 585–593.

Wilkins, L. & Richter, C. P. 1940. *A great craving for salt by a child with cortico-adrenal insufficiency.* Journal of the American Medical Association, 114, 866–868.

Mark S. Blumberg

■ Behavioral Physiology
Turtle Behavior and Physiology

Turtles display some of the most remarkable behavior found in nature. These behaviors include the oceanic migrations of sea turtles, where newly hatched animals locate faraway feeding grounds never before visited, and females return across entire oceans to the beaches of their birth to lay eggs. Other turtles can survive winter in cold forests by freezing into icy blocks, then unthawing the following spring with no apparent harm. Winter hibernation in some turtles involves submergence under water where they may not come to the surface to breathe for many months. Underlying these behaviors in turtles are extraordinary physiological abilities whose mysteries scientists are just now starting slowly to unravel, often with the help of emerging technologies.

Satellite telemetry, for example, has provided a wealth of new information on the long-distance movements of sea turtles, providing details on their exceptional navigational abilities. Recent tracking studies of green turtles (*Chelonia mydas*) have shown that these animals can find small oceanic islands with straight-line precision in journeys covering thousands of miles. How do turtles find their way so well across, what seems to us, a featureless expanse of ocean? Many hypotheses have been advanced to explain sea turtle orientation, including their use of the earth's magnetic field, ocean currents, wave direction, and visual and chemical cues. While much about turtle orientation remains enigmatic, experimental and field work suggest that turtles use multiple cues to guide them. Navigational mechanisms may also change during different parts of the turtle's migratory trip.

Even more impressive than the migrations of sea turtles may be the capacity to withstand freezing shown by hatchling North American painted turtles (*Chrysemys picta*), the highest vertebrate that can tolerate the natural freezing of its extracellular body fluids during winter hibernation. These animals spend their first winter inside shallow nests in the ground where temperatures are frequently life threatening. Here they use unique strategies to survive the cold by supercooling their bodies. This is done in one of two ways. First, the hatchling can avoid freezing through the development of a freeze-resistant *integument* (outer covering) and elimination of materials that favor the formation of ice. Alternately, it can freeze around half of its body, but protect tissues through cryoprotective agents such as glucose and glycerol. Understanding how these turtles survive freezing is of great interest to scientists trying to develop effective methods of cryonic preservation of human organs and tissues.

Winter hibernation in the adult painted turtle is also quite extraordinary due to its ability to remain submerged for incredibly long periods, a capacity shared by many turtle species living in both freshwater and marine environments. While even the most accomplished diving animals, such as marine mammals, can usually only hold their breaths for minutes at a time, painted turtles can remain under water continuously for months during winter, where they may be trapped below ice covering ponds and streams. Being air-breathers, how are these turtles able to do this? The answer lies in the workings of their metabolism. As cold-blooded animals, the metabolic rate of turtles varies with body temperature. Consequently, the amount of oxygen consumed by the animal is very low when it is cold during winter hibernation. Moreover, turtles have an unusually high tolerance for the absence of oxygen, or *anoxia*, under which conditions anaerobic metabolic pathways will be used, with potentially poisonous end products being taken up by the turtle's shell. This capacity for anoxia, combined with its low metabolic rate at cold temperatures, allows the hibernating painted turtle to drop its metabolic rate to over 10,000 times lower than a similarly sized mammal resting at its normal body temperature. Thereby, this turtle species can remain submerged for periods of time far and beyond what would be lethal to other air-breathing vertebrates.

In the past, physiologists also looked to the circulatory systems of turtles to help understand their exceptional diving abilities. Turtles were often featured as textbook examples illustrating adaptive cardiovascular responses to diving, part of the so-called "dive reflex" thought to extend dive duration in many air-breathing animals. This reflex, which includes a drop in heart rate and, in turtles and other reptiles, the shunting of blood away from the lungs made possible by their three-chambered hearts, was identified in early laboratory studies of diving. However, the advent of new technologies has allowed physiological measures to be made for the first time in voluntarily behaving animals. Thus, much of our conventional thinking about the dive reflex in turtles and other divers has proven to be wrong. More recent research has revealed that when turtles and other animals dive on their own, their cardiovascular responses may not resemble the standard dive reflex, and that results obtained in earlier diving studies were likely artifacts of forced diving, when animals were primarily fearful. New findings on freely-diving turtles suggest that the important variable determining physiological reactions may be the level of activity shown. Turtles who are swimming or otherwise moving while voluntarily submerged show cardiovascular responses akin to those found in exercise generally, despite holding their breaths. While resting, either in air or submerged, turtles display similar cardiovascular patterns, including a much reduced heart rate. Therefore, under normal circumstances the "dive reflex" may be more aptly called the "rest reflex" in turtles.

Instrumentation of freely-behaving turtles has yielded new and unexpected insights to their inner states as well, challenging many stereotypes about them. Data have been collected showing that unlike the slow, lumbering animal that its external behavior might suggest, turtles exhibit physiological responses to their environment that can be very sensitive. Rapid and profound changes in cardiovascular variables, such as heart rate, can occur in response to handling or even small visual cues of impending danger. The sensitivity of these reactions may be adaptive in allowing the turtle to withdraw into its shell effectively for protection. Such studies may provide strong evidence that turtles and other cold-blooded animals do not deserve the reputation of being unfeeling creatures, and should be afforded the same protection for their welfare as the so-called "higher vertebrates," birds and mammals.

See also Hibernation
Navigation—*Natal Homing and Mass Nesting in Marine Turtles*

Further Resources

Jackson, D. C. 2002. *Hibernating with oxygen: Physiological adaptations of the painted turtle.* Journal of Physiology, 543 (3), 731–737.

Krosniunas, E. H. & Hicks, J. W. 2003. *Cardiac output and shunt during voluntary activity at different temperatures in the turtle,* Trachemys scripta. Physiological and Biochemical Zoology, 76(5), 679–94.

Packard G. C. & Packard, M. J. 2003. *Natural freeze-tolerance in hatchling painted turtles?* Comparative Biochemistry and Physiology A, 134(2), 233–246.

Papi, F., Lutschi, P., Akesson, S., Capogrossi, S., & Hays, G. C. 2000. *Open-sea migration of magnetically disturbed sea turtles.* Journal of Experimental Biology, 203, 3435–3443.

Egle H. Krosniunas

■ Behavioral Physiology
Visual Perception Mechanisms

Being able to "see" the world is a great advantage to animals. It allows valuable sensory information to be detected at a distance, without the close physical contact required by the senses of touch and taste. The value of seeing is underscored by the variety of different visual systems that evolved in animals, ranging from elementary photoreceptors that only discriminate light and dark to more advanced systems capable of discriminating surfaces and objects.

Because of our own strong reliance on sight, it is easy for humans to underestimate the demands that processing visual information places on a brain. After all, we just open our eyes and the visual world is "right there." But computer scientists have found the challenges of vision to be many and vexing when trying to program computers to "see." For just one example, how is the three-dimensional impression of depth reconstructed from the two-dimensional information projected on the retinal surfaces within the eyes? The rapid and seemingly effortless solution of this kind of "visual" problem makes the brain as vital an organ of visual perception as the eye.

The physical stimulus for visual perception is the light of differing wavelengths that is reflected by the surface of all objects. Humans perceive light wavelengths between 400 and 700 nanometers; these different wavelengths are responsible for our impression of color. Interestingly, other animals can sense wavelengths outside of this range. Bees can detect ultraviolet light, which allows them to see distinct patterns in the petals of flowers that we cannot. Homing pigeons can see patterns of polarized light in the daytime sky that are invisible to us, providing them with an additional source of information for homing. As such, it is important to remember that each animal inhabits a sensory world that is appropriately tuned to its own natural behavior. Dogs are highly sensitive to smells because their behavior is importantly determined by scent. Likewise, songbirds have exquisite auditory systems. Collectively, the German word *Umwelt* is often used to describe the differing perceptual worlds of each species of animal. As such, one important lesson is that one's own visual experiences are only a rough, and probably biased, guide to what other animals see.

Comparative psychologists have made substantial progress in understanding visual perception and its mechanisms in animals, making a comprehensive review of this information impossible. Thus, the current essay is limited to visual perception in birds. Nevertheless, it will provide a good overview of many of the general issues and techniques used in the behavioral study of animal perception.

Humans see a world of stable, meaningful, and unified objects that can be detected, grasped, sought, and avoided with little effort. Pigeons behave as if they too perceive a stable and object-filled world. If this interpretation of pigeons' visually guided actions is correct, then an interesting paradox arises. Because of the demands imposed by flight, pigeons have been subjected to strong evolutionary pressure for the last 200 million years to keep their overall size, especially their brains, to a minimum—heavy heads would make flight difficult or impossible. Although a large portion of the avian central nervous system is indeed devoted to visual processing, the pigeon's brain is still only 1/1000th the size of a human's. Given this tiny brain, how are pigeons able to see so well? Answering this question not only illuminates avian visual cognition (Cook 2001), but extends our general understanding of visual cognition in all species.

Just how similar are the visual worlds of birds and mammals? To answer this question, behavioral studies have focused on how pigeons and humans perceive a variety of visual stimuli. If they see visual information in the same way, then pigeons' reactions should be

Vision, Skull Shape, and Behavior in Dogs

Paul McGreevy

The domestic dog, *Canis lupus familiaris*, demonstrates striking morphological diversity, especially when one compares doliocephalic (long-skulled) and brachycephalic (short-skulled) breeds. However, such diversity is minimal in the dog's ancestor, the grey wolf, *Canis lupus*, even though it has almost identical mitochondrial DNA. This highlights the profound influence of human selection on canine phenotypes.

The *doliocephalic* dogs with the longest skulls (relative to width) comprise a group known as the sight-hounds, which primarily use vision in hunting, in contrast to, say, the scent-hounds, which use olfaction. Examples of sight-hounds include greyhounds, whippets, borzois, salukis and Afghan hounds. Their behavior in the presence of moving prey is radically different to that of dogs at the other end of the skull-shape spectrum, as exemplified by pugs, mastiffs, boxers, and Boston terriers.

A recent study in various dogs of different shapes and sizes has shown that variation in body shape is associated with a wide variation in the structure of the retina, notably the distribution of ganglion cells. This variation is not random but is highly correlated with a change in skull length. The ratio of peak ganglion cell density in the area centralis to visual streak correlates inversely with skull length and positively with cephalic index (skull length/width). So, the distribution of ganglion cells in the retina varies from a horizontally aligned visual streak of fairly even density across the retina in long-skulled doliocephalic dogs (and also the wolf) to a strong area centralis with virtually no streak in short-skulled brachycephalic dogs. Such tremendous variation in ganglion cell density within a single species is unique.

These results suggest that we cannot expect dogs of different skull shapes to see the world in the same way as one another, and this may explain some differences in their behavior. For example, dogs bred with short faces may show behavior different from the sight-hounds, being less likely to act like running predators and hunt in packs and more likely to focus on human faces.

Developmental studies involving fetal retinae in other species with an area centralis have shown that ganglion cell density gradients arise later in development from an originally even distribution. In contrast, in those species with a strong visual streak, the visual streak is present from the earliest stage of development and then becomes more pronounced during development. Fetal dogs have yet to be studied to explore whether these two different mechanisms for producing cell density gradients apply to canids. However, recent evidence is that shortening of the skull in dogs is correlated with more frontally placed eyes. This may be of importance when one considers the development of retinal differences at the level of the embryo's optic vesicles. It is possible that variation in the vesicles' shape is influenced by the angle at which they project from the brain and that this angle is linked to nose length.

Further Resources

McGreevy, P. D., Grassi, T. D. & Harman A. M. 2004. *A strong correlation exists between the distribution of retinal ganglion cells and nose length in the dog.* Brain, Behaviour and Evolution, 63 (1), 13–22.

Newby, J. 2004. *New findings—Do different dogs "see" differently?* Bark Magazine, Spring 26, 36.

Robinson S. R., Dreher B., McCall M. J. 1989. *Nonuniform retinal expansion during the formation of the rabbit's visual streak: Implications for the ontogeny of mammalian retinal topography.* Visual Neuroscience, 2, 201–219.

similar to ours. On the other hand, if they see visual information in a different way, then pigeons' reactions should be different from ours.

For instance, Blough taught pigeons to discriminate among letters of the alphabet by training them in an operant conditioning chamber in which he could project and the pigeons could peck at the letters (Blough 1982). The pigeons' task was to discriminate among the letters to get food reward. Blough found that the pigeons made the same kinds of letter confusions that we make (e.g., confusing similar letters like "O," "Q," and "D"). This behavioral similarity suggests that, despite the considerable differences in the size, organization, and natural history of the mammalian and avian brain, we perceive these letters similarly. By using this general strategy, but testing more theoretically designed stimuli, one can begin to isolate and measure the different stages of the perceptual process leading to such visual impressions.

Perceiving Edges and Surfaces

One of the important main goals of early visual processing is to detect and organize visual information into perceptual edges and surfaces—the building blocks for the later recognition of objects. Studies using *texture* stimuli have found that our early visual system can quickly group similar color and shape features into regions and then rapidly segregate them at their boundaries or edges. For instance, you should have no problem segmenting the top two stimuli of the first figure shown into two distinct areas and then locating the smaller bounded area within each one. We easily segregate such *feature* stimuli into regions when there are consistent groupings of either color (left stimulus) or shape (right stimulus) features—perhaps because features that are perceptually grouped by separate dimensional channels which violate this dimensionally consistent organization are much harder to segregate into regions visually, such as the texture stimulus in the lower left of the figure. Believe it or not, this stimulus has a small odd region just like the above two stimuli, but it is much harder to see because the edges of this target are still not easily detected or connected by such grouping mechanisms (the light circles and dark triangles in the lower left part of the display). Instead, a slower, more focused attentional process is required to process these kinds of *conjunctive* stimuli accurately.

Would pigeons show the same superiority of discriminating feature displays relative to conjunctive stimuli? Using stimuli similar to those in the figure, pigeons learned to locate and peck at the odd region in such texture stimuli to obtain food reward (see diagram of operant chamber in lower right of the figure). When tested with the same types of features and conjunctive arrangements, pigeons were far better at locating targets in feature displays than in conjunctive displays, just like humans (Cook 1992; Cook, Cavoto & Cavoto 1996). By manipulating factors such as the number of surrounding elements and the presence or absence of irrelevant dimensional variation, we established a number of additional similarities in how humans and pigeons process these displays. Such outcomes have suggested that the mechanisms of perceptual grouping and search in humans and pigeons share many similar properties that allow each species to detect and identify the edges and surfaces of objects in the visual world rapidly.

Perceiving Objects

Prior to the development of representational art and photography, humans only rarely needed to discriminate anything other than three-dimensional visual stimuli. Nevertheless, adults and even very young children adeptly discriminate such two-dimensional representations of the real world. Given that we know of no other animal that spontaneously

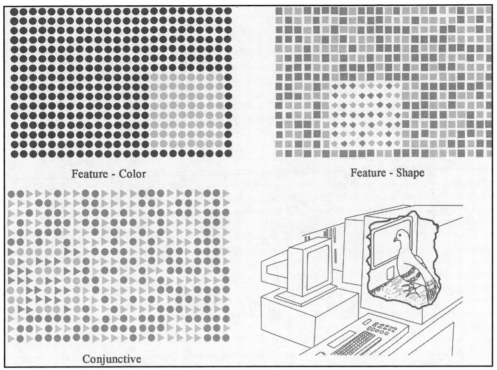

Feature - Color

Feature - Shape

Conjunctive

Representative examples of feature–color, feature–shape, and conjunctive–texture displays used to test pigeons and humans. These displays are made up of four elements composed from combinations of two shapes and two colors. The feature–color and feature–shape labels refer to displays where the two regions are grouped according to one of the two dimensions (color or shape). The conjunctive–texture display is made from similar combinations of elements, but designed to make both dimensions (color and shape) relevant to finding the target. The colors and shapes used to make these displays were randomly changed in each trial when testing the pigeons. The bottom right panel shows a cut away drawing of a typical operant chamber used to test the pigeons. The birds view a computer screen through a glass window in the front of the chamber. Pecks to the stimuli presented on this screen are detected with a touchscreen that can tell us where and when a pigeon pecks on each trial. A food hopper, located just below the viewing window, is used to deliver mixed grain to the pigeons as a reward for correct responses.

draws or otherwise reproduces such visual representations, possible differences in picture perception might exist among animals. What do pigeons perceive when they look at a picture?

A recent theory of visual perception by humans represents a significant forward step for studying picture perception by pigeons. Biederman's (1987) theory of recognition by components is a model of the perceptual process which assumes a series of stages of information processing that extract both specific and general information from complex visual stimuli. Biederman hypothesizes that the visual system divides complex visual stimuli into elementary components or *geons*, whose spatial arrangement defines virtually any object. Using this theory as a guide, experiments with pigeons have focused on whether they process geons in a fashion similar to that established in humans.

The line drawing stimuli used in Experiment 2 performed by Wasserman and his colleagues. The left column depicts the original training stimuli; the center column depicts the complementary test stimuli; and the right column depicts the spatially scrambled versions of the complementary test stimuli.

Copyright 1992 by the American Psychological Association. Reproduced with permission.

The first investigation (Van Hamme, Wasserman & Biederman 1992) examined the possibility that picture perception might transfer reliably from one portrayal to another, even though no exact pattern match was possible between the different visual portrayals. Some of the line drawings used in this experiment are shown in the second figure. These illustrations originated from complete line drawings of common plants, animals, and human-made objects; they were created so that only half of the line contours were present in each of the drawings in the left and center columns of the figure. The theory of recognition by components predicts that objects can be recognized even when visual noise is added or defining information is eliminated, if sufficient information for geon extraction and organization is available. The partial objects in the left and center columns of the figure were created so that adequate information to permit object recognition was maintained, despite half of the contours being deleted. According to the theory, such complementary halves should support equivalent picture recognition.

Pigeons learned a four-choice task to discriminate among the four objects in this figure. Following the presentation of each object, the pigeon had to choose the correct one of four buttons for food reward. After learning, the pigeons were tested with the complementary picture of each object. Although there was a slight reduction in accuracy, the pigeons quite successfully recognized the objects in the complementary test stimuli. Discriminative performance was clearly supported by drawings of objects that did not match the training pictures.

Testing was further conducted using the line drawings in the right column of the figure. These drawings entailed the same line segments as the complementary drawings in the center column, but those segments were spatially scrambled to prevent geon recovery and possibly picture recognition. Here the pigeons' performance dropped to just above chance proving that these fragments themselves could not support accurate responding: Only when the line segments were arranged to permit geon recovery could the pigeons do the task.

A second investigation (Wasserman, Kirkpatrick-Steger, Van Hamme & Biederman 1993) studied the role that geon arrangement plays in picture perception. This investigation was inspired by an earlier proposal that the pigeon's recognition of line drawings depends only on the presence of certain features, not on the spatial organization of those features. Biederman's theory, on the other hand, relies critically on organisms being able to recognize the spatial organization of the geons. To test these ideas, pigeons were trained to discriminate the four line drawings in the left column of the third figure. As can be seen, the desk lamp, the iron, the watering can, and the sailboat each comprised four geons. After training, the pigeons were tested with novel stimuli composed of the same four geons, but in new spatial arrangements. The center and right columns of this figure show the test configurations for the desk lamp. These test configurations preserved the overall height and width of the original drawings as well as the orientation of each of the component parts; those in the center column preserved the exact line contours of the original drawings (creating the small notch in the base of the lamp), whereas those in the right column preserved the uninterrupted contours of the underlying geons (eliminating the notch in the base of the lamp). In each case,

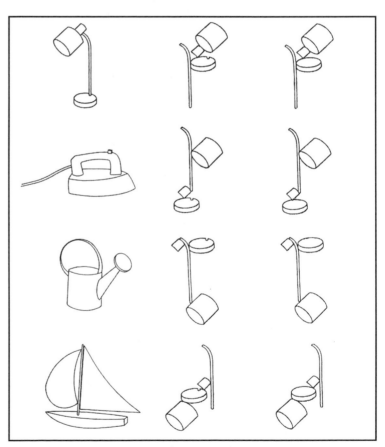

The leftmost column shows the line drawings used during the original training phase of Experiment 1 performed by Wasserman and his colleagues. All of the spatial scramblings of the desk lamp were used during testing in Experiment 1. The center column depicts the "notched" versions and the right column depicts the "smooth" versions. Spatial scramblings of the other three training stimuli were similarly created.

there was a drop in accuracy whenever the geons were scrambled into new arrangements. These drops in accuracy suggest that spatial organization of an object's elemental parts is an important ingredient in the recognition of complex visual stimuli by pigeons, just as it is for humans. This evidence and other research showing that, like humans, pigeons can recognize pictures of familiar objects that are rotated in depth suggests that the pigeon's analysis of complex visual stimuli may proceed along lines that are described by the process of recognition by components.

The Comparative Puzzle

There are behavioral and biological similarities between pigeon and human visual perception and cognition; however, it is important to keep in mind that empirical differences do exist and must be understood and incorporated into any final accounting of this species comparison. Pigeons are tetrachromatic, for instance, and show sensitivities to ultraviolet and polarized light that may play crucial roles in homing and mate selection. Pigeons integrate information from a wider visual field than do humans; and, they do so using two separate specialized fovea-like areas within each eye (we have only one in each eye). Some behavioral experiments have suggested that pigeons and humans may not use the same set of primitives or features in processing visual information. For example, pigeons do not show search asymmetries for the same features as humans (Allan & Blough 1989), may tend to look at smaller features prior to larger features in contrast to humans (Cavoto & Cook 2001), and may process configural information about shapes differently as well (Kelly & Cook 2003). Although still debated, pigeons may even be superior to humans in their capacity to rotate visual information mentally, exhibiting little or no slowing in reaction time as a function of angular orientation (Hollard & Delius 1982).

Much remains to be done to complete this comparative puzzle. Nevertheless, consider for a moment the implications of the above evidence suggesting that human and avian visual cognition share functional and operational properties. In conjunction with comparative research on insects, amphibians, mammals, and computers, the study of avian visual cognition helps us to identify the generalized information processing principles used by many species as well as the biological conditions responsible for their development. That study may also uncover divergent, but efficient alternatives to solving common visual problems. More speculatively, research on avian visual processing may also help us further understand the interrelation between sensation, perception, and consciousness. Besides these interesting psychological implications, just as with the inspiration for and solution of human flight, birds may also hold the practical key to our engineering of small, self-guided, adaptable robots and the ongoing development of compact prostheses for the visually-impaired.

Further Resources

Allan, S. E., & Blough, D. S. 1989. *Feature-based search asymmetries in pigeons and humans*. Perception & Psychophysics, 46, 456–464.

Biederman, I. 1987. *Recognition-by-components: A theory of human image understanding*. Psychological Review, 94, 115–147.

Preparation of this essay and its empirical contents was supported by grants from the National Institute of Mental Health (EAW) and the National Science Foundation (RGC).

Blough, D. S. 1982. *Pigeon perception of letters of the alphabet.* Science, 218, 397–398.

Cavoto, K. K. & Cook, R. G. 2001. *Cognitive precedence for local information in hierarchical stimulus processing by pigeons.* Journal of Experimental Psychology: Animal Behavior Processes, 27(1), 3–16.

Cook, R. G. 1992. *Dimensional organization and texture discrimination in pigeons.* Journal of Experimental Psychology: Animal Behavior Processes, 18, 354–363.

Cook, R. G. 2001. *Avian Visual Cognition.* www.pigeon.psy.tufts.edu/avc/

Cook, R. G., Cavoto, K. K., & Cavoto, B. R. 1996. *Mechanisms of multidimensional grouping, fusion, and search in avian texture discrimination.* Animal Learning & Behavior, 24, 150–167.

Hollard, V. D. & Delius, J. D. 1982. *Rotational invariance in visual pattern recognition by pigeons and humans.* Science, 218, 804–806.

Kelly, D. M., & Cook, R. G. 2003. *Differential effects of visual context on pattern discrimination by pigeons* (Columba livia) *and Humans* (Homo sapiens). Journal of Comparative Psychology, 117, 200–208.

Van Hamme, L. J., Wasserman, E. A., & Biederman, I. 1992. *Discrimination of contour-deleted images by pigeons.* Journal of Experimental Psychology: Animal Behavior Processes, 18, 387–399.

Wasserman, E. A., Kirkpatrick-Steger, K., Van Hamme, L. J., & Biederman, I. 1993. *Pigeons are sensitive to the spatial organization of complex visual stimuli.* Psychological Science, 4, 226–341.

Robert G. Cook & Edward A. Wasserman

■ Behavioral Plasticity

Environmental change is a fact of life for most animals. Animals often respond to these changes by altering their *phenotypes* (the expression of physical and behavioral traits in response to the environment). *Behavioral plasticity* is a form of phenotypic plasticity that applies only to animals. It is defined as an adaptive change in behavior in response to an environmental change. An adaptive behavioral change increases the individual's chance of surviving and reproducing. The "environmental" includes factors such as population density, resource availability, predation pressure, and abiotic factors (e.g., climate). There are two main types of behavioral plasticity: (1) differences in the development of behavior between different individuals and species, and (2) adjustments in behavior through learning.

Amazingly, the population of coyotes responds in inverse proportion to efforts by humans to control it.
Courtesy of Corbis.

The expression of behavioral plasticity is poorly understood because there are a number of possible mechanisms guiding its development. First, specialized genes may be activated in response to an environmental change and increase the plasticity of other traits when they are expressed. Second, plasticity may be the result of many genes, each related to a single environment. Each gene would allow one behavior to be expressed per environment, but will stop being expressed when the environment changes. A new gene will then express a new behavior, which is best adapted to the new environment. There is evidence for both of these mechanisms in the expression of behavioral plasticity. However, it is important to remember that behavioral phenotypes are not the result of genetic differences between individuals or populations.

How Is Behavioral Plasticity Adaptive?

Behavior (self-directed movement) is one of the most plastic categories of traits expressed by animals. Because behavior is the primary method animals use to respond to changes within or between environments, behavioral plasticity allows species to respond quickly and efficiently to stimuli. Behavioral change is useful in a changing environment because changes can be quick and often reversible compared to other phenotypic traits. It is important that behaviors change in an adaptive manner rather than at random because the chances of a random change selecting the right phenotype are low.

Species that experience more change throughout their lives are expected to express greater amounts of behavioral plasticity. For example, species that have generalized diets (as opposed to feeding on one thing) tend to display more plasticity. They also tend to have larger geographic distributions and live in larger social groups (e.g., ravens, primates, and coyotes) compared to species that do not have these traits. Therefore, animals living in changeable environments need changeable behaviors to survive. This would provide them with the opportunity to exploit a wider range of habitats than a more specialized species.

A raven scavenging on a road in Denali National Park carries a road-kill arctic ground squirrel away for food.
© *Alissa Crandall / Corbis.*

The foraging behavior of ravens (*Corvus corax*) demonstrates behavioral plasticity over a wide geographic range. Ravens are the largest member of the bird family *Corvidae*—which also includes crows and jays—and are found throughout Asia, Europe, and North America. Having such a large range means that a local population of ravens is unlikely to have the same habitat (or diet) available as birds in other areas. Therefore, each population needs to adopt tactics suited to its own environment. As expected, ravens adapt a number of different foraging styles depending on their local habitat. In costal areas, ravens eat shellfish and even attack seal pups; in orchards and market gardens they eat fruit; and in farmed areas they will eat carrion, insects, and small lizards. In order to forage in this generalized manner, ravens need to know how to recognize and exploit foraging opportunities for survival. Hence, a jack-of-all-trades may often be master of none. As a result, ravens are not as efficient at exploiting shellfish, for example, as oystercatchers, which specialize on this prey item.

Costs and Benefits of Behavioral Plasticity

The advantages that plasticity confer could theoretically allow species to be unlimited in their plasticity (i.e., express the best behavior in every environment). The differences in plasticity observed between species indicate that there are costs and constraints that limit the expression of plasticity among species. There are several costs to the unrestricted evolution of behavioral plasticity:

1. There may be fitness costs for plastic species. If a species inhabits many environments, it may not adapt well to any of them, and so may not be able to utilize resources as efficiently as a specialist species.

2. Some species may be unable to evolve plasticity because it is too costly. For example, one form of behavioral plasticity is adjusting behavior through learning. However, learning

might require a larger than average brain for its body size to enable the animal to remember all of the required environmental cues and responses. Brain tissue is energetically expensive to grow and maintain, meaning this type of plasticity might only evolve in species with access to rich resources. However, I am not suggesting that plasticity might be more common in vertebrates. Even minute insect brains are able to display high rates of plasticity.

3. Plasticity might be expensive to evolve and maintain. If any of the above costs are greater than the benefits accrued by plasticity, then behavioral plasticity will not evolve because fixed-strategy species will out-compete plastic species in an unchanging environment.

Correlates of Behavioral Plasticity

The main prediction of evolutionary theory regarding behavioral plasticity is that plasticity should increase with increased environmental variability (see raven foraging example above). There is also evidence that behavioral plasticity increases with learning ability (and relative brain size). However, learning is not a prerequisite for behavioral plasticity. Longevity may also increase the need for plasticity. Individuals of long-lived species may have an increased chance of experiencing environmental changes throughout their lives.

The distribution of food in the environment can be an important determinant of social behavior, which can then affect behavioral plasticity. For example, compare the mating systems of pronghorn (*Antilocapra Americana*) and bison (*Bos bison*). Both species have the same geographic ranges and have similar demographic and ecological properties. However, the social mating behaviors of these two species are very different. Male bison will defend only one receptive female at a time, whereas pronghorn will adopt many different tactics including defending harems, defending territories, and pursuing groups of females. These differences come from the fact that bison specialize in feeding on vast grasslands in which it is difficult for males to monopolize females since they can scatter over larger areas. In contrast, ponghorns are more specialized in their foraging needs, and this means that groups aggregate in areas rich in food. This allows dominant pronghorn males to exclude lower ranked males from females and allows them to form harems.

Plasticity is an adaptive response that enables individuals to cope with environmental change. Unfortunately, research on behavioral plasticity has lagged behind that of morphological plasticity. However, there is an increasing amount of research now being conducted on behavioral plasticity. This is partly because biologists are realizing that it is an exciting field of study. In addition, human-induced changes in the environment (e.g., global warming and habitat modification) will also have an increased impact on animal behavior.

See also Development—*Adaptive Behavior and Physical Development*
Development—*Behavioral stages*
Development—*Embryo Behavior*
Development—*Intrauterine Position Effect*
Feeding Behavior—*Social Foraging*
Feeding Behavior—*Social Learning: Food*
Learning—*Evolution of Learning Mechanisms*
Learning—*Social Learning*

Further Resources

Bekoff, M. 2001 *Cunning coyotes: Tireless tricksters, protean predators.* In: *Model Systems in Behavioral Ecology: Integrating Conceptual, Theoretical, and Empirical Approaches.* (By L. A. Dugatkin), pp. 381–407. Princeton, NJ: Princeton University Press.

Heinrich, B. 1999. *Mind of the Raven.* New York: HarperCollins.

Jackson, R. J. & Wilcox, R. S. 1998. *Spider-Eating Spiders.* American Scientist, 86, 350–357.

Komers, P. E. 1997. *Behavioral plasticity in variable environments.* Canadian Journal of Zoology, 75, 161–169.

Craig Barnett

■|Behaviorism

"Animals are little people in fur coats." This phrase appears below the cherubic faces of a dog and a cat on a small refrigerator magnet that professional friends gave me as gag gift a few years ago.

Surely, dogs and cats are not little people in fur coats; nor are quail and goldfish little people clad in feathers and scales. Each species of animal is a distinctive group of organisms with a unique evolutionary history that inhabit a highly particularized ecological niche. Assuming that one species is just like any other appears to be a serious biological error. Indeed, assuming that nonhuman animals are just like people is such a blatant interpretive error that it has been given a special name—anthropomorphism.

The manufacturers of such refrigerator magnets might be dismissed as pandering capitalists who brazenly exploit the animal lovers among us. The problem of anthropomorphism, however, is one that has for nearly a century vexed scientists interested in animal behavior. No less than the Nobel laureate Ivan P. Pavlov once adopted an anthropomorphic approach to understanding the conditioned reflexes that he and his coworkers discovered in their studies of canine digestion.

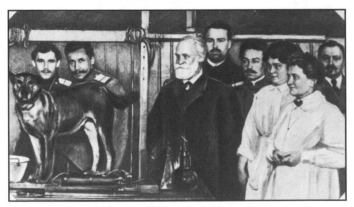

Pavlov and his early experiments with conditioned reflexes in dogs is an excellent example of people's tendency to anthropomorphize and led to a more objective approach to study.
© *Bettmann / Corbis*

The beginning of that story is familiar enough: Pavlov and his Russian colleagues serendipitously observed that hungry dogs not only salivated to food in the mouth, but also to stimuli that were repeatedly paired with food, like the sight of the experimenter entering the room holding a food bowl. The end of the story is also well-known: Pavlov vigorously insisted that natural scientific laws of association formation could be experimentally established that linked—via neural machinery—temporally contiguous stimuli, like the sight of the food bowl with food in the mouth. What is missing from most textbooks is an account of the extraordinary difficulty that Pavlov and his collaborators had in deciding just how to go about investigating and interpreting their groundbreaking observations.

Amusing Tales of Animal Minds

One result of mentalistic musings is entertaining literature. In a recent romp of a read, Boffa (in *You're an Animal, Viskovitz*. New York: Knopf, 2002) projects his own irreverent view of the world into different animal species. His protagonist, Viskovitz, is variously a dormouse, a snail, a parrot, a lion, a chameleon, and many others. As Viskovitz inhabits each of these different species, Boffa has us explore the human condition from a wealth of different perspectives: science, religion, philosophy, politics, and so on.

Infecting many of these metamorphoses is a keen sense of humor, laced with sexual and scatological innuendo. As a young mantis, Viskovitz asks his mother, "What was Daddy like?" Her wry retort, "Crunchy, a bit salty, rich in fiber." As a young dung beetle, Viskovitz learns the true meaning of existence from his father, "The only thing that matters in our life is. . .well, look. . .it's shit." Still other metamorphoses find Viskovitz in a more metaphysical mood. As a dog, he is, "focused in the greatest meditative absorption, close to a cessation of all activities, to a sublimated state of consciousness." And, as an earthworm, Viskovitz has his ego battered by his aloof object of desire. "Ljuba, why don't you love me?" Viskovitz asks. "Because you're a worm, you're vile, you don't have a spine, and you don't have any guts," she responds.

As literature, these stories about Viskovitz are most entertaining tales. But, can they in any useful way serve as scientific accounts?

In the 1928 book chronicling his first 25 years of conditioning research, *Lectures on Conditioned Reflexes*, Pavlov describes this fascinating story as involving two opposite paths to comprehending conditioned reflexes: the anthropomorphic approach and the scientific approach.

According to the anthropomorphic approach, one should be mainly interested in the internal or subjective world of the dog rather than in its overt actions. This anthropomorphic approach assumes that the internal world of the dog—its thoughts, its feelings, its desires (if it has any)—is analogous to ours. Pavlov and his colleagues actually entertained this approach prior to 1903 in order to understand the then-called "psychical" secretions of their dogs to signals for food. Using the anthropomorphic approach, the researchers tried to explain their findings by "fancying the subjective condition" of their dogs. But, all that resulted from these many musings were endless controversies and unverifiable personal opinions (see the sidebar "Amusing Tales of Animal Mind"). This interpretive breakdown forced the researchers to abandon what Pavlov suspected was an inborn inclination for people to adopt an anthropomorphic interpretation and to promote a less familiar, but more productive objective approach. This analytical transition from anthropomorphic interpretation to a natural science approach was not an easy one to make; indeed, Pavlov described the process as involving persistent deliberation and considerable mental conflict.

"Can I see it? Can I measure it? Can I repeat my results?" These three questions were to become a veritable mantra to Pavlov and his students, reminding them of their initial, futile attempts to apply an anthropomorphic interpretation to conditioned reflexes and cautioning them to stay the course of an objective scientific approach. This new natural science approach was one in which the world of reality would replace the world of fantasy. Here, the investigator—as a "pure physiologist"—would focus specifically on the relationship between external stimuli and the overt behaviors of the animal and never again stray into internal, subjective territory.

This then is the interesting story behind the story of Pavlov and conditioned reflexes. But, what exactly has it to do with the school of psychological science known as "behaviorism?"

Three points can be made in answer to this question. First, Pavlov—the pure physiologist—did not actually study the neurophysiological mechanisms of conditioned reflexes. He and his colleagues conducted purely behavioral investigations of what they then called "higher nervous activity" or what we now call *cognition*. Pavlov and his associates did indeed use the concepts and terms of the central nervous system; but, their studies are best considered to have been behavioral, not physiological.

Second, Pavlov was not interested in animal behavior for its own sake; rather, he studied salivation in dogs as a convenient model preparation for understanding the characteristics of complex behavioral adaptation. His aim was to uncover the behavioral and biological laws of learning and cognition—in animals and people. Indeed, Pavlov deemed the ultimate triumph of the human mind to be its determining the laws and mechanisms of human nature.

Third, Pavlov ardently believed that the true path to understanding human nature was not from within, but from without. An introspective approach which focused on an individual's private thoughts, wishes, and sensations was a scientific dead end; it was doomed to result in fruitless argument and useless evidence. An objective approach would instead be necessary, in which scientists would apply to the study of human nature the same methods that had so successfully illuminated other biological processes, such as the circulation of the blood, neural conduction, and hormonal regulation.

The focus on behavior, the use of animal models, and the deployment of natural scientific methods are hallmarks of *behaviorism*. Pavlov's behavioristic program of research was not only cogently conceived, but it was well-established by 1906, several years before the American psychologist John B. Watson is generally credited with having launched the behaviorist movement with his famous 1913 paper, "Psychology as the behaviorist views it."

Also predating Watson's polemical paper was publication of the work of H. S. Jennings, an American zoologist who studied the activities of single-celled animals. In his 1906 book, *Behavior of the Lower Organisms*, Jennings outlined a coherent behavioristic approach to the study of adaptive action. Many of his points reinforce those of Pavlov. Still others more fully explain and justify an objective approach to the study of behavior.

As did Pavlov, Jennings took a decidedly cautious stance toward the private world of thoughts and feelings. He recognized that the conscious aspect of behavior is of greatest interest to many theorists; but, he saw no way to delve into this private world with the natural scientific methods of observation and experiment. Therefore, he suggested that any claims about consciousness in animals may be inherently unverifiable. Jennings and later behaviorists do not deny the existence or possible importance of consciousness; instead, they simply claim that private data cannot be the subject of public science.

Jennings was more concerned with explicitly comparing the behavior of humans and animals than was Pavlov. Here, Jennings noted a glaring inconsistency. The behavior of animals is often described in objective terms, whereas the behavior of humans is often described in subjective terms; these very different kinds of descriptions give the impression of a complete discontinuity between human and animal behavior. Jennings believed that this view of the seventeenth century philosopher René Descartes was stale and erroneous; a fresh and proper twentieth century answer to the question of whether humans differed fundamentally from all other animals required examining their behavior from a common objective vantage point. Only by comparing such "objective factors" did Jennings believe that we can determine whether there is a continuity or a gulf between the behavior of animals and humans.

Based on that objective evidence, Jennings agreed with Charles Darwin and his theory of evolution by natural selection that there is no difference in kind, but complete continuity

Concept Learning in Pigeons

Edward A. Wasserman

People learn concepts. We equivalently respond to varied examples of related stimuli. For instance, we learn to use the words cats, flowers, cars, and chairs to denote highly individual examples from these four different conceptual categories. What about animals? Do they too learn concepts?

To find out, my colleagues and I first analyzed the familiar case of an adult teaching a child to name the pictures in a book. In this "name" game, the adult opens the book, points to one of its illustrations, and asks the child, "What is it?" If the child says the correct word, then positive social reinforcement is provided—"good." If the child says the incorrect word, then no reinforcement is provided; instead, the adult may ask the child to try again; and, if this request also fails to produce the correct word, then the adult may have to supply it.

We tried to extract and to apply the essence of the "name" game to pigeons to see if they too can learn to categorize stimuli from four different kinds of objects (Bhatt, Wasserman, Reynolds & Knauss 1988; Experiment 1B). We could not request verbal behavior from the birds, so we asked them to report members of four different categories—cats, flowers, cars, and chairs—by pecking four, corresponding circular keys around a central viewing screen.

On each of 40 daily trials, we showed the pigeons color slides of 10 different examples from each of the four categories. Within each category, the slides differed in the number, size, color, brightness, orientation, location, and context of the object. After 30 pecks to the viewing screen (to insure that the pigeon had seen the slide), each of the four report keys was lit with a different color (to help the pigeon distinguish its response options) and a single (choice) response was permitted. If the response was to the correct key, then the pigeon was given food reinforcement; if it was to any of the three incorrect keys, then no reinforcement was given and a correction trial immediately followed. An individual pigeon might have to peck the top left key in response to pictures of cats, the top right key in response to pictures of flowers, the bottom left key in response to pictures of cars, and the bottom right key in response to pictures of chairs.

Learning was quite orderly. Responding rose from the chance accuracy score of 25% correct to an average of about 80% correct after 30 days of training. Immediately after reaching the 80% level, the pigeons were tested with 10 brand-new snapshots of objects in each of the four categories. Accuracy averaged 81% for the old slides and 64% for the new ones. Thus, the pigeons had learned to categorize a set of highly complex and lifelike stimuli that they had seen 30 times before and still other stimuli that they had never seen before. We believe that this experiment represents a very clear illustration of concept learning in animals.

So, too, does Griffin (2001). But, he criticizes us because we are reluctant to interpret these behavioral results as proof that pigeons think consciously about the categories they learn to distinguish. Our question to him and to you is simple: What do we gain by speculating about the pigeon's conscious awareness?

One remark that Griffin makes in defense of his own anthropomorphic interpretation of our experiment is that is this kind of account is now "back in style." But, scientific theories are not like bellbottom trousers—drifting in and out of fashion. We need much firmer grounds for deciding between rival theoretical accounts. Predicting and controlling behavior seems to be a more worthwhile aim than conforming with popular fashion, no?

between the behavior of lower and higher animals—including human beings. Indeed, many years of research convinced Jennings that, if amoeba were a large animal and its actions were to come within the everyday experience of human beings, then its behavior would readily lead people to attribute to it states of pleasure and pain, hunger, desire, and so on.

Jennings' appeal for us to limit our consideration of both human and animal behavior to objective factors underscores the key imperative of behaviorism: to explain behavior in terms of matter and energy, thereby rendering unnecessary any psychical or mental implications. Mentalism was to play no part in this new behavioral science of the twentieth century, a science which remains prominent to this day.

Nevertheless, mentalism has staged a surprising comeback in the form of "cognitive ethology," a field founded by the biologist Donald R. Griffin. The goal of *cognitive ethology* is "to learn as much as possible about the likelihood that nonhuman animals have mental experiences, and insofar as these do occur, what they entail and how they affect the animals' behavior, welfare, and biological fitness" (Griffin 1978, p. 528). To Pavlov, Jennings, and Watson, this goal of studying animal consciousness falls outside the scope of a scientific psychology that has struggled for a century to avoid such analyses of subjective experience.

This contest between subjective and objective analyses of behavior is obviously an important one that has yet to be decided. The second century of behaviorism will have to prove to its many opponents that a natural science account of animal behavior and intelligence eclipses subjective and mentalistic interpretations.

One area where this contest will surely be waged is the study of more advanced forms of animal cognition, like memory and conceptualization. It might be easy to dismiss salivary conditioning as a rather mindless form of association formation and to grant behaviorists this narrow realm of behavioral adaptation. But, it is not going to be so easy to dismiss an objective analysis of abstract conceptual behavior. The battle is joined. The resolution will have important implications for our understanding of animal and human behavior.

See also Anthropomorphism
        Cognition—*Animal Consciousness*
        Cognition—*Cognitive Ethology: The Comparative Study of Animal Minds*
        Cognition—*Social Cognition in Primates and Other Animals*
        Cognition—*Theory of Mind*
        Comparative Psychology

Further Resources

Bhatt, R. S., Wasserman, E. A., Reynolds, W. F., Jr. & Knauss, K. S. 1988. *Conceptual behavior in pigeons: Categorization of both familiar and novel examples from four classes of natural and artificial stimuli.* Journal of Experimental Psychology: Animal Behavior Processes, 14, 219–234.

Blumberg, M. S. & Wasserman, E. A. 1995. *Animal mind and the argument from design.* American Psychologist, 50, 133–144.

Griffin, D. R. 1978. *Prospects for a cognitive ethology.* Behavioral and Brain Sciences, 4, 527–538.

Griffin, D. R. 2001. *Animal Minds.* Chicago: University of Chicago Press.

Jennings, H. S. 1906/1976. *Behavior of the Lower Organisms.* Bloomington: Indiana University Press.

Pavlov, I. P. 1928. *Lectures on Conditioned Reflexes.* New York: International.

Ristau, C. A. (Ed.) 1991. *Cognitive Ethology: The Minds of Other Animals*. Hillsdale, NJ: Erlbaum.

Roberts, W. A. 1998. *Principles of Animal Cognition*. New York: McGraw-Hill.

Wasserman, E. A. 1993. *Comparative cognition: Beginning the second century of the study of animal intelligence*. Psychological Bulletin, 113, 211–228.

Watson, J. B. B. 1913. *Psychology as the Behaviorist Views It*. Psychological Review, 20, 158–177.

Edward A. Wasserman

■ Burrowing Behavior

What Are Burrows For?

Animals build structures for essentially one of three reasons: to live in, to catch prey or as communication devices. By far the most common reason is as somewhere to live; this is true of burrowers as well, although for them we can add a fourth function, burrowing in search of food.

Homes may be differentiated structures, with different parts doing different things. It is common in mammal burrows, for example, for narrow tunnels to connect wider chambers that are used for sleeping or caring for the young.

Burrows may contain other features as well, for example a grain store. Some species of desert-dwelling rodents and ants feed on seeds, but the occasional rains result in no more than one crop a year. Kangaroo rats (*Dipodomys* species) and some ants of the family Myrmecinae overcome this problem by storing seeds in underground granaries. Leaf cutter ants (*Atta*) in tropical America, and African termites of the genus *Macrotermes*, excavate enormous quantities of soil to create their underground nests, which include gardens in which they cultivate the fungus on which they live using collected plant material as compost.

Constructed traps are rare among animals, and nearly always consist of nets made of a material secreted by the builder itself, a spider web being a good example. No burrow system appears to be designed to clearly act as a trap. A burrow that assists communication is, however, known and quite well understood. The singing burrow of male mole crickets (*Gryllotalpa*) consists of a bulb-shaped chamber, a constriction, and paired horns opening at the ground surface. The male positions himself in the constriction, facing into the bulb. The dimensions of these burrow features are such that the frequency of the sound he produces by rubbing his wings together resonates in synchrony in the bulb and the horns, very efficiently broadcasting his presence to nearby females.

Which Animals Burrow?

Burrowing animals are generally small. There may be some energetic advantage to smaller burrowers; however, burrows of small diameter are less in danger of collapse than larger ones, and larger animals have less need of burrows because they have a smaller range of organisms that prey upon them. The majority of burrows are therefore dug by invertebrates and, in particular, by *arthropods*: crustacea in aquatic habitats, and insects in terrestrial ones. Among the vertebrates a few species making simple burrows are found among the fish, amphibia, reptiles, and birds, but it is only the smaller mammals that show any number or variety of burrowing species.

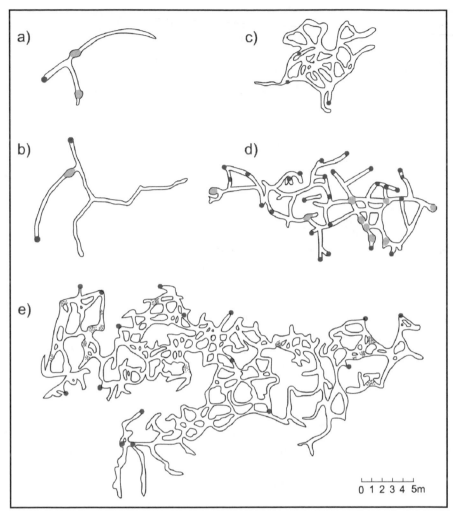

Large burrow system of the European badger (Meles meles). This may be occupied by no more than four or five badgers, but bedding material (stippled areas) identifies at least 15 sleeping places. Dark circles mark the presence of 16 entrances.
Courtesy of MAFF.

Among the insects, the solitary bees and wasps illustrate simple burrows excavated to raise their larvae. The subterranean nests of leaf cutter ants (*Atta*) illustrate the other extreme of burrow complexity in the insect order Hymenoptera.

Leaf miners are insects that make foraging burrows; however, not in the ground but inside a leaf. They mine the tissue between its upper and lower surfaces. This way of life is found in the larvae of some sawflies (Hymenoptera), flies (Diptera), beetles (Coleoptera), and moths (Lepidoptera). Life inside the almost two-dimensional world of the leaf probably provides some security and abundant food, but limits these insects to a very small size.

The mammal burrowers comprise species such as deermice and oldfield mice (Peromyscus) that dig simple burrow systems just as secure homes, and species that search for their food by burrowing. The latter includes vegetarian rodents such as mole rats and insectivorous moles, that make complex, extensive foraging burrows and are highly modified for perpetual subterranean existence.

Nature and Nurture

The burrowing patterns made by leaf miners are varied between, but often consistent within species, suggesting strong genetic control over the behavior. We have very little experimental evidence of the extent of the contribution of genes or experience to the development of burrowing behavior. However we do have some insight into genetic influences on burrowing in oldfield mice (*Peromyscus polionotus*) with deermice (*P. amniculatus*), from experimental hybrids between the two. The latter dig a simple chamber at the end of a short burrow, whereas the former have a burrow system where more than one exit tunnel connects to the nest chamber.

Both species of mice dig species-typical burrows, even after 20 generations reared in laboratory cages. Hybrids between them dig burrows characteristic of oldfield mice. These F1 hybrids, back-crossed to the recessive deermice, produce offspring with burrows showing varying mixtures of the characters of the two species, indicating determination by a number of gene loci. None of this, of course, excludes the possibility that experience plays a part in the development of the burrowing behavior.

Costs

The evident benefits of burrowing (security against predators and climatic extremes) must be set against its costs. Of particular significance is the energetic cost of digging compared with travelling the same distance above ground; this has been calculated at between 360 to 3,400 times greater for a terrestrial species, depending on the hardness of the soil. This has some important consequences. One of these is that burrow owners will not readily abandon their valuable homes, and so are dependent in finding all they need (food, water, etc.) near to their home burrow—so-called central place foragers.

The costs of burrowing can be minimized by creating a burrow that is barely wider than the burrower itself. This, however, may create problems of limited oxygen supply. Where the burrowing substrate is sufficiently permeable to air or water movement, this may not be a problem, but some species do ventilate their burrows, either by fanning or pumping behavior (suitable for water but not air), or by ventilation currents induced by the burrow architecture.

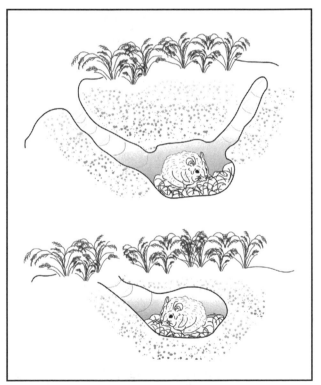

*The burrow system of oldfield mice (*Peromyscus polionotus*) (above) typically has more than one exit tunnel from the nest chamber to the surface, while that of deermice (*P. maniculatus*) (below) has a single short tunnel. Hybridization experiments demonstrate a genetic basis for these differences.*
Adapted from Hoffman. 1994. Animal Architecture. *Nature, 585, 52–85*

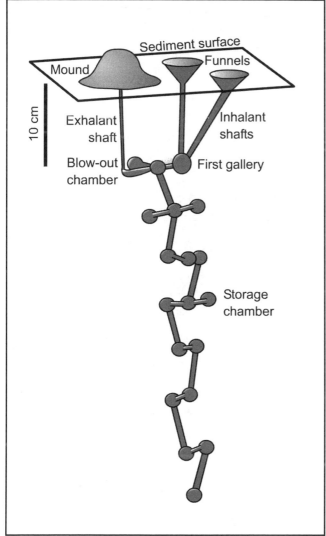

A diagrammatic representation of the burrow system of the mud shrimp Callianassa truncata *shows a complex arrangement of tunnels and chambers that penetrate deep into the sediment. Ventilation of the system is assisted by induced flow of water entering through the funnels and leaving through the mound.*

Adapted from Zeibis, W. et al. 1996. Complex burrows of the mud shrimp . . . Nature, 382, 619–622.

The marine polychaete worm, *Chaetopterus* pumps water through the U-shaped tube, which it excavates in marine mud, using fan-like modifications to some of its segments. This allows it to obtain oxygen, and to feed by filtering particles from the water with a fine mucous net.

Induced current flow is illustrated by the burrow design of the prairie dog, which exploits the pressure difference between two apertures linked by a tunnel. The prairie dogs create this difference by building a small mound around one burrow entrance, whereas the other is at normal ground level. This results in air moving faster over the top of the mound entrance, partly because it is raised a little above the slow moving boundary layer at ground level, and partly because the mound causes the air making contact with it to speed up in order to pass over it. The overall result is a lowering of the air pressure over the raised entrance, drawing air into the lower entrance. The same principle is used by aquatic burrow dwellers, such as the marine goby *Valencienne longipinnis* and the burrow of the mud shrimp (*Callianassa* species).

The energetic investment in creating a burrow system makes burrowers reluctant to pay the price of moving; however, the price of staying may be an accumulation of nest parasites. The complex burrow systems seen in the badger (*Meles meles*) and in Brants' whistling rat, both of which contain several sleeping chambers, are apparently in part an adaptation to controlling nest ectoparasites. A badger will change its sleeping chamber within the burrow every two or three days, but if treated to remove fleas, ticks and lice, will change places less often. A similar behavior change is shown by a treated Brants' whistling rat.

Digging Behavior and Anatomy

To dig a burrow, an animal needs to be able to do two things: First, dislodge the material from the workface, and second, remove the loosened material from within the burrow. It is a surprising feature of animal building behavior generally that, although quite complicated structures are built, it is the animal's behavior rather than its anatomy that is specialized for the job. It is impossible for example, to tell from looking at a bird what sort of nest it builds, or whether it builds a nest at all. The exception to this rule is animals that construct burrows. They often have very obvious adaptations to burrowing: powerful instruments for digging and sometimes also special anatomy for moving the dislodged spoil.

A number of orders of aquatic insects create burrows in the sediment. In some of these there is evidence of anatomy adapted for digging and/or spoil removal. In the mayflies (Ephemeroptera) comparison between specialist burrowing species and ones that cling to stones in the water current shows massive digging apparatus associated with burrowing. The mayfly larva *Pentagenia vittigera* is a digger in compacted clay. It shows quite stereotyped and repeated digging movements, using upward sweeps of its short, serrated tusks, as the main instrument to dislodge the compacted clay. Its broad front pair of legs push the loosened substrate back, at the same time widening the burrow with the spurs and stout bristles located on the forelegs and sweeping the loosened material backwards.

The fossorial rodents, (i.e., those species that live and feed entirely underground) are found in at least six families of rodents: Between them they have evolved two different specialist digging tools, the front limbs and the incisors. To remove excavated soil, rodents use a variety of techniques: with the hindfeet (e.g. *Heliophobius*, Bathyergidae), by turning round and pushing with the head and forefeet (*Tachoryctes*, Rhizomyidae), or by bulldozing with the head alone (*Spalax*).

In specialist limb-digging species, the front feet, forelimbs and pectoral girdle show substantial modification for powerful scoops into the soil at the workface. *Arvicola* (Arvicolidae), and *Heliophobius* both excavate with bites of their enlarged incisor teeth, moving spoil back with simultaneous kicks of their back feet.

Burrowers Change Habitats

Burrowers, by penetrating into the ground or aquatic sediments, may displace large amounts of material which can substantially alter the chemistry and consequently the biology of the habitats where they are active. Where marine burrowers make spoil heaps on the mud surface, estimates have been made based on observed burrow densities of the amount of sediment that is being disturbed over time. It has been calculated that a marine sediment burrower, the mud shrimp (*Callianassa subterranean*), alone can displace 11 kg (24 lb) (dry weight) of sediment m^{-2}, $year^{-1}$. Another mud shrimp species (*C. truncata*) although only 20 mm (.79 in) long, creates a burrow system that penetrates about 500 mm (20 in) down. Water circulates through the burrow system, partly because of the ventilation behavior of the mud shrimp and partly induced by the entrance designs. The result is to increase measurably the oxygen level almost to the bottom of the burrow. With a density of 120 mud shrimp mounds per m^2, it is apparent that these burrowers can alter this habitat substantially.

Pocket gophers (*Thomomys*) are fossorial rodents, which can also occur at high densities. The amounts of soil they move can also be very large, ranging from 3.4 up to 57.4 m^3 ha^{-1} yr^{-2}. Burrows can underlay 7.5% of the ground surface and the mounds created by spoil removal cover 5–8% of the area. These rodents are territorial, resulting in an even distribution of mounds and their associated burrow systems. The result on the surface is a mosaic of patches at a density of hundreds per hectare. The shifting pattern of soil disturbance over time maintains ecological diversity in the habitat through repeated plant colonization and succession.

This pattern of grassland plains covered by rather evenly spaced mounds, generally several meters wide and 1 m (3.3 ft) or more high is found in various parts of the world and referred to as *mima prairie*. Examples are found in North and South America, and

Africa and, although climatic explanations for their occurrence have been put forward, such as frost action or water erosion, the evidence generally has supported burrowing activity of rodents or termites as the cause.

Banner-tailed kangaroo rats (*Dipodomys spectabilis*) are burrowing rather than fossorial rodents, so competition between neighbors for food takes place above ground. The burrow system below ground is, however, quite substantial and spoil from it leads to the accumulation of a mound 2–3 m (6.5–10 ft) wide and over 40 cm (16 in) high within 2 years, while the territorial behavior leads to their even distribution.

Similar landscape effects have been shown to result from the activities of termites, alone or in combination with the activities of small mammals. Mounds with a mean diameter of 28 m (92 ft) in Cape Province, South Africa, have been shown to have a distribution that is quite regular. It has been suggested that colonies of the soil termite *Odontotermes*, through their activities over long periods of time, are responsible for some very large landscape features in southern Africa. These are bands of ridges and gullies which, although only about 2 m (6.5 ft) high, are up to 1 km (.6 mi) long.

See also Communication—Tactile—*Communication in Subterranean Animals*

Further Resources

Hansell, M. H. 1984. *Animal Architecture and Building Behaviour*. London: Longman.
Hansell, M. H. 2000. *Bird Nests and Construction Behaviour*. Cambridge: Cambridge University Press.
Kinlaw, A. 1999. *A review of burrowing by semi-fossorial vertebrates in arid environments*. Journal of Arid Environments, 41, 127–145. (Online at: www.idealibrary.com.on)
Zeibis, W., Forster, M., Huettel, M. & Jorgensen, B. B. 1996. *Complex burrows of the mud shrimp* Callianassa truncata *and their geochemical impact in the sea bed*. Nature, 382, 619–622.

Mike Hansell

■ Careers
Animal Behavior and the Law

We typically learn early in life about some animals' behaviors—the family dog's lick or the cat's purring, a wild bird's caution, a lizard's ability to remain motionless. It is only later, as we grow older, that we are introduced to the ways in which our human engagement with nonhuman animals is fraught with complex images of these animals, including some of the special social rules we have about them which we often make into official laws of one kind or another.

This order of learning—behavior first, rules later—might be used to suggest that our laws must surely reflect accumulated wisdom about animal behavior. But in only some societies is this true. In most indigenous peoples' societies, for example, adults continue to be very knowledgeable about and respectful of nonhuman animals' abilities. Strangely, though, laws in modern industrialized countries fail miserably to reflect much of what we knew as children, namely, that many nonhuman animals are, like us, sentient and complex individuals with personalities, emotions, and the ability to suffer. Instead, under modern legal systems of many countries (such as the United States, Japan, Canada, France, and Australia), dogs, cats, dolphins, elephants, chimpanzees, and all other nonhuman animals are isolated into the single category "legal things," while each and every human is placed in a separate category known as "legal persons."

The tension between these two notions—nonhuman animals as mere "legal things" without any legally recognized interests *versus* nonhuman animals, such as a chimpanzee or a dog, as complex individuals exhibiting specific behaviors that reflect sentience and unique interests—is what motivated me to look into the history of laws and our general cultural thinking about our cousin animals. Why, I asked, don't lawyers, judges, and legal philosophers notice what any careful, humble observer notices about the actual behavior of the animals around us? This is not only a legal question, but also an ethical question—many philosophers have noted that our ethics will be determined by the beings we are trained to notice and take seriously. "Training" of this kind can come in the form of religion, consumerism, education, or perhaps just our parents' way of talking about other living beings.

The bottom line today is that most legal systems take *only* human individuals seriously, thereby advancing a serious historical prejudice rather than an ethical vision of compassion that is tied to our fellow beings' actual lives. Because of this inherited prejudice, today's legal systems are surprisingly blind to nonhuman animals' behaviors—that is, the *actual* realities of their day-to-day lives—except in some minor ways (such as whether a dog has bitten someone before). The most basic facts about nonhuman animals—their intelligence, emotions, and sentience—remain, in the law's eyes, unimportant when it comes to determining any nonhuman animal's legal status.

Despite our current legal system's ignorance-driven arrogance about the lives of nonhuman animals, some lawyers today seek to have both science and ethics work together to guide the evolution of law. This is one of the principal themes in the "animal law" classes of the kind that I and others have taught at Harvard Law School, Yale Law School, and Boston College. Traditionally, law school education trained law students *not* to notice or take seriously animal behaviors. No wonder, then, that today's lawyers, judges, and legislators see them as mere legal things. Those of us who are lawyers and who care about animals' actual lives learned about the significance of animal behavior, not from our teachers in law school, but, instead, from the nonhuman animals we observed in the world—for me, these included bottlenose dolphins, gray and humpback whales, deer in the local woods, coyotes, and dogs and cats that were integral members of families, including my own. These nonhuman individuals trained me to ask the questions that drive the courses that I now teach—law and animals, religion and animals, and religion and ecology.

Pursuing such a common sense approach has taught me that, as Voltaire once said, "If we believe absurdities, we will commit atrocities." Today's legal systems are, in fact, premised on an absurdity—that the legal standing of a chimpanzee or any other nonhuman animal is not affected by that individual's scientifically verifiable abilities. Today's legal system is thus responsible for many atrocities. Tomorrow's legal systems can avoid such problems, and instead develop justice for all animals, human and nonhuman alike, if they are premised on a more realistic, ethical, and scientifically informed view—that other animals' realities should guide us in how we treat them.

It is law students who have pushed many law schools to offer animal law courses. It will likely be students who prompt other parts of our education system—high schools, colleges, other professional schools—to offer realistic engagements with all animals and their day-to-day realities. We, as human animals, clearly have the capacity to engage such realities. We now need to make modern legal systems, as well all levels of education, recall the basic lessons of our own childhoods—humans are animals, we are related to and can have special relationships with other animals, and we, as fellow earth creatures, have a special ability to notice and take other animals seriously. This is a closeness that so many children readily teach us today.

Further Resources

Animal Law. A journal published annually by the Northwestern School of Law of Lewis & Clark College, Portland, OR.

Curnutt, J. 2001. *Animals and the Law*. Santa Barbara: ABC-CLIO.

Francione, G. L. 1995. *Animals, Property, and the Law*, Philadelphia: Temple University Press.

Waldau, P. 2001. *The Specter of Speciesism: Buddhist and Christian Views of Animals*. New York: Oxford University Press.

Waldau, P. 2001. *Will the heavens fall? De-radicalizing the precedent-breaking decision*. Animal Law, 7, 75–118.

Wise, S. M. 2000. *Rattling the Cage: Toward Legal Rights for Animals*, Cambridge, MA: Merloyd Lawrence/Perseus.

Paul Waldau

■ Careers
Animal Tracking and Animal Behavior

Wild animals are often elusive. From tracker/hunters to modern scientists much of what we learn about secretive creatures results from our ability to "read" the sign left when the animal passed. Different societies call reading sign either tracking or trailing. Whatever the name used, this ancient skill was the foundation of modern science.

The first trackers were naturalists who named the first footprint, tested the first hypothesis, such as "I'll wait at the water hole and kill supper"; or developed the first correlations, "little foot, little animal; big foot, big animal"; and learned animal behaviors. As naturalists, the first trackers were the first scientists, although they did not call themselves scientists—they simply survived! The first hypothesis was tested, not with the goal of receiving a monetary raise or obtaining tenure, but with the need to continue the gene pool by feeding one's children. Some trackers were good. But the genes of the others are not present in the gene pool today.

Today, for most of us, tracking is not an everyday necessity guaranteeing our survival, but tracking still plays an important role in the success of scientists. I track to learn about animals, their identities, what they are doing, and their interactions with their environment—in short, to obtain the information we need for the sciences of behavior and ecology.

The tracking pyramid forms the perceptual basis for knowledge acquisition. The corners of the pyramid are track identification, sign identification, and gait identification. Knowledge gained at the foundation leads to the apex: story interpretation—the final version of what a given trail is telling the tracker.

For example, foot anatomy reveals the identity of the track maker. The stride (distance from a point where a foot touches the ground to where the same point on the same foot touches the ground again) shows the body size of a walking animal. The pattern of tracks on the ground reveals gaits other than walking, and in gaits such as trots, lopes, and gallops the stride now reveals relative speed. Taken with other signs, gait patterns explain animal behaviors.

The tracker might read the typical hunting sequence for a wolf pursuing an elk like this: First, along the walking trail of a wolf, the stride averages 90 cm (35 in) indicating that the mechanical distance from the hip to shoulder rotation points (that portion of the body directly responsible for movement) is 90 cm (35 in), big enough for an adult male.

Next, the stride shortens with the hind foot striking the ground posterior to the front footprint indicating the animal has slowed considerably. Nearby elk tracks reveal the wolf slowed to stalk an elk.

A burst of speed is revealed by groups of four footprints (two front, two hind) with a stride that measures about 600 cm (20 ft). The wolf has broken into full gallop and its trail converges with the trail of an elk. Scuffed ground and blood reveal a "take-down." A short distance away a pile of bones and hair bodes well for the wolves. Today, the wolves are well fed—tomorrow the elk may get away.

I often detect reproductive behavior by following the trails of animals. Footprints in the late January snow reveal a pair of wolves. While the trail passes on one side of a leafless deciduous shrub, yellow snow on the other side smells of urine.

Careful examination reveals two sets of three footprints melted deeper into the snow. In the melted footprints, each of the wolves stood longer than in the rest of the footprints made while walking. Three footprints indicates the wolves raised their hind legs to urinate.

The urine stain associated with one set of footprints is directly to the side of the melted hind foot indicating a female. The urine stain associated with the other set of three tracks is forward and out to the side of the melted hind foot, a male urinated here. In wolf packs only the top male and female wolves raise their legs to urinate, therefore this is the trail of an alpha pair.

While the urine stain appears to be on the other side of the shrub, its position reveals that the wolves were trying to urinate high on the shrub, but the shrub did not intersect much of the urine. The purpose of the attempt to place urine high is to allow the wind to waft the odor around advertising, "This is my territory."

Careful examination of the urine stains reveals a few small drops of blood in the stain made by the female. The female wolf may be in *estrus* (reproductively ready for mating), or she might be injured, perhaps by the kick of an elk. Further tracking to additional scent marks reveals the female does not have blood in every urine stain indicating early estrus and supporting the hypothesis that she is about ready to mate. A wounded female would probably have blood in every urine mark.

One full-moon night in January at West Thumb, Yellowstone National Park, I watched from a knoll. Looking across Yellowstone Lake, I spotted a coyote heading for my tent camp. Later I returned to my camp to discover my metal cooking pots, which I had carefully washed after supper and set by my tent, were gone.

Perplexed and gazing at the snow in the moon light, I spotted coyote tracks. Earlier I had given little thought to the night roamer approaching my tent, but evidence now suggested coyote mischief. The light of the moon was adequate to trail the canid out on to the lake ice.

As I followed the trail, it became obvious that the all the front feet were displaced from the trail to the left side. Quadrupedal animals have their eyes placed further to the side than humans. The side placement allows for greater peripheral vision and by turning the body slightly to the side, the animal can see where it is going and who is following it.

However, the frost forming in the tracks revealed that the coyote had left our camp soon after we did and, because of the time lapse, there was no reason to be looking behind at a pursuer. Why had it used a side gait? Then, I realized that holding my cooking pot in its mouth would interfere with its front feet. Turning to the side would allow the coyote to carry my pot with ease. My case against the coyote strengthened.

About a half mile out on the ice, I spotted my pot. Picking up my pot, I cussed the little mischief maker and went back for a good night's sleep.

The next morning, while cooking breakfast and reflecting on my night visitor, I smelled something acrid—urine, hot urine. The smell emanated from my oatmeal pot. The coyote had left another sign, a scent mark! It was the coyote's way of saying I was in his territory.

The link between tracking and behavior is strong. Knowing behavior helps reveal the stories animals write with their tracks and, conversely, following a trail reveals behavioral interactions of the track maker. From social interactions to reproduction to hunting to play, the trail holds a record available after the passing of the animal. The record would be lost were it not for the tracker, naturalist, behavioralist, ecologist, scientist. The record revealed by the trail may even persist millions of years as revealed by dinosaur tracks which tell about dinosaur behavior during a few minutes in the life of a creature that none have ever seen.

Tracking is a powerful technique, born of ancient times but which, as a modern science, still captures our imagination and provides scientific knowledge. Tracking is easy and takes but just a pencil, paper, and a ruler to become a tracker. Prepare ahead of time with a book or video, take good field notes, and soon animals will be revealing their stories to you.

Further Resources

Elbroch, M. 2003. *Mammal Tracks and Sign: A Guide to North American Species*. Mechanicsburg, PA: Stackpole Books.

Forest, L. R. 1988. *Field Guide to Tracking Animals in Snow*. Mechanicsburg, PA: Stackpole Books.

Halfpenny, J. C. 1986. *A Field Guide to Mammal Tracking in North America*. Boulder, CO: Johnson Publishing Company.

Halfpenny, J. C. 1997. *Tracking: Mastering the Basics*. (Video.) A Naturalist's World, www.tracknature.com, PO Box 989, Gardiner, MT 59030.

Halfpenny, J. C. 1998. *Scats and Tracks of the Rocky Mountains*. Helena, MT: Falcon Publishing Inc.

Muric, O. 1952. *A Field Guide to Animal Tracks*. Peterson Field Guide Series, no. 9. Boston: Houghton Mifflin Company.

Rezendes, P. 1992. *Tracking and the Art of Seeing: How to Read Animal Tracks and Sign*. Charlotte, VT: Camden House Publishing, Inc.

Seton, E. T. 1958. *Animal Tracks and Hunter Signs*. NY: Doubleday & Company. (Also available as the First Laurentian Library edition 1978. Macmillan Company of Canada, 70 Bond Street, Toronto, Ontario M5B 1X3.)

Jim Halfpenny

■ Careers
Animal-Assisted Psychotherapy

Sam spit on the ground again and as he did the young male llama walked away keeping a wary eye on the young man who had just spit. Completely oblivious to this movement of the llama walking away, Sam (not his real name), said in a loud, gruff voice, "This book I'm writing, Cheetaman is evil!! He goes after everyone!" Sam starts wildly swinging around the lead rope he is holding, trying to tell me yet another violent episode of his book. The llama stands in the corner of the paddock, and inches his way to the part of the fence that separates him from the male alpaca, who stands on his side of the fence. They touch noses. Sam is completely unaware of the llama and how his own actions are having an impact on the animal.

I ask Sam to describe to me what he sees going on in this paddock. Sam looks around and says, "Whatyda mean?" I repeat the question and he looks around, laughs, and says in an exasperated tone, "We're standin' in here and there's the llama."

"Yes." I say. "What might the llama be aware of right now? Do you think he is happy we are in here?" Sam looks at the llama who is still near his buddy, the alpaca, but facing us.

"I don't know," Sam says as he begins walking over to the llama, who quickly walks away from him. Sam begins to follow the llama, with his hand outstretched as if he had food or a treat in his hand. Raisin, the one year old llama, keeps walking away and Sam, a lanky 13-year-old, begins to slowly chase the llama around the paddock at a fast walk, attempting to touch it.

"What's happening now?" I ask.

Sam, who is basically friendless, continues to follow the llama around with one outstretched hand and the other in his pocket.

He states, "He won't let me pet him."

Sam runs a little at the llama, stops, and spits, then slowly swaggers over to me. He spits again. The llama moves as far away from Sam as he can. Raisin continuously looks at his buddy, the alpaca, who is on the other side of the fence, wanting to get closer to him for protection. But, to do so, would mean moving closer to Sam.

"Can we work with another animal? This one doesn't like me!"

"Why do you think he doesn't like you," I ask, attempting to see if Sam can see that his own behavior is pushing the llama away.

"Hell, who knows," says Sam and spits again.

Editor's Note: This entry describes how knowledge of animal behavior has helped to change the behavior of a 13-year-old boy who has trouble building relationships with his peers. The author is a clinical psychologist and former livestock manager who now works at a residential treatment center for emotionally disturbed and learning-disabled children.

Llamas have proven to be helpful to humans in many ways, including psychotherapy.
© Corbis.

Sam has very few friends. He is unable to see that his behavior actually pushes people away from him. He is obsessed with violent video games and movies. He has been referred to me because he has experienced some traumatic situations in his home and has been verbally aggressive toward some animals. Sam seems unable to see that how he expresses himself can actually determine whether people move toward him or away from him. We hope that the behavior of the llama, Raisin, included in a clinical session with me, the psychologist, will teach Sam something about himself. Our psychotherapy session is in the llama's paddock with Raisin.

Sam slumps against the fence and picks up a stick of hay and puts it in his mouth, sucking on it.

"You attempt to get closer to the llama and all he does is walk away," I say in a slow and calm way. Sam says nothing and starts to walk toward the llama, again in a fast and purposeful manner, making direct eye contact with the llama. The llama begins to move away quickly, and again Sam begins his slow chase around the paddock. Sam gets angry and scares the llama by lifting up his arms aggressively and shaking them at the llama, who quickly moves away. The llama stops and looks at Sam. They stand looking at each other, neither moving, facing off. They continue to do this when Sam's body begins to relax. Sam looks down and as he does Raisin stretches out his neck toward Sam.

I say to Sam, "Did you just see what Raisin did?"

Sam spits, Raisin moves away and Sam says, "Yea," in a so-what tone of voice.

"He tried to reach toward you a little," I said. "At first you were pushing him away by your energy, then what happened?"

Sam spit and said, "I stopped running after him."

"Yea that's right," I say, "and do you know what you did to let Raisin feel less scared?"

"No." he says. I role model what I saw.

"When you stopped chasing him, and just stood looking at each other, what were you feeling?" I asked.

"I don't know," Sam says. "I wanted to get him to let me touch him."

"Yea," I said, "I saw that the 'I want to touch him' had a lot of AGGRESSIVE I WANT TO TOUCH HIM ENERGY, right?"

"Yea," he says.

"Then I saw you change that," I said. "Did you see it yourself, what you did"?

Sam says, "I just stopped running after him, I was tired. It was going nowhere."

"That's right," I say. "What were you feeling though?" Sam shifts his position and looks at Raisin out of one eye with his head cocked. Raisin is now nibbling hay a short distance away.

"I wasn't angry anymore; I didn't care if I touched him or not."

"So, something as small as just letting go of anger in your body, relaxing a little, allowed Raisin to reach out a little to you! Let's see if Raisin would let you touch him or get close if YOU were more relaxed and didn't have that AGGRESSIVE I WANT TO TOUCH YOU ENERGY."

Sam spits, and acts like he is bored.

"Try not to make eye contact, and see if you can approach him relaxed and at an angle toward his withers (high point of the animal's back)," I say.

Sam starts to walk slowly to Raisin. Raisin stops eating and looks up. Sam continues to walk toward Raisin approaching near his withers and Raisin begins to walk slowly away.

"What is Raisin's behavior telling you now?" I ask.

"He still doesn't like me," Sam says.

"How do you know that?" I ask.

"Because he walks away still."

"OK." I say. "Good read of the behavior; he does walk away. Could it be Raisin might be feeling something else besides not liking you? What did his behavior tell you? Think about how you were walking toward him for most of this hour."

Sam says, "Maybe I scared him?"

"Yea, could be," I say.

Sam seems to become a little interested, challenged maybe. He is beginning to see that he has some control over how the llama behaves around him. Sam again tries to walk toward Raisin who is again nibbling at a pile of hay in the corner of the paddock. Sam walks very slowly, almost nonchalantly toward the back of the llama not making eye contact. Raisin looks up and Sam stops walking.

"Why'd you stop, Sam?"

"I thought it might scare him if I kept walking," Sam replies.

"Good for you!" I say. "Good read on what Raisin might feel or need from you."

Sam has not spit on the ground for at least fifteen minutes. Raisin keeps looking at Sam as Sam approaches slowly. Sam stops and just stands there looking more relaxed. He begins to talk to Raisin like you might talk to your dog.

"Good boy, Raisin!" he says. "Come on, boy; come on." Raisin just stands by his hay. Sam stops about three feet before Raisin, who has not moved away, nor eaten any food. Sam looks up slowly and Raisin slowly reaches out his neck to sniff Sam. I see Sam become excited and he quickly reaches to pet Raisin. Raisin quickly moves his head away, but does not run away.

"AHW!!!" Sam stomps over to me. "See it doesn't work!!! Stupid llama, who cares about this stupid llama anyway!!!!!!" Sam spits and comes over to me. " Come on, let's get out of here, I want to go pet something!!"

"What did you see happen?" I ask. "You did great!"

"Stupid llama wouldn't let me touch him . . . who cares anyway?"

"You do!" I say. "Did he run away from you?"

"No, but he's playing with me, making me look like a fool."

"I saw him feel more comfortable with you. Up until the end, you were doing great," I say. Sam spits.

"He's playin' with my head," he says. "He lets me come close and then won't let me touch him."

"Llama's can't play with your head," I say. "But their behavior can teach us something about ourselves if we really listen to it and understand it. You did great right up to the end. What happened at the end?"

"I thought I could just pet him," he says.

"Yeah, and in the intense energy of 'I want to touch him now,' what happened?"

"He moved away."

"That's right. Try it again and at the end keep the same slow energy you had prior to just touching him."

Sam spits and eyes Raisin, who is eating again. Sam slowly walks over to Raisin. He walks at an angle toward the llama's withers without making eye contact. Raisin looks up. Sam stops. Sam then begins to walk very slowly to Raisin. Raisin lifts his head and lets Sam touch the tip of his nose. Raisin sniffs his hand. Sam just stands there with his hand out. Raisin sniffs again and puts his head down to eat more hay still looking up at Sam while he chews. Sam slowly backs away and then grins at me.

"HEY!!!!!! Good work!!!!!!" I say. "What made it work this time?" I ask.

Sam looks down and says, "I didn't try to scare him."

"Yup," I say, "but what behavior did you have that wasn't scary?"

"I was slow, I didn't make eye contact. I know THAT'S aggressive!" he says.

"Raisin wanted to reach out to you then," I said.

"Yeah . . ."

"Sam, with other kids your age, do you think you come on too strong?" I ask.

"Ya mean, do I try too hard," he says.

"Yeah, what makes it hard for you to make friends, what might you be doing that pushes people away, like it did Raisin in the beginning?"

"I don't know," Sam says. "Maybe I scare them, too."

"How might you do that? What happened with Tom the other day?" I ask.

Sam thinks. "Tom thought I was going to hit him cause I was mad at something."

"Yeah, that's right. I bet if you could deal with your anger better, you wouldn't have kids push you away all the time. Do you see, how, when you changed your behavior with Raisin, he let you touch him? If you can be less aggressive to your peers I bet they wouldn't exclude you so much. Why don't you try it?"

Sam and I have continued to work with Raisin. Sam began to really learn llama behavior and to see what of his human behavior makes Raisin come close or move away from him. In working with the llama over time, Sam is able to translate what he is learning about his own behavior back into his peer relationships.

Further Resources

Green Chimneys Farm and Wildlife Center
http://www.greenchimneys.org/
> Green Chimneys Farm and Wildlife Center in Brewster, New York, is a pioneer in animal assisted therapy and provides innovative services for children and adults. By using farm animals and wildlife-assisted activities, Green Chimneys works to help emotionally injured children reclaim their youth.

North American Riding for the Handicapped/Equine Assisted Mental Health, P.O. Box 33150, Denver, Colorado, 80233, http://www.narha.org

Our Farm, Taylor, Texas: http://www.ourfarmschool.org/

Susan Brooks

Careers
Applied Animal Behavior

The field of applied animal behavior study employs the knowledge of animal behavior that we gain through observation and experimentation for a particular purpose. This purpose may be to benefit humans or the animals. In many circumstances applied animal behavior can be used to mediate conflicts or problems that exist between humans and animals in a way that does not require killing or hurting animals. Applied animal behavior is often the ultimate test of our theories on how the behavior of animals is motivated and maintained. If a theory is robust, and the variables that it identifies are important in the control and management of behavior, we should be able to manipulate those same variables in a real world setting to obtain a desired result. Since the applied animal behaviorist works outside the protective walls and controlled conditions of the laboratory, their interventions need to have an impact that cuts through the *noise* of the many extraneous events and stimuli that the real world presents.

The applied animal behavior field is growing as people seek a more harmonious relationship with animals. It requires a broad knowledge of animal behavior and the ability to translate that knowledge into practical methods that reduce the conflicts that people may have with animals, or to enhance the enjoyment people can have living near or with animals. There are a variety of career fields in applied animal behavior. Applied animal behaviorists may study and research animal behavior, practice as veterinarians who specialize in animal behavior, or work as animal trainers.

Although applied animal behaviorists may have been trained in a particular theoretical tradition, more often than not they are required to combine and apply the theories and approaches to an understanding of animal behavior developed through several different approaches and models. In the early days of modern animal behavior study two primary schools of thought dominated the field. Comparative psychology utilized carefully controlled conditions and stimulus presentations to develop standard models of how animals received, processed and responded to stimuli. Within this context *Learning Theory* built models of how the previous experiences of animals could modify subsequent behavior. *Ethology* emphasized the manner in which members of a species inherited specific behavioral adaptations to a particular environment, providing them with the advantages needed to succeed through natural selection.

Applied animal behavior shares, with experimental and theoretical efforts, the need for careful observation and documentation of behavior, identification of important variables, and the evaluation of behavioral changes following the systematic manipulation of identified variables. Rather than measuring success with a statistically significant test, applied animal behaviorists measure a significant result as the successful resolution of the original problem or concern. Most cases presented begin with the development of a case history that includes information on the animal, including species, age, sex and reproductive status, when the problem behavior first appeared, its frequency and circumstances; and the presence of antecedent stimuli, the apparent target of the behavior and the result.

Once the case history is developed, the applied animal behaviorist can develop a theory of why or how the behavior developed and the manner in which it is maintained, and design an intervention that can reduce or eliminate the problem. At this point the two common approaches to animal behavior can come into play. If the case history suggests that the animal's behavior is the expression of a normal part of its behavioral repertoire, or *ethogram*, but in the wrong place or time, one intervention solution would be to provide the animal with an alternate, appropriate context for the behavior. One example would be a kitten or cat that chases and bites at people's ankles. This is an expression of the normal predatory behavior of felines. In this case, an appropriate intervention would be to provide the cat with a variety of toys to redirect the predatory behavior, and scheduled play sessions to dissipate the cat's predatory play drive.

In other circumstances the case history may suggest that the behavior is the result of previous experience. For example, if a young dog is frightened by a loud, surprising noise, it may develop a life long fear or phobia of loud noises such as thunder or firecrackers. One treatment intervention would be to help the dog "unlearn" its fearful association with loud noises. A common approach is to employ counterconditioning. The dog would be exposed to low volume recordings of thunder in the presence of desirable, positive stimuli such as food or treats. Over a period of successive presentations, the volume of the recording would be slowly increased, all the while providing positive stimuli. In this way, the dog would learn to associate the loud sound not with surprise and fright, but rather with a pleasant experience.

At other times, it may be necessary to combine these two approaches. Dogs often jump on people as a part of their greeting. This is not surprising if you have ever watched two dogs that know one another bounce around in excitement when they get together. A frequent intervention used here is to teach the dog an alternative form of greeting. A dog that has already learned a good "sit–stay" in obedience training can learn to transfer the behavior to greetings at the front door. People can come in the door with treats to reward a "sit" when they come in. Another strategy would be to have the person enter and toss a few treats on the floor when the dog approaches, diverting his exuberant greeting just before he is ready to leap up. The dog will typically keep all four feet on the ground, looking for the treats. In this case, the dog will eventually learn that good things come from staying grounded when someone arrives, not by jumping up.

Behavior problems are the most common reason given for the surrender of companion dogs and cats to animal shelters in the United States. Many animal shelters now provide programs that include behavioral evaluations of animals in the shelter, behavioral enrichment and rehabilitation for the animals during their stay in the shelter, and behavior helpline support for new adopters.

In addition to companion animals, applied animal behaviorists work with farm animals, laboratory animals and wild animals, both in the wild and in captivity. Applied animal behaviorists can have a significant impact on the welfare of animals by providing opportunities to reduce stress that results from the frustration of natural behaviors, or training alternate behaviors that provide mental stimulation and enrichment. Temple Grandin has dramatically reduced the stress experienced by cattle brought to slaughter by modifying the chutes and ramps employed to move the animals from place to place. In zoo settings animals may show stereotyped, repeated behaviors as a result of boredom and limited opportunities to engage in species-typical behaviors. Behaviorists that work at zoos and aquariums have developed a wide variety of protocols to provide animals with opportunities to express species-typical behaviors. Food for primates can be hidden in puzzle boxes or scattered and covered with straw or hay. This allows the primates to engage in their normal

Captive bighorn sheep in a naturalized exhibit. Rocks and sculptured concrete suit the bighorn sheep (Ovis canadensis) *at the Arizona-Sonora Desert Museum in Tucson, an innovator in naturalistic zoo exhibits.*
© *Michael Nichols/National Geographic Image Collection.*

food searching and gathering behaviors. A well-known example used for bears is to place a device into a tree in the enclosure that "leaks" honey at random intervals for the bears to find and lick. A dramatic example of enrichment has been used with cheetahs: A food item is hung by a cable from an overhead pulley and rapidly dragged across their enclosure stimulating a high-speed predatory chase.

Given the broad range of activities that applied animal behaviorists pursue, there is a complementary wide range of educational options. A common strategy would be to pursue an undergraduate degree in biology, psychology or animal science, and then an advanced degree in animal behavior. While few universities have an animal behavior graduate program, most offer graduate degrees in either biology or psychology, and a concentration in animal behavior is possible. As noted above, approaches in applied animal behavior may require the integration of both the biological–ethological and the psychological–learning/conditioning approaches. For this reason, training for applied animal behavior should include coursework and experience in each of these two disciplines. The Animal Behavior Society has a program for the certification of applied animal behaviorists (CAAB). Individuals are required to have either a Ph.D. or Masters degree (for certification as an associate applied animal behaviorist), 5 years of experience in the field, and a record of professional accomplishment and contribution to the practice of applied animal behavior. Additional details on the certification process are available from the Society (*www.animalbehavior.com*).

A behavior specialty has also developed within the practice of veterinary medicine. Following completion of a veterinary degree, it is necessary to complete a 2-year residency under the supervision of a veterinary behaviorist. The American College of Veterinary Behavior requires successful passage of an exam as part of the requirements to become a diplomat of the college (i.e., board certified veterinary behaviorist). In addition to the use of behavior modification and environmental management employed by CAABs, veterinary behaviorists may prescribe drugs to treat some behavior problems in animals. It is not uncommon for CAABs to work in partnership with veterinarians to help animals with behavior problems.

Dog trainers practice a third branch of applied animal behavior. While they may not have advanced degrees in animal behavior, they will generally have a strong knowledge base in dog behavior and learning theory. The educational path for dog trainer is not as formal as that for CAABs or veterinarians. It is not uncommon for dog trainers to serve an apprenticeship with an experienced trainer before working on their own. Continued profes-

Helen Keller with her seeing eye dog. Just one example of many how applied animal behavior is used.
© *Bettmann/CORBIS.*

sional development is available through a wide variety of conferences and training seminars offered through groups such as the Association of Pet Dog Trainers (APDT). The Certification Council for Pet Dog Trainers offers certification for dog trainers following successful passing of an exam.

Dog trainers may work with groups of people who are training their own companion dogs, with individuals and their dogs, or directly with someone's dog. Very experienced trainers may also work with dogs that have specific behavior problems. There are a number of specialty areas such as training for assistance dogs, search and rescue, law enforcement and various competition fields such as agility and obedience.

Further Resources

Carlstead, K. & Shepherdson, D. 2000. *Alleviating stress in zoo animals with environmental enrichment.* In: *The Biology of Animal Stress* (Ed. By G. P. Moberg & J. A. Mench), pp. 337–354. New York: CABI publishing.
Grandin, T. 1995. *Thinking in Pictures.* New York: Doubleday.
Hetts, S. 1999. *Pet Behavior Protocols.* Lakewood, CO: AAHA Press
Wright, J. 1999. *The Dog Who Would Be King.* Pennsylvania: Rodale Press.

Stephen Zawistowski

■ Careers
Careers in Animal Behavior Science

Most of us who are employed in the field of animal behavior have one thing in common—we are all fascinated with and love animals. Many of us also like to ask questions about animals. Why is that animal doing that? How does that animal show that behavior? What is it like to live that way? Additionally many of us weren't sure that we could take our interest in animals and use it to make a career. Luckily for us (and you!) the careers available for people interested in animal behavior are diverse, and these different jobs require different levels of education. One thing common to all careers in animal behavior is the first step—learning as much you can about animals.

The study of animal behavior is undertaken by people in a variety of fields, including biology, chemistry, psychology, anthropology, as well as other areas. People who study animal behavior may be called behavioral ecologists, animal behaviorists, ethologists, animal psychologists, or even applied animal behavior consultants. As you might expect, the salary one can earn in the field of animal behavior is across-the-board, usually correlating with how much formal education someone has. All jobs in animal behavior require substantial knowledge about animals, and a formal education is a good way to get this knowledge. Some jobs in animal behavior only require a Bachelor of Science (B.S.) or a Bachelor of Arts (B.A.) degree. These degrees are usually earned in 4 years of study at a college or university, and many people employed in the field of animal behavior have advanced degrees. Advanced degrees include a Master of Arts or of Science (M.A. or M.S.), Doctorate of Philosophy (Ph.D.) or Doctorate of Veterinary Medicine (D.V.M.).

Below are outlined some careers in animal behavior, based on the working environment. Remember that interest, passion and knowledge are the key to getting a career that you want!

Zoological Parks and Aquariums

Because zoos are likely the first place you have seen animals other than your pets, it is one of the first places to envision a career in animal behavior. There are many different types of jobs available at these places. Over the years, the animal displays and programs seen at zoological parks and aquariums have become technical and specialized. Since the early 1990s many entry-level keeper positions now require a 4-year college degree, with an emphasis on animal science, zoology, marine, or conservation biology. A 4-year degree, coupled with years of on-the-job training, is now a standard requirement for caring for the more exotic animals found at aquariums and zoos.

Curators and researchers usually have Ph.D. or D.V.M. degrees. They also have training in animal husbandry, ecology, or in one of the specific areas such as herpetology or ornithology. A very few places offer curatorial internships, which are designed to provide practical experience. Researchers study the behavior of the animals at a given zoo whereas curators are responsible for collecting, keeping and displaying an array of animals, some of which are rare and endangered. They are also involved in collaborative breeding and animal management programs. Researchers at zoos are primarily concerned with improving the health and well-being of the animals on display. Studies on boredom shown by animals in captivity have resulted in a new area of study, enrichment, and people employed in this field have implemented a wide variety of new approaches for keeping animals active and healthy in these settings. Animals have been given food in a more natural way, and objects have been placed in many exhibits for animals to play with when they are visible to the public, as well as in their off-exhibit areas.

Some zoological parks and aquariums hire people that specialize in education. Educators inform the general public about animal behavior by giving tours, workshops, and lectures, and by designing the educational displays in exhibit areas. Educators hold a range of degrees, from a high school diploma all the way through to a Ph.D. Experience and formal knowledge are important in getting a job in these fields. Sometimes working at a zoo or aquarium may not pay very well, but the pay is steady and the pleasure of working with animals to make their lives in captivity better is rewarding.

Research and/or Teaching at a University, College or K-12

Students interested in animal behavior might also imagine working, like their teachers or professors, at high schools, colleges or universities. There are lots of opportunities for animal behaviorists in academia. Working as a college professor offers both advantages and disadvantages. The advantages include a relatively high salary and a fair amount of freedom and flexibility. Professors teach courses about things they know well, go to conferences, and design and carry out their own studies. Some disadvantages are the length of time to become qualified (usually a Ph.D. takes 5 or 6 more years after a B.A. or B.S. degree), and the high level of competition. (Only the best succeed.) It is often necessary to write grants to gain research money, and there are many other tasks, besides studying the animal behavior, that must be done.

Professors vary the amount of teaching and research that they do, depending on the type of institution where they work. There are many more jobs at colleges that emphasize teaching than there are research positions at larger institutions. The average professor spends most of his/her time in the classroom and working with students during the 9-month school year and as much time as possible working on research during the summer. This is particularly true of field research. Teaching at the college and university level can be very rewarding—you have the brightest people out there, and they want to learn what you have to teach. Teaching your specialty flows naturally into giving students the chance to work with you. Many students become involved in field courses, lab assistantships and data analysis, and you may be the person who starts scientists in careers in animal behavior. However, there never seems to be enough time; professors commonly work at least 50-hour weeks.

Most scientists directly involved in animal behavior work in one of four broad fields: ethology, comparative psychology, behavioral ecology, or anthropology. These disciplines overlap greatly in their goals, interests, and methods. Historically, psychologists and ethologists have been concerned primarily with the regulation and functions of behavior, whereas behavioral ecologists have focused on how behavioral patterns relate to social and environmental conditions. Ethologists and behavioral ecologists usually are trained in biology, zoology, ecology and evolution, entomology, wildlife, or other animal sciences. Most comparative psychologists are trained in psychology. Behaviorists specializing in the study of human behavior are usually trained in anthropology, psychology, or sociology. There are many veterinary schools across the United States that employ animal behaviorists. Animal behaviorists are in a variety of fields in these schools, from training applied behavior students to developing, testing, and supervising the production of drugs and biological products for animal use.

Another area of work in colleges and universities is as technicians or laboratory assistants. Some institutions have full-time technicians, whereas many others use graduate students to do the work as part of their training and to earn money to pay graduate fees.

With an interest in animal behavior, you could become a high school science teacher, most likely biology. For this you would need a wide background in the sciences in order to be able to teach some general science as well. People who teach are usually interested in working with people and like helping others. Teachers need an undergraduate degree in their area and then a year or more of teacher training, and often do courses in summertime. The pay for schoolteachers isn't the highest but the job is a steady one, with the benefit of shaping the lives of young people. Teachers spend a lot of time preparing for their classes and grading student work. Teaching can be a very rewarding career, with the advantage of having weeks off in the summer to pursue hobbies and to spend time with family.

Applied Animal Behaviorists at Zoological Parks, Veterinary Clinics, Hospitals and Animal Shelters

People who work in applied animal behavior use the knowledge of behavior that they gained in training to work out real-life problems with animals. They help with the diagnosis and treatment of behavior problems of companion animals (usually dogs and cats), farm animals, laboratory animals, and animals found in zoological parks. In fact, zoological parks probably employ more applied animal behaviorists than they do researchers. A background in comparative psychology would be very useful here, combining an understanding of nonhuman animals with the study of the reactions of their human companions. Training handlers and owners to work easily with their pets or farm animals, fixing a problem behavior that an animal shows, is a big part of applied animal behavior. Such work might be done in a not-for-profit organization such as the Society for the Prevention of Cruelty to Animals or working independently as a consultant in private practice, or as a specialty veterinarian in a hospital or private practice. Stability and variety of situations to deal with are the advantages of working in an

Zoologist Roland Kays releases a kinkajou from a cage after observation. Kays releases kinkajous back into their habitat after they are fully alert and recovered from being tranquilized.
© Mattias Klum/National Geographic Image Collection.

organization. The person in independent practice must be an entrepreneur, self-organizing, and willing to work long hours.

Both inside and outside of an organization, people can qualify for certification as applied animal behaviorists through several organizations. The American College of Veterinary behaviorist program or the Animal Behavior Society's applied animal behavior specialty programs are the two major academic associations providing certification. The University of Georgia actually has a degree program in applied animal behavior. It takes a long time to achieve the necessary credentials for certification. Usually one must take undergraduate training in biology or psychology, and then either complete a Master's or a Ph.D program in animal behavior or a veterinary degree (DVM). After that, several years of experience in the field and a record of accomplishment and contribution are also required. For more information, see the details at the Animal Behavior Society site (*www.animalbehavior.com*) or the American College of Veterinary Behaviorists (*www.var.vet.uga.edu/behavior/html/ACVB.htm*). (Look also at the Applied Animal Behavior entries in this encyclopedia). The value of a career in animal behavior is the chance to make a difference by helping people and their animals because many owners give up pets because of behavior problems.

Writers in Animal Behavior

Writing is a basic way of communicating with other people, and animal behavior is one of the foundations of our knowledge about animals, so it is not surprising that the two together make an opportunity. Writing about topics such as animal behavior is an important part of many different occupations. Skilled writers write children's books, popular adult and technical works. People interested in animal behavior also contribute articles to magazines such as *Discover, Natural History, Zoo Book,* and *Audubon*. In addition, writers are needed for technical manuals, public relations and media stories, to work in the film industry, and for journals and magazines.

Editors are needed to find people to write books or articles, help them through the writing, get reviewers, and manage the finalization of a publication. They need to be patient, like people, and above all be organized.

Writing about animal behavior requires a broad knowledge, rather than a narrow focus on one or a few problems. It can be a part-time job or one that you do from home, giving flexibility to combine working and raising a family.

Unlike the other areas described here, writing doesn't require a lot of formal education in animal behavior, although it would help. Instead, writers often gain broad knowledge from experience as well as from formal classes (think about what birdwatchers know about birds). Like any other craft, writing depends partly on ability; those who can write well often know it from childhood. But it also relies on practice, and the person who wants to write about animals will have to try, try and try yet again. Writing for any publication that will accept your writing is necessary, so that you can make a portfolio of "clips" or clippings from your published articles. Internships, media fellowships and science journalism training are good ways to get this experience. One problem is that although the public wants a lot of writing about animals, many people want to write about them. The area is full of competition, and you must try every opportunity, juggle several pieces of writing at once, and read everything you can find about your area. As well, although technical writing and public relations positions can be steady jobs, much writing is freelance and does not provide good job security. The freelancer must take every opportunity that comes up.

Running Your Own Business

An interest in animal behavior can lead to a career in a behavior-related business. Think of being an entomologist and growing it into a beekeeping specialty, of taking your love of domestic animals to a job as a veterinarian or veterinarian assistant, of starting in 4-H membership and moving to specialty farming, of learning all about fish and running an aquarium supply store, or even of taking a general interest in animals to a career in television. Perhaps a training ability for pets can grow into a career as an animal trainer for film. One thing most of these jobs require is a different personality than the teaching and helping jobs. You should be a bit enterprising, the kind of outgoing, energetic, and persuasive person who wants to work with people and convince them to help you accomplish your goals. You must be a leader, willing to follow your ideas wherever they lead, including some places you never thought of going.

There isn't a specific educational route for animal behavior entrepreneurial jobs; they are too varied. No specific education, not even a university degree, will get you into one of these specialized niches. However, a good understanding of animals is critical, so the more you know the better. It is also important to be trained in principles and techniques of handling animals (maybe by an applied animal behaviorist) if you are going to work directly with them. If you plan to set yourself up as an independent career person (and small businesses are the fastest growing area of the economy), then you need some training in small business management—either in community college or university programs. The most necessary ingredients are many years of learning and an absorbing interest that means you really know everything about your specialty. Because these are all very different jobs, there's no clear guide to what you will earn. In a big company you'll have a good salary, stable position and clear hours of work. If you set up in business for yourself, you will have much less financial security though much more flexibility to do what you want. Be aware that people who advise small business owners say that the hours are usually very long—but that doing what you want for yourself is its own reward.

Government, Private Research Institutions, and Industry

A growing number of animal behaviorists work in government laboratories or in private business and industry. Many of these jobs involve health-related research. For example, drug companies or government laboratories may hire animal behaviorists to conduct research on the behavioral effects of new drugs, to examine the links between behavior and disease, or to evaluate the well being of animals under their care. Many state and federal government agencies responsible for natural resources management sometimes hire animal behaviorists to work in their wildlife programs. Increasingly, private environmental consulting firms are employing behaviorists to examine the effects of habitat alteration on foraging patterns, spatial dispersion, and reproductive processes in animals. A career in a state or federal agency provides a stable income, but there is a lot of red tape if you want to do something unique or different. The pay scale in governmental jobs is often determined by the number of years you work in a given field or your educational degree. A job in industry often pays more than a government job, but sometimes offers less job stability. Companies that use animals for drug testing often need animal keepers and even trainers for laboratory work or to care for animals.

Examples of Careers in Animal Behavior

Zoo/Aquarium

Roland Anderson is a biologist at the Seattle Aquarium in Seattle, Washington. His job ties in well with his degree in marine biology and with his interests in beach combing and SCUBA diving. He plans, organizes and collects animals for exhibits, does routine maintenance, communicates with the public and with staff at other aquariums, and supervises volunteers. In addition he carries out research, writes scientific publications, as well as contributes to the semi-professional "grey press" on mollusks. He is also active in his local shell club.

University

Jennifer Mather is a professor in the Department of Psychology and Neuro-science at the small (7,000 students) University of Lethbridge in Lethbridge, Alberta, Canada. She teaches two courses a semester and has several indepen-dent study students and a couple of long-distance graduate students in Vienna. She spends a month each summer doing field research in the Caribbean, and also has an interest in the role of women in science.

Writer

Tracey Sanderson did her graduate work in London specializing in insect feed-ing behavior. After a couple of postdoctoral positions she decided she wanted something with more stability and joined Cambridge University Press (CUP) as an editor in 1992. She did a combination of managing projects, working with authors through the writing process, and looking for new authors who might write books for CUP. In 2003, she decided to downsize her commitment so she could spend time with her son and has moved to Derbyshire working 2 days a week out of her home. As well as editing, she plans to write a variety of materials, from books to technical reports, as well as journal articles.

Applied Animal Behavior

Steve Zawistowski is vice president of the American Society for the Preven-tion of Cruelty to Animals, based in New York City. He investigates situations with problem animals, supervises training of pets and pet owners, travels to various branches of the ASPCA to supervise and set up educational programs. He travels to conferences and speaks often about animal welfare issues and is coeditor of *Journal of Applied Animal Welfare*, which publishes articles on these issues. He holds Applied Animal Welfare certification.

Naturalist

Jerry Ligon taught biology for 15 years in Colorado before discovering the island of Bonaire in the Caribbean. He works there for the Sand Dollar Dive Center as a naturalist. He takes the normal shift leading dives to the many dive sites on the island, but has other specialties, too. He has become the island's authority on the birds and regularly takes birders out on expeditions. He's also attempting to document the island's fish fauna and dives twice weekly to conduct a census. He posts an annotated bird list and a list of fish species seen on the web, and consid-ers himself to be a naturalist with a year-round warm one-island laboratory.

Where Can I Get More Information?

For more information about the science of animal behavior, begin at your local public or college library. Many books on animal behavior have been published in recent years. A web search or a librarian can help you locate them. You can also obtain more information from magazines such as *Science News, Discover, Natural History,* and *Audubon* or from the Animal Behavior Society (*www.animalbehavior.com*).

What Can I Do to Further My Interest in Animal Behavior?

As a budding scientist, you might be asking yourself: What things might I do to determine if an animal behavior career is right for me? Here are some suggestions.

Many animal behaviorists started their careers by just observing the animals around them. Keep a journal that charts the arrival of the birds into your neighborhood each spring. You could start by watching a bird feeder. Do certain birds select different seeds? Do certain birds feed only in the morning? Get a stopwatch and record how long it takes for a bird to feed or time between trips to a feeder. Observe a spider making its web. What size insects does it catch? Go to a park and observe the behaviors of squirrels, rabbits or ducks on a lake. If you live in an urban area you can observe how dog owners control their dogs in a dog park.

Buy an aquarium and place fish in it one at a time. Keep track of where a fish spends its time when it is all alone. Now add another fish. How does the first fish respond? Did the first fish change where it hangs out in the tank? Did you see any indications of behavior changing, like raising fins up or chasing the introduced fish around the tank?

There are other things you can do, besides animal-watching, to start your career. Visit nature centers, watch the Discovery channel, or even volunteer at a local animal shelter.

Further Resources

Here is a sampling of a number of well-written and engaging books about people who study animal behavior.

Bekoff, M. 2002. *Minding Animals: Awareness, Emotions, and Heart.* New York: Oxford University Press.
Douglas-Hamilton, I. & Douglas-Hamilton, O. 1975 *Among the Elephants.* New York: Viking Press.
Frisch, K. von 1967. *The Dance Language and Orientation of Bees,* Cambridge, MA: Belknap Press of Harvard University Press.
Goodall, J. 1996. *My Life with the Chimpanzees, Through a Window.* New York: Pocket Books.
Heinrich, B. 1999. *Mind of the Raven.* New York: Cliff Street Books.
Lorenz, K. 1952. *King Solomon's Ring.* New York: Crowell.
Moss, C. 1988. *Elephant Memories.* New York: W. Morrow.
Payne, K. 1998. *Silent Thunder: In the Presence of Elephants.* New York: Simon & Schuster.
Poole, J. 1996. *Coming of Age with Elephants: A Memoir.* New York: Hyperion.
Schaller, G. 1963. *The Mountain Gorilla.* Chicago: University of Chicago Press.
Tinbergen, N. 1953. *The Herring Gull's World.* London: Collins.
Weiner, J. 1994. *The Beak of the Finch.* New York: Knopf.

Mary Crowe & Jennifer A. Mather
On behalf of the Animal Behavior Society Education Committee

■ Careers
Mapping Their Minds: Animals on the Other Side of the Lens

We are well-accustomed to the adage, "A picture speaks a thousand words," and that, "The eyes are a gateway to the soul," but the feet? In the short film produced as a companion piece to my book *Spirit of the Rockies: The Mountain Lions of Jackson Hole*, a cougar gazes down from the boughs of an evergreen tree, shifting her weight tentatively from left to right foot before finally settling onto her haunches. A sharp branch, jutting up out of the main limb, disappears into the animal's tender paw pad. A single thread of drool hangs from her closed mouth. The cat, nearly motionless, sits trapped and treed by a hunter and a pack of dogs just 20 feet below.

In spite of the fact that the film eventually depicts the same cougar being shot from close range as a sitting target, it is this behavior—the expression of the cat's feet—that the vast majority of the audience responds to most viscerally.

Perhaps it is as simple as the empathetic visual of pain to the feet, be they animal or human, impaled on a sharp object. Or it may be what the gesture represents, a symbol of something more than physical discomfort, one that crosses over into the expression of being truly ill-at-ease. Who can't remember standing on stage during that first school play, nerves running wild trying to remember a cue or line, or that uncertain time of childhood, needing to go to the bathroom and not knowing where to go?

Archetypically speaking, it is also the image seared into our imaginations as children hearing Bible stories, the combination of curious, malleable minds, roving eyes fixing on religious statues and relics during Sunday School or church: Jesus nailed to the cross by outstretched arms and crossed feet. This is perhaps the most compelling image of vulnerability, one that crosses over from the spiritual into the literal world of pain. Nailed, we become exposed; exposed, we become vulnerable.

Our feet are what secures our path to this earth, grounds us in the experience of being alive. They are the overlooked warriors of life. Without them, we cannot get around on our own without either the help of others or using a device designed to help us do so. Animals in the wild, on the other hand, do not have this choice. Without their feet, they perish.

Leghold traps designed for canids—coyotes and wolves—achieve just this. By seizing their freedom of mobility, they perish. The end comes into focus, the inevitable closes in.

And with this knowing, the animal mind arrives at a frantic place, compelling it to acts of desperation. Facing entrapment, it chews off its foot at the source, committing its life to perhaps the greatest handicap a wild animal could bear, one of hobbling on odd-numbered limbs, but committing to life nonetheless.

A story of a mountaineer echoes this same innate capacity to comprehend the slow death, the

The expression and emotion captured in this shot of a cougar in the treetops speaks more than a thousand words.
Courtesy of Cara Blessley Lowe

long-suffering anxiety the entrapment lays before the entrapped. Faced with his wrist inextricably wedged in an impossible crack in a rock, the climber instead chooses to sever his own limb with a Leatherman tool, sawing through flesh, bone, blood, vein. The price of freedom.

Where the animal and human mind meet—in the far-flung scientific arena called "animal behavior"—we recognize the baseline of behavior that both share in common. In looking at the other, we perceive ourselves, mirrored. An inescapable fact of this field of study is that any behavior documented of animals' lives is first and foremost informed by our own two eyes, by what and how we elect to see.

Film is such an effective tool for communicating this that it is no wonder entire cable networks have devoted their content to stories of the animal world. And many times, it is through the steadfast eyes of the filmmaker that these animal worlds come into focus. These individuals who devote hours of field time to bring forward the lives of animals to the human world become nothing short of scientists themselves. What is a scientist if not one who has the curiosity to ask the question, the patience to wait for the answer?

These scientists, these cultivators of story, set out with a premise and go and sit quietly in the natural world to uncover what they believe to exist, to be true. In doing so, they open themselves to a previously unimagined miracle. And so it is with animal behaviorists as well.

As with all kinds of research, often times the hypothesis is rewritten when a new or different discovery is revealed. Findings assembled, the story becomes one that the observer had hardly dreamed it could be. In the realm of animal behavior studies as with natural history filmmaking, it is being aware of the potential to perceive the possibility of a new story, or the same story told a new way that cleaves us open, enabling us to receive the stories the animals have to tell.

Animals choose us. In choosing to see them for who they truly are and what they are capable of, we, in turn, choose them. A marriage of vision, a partnership, co-collaborators. Observer and observed join as one in the world of story.

Scientists, filmmakers, photographers, writers bring these stories to the forefront of society. What is required of these recorders of story is a consciousness and subsequent liberation from the leanings that they take with them into the field. Hypotheses, by their very nature, represent a bias—there must be some starting point to rub up against to allow the true patina of story to shine through.

"Kill your darlings," Faulkner's advice for would-be writers, serves as a dark but necessary reminder to grant the grace distance affords animal behaviorists and filmmakers alike from their most beloved, and therefore, most dangerous, ideas. Science in general and animal behavior studies in particular have long been trapped, much like the coyote in the leghold trap, in a static world of absolutes. Several pioneering souls, beginning most prominently with the work of Dr. Jane Goodall, dared to escape the trappings of a field that favored drawing a line in the sand separating man from animal. Under the guise of being "scientific," this separation casts more prejudice than admittedly seeing with human eyes and then seeing beyond what it means to be human, biases and all.

Similar to many religions that declare that God cannot be known but through the vehicle that is dogmatic practice, science has traditionally maintained that nothing can be known until witnessed—often repeatedly, often by many—and proven as truth. Some scientists, in a restrained frenzy to dance around what could be perceived as bias or innaccuracy, often speak and write as though they were the cougar on the limb, unsure of their footing, wary of being blown away or dismissed. Conclusions—the illicit handmaiden to animal behaviorists—are conventionally shunned.

Cougars take to trees or cliffs or high precipices to flee from pursuers. On the historical ecological timeline, this amounted to only wolves, cougars' main and sole archenemy, carnivorous alter-ago, fellow meat-eater, canine predator. The cougar that was filmed in the tree fled from her modern day pursuers; in fleeing, she took flight. Abandoned the earth. What a human may deem, "A leap of faith," the cougar's behavior spoke of the hope that in leaving the ground, she might find safety.

Cougars rarely take to trees, seeking them only as a last resort.
Courtesy of Cara Blessley Lowe

Once in the tree her feet, unsteady in this state of limbo, shifted in nervous anticipation at this place of separation between earth and sky. Unlike leopards who frequently haul killed prey into trees to conceal it from marauding lions and hyenas, cougars rarely take to trees except to seek them as a haven, a last resort. So it is of this that the dance of feet on limbs speaks: the memory of earth and solid ground and freedom and life so immediately a distant memory.

Observation, imagination and personal experience form the bridge between animal behavior and the human mind. The koan of science asks us to penetrate the obscurity of our subjects and revel in the unknowing, emerging from the other side not necessarily with knowledge, but with another question. In its essence, science never quite answers the questions it seeks; it is never truly complete.

Cara Blessley Lowe

■ Careers
Recording Animal Behavior Sounds: The Voice of the Natural World

Some say that the work I do is perilous. I think it's magical. I record the sounds of all kinds of creatures—animals as large as whales and those so small they can only be seen with a microscope. To capture these mortal voices I travel everywhere on the planet, from the Arctic to the Antarctic, the rainforests and oceans at or near the Equator, and temperate regions in between. My field is called *bioacoustics*. The term comes from two sources: the Greek word *bios*, meaning "life," and "acoustics," which refers to the science of perceived sound. When the two word segments are combined, the term means the study of sounds produced by various living organisms.

Every living organism makes some kind of sound—even viruses. These sound signatures are unique to an organism's characteristics, what it does, and how it lives out its respective life. Humans have a primarily visual culture—we try to understand the world around us mostly from what we see. As a result, vocal behavior of critters is often overlooked or not considered as seriously as it might be. However, a few remaining human populations and nonhuman creatures who live much more closely connected to the natural world, understand and rely on all of the senses to inform them of their surroundings. This is especially true of sound. Because of the nature of my work in the wild, I have learned that creature sounds are among the most essential forms of communication behavior we know of. Drawn to the music of the nonhuman animal world, I have spent most of my adult life demonstrating the impact of that awareness to researchers and ecologists in the field of the natural sciences.

I wasn't always a bioacoustician. For the first half of my life, I was a well-respected and highly paid musician—playing guitar and synthesizer with many famous groups and stars of the 60s and 70s. My music can also be heard on many major feature films. At one point, tired of being indoors—working day and night in an air-conditioned recording studio— I began to study marine biology, learning about whales and other sea mammals.

During that same period, I was writing and producing my first record album called *In a Wild Sanctuary*, created with my late musical colleague, Paul Beaver. The theme was unique because it was the first music album to address the subject of ecology. As such, it required us to venture outside of our normal safe haven, the studio, and trek into the field to record natural soundscapes—something I had never done before. I grew up terrified of animals. We never had a dog or cat in the house during my childhood. A pet fish was considered dangerous by my parents. Even as a young adult, my idea of the wild in the late 60s was a popular redwood park just north of San Francisco, or the zoo. Although these weren't exactly the most primitive places on the planet, the experience of sitting alone outside in the woods with a recorder, listening through headphones to the miraculous sounds picked up by microphones, changed my life. And *In a Wild Sanctuary* became the first album to use natural sounds as a major component of musical orchestration. Recording the natural sounds was not easy. I realized that we were shutting out these delicate sounds with all the human noise around us, and in trashing our natural habitats, we were silencing a special collective voice that provides one of the most important and enchanting encounters we might ever have.

As the studio and performing life of music became tedious, I was drawn to places far distant from the incessant racket of cities so that I could record the wild animals I was beginning to love and know. In order to get sounds on tape, it was necessary to find places with no human noise, and then to be extremely quiet for long periods of time. The birds, mammals, insects, amphibians and reptiles taught me great patience because they didn't always perform in expected ways. Sometimes that meant sitting for 30 hours in one place without moving in order to capture the special song of a single bird or frog.

In order to work effectively, I had to acquire new skills of listening in ways I had never thought possible. The first objective was to understand the nature of the *soundscape*. The soundscape is the entire body of sound we hear in any habitat—everything from a city to a tropical rainforest. It consists of three elements in various combinations: (1) *biophony*, or non-human creature sounds singing together in a given habitat, (2) *geophony*, or non-biological natural sounds in a habitat such as wind, rain or streams, and (3) *anthrophony*, the sounds introduced by humans such as automobiles, motorcycles, leaf blowers, boom boxes, and 808 subwoofers in cars.

The primary focus for this discussion is biophony. Here is an example: Most of us think that birds in spring, or insects or frogs on midsummer nights create quite a racket—a jumble of unrelated noise. I certainly thought so at one time. But then, late one evening, while recording at a location in Kenya in eastern Africa, my musical training took over and I began to hear distinct patterns expressed within the collective natural soundscape, designs that suggested unique relationships rather than discord. When I returned to the lab and tested the idea on a computer, I found that the animals in that particular habitat were all vocalizing in *relationship* to one another, just like instruments in an orchestra. I then examined all of my other recordings from tropical and temperate rainforests, alpine meadows, coral reefs in the far reaches of the ocean, and even tide pools along the shore. All of the tests revealed distinct patterns unique to each habitat. The animal voices had their own frequency slot (or niche)—some with very high-pitched voices, some low-pitched, and some right in the middle. Also, a number of critters that resided in each habitat seemed to sense when it was an appropriate *time* to vocalize. The high-pitched voices allowed the lower-pitched voices to be heard at the same time because they naturally evolved to stay out of each other's way in terms of their respective frequencies. On the other hand, the selective timing of the other animal vocalizations allowed recognition because those with the same range voices would avoid competition from others when singing at the same instant. This animal orchestration—or symbiotic vocal behavior among all the species—is what I call a *biophony*, with *bio* referring to life, and the Greek word *phon*, meaning "sound." Biophony is different from the more general scope of bioacoustics because it refers to the specific ways in which collective groups of creatures vocalize in a distinctive relationship to one another in a given habitat.

Another reason the perspective of biophony is important is because it communicates to us the ways in which various species of animals are relating to one another within their habitat, otherwise referred to as a holistic overview. If they are under stress because of human noise or because the habitat has been altered, that will become evident when the collective expression of the natural soundscape is examined. Older methods of recording and study—where one kind of bird, or mammal or insect is singled out and captured on tape—reveals nothing about the health of that particular environment or the manner in which these creatures relate to one another. Their combined vocal behavior, however, sheds a great deal of light for those of us willing and able to listen.

My colleagues and I have also found that biophonies define the territory. Humans, with all of our precise equipment and logic, tend to map and define territories in grids of space such as 100 square feet or yards or miles, or by visual cues such as common vegetation or geography. However, if we want to learn about how animals define their turf, we only need to do a simple sound survey. A *biome*, or a naturally occurring community of plants and animals, can easily be understood and delineated by the continuity of creature sound within its space. The point at which the natural soundscape begins to change as one walks about and listens is a zone that becomes transitional (otherwise known as the *ecotone*), and the boundary in that region is established. However, the territories defined by the biophony will likely be very different from the very straight and rigid boundary lines established by human visual observation. We tend to convince ourselves about one thing based on a little knowledge. The animals, however, tell us quite another story. Within the natural world, I tend to rely on the critters to reveal certain truths about life around us.

Once, when we were doing an informal study to see why certain migrating warblers that fly up and down the east coast of the United States as far south as the tropical forests of Venezuela were disappearing, we saw and heard something impressive. These particular warblers, able to learn only one song in their lifetime, needed to find safe territory that allowed their voices to be heard at the end of their migration. Their "safe place," of course, was determined by the biophony. As they flew over the canopy of the rainforest listening for an appropriate place to land, they passed over many grids of sound until they found one that matched their voices thus allowing it to be heard. When and if that part of the forest was cut down, the birds no longer had a place to return to that was considered safe. Their voices couldn't be heard in other locations because they were masked by the unique properties of the unwelcoming biophony. We now believe that the disappearance of a relevant biophony is a contributing factor to the reason they became lost and disappeared.

When you are listening, remember that the biophony is determined by many natural conditions. Among them are, first, the time of day—during spring and summer the biophony is usually louder and more active at dawn and dusk; second, the season—spring and summer as contrasted to winter and fall; and third, the weather (the effect of *geophony*).

After presenting talks on my work, students often ask me three types of questions. First to fifth graders generally bring up two issues: (1) How can I do what you do? and (2) What's the most dangerous animal you've ever recorded. From the sixth grade to college graduate level, it usually boils down to question number (3): How much money do you make? and the altruism all but disappears, buried under the debris of the practical.

To answer the first, I usually respond that I can't imagine doing anything more engaging, fun, or that connects me more personally, spiritually, or emotionally to life then working within the wild natural listening to the non-human world around me sing. To me, these voices represent God speaking. The forest is my church, or temple, or mosque. Being present in that environment makes me feel especially healthy and alive. And I feel delighted to have found a way to enjoy the natural world without injuring it or feeling as if I am in conflict with it. After all, it was the music of the animals that taught us to dance and sing and drum in the first place.

Bernie Krause records the sounds of mountain gorillas in Rwanda, 1988.
© *Nick Nichols / National Geographic*

As for the second question—dangerous encounters—the following are some events that were memorable: A mountain gorilla high in the Virunga Mountains along the Uganda/Rwanda border in Africa once grabbed my shoulder and tossed me and 50 pounds of recording equipment strapped to my body 15 feet through the air into a patch of stinging nettles. I was new at the late Dian Fossey's research site, Karisoke, and didn't quite know the social rules of gorilla behavior. As I mistakenly stepped between two males fighting over a female (no matter what the species, you want to be very careful not to do that), I learned my first lesson in mountain gorilla etiquette.

One time, while recording bowhead whales in late spring on the North Slope of Alaska, I was camped alone for two weeks on the shore. About 10 days into the trip on a very cold morning at first light, I heard a crunching noise on the ice outside my tent. Surprised by the sound, I unzipped the tent flap to find a polar bear standing about 20 feet away and moving toward me. Except for a flare gun meant for emergencies, I had no weapon (I generally choose not to carry them). The batteries on my radio were dead. Remembering that flares contain magnesium which burns very hot, I scrambled out of my tent and when the bear came so close I couldn't miss with the one shot I had in the chamber despite my shaking hands, I aimed at it's chest and squeezed the trigger. The flare didn't injure the animal, but the smoke from the projectile singed its fir and frightened him enough so that he ran off, never to return.

Recording killer whales in the Antarctic, I was sitting on a rock ledge overlooking the water in a dry suit (used for scuba diving but, in this instance, designed to keep me warm in −40° F weather), recording a pod of over 100 orcas. One large male, in particular, was spy-hopping (sticking his head above the surface looking around for something interesting) when suddenly it leaped out of the water onto the 5 foot ledge not 10 feet from where I was perched, and hauled his enormous body about 50 yards overland to a rookery of emperor penguins. After chasing one down and grabbing it in his mouth, the whale reversed direction and returned to the water where it devoured its prey. I never wore a black wet suit again while recording killer whales.

In the Amazon rainforest very late one night, my colleague and I smelled the unmistakable scent of a nearby jaguar as we were making our way down a trail to find a likely spot to record. After a mile or two, we still smelled the musky odor, but hadn't heard or seen the cat. At one point my colleague and I split up in different directions. Alone on the trail, I set up my microphones and then sat down about 30 feet away listening intently through my headphones. A few seconds later, the jaguar that had been following us stepped right up to the mic and began to chuff and growl, just like a singer in the recording studio. (For the jaguar recording, listen to *Amazon Days/Amazon Nights*, Cut 2, a CD of that adventure.) What took only 2–3 minutes, seemed like hours as I sat there in the dark, afraid to shine my light and startle it.

And finally, the National Park Service hired us to do some recording in Sequoia National Park. We had just set up our microphones at about 9:30 one evening and were sitting about 50 feet away, when we heard the sound of a bear padding through the forest in the direction of where we had placed the equipment. Sure enough, when we listened to our recording later that night, we found that the bear had engulfed the entire microphone apparatus between his jaws. As a result, we have the only stereo recording of what it is like to be *inside* a bear's mouth, and a large hole in one of our mics.

A young fourth grader once asked me "the most dangerous animal" question and, without hesitation, I responded "Man." "Whaddya mean?" he sputtered from the back of the room, his face red with the rage of challenge and youth. "My dad says that a polar bear is the most dangerous animal. And my dad knows everything!" When the boy finally calmed down a little, I asked, "Has your dad ever seen a polar bear with a high-powered rifle in its hand?"

Unlike my experiences in large populated cities, I have never felt in danger or threatened in the wild. I am always very careful to be aware of my surroundings and do not take unnecessary chances. I learn as much as I can about where I'm going and what I'm likely to encounter and never go anywhere without letting others know where I'm headed and when I should return.

The last question—"How much money do you make?"—brings all kinds of things to mind. First, I create several basic products from my recordings: CDs of soundscapes from rare habitats of which there are now 20 titles; natural and cultural soundscapes for museums and other natural history public spaces; sounds for many feature films (*Shipping News, Castaway, Perfect Storm*, and others); and books, articles and lectures on the subject of wild soundscapes. Our group also does biological habitat studies for various local, regional, state and federal government agencies such as the National Park Service, using biophony as the basis of our work. Second, my work allows me to do exactly what I want and need to do at any time I choose. Third, I think of money as only one form of payment. I feel blessed to be paid in many ways—by the fact that I love what I do and have the friendship and support of the folks I do it with. I have been able to sustain a decent living, especially since I realized the importance of getting an excellent education and earning a doctorate even though I was 40 years old at the time. I love working with animals—living and recording in the wild. That's almost payment enough. When you love what you're doing and you're doing something that is honest, utilizes few resources, and provides something others truly need, you'll have everything you need in this life.

Further Resources

Eiseley, L. 1969. *The Unexpected Universe*, NY: Harcourt Brace.

Krause, B. 1989. *Habitat Ambient Sound as a Function of Transformation for Resident Animals and Visitors at Zoos, Aquaria, and Theme Parks.* In: AAZPA 1989 Annual Conference Proceedings. (no editor noted)

Krause, B. 1998. *Into A Wild Sanctuary.* Berkeley, CA: Heyday.

Krause, B. L. 2002. *Wild Soundscapes: Discovering the Voice of the Natural World.* Berkeley: Wilderness Press. (Book & CD)

　　　(CDs of different soundscapes can be found on our web site: http://www.wildsanctuary.com)

Shafer, R. M. 1977. *Tuning of the World*, New York: Knopf.

Shepard, P. 1996. *The Others: How Animals Made Us Human.* Washington, DC: Shearwater.

Turner, J. 1996. *Abstract Wild . . .* Tucson: University of Arizona.

Wild Sanctuary

http://www.wildsanctuary.com/

　　　The Wild Sanctuary, sponsored by Bernie Krause, is a resource for natural sounds, with a collection of 3,500 hours of recordings, which represent 15,000 species. At this website, you can "travel the world by sound safari and listen as the most exotic creatures sing" by selecting a location and clicking on it to hear the voices of the animals who live there.

Bernie Krause

■ Careers
Significance of Animal Behavior Research

Animal behavior is the bridge between the molecular and physiological aspects of biology and the ecological. Behavior is the link between organisms and their environments and between the nervous system and the ecosystem. Behavior is one of the most important properties of animal life. Behavior plays a critical role in biological adaptations. Behavior is how we humans define our own lives. Behavior is that part of an organism by which it interacts with its environment. Behavior is as much a part of an organism as its coat, wings, and so on. The beauty of an animal includes its behavioral attributes.

For the same reasons that we study the universe and subatomic particles, there is intrinsic interest in the study of animals. "Animal Planet," popular on cable and public television, devotes significant amounts of time to animal films. People spend a lot of money on nature books. There is much more public interest in animal behavior than in neutrons and neurons. If human curiosity drives research, then animal behavior should be near the top of our priorities.

Research on animal behavior and behavioral ecology has been burgeoning in recent years despite below inflation increases (and often decreases) in research funding. The major journals have increased in frequency of publication and number of pages, and rank near the top of citation lists in both behavioral and biological sciences. Animal behavior research is an active and vital field.

Although the study of animal behavior is important as a scientific field on its own, this science has made important contributions to other disciplines with applications to the study of human behavior, to environment and resource management, to the study of animal welfare, to neuroscience and to the education of future generations of scientists.

Animal Behavior and Human Society

Many of the problems in human society are related to the interaction of environment and genetics on behavior. The fields of sociobiology and animal behavior deal with the issue of environment and genetic interactions on behavior both at an evolutionary level and a proximate level. Increasingly, social scientists turn to animal behavior as a framework in which to interpret human society and to understand possible causes of societal problems. A new field of evolutionary psychology looks at human courtship, parenting, aggression, violence, and cognitive ability based on an evolutionary analysis developed from studies on animal behavior.

Research on chimpanzees, monkeys, and many other species has illustrated the importance of cooperation and reconciliation in social groups. This work provides new perspectives for interpreting and reducing aggressive behavior among human beings.

The observational methods to study animal behavior have had a tremendous impact in psychology and the social sciences. The child psychologist Piaget began his career by studying snails, and he applied careful behavioral observations and descriptions to his landmark studies on human cognitive development. Other major scientists of human psychology began by studying animal behavior. Experimental designs, observation methods, and attention to nonverbal communication signals were developed in animal behavior studies before their application to human behavior. The behavioral study of humans would be much diminished today without the influence of animal research.

Charles Darwin's work on emotional expression in animals has had an important influence on modern psychologists who study human emotional behavior and emotional disturbances.

Research on the importance of social attachment in rhesus monkeys has dramatically changed how we take care of our own infants. The long lasting behavioral disorders of adopted children from Eastern European orphanages can be best understood in the context of work on attachment in monkeys. Animal studies on learned helplessness has led to new ways to understand and treat depression.

The comparative study of behavior over a wide range of species can provide insights into human behavior and how to change it for the better. For example, the woolly spider monkey in Brazil displays no overt aggressive behavior among group members. We might learn how to minimize human aggression if we understood how this species of monkey avoids aggression. If we want human fathers to be more involved in infant care, we can study the conditions under which fathers care for infants in other species like the California mouse or in marmosets and

Demonstrations on how dolphins use sonar to locate objects has led directly to the application of sonar imaging techniques from the military to medical diagnostics.

© Reinhard Dirscherl / Visuals Unlimited

tamarins. Models of the development of communication in birds and mammals have had direct influence on the development of theories and the research in the study of child language. The richness of developmental processes in behavior, including multiple sources and the consequences of experience are significant in understanding processes of human development.

Understanding how species that can live in a variety of habitats differ from those that are restricted to limited habitats help us understand how humans might adapt as our environments change.

Research on animal sensory systems has led to practical applications for extending human sensory systems. Demonstrations on how bats and marine mammals use sonar to locate objects has led directly to the application of sonar imaging techniques from the military to medical diagnostics.

Studies of apes and parrots using language analogues have led to new technologies that have been applied successfully to teaching language to autistic and retarded children.

Research on circadian and other endogenous rhythms in animals has directly helped humans in areas such as coping with jet lag or changing from one shift to another.

Animal behavior research developed many of the important concepts relating to coping with stress; for example, the importance of prediction and control on coping behavior.

Animal Behavior and the Environment, Conservation, and Resource Management

The behavior of animals often provides the early warning signs of environmental degradation. Changes in sexual and social behavior appear much sooner and at lower levels of environmental disruption than changes in reproductive outcomes and population size. When we wait to see if numbers of animal populations are declining, it is often too late to save the environment. Studies of natural behavior of wild animals are vital to provide baseline data for future environmental monitoring. For example, the Environmental Protection Agency uses disruptions in swimming behavior of minnows as an index of pesticide pollution.

Basic research of salmon migration to their home streams, started more than 50 years ago, has taught us not only about the mechanisms of migration, but has been vital in preserving the salmon industry in the Pacific Northwest. Understanding salmon migration led to the development of a salmon fishing industry in the Great Lakes. Animal behavior research has important economic implications.

Animal behaviorists have described insect reproduction and host plant location leading to the development of nontoxic pheromones for insect pest control that avoid toxic pesticides. Understanding predator–prey relationships can lead to the introduction of natural predators of prey species.

Knowledge of honeybee foraging behavior is directly relevant to mechanisms of pollination, which in turn is important for plant breeding and propagation.

The foraging behavior of animals is also important in forest regeneration. Many animals serve as seed dispersers and are necessary for the propagation of tree species and therefore habitat preservation.

Conservation of endangered species requires that we know enough about natural behavior (migratory patterns, home range size, interactions with other groups, foraging demands, reproductive behavior, communication, etc.) to develop effective reserves and

Unconventional Uses of Animal Behavior

Arnold S. Chamove

Other than conventional careers in animal behavior, there are those that reflect a use for animal behavior in areas not normally considered, based on the premise that animal behavior exhibits some of the biological bases of human behavior. With that in mind, it is not difficult to postulate areas that might benefit from knowing about the biological bases of some behaviors. Four examples follow:

Knowledge of animal behavior has been used in a clinical psychology practice, where the study of animals was thought to be beneficial in treating anorexia and anxiety. Both can be viewed as if the individual were a subordinate in a social group. Then ways of increasing dominance within the group or forming new social groups where the individual occupies a more dominant rank are beneficial.

Direct knowledge of primate behavior also may help in dealing with those corporate problems that are not normally the focus or expertise of those in human resource or consultancy positions. When forming new groups of animals, it is helpful to determine the goals of those individuals—access to resources, preserving dominance, protecting offspring. When new groups form in a business setting, problems arise as individuals seek to maintain or improve access to their goals. These goals are hidden to those wrapped up in the conflict with the new business setting, although they are quite clear to a trained observer of animals, one trained to look for goals, look for feelings, look at postures independent of language.

Knowledge of animal behavior can be used to improve the design of equipment used for both animals and humans. Imagine an electric fence. Neither humans nor animals know it is electric until they touch it and then must learn that certain tapes or wires are dangerous and certain ones are not. Knowledge of animal warning patterns suggests that if a fencing has warning patterns on it, children and animals will have an inherent aversion to the fence, can learn it is dangerous more quickly and remember it is aversive for longer than one with no pattern. This is the basis of a patent used in electric fencing tape. The invention is beneficial to humans because it has been found that animals avoid the tape even if the electricity is off. And it benefits both humans and animals in that it warns them before shocking them.

When humans go into the ocean, there can be problems with sharks. When they scuba dive, humans usually wear dark colors, which are the colors of a shark's major food item—seals. And surf boards have a white underside, much like the belly of a seal seen from below. Based on a knowledge of animal behavior, what modifications could be considered to reduce the frequency of shark attacks to those two groups? Answering questions like this can be useful in certain careers.

People with training in animal behavior make good observers of human behavior, and most careers outside animal behavior involve observing human behavior. Such training leads one into new insights if trained as a counsellor, family therapist, or clinical psychologist; into novel ways of viewing and modifying corporate behavior; and into new ways of designing more friendly items for everyday use.

effective protection measures. Relocation or reintroduction of animals (such as the golden lion tamarin) is not possible without knowing a species' natural history. Habitat preservation programs and human management of populations of rare species, both in captivity and in the natural habitat, require animal behavior research. Many of the world's leading conservationists have a background in animal behavior or behavioral ecology.

Studies on reproductive behavior have led to improved captive breeding methods for whooping cranes, golden lion tamarins, cotton-top tamarins, and many other endangered species. Captive breeders who are ignorant of the species' natural reproductive behavior are usually unsuccessful.

Animal Behavior and Animal Welfare

As a society we have placed increased emphasis on the welfare of research animals, exhibit animals, and companion animals. United States law now requires attending to exercise requirements for dogs and the psychological well-being of nonhuman primates living in these roles. Thoughtful animal welfare requires knowledge of behavior. Animal behaviorists understand the behavior and well-being of animals in lab and field. They provide expert testimony to develop reasonable and effective standards for the care and well-being of animals.

Zoological parks and wildlife reserves recognize the need for research-trained animal behaviorists to be permanent staff members to design better environments, evaluate animal health and well-being through behavior, and to interpret animal behavior for zoo and reserve visitors.

Many people experience behavioral difficulties with their companion animals, and applied animal behaviorists can help people train their animals appropriately and provide solutions to behavioral problems.

Improved conditions for farm animals, breeding of endangered species, and proper care of companion animals all require a careful understanding of behavior.

Animal Behavior and Neuroscience

Sir Charles Sherrington, an early Nobel Prize winner, developed a model for the structure and function of the nervous system based on careful behavioral observation and deduction. One hundred years of subsequent neuroscience research has completely supported the inferences Sherrington made from behavioral observation.

Neuroethology is the integration of animal behavior and neuroscience and provides important frameworks for hypothesizing neural mechanisms. Careful behavioral data allow neurobiologists to narrow the scope of their studies and to focus on relevant input stimuli and attend to relevant responses. The use of species-specific natural stimuli has led to new insights about neural structure and function that often contrast with results obtained using nonrelevant stimuli.

Animal behavior research has demonstrated a downward influence of behavior and social organization on physiological and cellular processes. Variations in social environment can inhibit or stimulate ovulation, produce menstrual synchrony, or induce miscarriages. The quality of the social and behavioral environment has direct effects on how the immune system functions. The social and physical environments have direct effects on gene expression. The nature of early social experience can rewire the structure of parts of the brain.

Neuroscientists and immunologists need to understand these behavioral and social influences to properly control and interpret their own studies.

Animal Behavior and Science Education

Many in our society are concerned with the lack of scientific literacy, the low level of interest that students have in science, and the fact that women and minority groups are underrepresented in science. Courses in animal behavior and behavioral ecology serve as hooks to interest students in behavioral biology. At many universities courses in animal behavior and behavioral ecology are extremely popular for anthropology, biology, and psychology majors as well as non-majors.

For many students, these courses often are the first introduction to behavioral biology. Many female undergraduates develop interest in attending graduate school and research careers after taking these courses. As more high schools provide teaching in animal behavior many students begin college already motivated. A good proportion of students enrolled in animal behavior courses become motivated for research careers. Career options are somewhat limited by funding and job opportunities, but a highly intelligent, hard-working student, passionate about animal behavior, can find opportunities in academic settings (teaching and doing research), in zoological parks and aquariums (designing habitats and evaluating the behavior of exhibit animals), as an applied animal behaviorist (working with companion animals and their owners) and in conservation agencies (developing management plans for preserving wild populations and preparing animals for reintroduction). As with any other career, determination, competence and motivation are important variables leading to success.

Charles T. Snowdon

■ Careers
Veterinary Practice Opportunities for Ethologists

The Importance of Behavior Counseling in Veterinary Practice

Many new pet owners are misinformed or at least naive when it comes to understanding the behavioral development and training of pets. Since most pet owners seek veterinary attention shortly after obtaining their pets, the veterinary practice should serve as a primary resource for behavioral advice, in much the same way a new parent might seek child care advice from a family doctor or pediatrician. Therefore, the first few veterinary visits should be devoted not only to the nutritional and health care needs of the pet, but also to the guidance needed to prevent behavioral problems. Owners should be counseled on normal species-typical behavior and how to create a home environment and schedule that meets all of the pet's needs. Most importantly, by reviewing basic learning principles, owners can be set on a path of shaping desirable responses rather than punishing undesirable behavior. Studies by Patronek, Glickman and Beck have shown that the risk for relinquishment is lowest for dog owners who receive behavioral guidance during their first veterinary visits and for those owners who attend obedience classes; similarly they found that cat relinquishment might be reduced by providing sufficient counseling or reading material, and advising cat owners on how and why to house cats indoors. Most pets relinquished at shelters in North America are a result of a poor pet–owner bond, often arising from undesirable behavior. In one Canadian study, Gorodetzky estimated that 11.5% of cats and 13% of dogs euthanized at veterinary clinics were due to behavioral reasons, where as in a recent study involving U.S. veterinarians, Patronek & Dodman (1999) found that approximately 224,000 pets each year were euthanized at veterinary clinics for behavioral reasons.

Veterinary Roles in Behavior Medicine

In veterinary practice, the discipline of behavior medicine focuses on both preventive counseling for new pet owners (or for any species that might be kept in captivity), and on the treatment of behavior problems in these species. The diagnosis and treatment of complex behavior problems in animals, in most jurisdictions in North America, is considered to be the practice of veterinary medicine, and as such must be performed by a licensed veterinarian. Consider the very definition of "veterinary practice" offered by the AVMA (American Veterinary Medical Association) Model Practice Act: "Practice of veterinary medicine means: to diagnose, treat, correct, change, relieve, or prevent animal disease, deformity, defect, injury, or other physical or mental conditions; . . . or to render advice or recommendations with regard to any of the above." (2003, p. 334) Since virtually any medical condition might cause or contribute to changes in behavior, a veterinary assessment is the first place to start. Veterinarians also play a unique role in behavior medicine since some problems benefit from the use of prescription drugs, supplements, appliances, or surgery to achieve the most successful and humane resolution.

Ethologist Roles in Veterinary Practice

The Oxford dictionary defines *ethologist* as "one who studies the science of animal behavior." There are many roles in clinical practice that ethologists can fill to help expand veterinary services, particularly in areas such as client education, preventive behavioral counseling, and training. The ethologist might be employed as part of the practice staff or on a contracted basis. This will vary with the behavior expertise within the practice, as well as the size of the practice, the facilities available, and the services that they wish to provide.

Pet Selection

To help owners choose a pet that is best suited for their home and lifestyle, one valuable service that could be offered by an ethologist in the veterinary practice setting is pet selection counseling. This may be as simple as providing handouts or internet links. However, the optimal approach is to utilize the expertise of a veterinarian or contracted ethologist in a visit that guides the owner through the selection process on what is most appropriate for their household. This should include: consideration of species, age, sex, and breed differences; how to find a good source for the pet; and how to assess the kennel, breeder, and individual pet.

There is little evidence that the assessment of young puppies and kittens is a valid means of predicting adult behavior. However, if the prospective pet owner is considering an older puppy or kitten, or especially an adult pet, some guidance should be provided as to how to best evaluate these animals and which temperament tests are most likely to be predictable.

The pre-selection visit can provide the client instruction on how to introduce the pet into the home and advice needed prior to the pet's first veterinary visit. Finally, the ethologist could consider traveling with the client to aid in the selection process and/or make a housecall to help evaluate and set up the home for the pet's arrival.

New Pet Owner Counseling

Ideally, once the pet is obtained, the next step is preventive behavior counseling. This might include: reducing fear through effective socialization; reward based training; insuring that the behavioral needs of the pet are effectively met; and, adapting the household to prevent problems (i.e., setting the pet up to succeed). The ethologist might work with the veterinary clinic as a staff member who counsels pet owners at the clinic, who visits the home for hands-on advice, or who teaches community behavior seminars for new pet owners after hours in the veterinary lobby.

Puppy and Kitten Classes

Advising pet owners to keep their pets away from other pets until after all vaccinations are complete may be an excellent way to avoid contagion, but is counterproductive at a time in the pet's life where primary socialization and habituation to new stimuli is critical. Therefore, one compromise is to introduce and socialize puppies to other pets and people in a controlled environment such as the veterinary clinic. This has the added benefit of multiple enjoyable visits, bonding the puppy and owner to the veterinary facility. Therefore, an increasing number of veterinarians are offering socialization classes for new puppies (e.g., a 4–6 week course including an initial seminar followed by a few weeks of puppy classes in which the owners receive guidance on training, play, and socialization). Ideally, these classes provide reward-based obedience training demonstrations, an opportunity for supervised social play, as well as a variety of novel stimuli such as umbrellas, crutches, balloons, etc. In this way the puppy

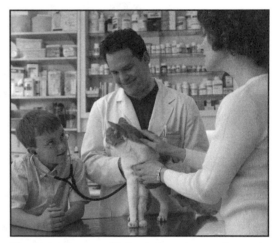

The importance of learning proper pet care is critical. The first trip to the vet should include a lesson.
© *Jim Craigmyle/CORBIS.*

can be exposed to a wide variety of social and environmental stimuli in a safe, controlled environment, while avoiding health risks associated with exposure to public parks and other dogs with unknown vaccination status. Similar sessions might be offered for kitten socialization, perhaps with an initial seminar followed by a second week where the kittens are brought in for socialization, play with appropriate toys, and basic training demonstrations. The ethologist might be employed by the veterinary clinic to provide these socialization classes.

Canine Day Care Supervision

Some veterinary practices may wish to expand their services to include day care for canines. An ethologist working with the practice could contribute by evaluating new dogs, integrating them into the facility, supervising play groups, and designing the daily program to maximize the benefit to pet and owner. The pet's enjoyment and intelligence might be stimulated by a variety of new sights, sounds, smells, and experiences. Certainly, the average canine will develop improved social skills resulting in an increased opportunity to accompany the owner in public. Finally, the play, exercise, and daily routine may make for a more contented dog with less likelihood for the development of behavior problems.

Early Intervention

Another type of preventive counseling might be described as early intervention. This is defined as counseling a pet owner (with a healthy pet) who has a concern due to a lack of knowledge about *normal* species behavior, training, or other behavior management issues. This might include advice on emerging concerns such as housesoiling, play biting, play destruction, or overexuberant play. This type of counseling might be delegated to a staff ethologist or behavioral technician who could deal with the specific issues in the particular household. Some veterinary clinics might also wish to offer their own obedience training classes or private obedience sessions as auxiliary services. A staff ethologist with sufficient knowledge of animal behavior and reward-based training might be contracted to offer these services on the premises, in a separate training location, or in the case of private sessions, in the owner's home.

Behavior Problem Counseling

The diagnosis and treatment of complex behavior problems is a process that requires veterinary input. However, within the veterinary consultation, there is often a need for a behavioral coworker. This person might be a behavioral technician, a trained staff member, or a staff ethologist. The role might include some or all of history taking; in-home evaluation; animal restraint during clinical assessment; helping demonstrate procedures or products that the veterinarian has recommended; providing appropriate handouts and reading material; insuring adequate case follow-up; and/or helping implement the suggested behavioral program. Following the consultation, many clients may be in need of ongoing guidance to help implement the program, especially with respect to reward-based training, head halter use, ongoing assessment and review, and the control and introduction of stimuli for desensitization and counterconditioning.

Consulting Regarding Other Species

Although most veterinary behavior consulting is for canids and felids, behavioral consultations and management advice might be requested for a multitude of other species such as ruminants, equines, laboratory animals, avian or zoo animals. An ethologist with training in these species might collaborate with a veterinarian who consults on or manages any of these species.

Pet-Assisted Therapy Training and Assessment

With the increase of animal assistance programs for humans with special needs comes the need for pet training and assessment of suitability for pet-assisted therapy. For pet dogs, temperament testing such as the AKC (American Kennel Club) Canine Good Citizenship or TDI (Therapy Dogs International) tests might also be offered by the practice. With proper training and in some cases certification, these are procedures that can be offered by a trained staff member or contracted ethologist.

Educating Veterinary Staff

Veterinarians and their staff require sound knowledge of normal behavior for each species they offer to treat. This serves as the basis for the study of humane restraint and handling, health management, husbandry, housing, and training. The safety of the pet owner, veterinarian, and veterinary staff is at risk without understanding normal behavior of each species such as vocalization, facial expressions, body language, or posturing. Since abnormal behavior might be a result of an underlying medical problem, pain, or emotional distress, ethology education to differentiate normal and abnormal behavior provided to a veterinary staff is helpful. Even with a board certified veterinary behaviorist on staff, his or her time and expertise might be best utilized by focusing on clinical aspects such as the diagnosis and treatment of behavior problems, case consulting, and training residents. Another option for the ethologist who might wish to further his or her own education, would be to pursue a veterinary degree and then to use the combination of ethology and veterinary training to proceed into a residency program to become a veterinary behaviorist, who, with training in both disciplines, would add a unique perspective to the field.

Educational Opportunities and Clinical Studies

Over the past few decades, the American Veterinary Society of Animal Behavior (*www.avma.org/avsab*) has been established to help promote an increased veterinary involvement in behavior. In the mid 1990s, the American Veterinary Medical Association granted board certification status to the field of veterinary behavior as diplomates of the American College of Veterinary Behaviorists (*www.dacvb.org.*) However, with only 36 certified veterinary behaviorists as of 2004, and many of these working in either clinical practice or industry, there are insufficient numbers to fill the positions that would be needed if every veterinary school, technician college, and continuing education seminar were to require a board certified behaviorist. Therefore, the ethologist can play a important role in the training of the veterinary profession. The optimum situation, however, would be for a veterinary behaviorist and ethologist to both be employed at the veterinary teaching institution. In this way, the expertise from each discipline could be used most effectively to train undergraduate and graduate students, as well as to design and implement research that will further advance the field.

Ethologists have increasing opportunities in veterinary practice, as well as in the education and training of veterinarians and their staff. The greater the cooperation between disciplines, the better the services and expertise available to the veterinary clients and patients.

See also Applied Animal Behavior—*Social Dynamics and
Aggression in Dogs*

Further Resources

American Kennel Club. (www.akc.org)
Animal Behavior Network. (www.AnimalBehavior.Net)
    Founded by Rolan Tripp, DVM, the Animal Behavior Network (ABN) seeks to link trainers and behaviorists with veterinarians who wish to offer applied animal behavior counseling for pets
Gorodetsky, E. 1997. *Epidemiology of dog and cat euthanasia across Canadian prairie provinces.* Canadian Veterinary Journal, 38, 649–652.

Line, S. W. 1998. *Factors associated with surrender of animals to an urban humane society.* Proceedings of the American Veterinary Medical Association Annual Conference, Baltimore. 345–348.

Miller, D. D., Staats, S. R., Partlo, C., Rada, K. 1996. *Factors associated with the decision to surrender a pet to an animal shelter.* Journal of the American Veterinary Medical Association, 209, 4, 738–742.

Patronek, G. J., Dodman, N. H. 1999. *Attitudes, procedures, and delivery of behavior services by veterinarians in small animal practice.* Journal of the American Veterinary Medical Association, 215, 11, 1606–1611.

Patronek, G. J., Glickman, L. T., Beck, A. M., et al. 1996. *Risk factors for relinquishment of cats to an animal shelter.* Journal of the American Veterinary Medical Association, 209, 3, 582–588.

Patronek, G. J., Glickman, L. T., Beck, A. M., et al. 1996. *Risk factors for relinquishment of dogs to an animal shelter.* Journal of the American Veterinary Medical Association, 209, 3, 572–581.

Therapy Dogs International www.tdi-dog.org

2003 *American Veterinary Medical Association Resource Manual*, Model Practice Act, "Definitions," p. 334, Section 2, Article 8, Subsection (a)

Rolan Tripp & Gary Landsberg

■ Careers
Wildlife Filmmaking

I grew up in Wisconsin with a gun in one hand and a camera in the other and learned at a young age how to put meat on the table. Although I have since given up hunting, it was in my youth that I learned the skills that make a good hunter or photographer: how to stalk animals and the willingness to be patient and persevere. My father was an avid hunter and a very good amateur cinematographer who particularly enjoyed hunting—and filming—ducks. I spent many hours in the blind with my father learning the technical skills that make a "hunter" using a gun—or camera—successful and learning about my father's philosophy about hunting and life in general.

In high school, I went on several big game hunts with my father to northern British Columbia. These were long trips, which often required 21 or more days spent packing and hunting from horseback. It was during these trips that I started using a movie camera and discovered the importance of recording the story of a hunt. Of course, we also brought back a few trophy animals, but for me, the lasting influence of these trips was my genesis as a future wildlife filmmaker.

After I married, my wife Connie also joined me in the field, stalking and photographing animals. In 1965 we traveled up the unpaved Alaska Highway to Mount McKinley National Park in the center of Alaska. Before going on the trip, I read *A Naturalist in Alaska* by Adolph Murie. This book was very influential because it made me realize that observing animal behavior has its own rewards, separate from hunting or even filming. The book also provided an avenue to meet Adolph Murie who was still doing research in Mount McKinley National Park on grizzly bears

Wolfgang Bayer, wildlife cinematographer, produces, films, and directs his own wildlife films and television programs.
© *Ed Kashi / Corbis*

and wolves. Adolph, a very thoughtful conservationist and biologist, also had a sense of humor. Once when I commented on the "big heads" that Dall Sheep have, Adolph quipped that the rest of the body isn't bad either.

I shot my last trophy, a record book Dall Sheep, in 1969. My experiences hunting and the skills I learned would continue to serve me well as a photographer, but I discovered that I found it more satisfying to secure photographs of wild animals than "big heads." At first, I worked to take beautiful shots of still animals, much like a photographer, but in 1933, in McKinley Park, I filmed a grizzly sow with two spring cubs killing a caribou on the East Fork River, and my career was launched. At this time, predation sequences were extremely rare and the footage was in instant demand. I sold the rights to use this sequence to the British Broadcasting Corporation (BBC) and used the money to upgrade my equipment. My semiprofessional career in wildlife filmmaking was underway. There was little demand for wildlife footage in the '70s and early '80s so I paid for my filming habit by teaching high school math. In 1968, Connie and I moved to Billings, Montana, and I spent my weekends filming wildlife in Yellowstone National Park and my summers filming wildlife in Alaska.

In 1981, on the way home from filming for the National Park service in Denali, I was able to film Dall sheep in rut in dominance fights. Twelve years after killing my last animal, a large Dall ram, I now had a trophy far superior to the one that hung on my wall. My film sequences of dominance fights have been seen and enjoyed by millions, and the Dall sheep lived to fight another day.

In part, perhaps because of Adolph's influence, I concentrated on wolves in Alaska and their smaller brothers, the coyotes, in Yellowstone. After several weekends filming in Yellowstone, I had the opportunity to catch some unique coyote behavior on film—that of coyotes stealing fish from river otters. With my new partner, Dale Johnson, I produced a 25-minute film on coyotes titled "Song Dog" that was noted for the unusual behavior it included. A few years later, National Geographic Television commissioned me to produce an hour-long film on coyotes for television.

In 1993, I retired from teaching in order to devote myself full time to wildlife filmmaking. In 1995, wolves were reintroduced into Yellowstone National Park and the rest is history. More recent work includes four films: one about bison, one about the Yellowstone Lake ecosystem, one about wolf reintroduction in Idaho and the role that the Nez Pierce Indians played in that reintroduction, and a 2-hour DVD on wolf behavior. All of these films depend on viewing animal behavior. Although predation certainly plays a big role in the behavior of both predators, such as wolves, and their prey, such as elk, it's only part of the story. In my films on natural life cycles, predation is only one component. Young at play, social communication, interaction with related animals, such as pack or herd dynamics, and interactions with other animals, including rivalry between wolf packs and wolf–coyote interactions, are equally satisfying to film and watch.

In addition to films for television, I also work with researchers in the field by providing videotape, often uncut, of animal behavior. It's very gratifying to work with both established researchers and students working on their doctoral degrees by providing filmed sequences, which in turn provide the raw material for scientific papers on animal behavior.

Forty-three years after first filming a wild animal and 33 years after killing my last trophy, I realize that certain skills, such as the ability to locate and stalk wild animals and the perseverance to endure hardships, are necessary for both activities. However, in most cases, securing good wildlife behavior sequences on film is more difficult than hunting. Like us, most animals live mundane lives with occasional brief bursts of activity. The cameraman must be there when the activity occurs—thus requiring a lot of time in the field just watching.

When something happens, he must be close enough to get suitable images and there must be adequate light.

As humans, we relate our experiences as stories with beginnings, middles, and ends and so, too, wildlife sequences must tell a story. When a film is edited, real time is condensed, but the illusion of real time must be maintained. This is done using "matched cuts" which require images of different sizes (i.e., close-up shots, middle shots, and wide shots). These requirements can be very challenging because it means that the wildlife filmmaker has to see the necessary components while shooting an action scene. In Hollywood, the director has the luxury of simply asking actors to repeat a scene if the camerawork isn't perfect. In filming wildlife, however, there are no second chances, which is why it is so important to understand animal behavior and be able to anticipate a wild animal's next move.

Further Resources

Landis Wildlife Films
http://www.wolftracker.com/Landis/
>   Landis Wildlife Films provides information about Bob Landis's wildlife films and also links to the Yellowstone Wolf Tracker, which provides information on the gray wolf recovery in Yellowstone National Park.

Bob Landis

■|Careers
|*Wildlife Photography*

Over the years many individuals have asked me for advice on how to become a successful wildlife photographer. The first question is usually, "What kind of camera and equipment do you use?" followed by, "What kind of film, who does your processing, and who makes your prints?" Often after I answer these questions, the curious individual will still have a puzzled look on their face and say, "That's great. I use Nikon and Fuji films too." Then they add, "I've been to many of the same places. I've done the polar bears in Churchill and the lions in the Serengetti, but my pictures didn't turn out like yours. What's the name of your printer again?"

This exchange is usually followed by more technical questions on topics ranging from auto focus to matrix metering to the newest vibration reduction lenses. The person continues digging for answers, digging for "the secret" deep in the recesses of my mind. Surely, I must be hiding something.

When I tell them I own most of the latest and greatest equipment but am generally technically challenged, they look even more perplexed. I explain that I don't know how to take advantage of all the expensive built-in computer programming and haven't even read a camera manual in 15 years. Still looking for the secret, eh?

"Don't get me wrong," I say, "I obviously know about the basics of photography, such as focus, depth of field, and shutter speeds. And yes, I know something about art: subject, composition, light, texture, form, and light."

The "secret" for which they are searching is something that I've often taken for granted as being obvious. It is a thorough and intimate knowledge of animal behavior, which does not come easily. Much can be learned from research done on a species, but true knowledge, especially of an individual within a species, takes time, patience and keen observation. Even after 40 years of observing chimpanzees in Gombe, Tanzania, Jane Goodall is still learning new things about chimpanzee behavior and individuals she has known for nearly four decades.

My interest in observing animals and the desire to capture those animals and scenes in some sort of "decisive moment" led me to photography. Sharing those moments with others through photography has always been the driving force. Before I was 21, I had shot less than a roll of film. However, observing and mentally capturing images of the wild started for me at a very young age—probably when I was 4 or 5 years old.

Being exposed to wild places or animals affects each of us differently. These experiences may barely be noticed by one or may leave a lifelong impression on another.

Growing up on the Platte River in central Nebraska, my three brothers and I had essentially the same upbringing for the first 20 years of our lives by a father who was an avid sportsman, hunter, and fisherman. My father, Harold, wanted to instill in each of us his love of the outdoors—the pursuit of ducks, geese, and catfish. My brother Bill was never fascinated by birds or animals. He has lived in Hong Kong for 30 years and thrives on the "mass of humanity," as he calls it. David and Hal, my two younger brothers, are interested in wildlife, still live in Nebraska, and spend a few weekends hunting and fishing on the Platte, but most of their time is spent running the family-owned department store.

In contrast, I could never get enough of observing animals. I could sit all day at the table where I am now writing and stare out of the picture window. Watching the chickadees, goldfinches, and nuthatches at the feeder and the chipmunks coming and going from the woodpile, is my meditation. I am riveted by a kingfisher on a quaking aspen whose eyes are fixed on the glassy surface of the pond waiting for the rise of a tiny trout. In order to finish this essay I should drop the bamboo shade, but I can't.

During the winters in Nebraska, we made decoys and spent hours practicing our duck and goose calling. The summer months, prior to the opening day of duck season, we worked on the river. It was the preparation, the challenge, the "game" that drove my father—not the killing—though we relished the roast wild duck or pheasant on the table. Our late summers were spent cutting back the advancing brush and trees that were choking the Platte. The river's channels, once "a mile wide and an inch deep" continued to shrink due to the advancement of farming, of clearing more land, and of diverting more water from the Platte for irrigation. By mid-summer in the early '50s, the Platte "ran" dry, but by opening day of duck season, in early October, the water always returned. My father fought the irrigators and those who wanted to divert yet more water (only about 20% of the Platte's historical flows remained by the early '50s) for municipalities and cities, like Denver far upstream. These were my first lessons in conservation and in becoming an activist.

The great fall and spring migrations of the cranes, the greatest on the North American continent, were not to be taken for granted. Key to their survival, the Platte is a place of rest and rejuvenation on their long flight to their nesting and wintering grounds. I feared these subjects might one day disappear so I fixed them in my mind. I envied the painters Lynne Bogue Hunt and Richard Bishop whose paintings hung on our cabin's walls and captured precious moments of a flock of pintails taking flight or a gathering of Canada geese coming in for landing against the blue sky of Indian Summer, necks extended and webbed feet spread wide. There were no photographs hanging in the cabin. I didn't know anything about photography—only that my father would take the occasional picture—usually of us boys holding our dead ducks, geese, or catfish. Now, thinking about it, I don't remember a single snapshot "wasted" on a landscape. There were some images of the blind and decoys, and a few of flocks of geese coming into our setup. My dad had an old Mercury half frame view camera that shoots 72 frames on a 35 mm roll and sits on my bookshelf.

As the years passed, my knowledge of and passion for wildlife grew, and I gradually traded in my shotgun for the camera. The beauty, the abundance of life, and the plight of the Platte drove me to photography. I wanted to show others what we were foolishly risking.

Today, October 22, 2003, the Platte River, where my brothers and I waited every September nearly 50 years ago for the water to arrive, is completely dry. My brother David walked across the riverbed this morning and called to tell me there was no water in sight. The duck season will come and go, and the Platte, depending on snows in Colorado and Wyoming this winter, may not have water until spring.

In 1969, I graduated from Doane College in Crete, Nebraska, with an undergraduate degree in biology. I focused on biology thinking I would go into pre-med. But before gradu-

Tom Mangelsen in the field.
© *Images of Nature*

ating, I knew I wanted to learn more about wildlife, ecosystems, and animal behavior. So, I spent the following year studying zoology at the University of Nebraska under Paul Johnsgard, world famous ornithologist. The following year, I moved west to the Rocky Mountains near Boulder, Colorado.

I continued my studies in wildlife biology at Colorado State in Fort Collins and Arctic alpine ecology at the University of Colorado in Boulder. While in Boulder I met Bert Kempers, a wildlife filmmaker, who had just been contracted to do five animal behavior films for the University. He hired me, *with no motion picture experience*, as a cinematographer and editor. I was learning a lot during graduate school, but had to choose between finishing my masters degree or possibly having a career in wildlife

filmmaking and photography. The five animal behavior films solidified my choice. Being in wild places and photographing and observing birds and mammals year round made it difficult to think about stuffy classrooms, term papers, and theses. My thoughts and heart remained with the Platte and what was happening to it. I returned to the cabin often, especially during the fall and spring.

During the next 15 years, I spent many months documenting cranes and life on the river. I followed the cranes from their nesting grounds on the Arctic coast of Alaska to their wintering ground on the Gulf of Mexico. I shot more than 50,000 still frames and made three television films.

Going to Alaska introduced me to the Arctic wildlife, made me fascinated with the far north and obsessed with polar bears. From the mid-1980s to the end of the 1990s, I shot over 80,000 still frames of polar bears, published a book *Polar Dance*, and shot more than 100,000 feet of 16 mm film of the great white bears. To me, there is no animal on earth more beautiful or interesting.

However, on February 14, 1999, a family of cougars—a mother with three cubs, was spotted in a cave on a butte on the National Elk Refuge about 10 miles from where I live in Jackson Hole, Wyoming. I spent the following 42 days from dawn to dusk photographing the family. Having never seen a wild cougar before, it was a once in a lifetime experience. It was the first time a cougar family been observed in the wild for that length of time.

My knowledge of cougars was minimal until that time. I did know that nearly all cougar photographs are of trained animals kept in small cages at game farms. Along with many other species, such as wolves, bears, lynx, bobcats, tigers, badgers, and raccoons, hundreds of animals, at dozens of game farms, are rented out daily as "wildlife models" for photographers. The animals are forced or trained to perform for the camera. It is a cruel and inhumane business. Of course, the animals are not acting naturally. There is little skill or knowledge of animal behavior necessary, and in one way or another the resulting images are a lie.

The family of cougars was a rare gift, a serendipitous event that changed my life. I learned a great deal about cougar behavior during those 42 days. Most of what I learned was how very difficult and challenging it is for a female cougar to make a living for a family of three, even at a place like the National Elk Refuge, which has a bounty of elk, deer and big horn sheep. She risked her life every day pursuing prey 5 times her size. Other days, she left the safety of the refuge in search of prey and had to avoid cougar trophy hunters. Since the cougar family left the refuge following the elk herds in mid-March 1999, more than 40 cougars have been killed for trophies in their home territory. The likelihood of any of them still being alive is remote. There are more than 3,500 cougars killed by trophy hunters in the western United States every year with an equal number of orphan cubs that will not survive and are unaccounted for.

It was after the winter of 1999 that my passion for polar bears was necessarily curbed by a concern for cougars. In 2001, together with my partner Cara Blessley Lowe, we founded the Cougar Fund. Now, I spend more time trying to protect cougars than I do photographing.

So, the journey continues, and my fascination with wildlife that began on a prairie river in the middle of nowhere has consumed my life. I imagine my fight for cougars will be the longest and will likely take me to the end. Between battles, I will return to the Platte in the spring to see the cranes. I hope to be fortunate enough to go to some old and familiar wild places and to a few new and faraway ones to enjoy and learn more about the behaviors of our earth's amazing creatures.

Thomas D. Mangelsen

■ Careers
Writing about Animal Behavior: The Animals Are Also Watching Us

All I ever wanted to be was a writer. If you had told me in high school or college that one day I would share my literary career with whales and dolphins, wolves and grizzly bears, seals and horses, I would have said, "Yes, my childhood was spent in the wild with animals as my brothers and sisters, but literature and animal behavior, well, they're very separate fields."

After college, when I landed a job as an editorial assistant at the *New Yorker* magazine, the separation of city from nature, and literature from animal behavior studies, seemed as concrete and vast as the stone Manhattan canyons.

This is not to say that I didn't often observe artists or writers behaving like animals—following instinct, adapting to the complex society, using communication skills, and enduring the pecking orders of publishing. Among writers and editors, as among nonhuman animals, there was always display behavior, altruism, courtship rituals, and fierce competition.

I was somewhat prepared for this artistic career from my days in a high school symphony, one of the finest in the state of Virginia. I survived the highly stratified society of high school with its clan-like sororities and fraternities and with its symphonic challenges of first, second, and third chairs, by writing a weekly soap opera in which all the musicians were characterized as animals. I took my inspiration for this parody from George Orwell's *Animal Farm*—thus, the first flautist was a temperamental swan, the first oboe a rat, the trombone impresario an elephant, and the conductor a regal baboon. My favorite chapter of my soap opera was entitled, "Some pigs are more equal than others."

It seemed natural to me, raised on a national forest in the High Sierras and now finding myself in the wilds of an East Coast high school, to call upon my affinity for animals to understand such things as prejudice, class structure, and dominance struggles.

When I took up my first post-college job at the *New Yorker*, I diligently cultivated a mini-vegetable garden on the 22nd floor office, walking my Siberian husky dog in a park known for late-night muggings. While trapped in a city in which nature and animals are second-class citizens, I turned my memory and eye inward to create my first novel, *River of Light*.

Imagine me then, sitting in a dim office, watching the city soot settle over my miniature corn plants and cherry tomatoes, writing about the red mud, yellow rivers, and animals of backwoods Georgia. In my novel, a cow and a woman become a Greek mythic Io haunting the other characters, a snake tells the fall of the Garden of Eden, and a river overflows with the story of her people.

Critics may have called this literary license. But to me, wild-born in that Forest Service mountain lookout with only a handful of people amidst a million acres of wilderness, and with wild creatures as peers, it was not a far leap for all of my novels to portray the points of view of other animals, including dinosaurs in *Becoming the Enemy* and Mrs. Manatee in *Duck and Cover*.

It was when I was asked to collect some of my nonfiction essays on my chosen homeland, the Pacific Northwest, in my first nonfiction book *Living by Water*, that I actually made another evolutionary leap—I began to study and apprentice myself to the minds and lives of cetaceans—marine mammals. My essay on swimming with dolphins, "Animals as Brothers and Sisters," marks a sea change when I realized that I could be an artist *and* an activist. I could both artistically praise and scientifically help protect other animals.

So began more than a decade of writing about animals and, like Noah, carrying other species in the ark of my literary work. My memoir, *Build Me an Ark: A Life with Animals*, at last formally recognized both of my passions—*not just art-for-art's sake, but art for the sake of other animals.* At the same time, the field of "nature writing" began to truly flourish, both in the marketplace and within universities. As part of an Asian–American cultural exchange, I traveled to Japan and Hawaii to talk about the "culture" of cetaceans. The idea of animals having their own complicated cultures was being introduced by such scientists and visionaries as Jane Goodall, Hal Whitehead, Lindy Weilgart, and Marc Bekoff. Once we granted culture to other species, art was sure to follow. It took Western civilization several thousand years to return to what Native science and stories have always known—the First People are the animals. We learned from watching them; our hunters learned to stalk prey from Wolf, our families settled sensibly into caves abandoned by the Great Bear, our clans covered themselves with animal skins to take on the power and warmth of other great predators.

Modern physics teaches us that in the act of observing, we change what we see. Those of us who study animals, who spend our days in the field watching, noting, trying to understand their societies, must always remember this law of physics and humbly admit that just as we are observing, we are also keenly—with senses often far beyond ours—observed. And doesn't this then mean that we are changed, as well?

This realization almost overwhelmed me one morning just past dawn as I perched on a hillside in Yellowstone National Park with a biologist and a few others, our spotting telescopes dangling perpendicular from the frozen ground. We were watching the return of the first wild wolves to Yellowstone in 1995. Even from our distance, it was obvious the wolf pack was also keeping track of us—our scent on the wind, our bright specs on the hillside, and assessed as no great danger—for now. And then it happened. The telescope zoomed me in close to the alpha male who at that moment looked up from digging out a coyote den. A cool, wild gaze from deep yellow eyes pierced me so that I teetered unsteadily. I was seen just as I was seeing. For one, endless moment I was studied, deeply considered—then allowed, perhaps even accepted. Immediately the male was joined by his mate, the alpha female. For only a second longer they both stared up at our hillside, directly into our scopes. At last, they made the final decision—the male continued digging out the coyote den for supper, the female romped with her pups and a juvenile male. We observed a wild wolf pack going about its day for over 3 hours—a "rare and generous amount of field time," said our wolf biologist, who had studied gray wolves for over 20 years. A gift.

Turning to me, the biologist added with a grin, "Wolves look right through you, don't they?"

Animals do. The direct gaze of another species, the energy running between us is so real that it engages us for a lifetime of work. In my 20 years of following animals, especially marine mammals, I've been drawn again and again to their curiosity about us. One of my favorite Gary Larson cartoons shows three huge whales walking on their tails to follow tiny humans who are running away, screaming. The three whales comment on their human study animals: "Look! There are some now!" and "They seem so intelligent!" and finally "What beautiful sounds they make! Is it some sort of song?"

As I spent more time in the field—that liquid element that defines our earth as a water planet—I began to ask the same questions that animal behaviorists ask about cetaceans: "What is their world really like?" This called up in me not only the nature writer, but also the novelist, because it is through the imagination that we expand our own point of view to embrace the Other, including the "more-than-human."

The cetacean world is suffused with sound. Some dolphins, such as those in the flooded forests of the Amazon, the *boto*, are almost blind. To understand this acoustic world, we humans must imagine navigating with hearing as a primary sense. Over the years, as I listened in on hydrophones to this underwater conversation—rapid-fire bleeps and ricochets of dolphin talk, the eerie and mournful bass rumble of humpback lullabies, the blue whale's sonorous sonar scanning and perhaps mapping an abyss of underwater canyons—I heard other sounds. Disturbing and dangerous: Noise pollution from ship traffic that scientists have already documented as a cause of deafness in gray whales; under water seismic testing that has caused whales to veer from ancient migration paths; and most frightening of all—the new military sonar called LFA, or low-frequency active sonar, that the U.S. Navy and NATO plan to deploy in 80% of the world's oceans.

Through covering the issue of military sonar since 1998 as a nature writer and journalist, it became obvious to me that the government and scientists alone would not stop or limit this technology, which is dangerous to some animals. Around the world—from Greece to the Canary Islands, from the Bahamas to my home waters in the Pacific Northwest— mass strandings of whales and dolphins have been linked to military high-intensity sonar tests. I realized that only through telling the story of this sonar to a wide audience would there be a chance to make a difference. Think of what "Free Willy" did for captive orcas. So I began my fourth novel *Animal Heart* with its plot of everyday people rising up to protest and help stop the use of sonar in our oceans. Published in 2004, the book asks questions that are ethical as well as scientific. Commenting on the novel, Jean-Michel Cousteau wrote that "*Animal Heart* is full of factual information about the natural world; but more importantly, it is presented through a story which grabs our attention. I have always said, 'Science is not enough,' and feel information is most valid when it travels through the heart."

Science is not enough, nor is literature. What is equal to anything is the human heart. That is what I have discovered in my literary career intricately connected with my own and other species. There is something else that both science and storytelling share—dreams. During the 7 years that my Native American coauthor Linda Hogan and I followed the gray whale migration along the Pacific Coast for our National Geographic book, *Sightings: The Gray Whale's Mysterious Journey*, I discovered that Russian scientists had documented the fact that gray whales dream. So do birds, whose brain waves show the exact pattern during their dreams as when they are awake and singing. The notion that science has proved that animals dream was so exciting to me, because so much of my own literary work and stories have come from my dreams.

Dream time is a place that can't exactly be quantified; it exists alongside our days like one of those parallel universes that the new physics of "String Theory" posits. Add to this fact the statistic that for children, a full 80% of their dreams are devoted to animals; in adults that connection is reduced to only 20%. One has to wonder how much of animals' dreams is given over to humans? We will perhaps never know, but "in dreams begin responsibility," as the great Irish poet W. B. Yeats writes. And the responsibility of both artists and every great scientist is to listen to our intuition. As Einstein said, "The most beautiful thing we can experience is the mysterious." Perhaps intuition and full recognition of the many mysteries is where we all start, and search, whatever our discipline.

This notion led me as an editor to join forces with several coeditors to gather together scientists, naturalists, poets, and literary writers in the anthologies *Intimate Nature: The Bond between Women and Animals* and *Between Species: Celebrating the Dolphin–Human Bond*. This uniting of disciplines that often don't seem to connect, although they run on parallel lines, was very fulfilling. Sometimes navigating my way as an editor between "hard" scientists and poets I felt as if we were literally creating new synapses, new ways for specialized fields to communicate. During the long process of bringing these anthologies to life, we reached out to animal behaviorists, naturalists, and scientists—such wonderful witnesses as Birute Galdikas, Katy Payne, Roger Fouts, Sy Montgomery, and Elizabeth Marshall Thomas—who understood that only by telling vivid stories of their study animals could they truly engage a wide audience. At the same time, the poets and literary writers knew that by expanding our kinship systems to embrace the great dramas and narratives of other species, we become more than ourselves. We become, as scientists might say, "less anthropomorphic." But we also become more human, more humane. We remember, as indigenous peoples never forgot, that we are just one strand of an intricate and incandescent web.

Further Resources

Frohoff, T. & Peterson, B. (Eds.) 2003. *Between Species: Celebrating the Dolphin–Human Bond*. San Francisco: Sierra Club Books.

Hogan, L. & Peterson, B. (Eds.) 2002. *Sightings: The Gray Whales' Mysterious Journey*. Roanoke, VA: National Geographic.

Hogan, L., & Peterson, B. (Eds.) 2004, *Face to Face: Women Writers on Faith, Mysticism, and Awakening*, New York: Farrar, Straus, & Giroux.

Peterson, B. 1991. *Duck and Cover*. New York: HarperCollins.

Peterson, B. 2001. *Build Me an Ark: A Life with Animals*. New York: W. W. Norton.

Peterson, B. 2001. *Singing to the Sound: Visions of Nature, Animals, and Spirit*. Troutdale, OR: NewSage Press.

Peterson, B. 2002. *Living by Water: True Stories of Nature and Spirit*. Golden, CO: Fulcrum Publishing.

Peterson, B. 2004. *Animal Heart*. San Francisco: Sierra Club Books.

Peterson, B., Peterson, B., & Metzger, D. 1999. *Intimate Nature: The Bond Between Women and Animals*, Ballantine Books.

Brenda Peterson

■ Caregiving
Attachment Behaviors

Have you observed an infant cry and protest when removed from its mother only to stop when returned? Such behaviors indicate the infant is attached to its mother (an *attachment figure*). The evolutionary significance of attachment behavior is *inclusive fitness*. This means it is in the best interest for the survival of the infant to stay close to a primary caretaker (usually the mother) for food, warmth, protection from predators, to learn basic behaviors for survival, to learn species-specific behaviors (social and individual), and to learn to which species one belongs. All of these contribute to the potential reproductive success of the individual.

Attachment behaviors refer to a broad classification of behaviors such as clinging, crying, and approaching, as well as behaviors produced as a consequence of separation from the attachment figure. Since attachment behaviors vary across species, a number of criteria are used to identify attachment. These behaviors include:

1. preferences for an attachment figure with the ability to discriminate and respond differentially to that attachment figure;

2. responses to maintain proximity to the attachment figure;

3. differential responses to a brief separation from the attachment figure (e.g., protest) compared to other familiar stimuli. In addition, reunion with the attachment figure should diminish protest behaviors;

4. responses to extended separation from the attachment figure (e.g., despair); and/or

5. responses to the attachment figure as a secure base for exploration.

There is general consensus that the whole maturational sensorimotor organization must be taken into account when observing attachment behaviors. This organization involves at least two features. First, one or more sensory modalities must be capable of recognition and discrimination so that environments and objects become familiar. Second, the organism must be able to use the more complex "integrative" cognitive system (internal working model) to predict and relate to the world. Once these are understood, one can observe attachment behaviors unfold in a psychobiological and behavioral synchrony between the caretaker(s) and the infant.

In order to understand attachment behaviors, the observer must be fully aware of the nature of the organism they are observing. Two important factors are:

1. The sensorimotor functioning of the species under observation. Are they nocturnal or diurnal? Are they born with their eyes closed. In such cases odor, warmth, and tactile cues and possibly auditory cues (or ultrasounds) are usually the primary modalities for attachment. Later, other sensorimotor systems may come into play as they develop (e.g., vision, and visual discrimination).

2. Is the organism under observation altricial or precocial? *Precocial* (sensorimotor mature) infants (e.g., migrating herds) must attach immediately or get left behind—a sure death. If they do not attach immediately no one else will care for them. Licking by the mother insures attachment odors. The more *altricial* (sensorimotor immature) the infant the longer and more flexible the sensitive period is for attachment.

Mammals live in a diverse array of habitats and social structures. The basic unit of the family is the mother and infant. In fact, the class "mammalia" is named for her life-giving function. However, infant care varies across social structures. The infant may be cared for by the mother alone or by other conspecific (of the same species) caregivers involved to a greater or lesser extent than the mother. These caregivers can include father, siblings and/or peers, and other males and females (either relatives or nonrelatives). Infants will generally display attachment responses to those individuals that provide early nurturing contacts with the infant, even if the contact is minimal. While differences in habitats and social structures do not lead to precise predictions of social relationships, studies on attachment behavior must take these differences into account.

Attachment behaviors of primates (both human and nonhuman) have been studied the most. While development in primates differs considerably, there are some analogous features. Most sensory organs necessary for attachment are functional at birth or shortly thereafter and continue to develop and change (which is important for individual recognition). They are generally altricial, necessitating a long period of nursing. They generally only have one infant, which is nursed by the mother, although the father, siblings, or other conspecifics may become involved in infant care. Since brain development takes a long time, primates generally have fairly long periods of gestation and dependency during infancy. While some primates give birth in a nest and their infants stay with them until they are mobile, most primates carry their infants with them and the infants maintain contact by clinging to their mothers. This physical contact appears to be an important variable in the formation of primate attachments.

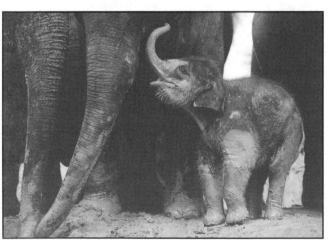

This baby Asian elephant will stick close to its mother for a good portion of its youth.
© *Michael Nichols/National Geographic Image Collection.*

There is considerable evidence that mothers who were atypically reared as infants may reject, neglect, abuse and/or kill their own infants. This is especially true of human and nonhuman primates, but can occur in a variety of mammals. Paradoxically, abused infants may display heightened attachment responses to abusing caretakers.

Mammals reared in isolation may show hyperagressiveness, hyperfearfulness, sexual dysfunctions, motor disintegration, social communication deficits, learning deficits, and many bizarre self-directed behaviors including self-injurious behaviors. Mammals reared in isolation generally show little affinity for conspecifics. Dysfunctional attachments in humans have been empirically tied to wide spectrum of clinical psychiatric disorders and stress related diseases.

One important outcome of conspecific attachments is species identity. Species identity simply refers to the fact that the infant affiliates with and prefers members of the species that are representative of the attachment figure. Normally, these would be conspecifics. However, when an infant is raised individually with members of a different species, it comes to prefer members of that species. For instance, if a kitten is reared with rats, it attaches to the rat species and won't kill rats similar to the ones it was reared with. A kitten reared with a dog litter prefers dogs to other cats. Dogs or cats reared alone by humans orient their social responses almost exclusively toward humans. Apes reared with humans prefer humans to conspecifics, and since apes also have self-recognition, they perceive themselves to be part of the human species and orient their social and sexual behavior exclusively toward humans. There is little doubt as to the evolutionary significance of normal attachment behaviors.

See also Caregiving—*How Animals Care for Their Young*
Caregiving—*Parental Behavior in Marsupials*
Caregiving—*Parental Care*
Caregiving—*Parental Care and Helping Behavior*
Caregiving—*Parental Care in Fish*
Caregiving—*Parental Care in Insects*
Caregiving—*Parental Investment*

Further Resources

Bowlby, J. 1969. *Attachment and Loss: Vol. 1. Attachment.* New York: Basic Books.
Gubernick, D. J. 1981. *Parent and infant attachment in mammals.* In: *Parental Care in Mammals* (Ed. by D. J. Gubernick & P. H. Klopfer), pp. 243–305. New York: Plenum Press.
Guyot, G. W. 1998. *Attachment in mammals.* In: *Comparative Psychology: A Handbook.* (Ed. by G. Greenberg & M. Haraway), pp. 509–516. New York: Garland Publishing, Inc.
Roy, M. A. (Ed). 1980. *Species Identity and Attachment: A Plylogenetic Evaluation.* New York: Garland STPM Press.
Scott, J. P. 1968. *Early Experience and the Organization of Behavior.* Belmont, CA: Wadsworth.

Gary W. Guyot

■ Caregiving
Brood Parasitism among Birds

If you were raised by a pack of wolves, would you grow up knowing that you were a human, not a wolf? In Rudyard Kipling's Mowgli stories, Mowgli needed some coaching by his bear and panther friends to recognize that he was human. Of course, it would be very unusual for humans to be raised by a pack of wolves or by any other species, but the "what if" question has intrigued people for quite some time through numerous jungle stories about characters such as Mowgli and Tarzan.

For some species of birds and a few other animals, being raised by other species is a regular way of life. Among birds, this occurs when the mother lays her eggs in other birds' nests. This peculiar way of reproducing is called *brood parasitism*; the *brood parasite* lays her eggs in other nests, and the *host* incubates the parasitic eggs and raises the chicks. Brood parasites are divided into two major categories: (1) species that parasitize other nests but may also raise their own young in their own nests—called *nonobligate brood parasites*, and (2) species that must parasitize other birds—called *obligate brood parasites*. Nonobligate brood parasites have the option of parasitizing their own species as well as other species. Obligate brood parasites, on the other hand, over time have lost the ability to build nests and incubate eggs; therefore, they are obliged to parasitize the nests of other species.

Almost 100 bird species (about 1% of all birds) are obligate brood parasites. These include some (but not all) cuckoos, cowbirds, African finches, honey-guides (related to woodpeckers), and one duck from South America. Some of these species parasitize only one or a

Warbling vireos are one of the brown-headed cowbird's favorite hosts. They rarely raise any young of their own in parasitized nests.
Courtesy of Catherine Ortega.

few species of hosts while others parasitize a wide variety of species. For example, some of the parasitic finches and some cuckoos specialize on only one or a couple hosts, whereas brown-headed cowbirds (*Molothrus ater*) in North America parasitize at least 226 hosts. Sometimes brood parasites "make mistakes" or lay in inappropriate nests. An inappropriate host is one that is considerably larger or smaller than the parasite, rejects parasitic eggs, or will feed the parasite an unsuitable diet. Brown-headed cowbirds have been known to occasionally lay their eggs in hummingbird nests and hawk nests, but neither of these birds would actually raise a cowbird.

So, how does a cowbird know it is a cowbird when it is raised by a yellow warbler (*Dendroica petechia*) or a red-winged blackbird (*Agelaius phoeniceus*)? As much as we would like to have the answer, we do not know how cowbirds or other brood parasites figure this out, but it is clear that they know exactly who they are—no matter who raises them. For instance, by the time a cowbird is about 60 days old, they join other cowbirds in flocks and migrate south with the adults.

How do parasites get away with it? In some cases, they probably dupe their hosts; in other cases, the hosts may be helpless to do anything about parasitic eggs in their nests. Hosts are faced with two problems when a parasitic egg appears in their nest: (1) They may not recognize that the egg is not theirs, or (2) even if they do recognize the odd egg, they may not be able to get rid of it because the egg is too large. Brood parasites, however, do not always get away with their behavior. Some potential hosts eject brood parasitic eggs from their nests either by picking them up whole in their bill (*grasp ejection*) or by puncturing the egg and flying away with the egg impaled on their bill (*puncture ejection*). Large birds, such as robins and jays, with large bills can grasp eject easily, but small birds, such as warblers, do not have bills large enough to pick up a parasitic egg.

Why don't they just puncture the parasitic egg? If they puncture the egg, the contents of the egg may spill onto their own eggs; this can cause their own eggs to adhere to the nest and not hatch. The odor may also attract predators to their nests. Additionally, the eggs of nearly all obligate brood parasitic species are very round compared with other eggs, and they have an extraordinarily thick shell—up to 30% thicker that other birds' eggshells. When hosts try to puncture the parasitic egg, their bills often slip off the egg, and they puncture their own eggs instead. Even though Bullock's orioles (*Icterus bullockii*) and Baltimore orioles (*Icterus galbula*) have bills large enough to grasp eject cowbird eggs, they puncture eject them instead—probably because it would be difficult to remove them whole from their deep hanging nests. Several experiments have been conducted to determine the cost of puncture ejection. Cowbird eggs and other more easily punctured eggs were added to oriole nests; in each study, the damage to oriole eggs was significantly higher in the nests with cowbird eggs. This may explain why very few hosts puncture eject parasitic eggs.

Some hosts are aggressive toward brood parasites and try to drive them away. Many experiments have been performed to determine if hosts can distinguish between brood parasites and other types of birds. Typically, stuffed brood parasites and controls (other birds, such as sparrows) are placed near nests, and host behavior toward the mounts is analyzed. Although the methods for experiments vary, and the experiments are far from perfect, it is clear that many hosts behave more aggressively toward brood parasites than toward other nonparasitic birds. However, aggressiveness is probably a poor indicator of how well hosts can prevent parasites from accessing their nests. Brood parasites can be extremely persistent and stealthy, waiting for just the right opportunity to hop on a host nest. Furthermore, brood parasites lay their eggs faster than any other birds—in as little as 30 seconds.

Most brood parasites are nomadic or partially nomadic, following their food resources, or they are closely tied to roaming mammals. Cuckoos wander around looking for outbreaks of caterpillars (one of their primary food resources), and parasitic finches wander the African savannas looking for seeding grasses. Honey-guides, from Africa and

A brown-headed cowbird egg (heavily spotted egg on the right) is incubated along with the four warbling vireo eggs. Brown-headed cowbirds almost always hatch 4–5 days before the vireos. *Courtesy of Catherine Ortega.*

Asia, are named after their behavior of guiding badgers to beehives, where they feed on pieces of honeycomb after the mammal has plundered the nest. Cowbirds are strongly associated with cattle and other large grazing mammals; they feed on insects stirred up by the mammals. Before Europeans brought cattle to North America, cowbirds followed migrating herds of buffalo, and in older literature, they are referred to as buffalo birds. Brood parasites are not tied down to a particular nest site as other birds are; therefore, they are at liberty to take advantage of these wandering food resources.

Most brood parasites have behaviors that enhance the probability of success in their hosts' nests—to the detriment of their hosts. Honey-guides have a dagger-like bill when they hatch, and they stab to death the chicks of their hosts (called *host siblings*); this helps to eliminate competition for food between the parasite and host siblings. Other behaviors also eliminate competition. For example, some cuckoo nestlings dump their host siblings or host eggs overboard by backing up to the chicks and/or eggs and forcing them out of the nest with a backward sweep of their wings. The adults of some adult cuckoo species remove a host egg from the nest when they lay their own eggs. Brown-headed cowbirds also sometimes remove host eggs, but apparently not all individuals engage in this behavior.

Brood parasites can affect the success of their hosts in other ways. They can cause some hosts to abandon their nests. They are also often (but not always) larger than their hosts, and it is usually the largest noisiest mouth that gets fed by the parents. More importantly, however, the incubation period (the time between egg laying and hatching) of most brood parasites is shorter than the incubation period of most of their hosts. This results in the brood parasite typically hatching several days earlier than the host chicks, and by the time the host eggs hatch, the brood parasite takes up most of the nest and most of the parent's attention. In these cases, often the hosts raise only the brood parasite and none of their own.

Many people do not like brood parasites because they find their behavior immoral. Some people even want to kill brood parasites or remove their eggs from host nests. However, it is not wise to impart human standards on the behavior of other animals because they are not human, and we should not expect them to act like humans. This behavior, as bizarre as we may find it, is just another fascinating way of reproducing.

See also Caregiving—*Brood Parasitism in Freshwater Fish*

Further Resources

Lowther, P. E. 1993. *Brown-headed Cowbird* (Molothrus ater). In: *The Birds of North America, No. 47.* (Ed. by A. Poole & F. Gill), pp. 1–23. Philadelphia, PA: The Academy of Natural Sciences; Washington, D.C.: The American Ornithologists' Union.

Lowther, P. E. 1995. *Bronzed Cowbird* (Molothrus aeneus). In: *The Birds of North America, No. 144.* (Ed. by A. Poole & F. Gill), pp 1–14. Philadelphia: The Academy of Natural Sciences; Washington, D.C.: The American Ornithologists' Union.

Lowther, P. E., & Post, W. 1999. *Shiny Cowbird* (Molothrus bonariensis). In: *The Birds of North America, No. 399.* (Ed. A. Poole & F. Gill), pp 1–24. Philadelphia: The Academy of Natural Sciences; Washington, D.C.: The American Ornithologists' Union

Ortega, C. P. 1997. *Social outcasts: Brown-headed Cowbirds are ingenious pests and a natural part of our ecosystem in Colorado.* Colorado Outdoors, 46(4), 22–25.

Ortega, C. P. 1998. *Cowbirds and other Brood Parasites.* Tucson: University of Arizona Press.

Robinson, S. K. 1996. *Nest Losses, Nest Gains.* Natural History Magazine, 105(7), 40–47.

Wyllie, I. 1981. *The Cuckoo.* New York: Universe Books,.

Catherine P. Ortega

■ Caregiving
Brood Parasitism in Freshwater Fish

Obligate brood (or social) parasites use hosts for providing parental care to the parasitic egg and/or young. Until recently, it was thought that this system occurred only in insects and birds. However, in 1985, a fish, *Synodontis multipunctatus*—the cuckoo catfish, that parasitizes mouthbrooding cichlid fish was found in Lake Tanganyika. Other than parasitic birds (e.g., cuckoos and cowbirds), the cuckoo catfish appears to be the only other known case of a vertebrate obligate brood parasite (i.e., total dependence upon a host for breeding). Brood parasitism has also been reported in other fish, but such cases are examples of facultative parasitism, where parasites occasionally use other hosts.

Mouthbrooding, in which one of the parents carry the eggs and fry in the mouth, is one of the most advanced parental care systems known among fish. Among Tanganyikan cichlids, mouthbrooding is well developed. It is effective in protecting young against predators. In the prototypal situation, a mouthbrooding pair circles in the breeding area, with the females dropping eggs. After oviposition, the eggs are taken into her mouth (buccal cavity) where they are brooded (oral incubation). Depending upon the species, fertilization might occur in the mouth, externally,

*Cichlid host (*Tropheus sp.*) with cuckoo catfish fry.*
Illustration by Alexander Cruz, Jr.

or a combination of both. The zygote undergoes development to fully formed juvenile, while using its yolk for nutrition. The amount of time in the mouth varies with species, but averages 18–21 days.

During a study of Tanganyikan cichlids, scientists collected 32 females of six species containing eggs or fry of the cuckoo catfish. In other lake areas, scientists found few parasitized females. However, under aquarium conditions, catfish have only successfully bred when using cichlid as hosts. In addition to hosts endemic to Lake Tanganyika, aquarists have also used hosts from different rift lakes (e.g., Malawi and Victoria). Thus, although the catfish has a very specialized reproductive mode, it appears to be a host generalist.

As the cichlids spawn, catfish enter the spawning area and simultaneously lay and externally fertilize eggs, while at the same time eating some of the cichlid eggs. Catfish are able to release small numbers of eggs (10 to 20 eggs at a time) and to repeat this 5 to 8 times during the host spawning, notwithstanding the cichlid's attempt to chase off the catfish. The same spawning female is able to release eggs over a period of 2 to 3 days, thus increasing the probability of successful parasitism. The end result is that the host collects both egg types, even though catfish eggs are 1/2 to 2/3 (3 mm vs. 4 to 5 mm) the size of host eggs and differ in appearance. The cichlid eggs are yellowish to light brown and opaque, while the catfish eggs are whitish and translucent. The catfish eggs develop more rapidly than the host eggs, and the yolk sac is absorbed 3 to 4 days after hatching. Following absorption, the catfish fry feed upon the host eggs, then the fry. Thus, the fry depend upon their hosts for food and protection, but also exploit almost their entire parental investment. The end result in many instances is the loss of the entire cichlid brood as a result of parasitism.

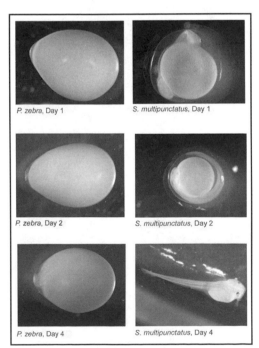

*Early sequence of development (day 1 to day 4) in cuckoo catfish (*Synodontis multipunctatus*) and host (*Pseudotropheus zebra*).*
Courtesy of Alexander Cruz.

In one study, the mean number of catfish young per cichlid brood was 6.7, and in one 3-month period, 31% of the broods were parasitized. Just hatched catfish are 6 mm (.24 in) in length and barbels are apparent, and within 4 to 6 weeks they resemble their parents.

In contrast to parasitic birds in which a host nest can be successfully parasitized for a period of a week, the window of opportunity is much less for the catfish, encompassing the time it takes for the cichlid to take the eggs into her mouth, a period averaging an hour. Close synchrony is essential for the parasite's success because there are just a couple of seconds between the host's egg release and the collection of eggs in her mouth. How the catfish synchronizes its breeding with the host is unknown, but observations suggest that is a combination of both visual and pheromonal cues.

Studies suggest that learning is involved in the host behavior towards the catfish. In a study comparing zebra cichlid (*Pseudotropheus zebra*) aggression in populations with and without previous catfish exposure, it was found that individuals that had experienced parasitism (37%) were more aggressive toward catfish than individuals that did not have any prior contact. Since zebra cichlids are from Lake Malawi, and thus have not coevolved with the catfish, we can rule out an evolutionary component. It was also found that hosts that had been parasitized several times by catfish decreased or stopped breeding.

The interactions between catfish and cichlids are an interesting example of a host–parasite system. Because of the close interactions between parasite and host, a study of such interactions can reveal much about coevolutionary mechanisms. The successful captive breeding of catfish with both Tanganyikan and non-Tanganyikan hosts makes it a model system to study host–parasite interactions.

Further Resources

Allen, B. 1995. *Breeding* Synodontis multipunctatus. Tropical Fish Hobbyist, 43, 146–162.

Barlow, G. W. 2000. *The Cichlid Fishes—Nature's Grand Experiment in Evolution.* Cambridge, MA: Perseus Publishers.

Fisher, R. M. 1987. *Queen-worker conflict and social parasitism in bumblebees* (Hymenoptera: Apidae). Animal Behavior, 35, 1026–1036.

Keenleyside, M. H. A. 1991. *Cichlid Fishes: Behavior, Ecology, and Evolution.* New York: Chapman Hall.

Ortega, C. 1998. *Cowbirds and Other Parasites.* Tuscon: The University of Arizona Press.

Rossiter, A. 1995. *The cichlid fish assemblages of Lake Tanganyika: Ecology, behavior, and evolution of its species flocks.* Advances in Ecological Research, 26, 187–252.

Sato, T. 1986. *A brood parasitic catfish of mouthbrooding cichlid fishes in Lake Tanganyika.* Nature, 323, 58–59.

Wisenden, B. D. 1999. *Alloparental care in fishes.* Reviews in Fish Biology and Fisheries, 9, 45–70.

Alexander Cruz, Janelle Knox, & Scott Pawlowski

Caregiving
Fostering Behavior

Fostering occurs when individuals other than biological parents provide care to offspring. The degree of care ranges from babysitting, which may or may not include feeding the young, to extended periods of complete care. Babysitting has been observed in a range of species (e.g., primates, canids, ungulates, and cetaceans) and usually involves one adult guarding the young of other group members so that parents may forage more efficiently.

Because babysitting is usually brief in duration, rarely includes feeding the young, and is generally reciprocated (i.e., other adults will later babysit the initial sitter's offspring), the cost to individuals is minimal. In contrast, providing more extensive parental care can be very costly in terms of time and energy. Since the latter type of fostering reduces the resources available for rearing one's own offspring, natural selection should minimize its occurrence unless fostering provides some other advantage to the providers. The cost should dictate the degree of benefits required for fostering to be maintained in a population.

Fostering has been observed in over 100 mammals and birds despite the fact that it reduces individuals' individual fitness (i.e., reduces their ability to rear their own offspring and hence pass on copies of their genes). Hypotheses proposed to explain the occurrence of fostering include:

1. *kin selection*, where individuals care for relatives' offspring and therefore increase their inclusive fitness (i.e., their help increases the number of copies of their genes passed on via relatives' offspring who also share their genes);

2. *parental experience*, where individuals improve their chances of rearing their future offspring by gaining parental experience caring for someone else's young;

3. *reciprocal altruism*, which can occur when long-term associations among unrelated individuals allow for an exchange of fostering care; and

4. *reproductive errors*, which occur when animals are unable to identify their young and so inadvertently expend time and energy caring for an unrelated offspring.

In the latter case, fostering is not evolutionarily adaptive, and should become less common if it exacts a high enough cost.

Among mammals, fostering is often preceded by a female losing her own offspring. For example, harbor seals, which breed in fairly dense colonies, foster only after losing contact with their biological pups. Separation from pups typically occurs because of storms, shark attacks or because the pups wander from where their mother left them when she went to hunt. Approximately 5–20% of females foster in a given breeding season (fostering duration ranges from less than 1 hour to more than 18 days, the normal lactation period). Although younger females are more likely to lose contact with their pups and hence to be available for fostering, the proportion of available females that foster is similar across all age groups. It therefore seems unlikely that fostering is maintained in the population solely due to advantages gained from additional maternal experience. Instead, fostering appears to be a consequence of a reproductive error—females being unable to recognize their own pups. In fact, once a female has adopted another's offspring, she may fend off her own pup if it approaches and attempts to nurse. Genetic studies indicate that females do not preferentially foster relatives' young (no direct kin selection), but colonies may represent groups of related individuals since animals tend to return to the colonies where they were born. Hence, it is possible that fostering in harbor seals is maintained because the cost of the error is offset by gains in inclusive fitness.

Fostering among mammals is also associated with cooperative breeding in social species. Cooperative social groups provide communal care of young in response to especially harsh or unpredictable environments (e.g., African wild dogs) or exceptionally crowded living conditions (e.g., Mexican free-tailed bat). Social groups may also permit cooperative foraging strategies (i.e., pack hunting) for large prey or more variable food sources (e.g., lions, coyotes, etc.). Cooperative groups are often composed of related individuals and hence, cooperative breeding can be the result of reciprocal altruism and/or kin selection (e.g., wolves, elephants).

A common example of fostering among birds is also due to kin selection. This fostering, usually referred as helpers-at-the-nest, is performed by the older offspring of a breeding pair of birds who stay and help their parents raise subsequent siblings. Helpers-at-the-nest are common when there is habitat saturation (i.e., all suitable territories are occupied) and so the younger, less competitive individuals are unlikely to be able to access the resources needed to breed successfully (i.e., mates, nesting sites, and food). Florida scrub jay adults will actually sabotage their offspring's breeding attempts so that the young will continue to help the parents (i.e., increase the number of biological offspring successfully produced by the parents, parents' individual fitness). The helpers offset the cost of their delayed reproduction by rearing related offspring (i.e., increase their inclusive fitness) and by gaining important parental experience.

Reproductive errors induced by brood parasites also account for a significant proportion of fostering among birds. Brood parasites, such as brown-headed cowbirds and European cuckoos, place their eggs in other species' nests and these hosts, if unable to detect the parasites' eggs, raise the parasites' offspring for them. Because the costs to hosts are not offset by gains in inclusive fitness or useful parental experiences, many species have developed improved abilities to recognize their own eggs or when extra eggs are present. Parasites, in turn, tend to have eggs that look similar to host eggs and/or to adopt useful behavioral strategies (i.e., stealthy approaches to host nests and removal of host eggs before despositing their own egg). In the United States, the parasites are often more successful than the hosts in this evolutionary arms race, and many species of songbirds have experienced dramatic reductions in numbers (i.e., populations are lost because adults don't raise enough biological offspring). Within-species parasitic egg dumping also occurs, especially in colonial breeding species. In species where females often deposit eggs in other females' nests, potential hosts may resort to infanticide to minimize the chances of unintentional fostering (e.g., gulls).

Fostering is generally restricted to species such as mammals, birds, and a small number of social insects that provide extended parental care. The frequency of fostering varies inversely with its cost to the providers. As reviewed above, costs can be offset by a variety of benefits including increased inclusive fitness, parental experience, and reciprocal altruism; and selection should act to enhance individuals' abilities to avoid unintentional fostering. Fostering behavior is frequently observed among captive animals; however, given the unusual selection pressures associated with captivity, this behavior may not reflect patterns present in wild populations. Extrapolations from data on captive animals should therefore be made with caution.

See also Caregiving—*Helpers in Common Marmosets*
　　　　　Caregiving—*Non-Offspring Nursing*
　　　　　Caregiving—*Parental Care and Helping Behavior*

Further Resources

Brown, K. M. 1998. *Proximate and ultimate causes of adoption in ring-billed gulls*. Animal Behavior, 56, 1529–1543.

Job, D. A., Boness, D. J. & Francis, J. M. 1995. *Individual variation in nursing vocalizations of Hawaiian Monk seal pups* Monachus schauinslandi *(Phocidae pinnipedia) and lack of maternal recognition*. Canadian Journal of Zoology, 73 (5), 975–983.

Roulin, A. 2002. *Why do lactating females nurse alien offspring? A review of hypotheses and empirical evidence*. Animal Behavior, 63, 201–208.

Cathy Schaeff

■ Caregiving
How Animals Care for Their Young

In order for a species to survive, sexually mature members of the species must produce a sufficient number of offspring that will themselves reach sexual maturity and reproduce. Otherwise, the species will become extinct. Species have evolved a variety of methods to insure that enough of their young survive to produce the next generation. For example, species vary considerably in terms of the care that parents provide for their offspring. Some parental care is directed toward the young animal, as when parents provide food, transportation, or warmth. Parental care may also involve behaviors that are not directed toward their offspring in as obvious a fashion, but nonetheless benefit the young animals. Examples of indirect parental care include the acquisition and defense of territory, the building of shelters, and the luring away of predators.

As humans, we are used to the notion that parents care for their children. Without extensive parental care, young humans would not survive. Human infants are an example of an *altricial* species, meaning that the young require a great deal of care. Other altricial species include dogs, cats, and altrices, a species of birds whose young remain in the nest for a comparatively long time (and the species from which the term altricial is derived). In contrast, the young of some species are *precocial*, meaning that the young can move about and perceive much of their world shortly after birth. Precocial species include ducks, elephants, and dolphins.

Regardless of whether young animals are altricial or precocial, their survival often depends on the care their parents provide. The type of parental care depends in a part on the species of the parents. For example, consider the many types of animals that produce eggs. These eggs will eventually hatch into young animals if they are not destroyed by the elements or eaten by predators. One might imagine that egg-laying animals would take great care to insure that their eggs survive. However, some species, such as the crown-of-thorns starfish, release large numbers of eggs into the sea at a time. The sheer volume of eggs insures that at least some offspring will survive. In contrast, many other egg-laying animals prepare nests before the mother lays the eggs. These nests help to protect the eggs from the elements and may also provide some degree of safety from predators. However, once they prepare their nests and lay their eggs, a surprising number of these animals depart and leave the eggs to their fate. For example, sea turtles create a nest for their eggs, lay the eggs, cover the eggs with sand, and then return to the water. Baby turtles must hatch themselves and then find their way to the water without being preyed upon by birds, foxes, or other predators. Other animals care for eggs until the offspring hatch. For example, male seahorses protect fertilized eggs by carrying them around in a special compartment called a brood pouch. However, once the baby seahorses hatch, they must find their own food and shelter. Other species, like the altrices, care for their young long after they have hatched. Thus, the range of parental care in egg-producing animals is vast. Some parents simple release their eggs when the time is right. Others build nests in which to lay their eggs, and then leave the eggs to fend for themselves. Still others care for the eggs until they hatch, whereas others care for their young after they have hatched.

It is not only egg-laying animals that exhibit a range of parental care. Many animals provide protection, transportation, shelter, and food for their offspring. Oftentimes, such care is provided only until the offspring are capable of obtaining their own food. However, the offspring of species such as geese, swans, human and nonhuman primates, and killer

Helpers in Common Marmosets

Augusto Vitale

An interesting topic of research for ethologists is the altruistic behavior. An altruistic behavior is described when an individual performs an act that increases the probability of survival of another individual, with a correspondent disadvantage for itself. A classic example is the behavior shown by some species of birds, nesting on ground, in defense of their nestlings. In this case, when a predator approaches the nest, the parent moves away from the nest, pretending to be injured. The potential predator then ignores the nest, and follows the supposedly injured bird that represents an easier prey. This behavior saves the chicks from the attention of the predator, but it is very dangerous for the parent who, in any case, takes off when the predator is distant enough from the nest. Therefore, the parent has temporarily decreased its probability of survival by attracting the predators to itself, and increased the probability of survival of its chicks.

This is an example in which parental care can be described as a form of altruistic behavior. Among the different forms of parental care, of particular interest are the behaviors labeled as alloparental care. In this case, all of the behaviors aimed at the growth and maturation of the young are shown not only by the real parents, but by other members of the social group, too. These latter individuals can be called, in ethological terms, "helpers." This behavior has been observed in different species of birds and mammals.

An interesting example of alloparental care comes from a small Brazilian monkey, the common marmoset, or *Callithrix jacchus*. This species lives in the Atlantic forest of the northeast Brazil in family groups composed by a minimum of 3 to a maximum of 15–16 individuals. In general, in each family, there is only one adult breeding pair, who are the parents of a series of offspring present in the family. The breeding female gives birth to twins, and the gestation time is about 5 months. The interesting aspect of the breeding behavior of the common marmoset is the communal care of the infants. This means that, in this species, not only the mother cares for the infants, but the father, the other adults and the older siblings care for the youngest members of the family, too. One of the reasons for the popularity of this kind of altruistic behavior in marmoset species can be indeed the regular presence of twins: Simply, the mother cannot care for two babies all of the time by herself. It requires too much energy from just one individual who has, among other things, to provide milk for the two infants. To care for the infants mainly means to carry them around, but observations both in nature and captivity have also shown that helpers actively feed the infants. This is observed especially when the infants are not yet able to capture live prey, such as insects.

The helpers are most active during the first 4 weeks of age of the little marmosets. Studies in captive colonies have shown that the older the helper, the more that helper cares for the infants. As a matter of fact, the adults of the family, beside the mother and the father, carry the infants for longer periods and more often than younger family members. Furthermore, they are also the first ones to intervene when the infants give signs of stress and cry out for help. In general, males appear to be more actively involved than females. However, despite these age- and sex-related differences, all of the members of a family of marmosets show a great interest in and excitement toward the infants. It also seems that social status affects the amount of care given to infants. In the family there is competition for carrying the infants and, usually, the individuals who have a higher social position within the group have privileged access to infants.

When it comes time to feed the infants, the marmosets are unique among the nonhuman primates in the amount of solid food that is offered to the youngest. As a matter of fact, in different marmoset species, infants are provisioned almost entirely when they begin to eat solid food, then gradually develop into independent feeders. Usually the mother does not feed the infants in this way, leaving it to the helpers in the family: The mother is already giving milk to the infants, and so she is not willing to search for extra food. Sometimes, the helpers do not actively give solid food to the infants, but, as it has been observed in the wild, they are very tolerant when the infants steal from their mouth very appetizing food items, such as insects.

An interesting question is: "Why do helpers help?" One hypothesis formulated by the researchers is that to be a helper is a convenient behavior. If you help you can remain within the family, and there are fewer possibilities to be sent away from the group when the food is in short supply for all of the members of the family. As a matter of fact, for a marmoset, to be alone in a forest is a very risky business. There are several type of predators that feed on marmosets: felids, eagles, snakes. If a marmoset remains in the group, it can take advantage of the protection against predators offered by the family. Furthermore, to be in a group makes it easier to find food, than being alone. Finally, the younger helpers gain an experience that can be helpful for their future life: It has been shown that the helpers will be better parents than others who had not had the opportunity to help their parents to care for the younger siblings.

See also Caregiving—*Fostering Behavior*
　　　　　Caregiving—*Non-Offspring Nursing*
　　　　　Caregiving—*Parental Care and Helping Behavior*

Further Resources

Clutton-Brock, T. H. 1991. *The Evolution of Parental Care*. Princeton, NJ: Princeton University Press.

McDonald, D. W. (Ed.) 2001. *The New Encyclopedia of Mammals*. Oxford: Oxford University Press.

Mittermaier, R. A., Rylands, A. B., Coimbra-Filho, A. & Fonseca, G. A. B. (Eds.) 1988. *Ecology and Behavior of Neotropical Primates* (Vol.2). Washington DC: World Wildlife Fund Press.

whales receive parental care long after they are able to forage or hunt for themselves. In such cases, parents continue to protect their young from predators, and sometimes teach or model important behaviors. Extended parental care may also provide supervised opportunities to interact with other members of one's species. For example, killer whale calves remain with their mothers for many years. Killer whale mothers protect their calves from predators and from overly aggressive whales as well as providing nourishment when the calves are young. In addition, killer whale mothers provide models of foraging behaviors, such as swimming up on a beach to catch a seal pup. They may also help their offspring learn proper modes of social interaction with other killer whales and so solidify their relationships with other animals in their social group. These latter abilities take time to learn, and so killer whale calves stay with their mothers long after they have become able to locate and catch their own food.

Thus far, we have considered parental care without much attention to which parent provides the care. For egg-laying fish, males are more likely to protect the eggs if fertilization takes place externally (the eggs and sperm are both released into the water). However, the mother is more likely to be the sole guardian if internal fertilization occurs (the male deposits his sperm into the female and the eggs are fertilized inside the female). The young of most bird species receive care from both parents, most likely because it takes two adults to provide sufficient food for the hatchlings. For some mammalian species, such as Titi monkeys, prairie voles, marmosets, wolves, and African hunting dogs, both parents help to raise the young. However, this is the exception rather than the rule. The mother is the primary caregiver for most mammals (e.g., elephants, horses, sheep, rats, rabbits, gerbils, cats, dogs, sea lions, seals, dolphins, whales, chimpanzees, monkeys, and gorillas) as well as certain insects, amphibians, and reptiles. For this incredible range of species, it is the mother that provides the offspring with food, shelter, and transportation (if necessary). She may also protect her young from predators and aggressive members of the same species, such as peers or adult males. The mother may also provide learning experiences that facilitate the acquisition of foraging and social skills. However, not all mothers are the same.

Maternal Styles

The underlying purpose of maternal care is the same for all species. *Maternal care* helps offspring to survive to maturity so that they may reproduce. However, the manner in which mothers accomplish this goal varies from species to species and from individual to individual within a species. Individual "maternal styles" have been found in a number of species (including humans). These include nonhuman primates (e.g., yellow baboons, rhesus monkeys, vervet monkeys, pigtail macaques), marsupials (e.g., red-necked wallabies and eastern grey kangaroos), and marine mammals (e.g., bottlenose dolphins). These differences affect the behavioral and social development of their offspring. Some of the best evidence for this comes from studies of human mothers. If they are sensitive and consistent in the care of their children, human mothers tend to produce children that seek their mother in times of illness, fatigue, fear, or stress. Children of such mothers are better able to deal with stressful situations than are children of neglectful mothers. Thus, maternal style has consequences for offspring.

For many species of primates (including humans), individual maternal styles fall along a continuum that ranges from *highly protective* to *severe rejection*. Maternal protectiveness involves behaviors that promote contact and proximity with infants. For example, a protective mother might restrain her infant in order to keep the infant near and out of harm's way. When the infant is not close, a protective mother might approach the infant and initiate contact with the infant. Protective mothers also engage in a variety of nurturing behaviors. For example, they may inspect or groom their infants, which promotes both the infant's physical health and the mother–infant bond. In contrast, mothers that are at the opposite end of the continuum are more likely to break contact, reject, or leave their infants.

There are also more subtle differences than the above polar opposites. For example, a group of yellow baboon mothers was characterized by two different maternal styles, neither of which involved the mothers rejecting the infants. Some mothers restricted their infants' attempts to leave the mother and explore the world. As a result, these mothers maintained contact with their infants for a longer period of time. Other mothers were more tolerant of infant-initiated separations and stopped restraining their infants at younger ages. These differences may reflect each mother's unique social status. Mothers that are lower in rank are often harassed by animals that have higher status in the social hierarchy. As a result, lower-ranking mothers may be more fearful and so more protective of their infants.

There are also species differences in regard to maternal protectiveness. For example, bonnet macaques allow other members of their group to hold their infants the first day of life. In contrast, rhesus macaques are much more possessive of their infants. These differences may be related to differences in the overall level of aggression of the two species. Bonnet macaques are less aggressive than are rhesus macaques, and so a rhesus mother may be more fearful that another monkey will harm or steal her infant. Although both within and between species differences exist, it is not yet clear how these differences affect the development of offspring. Part of the problem is deciding exactly what maternal differences mean. A yellow baboon mother that keeps her infant close at hand may produce an infant that is strongly bonded to her while also decreasing the chances that harm will befall the infant. However, a mother that allows her infant to explore the world may produce an infant that is more capable of taking care of itself if the infant survives its curiosity. What is clear is that a mother that rejects her infant places the infant in peril. But not all nonprotective behavior involves maternal rejection. There are more than two types of maternal styles.

For example, a number of maternal styles have been found in bottlenose dolphins. Some dolphin mothers maintain a proximity to their calves. They also monitor their calves' activities and are quick to intervene if the calf appears to be in some sort of trouble or strays too far from the mother. They will also discipline their calves if the calves wander too far or fail to return when called. In contrast, other mothers rarely remain close to their calves. These mothers also fail to monitor their calves' activities and rarely intervene on their calves' behalf or discipline the calf. Most mothers fall somewhere between these two extremes and alter their maternal behaviors as their calves mature. For example, a dolphin mother might insist that her young calf remain close by when the calf is young and discipline that calf when it strays too far. But as the calf matures, the mother is likely to allow the calf to roam more freely and to monitor the calf's behavior less diligently.

In most mammals, including dolphins, the mother is the primary caregiver. Maternal styles of dolphins fall on a continuum from very protective to permissive.
Courtesy of Stan A. Kuczaj.

Although the above example illustrates one way in which maternal style may change over time, some maternal styles remain remarkably stable. For example, some primate mothers maintain individual maternal styles throughout the entire childhood of their offspring. In addition, some mothers retain the same maternal styles for each of their offspring. An example of consistent maternal styles across offspring would be a protective mother of one offspring that was also protective of her next youngster. Consistent maternal styles across different offspring have been found for primate, marsupial and dolphin mothers. The stability of maternal styles is sometimes reflected in the transmission of mothering styles from one generation to the next. Mothers may pass their maternal style along to their daughters. For example, the daughter of a mother that rejected her may also reject her offspring. Dolphin mothers, like those of many other species, vary in terms of the stability of their maternal styles. Some dolphin mothers maintain the same maternal style throughout a calf's development, while others change their mothering behaviors as their calves mature. In addition, some dolphin mothers exhibit the same maternal style with each of their calves, whereas others change maternal styles as they have new calves. For example, one dolphin mother was very restrictive and protective of her calf during its first year of life. In stark contrast, she was much more permissive with her next calf that was born 5 years later. Although it is not clear why she changed her maternal style, there was a significant change in the social environment. When the first calf was born, an adult male was part of the social setting. The adult male was no longer present when the second calf was born. The adult male never attacked any of the calves in this social group, but the mother may nonetheless have been concerned about what the adult male might do, and so was more restrictive of her first calf. It seems likely that social context, environmental pressure, and maternal dispositions interact to produce the range of maternal styles that exist in the animal kingdom.

Consequences of Maternal Behavior on Offspring Development

Early research by Harry Harlow (1958) demonstrated the importance of maternal care for development. When rhesus monkey infants were deprived of their mothers, the infants proved incapable of coping with novel situations and also exhibited severe deficits in their development of social behavior. Some of the monkeys were never able to overcome these early separations from their mothers. These monkeys failed to acquire species-typical social behaviors and remained less likely to explore novel environments or objects. If orphaned monkeys were provided an inanimate "cuddly" surrogate mother, they behaved more like normal monkeys. They cuddled with the surrogate mother when they were tired or frightened, and used her as a secure base from which to explore their world, including novel objects. In addition, these monkeys were able to interact normally with other monkeys. Thus, even an inanimate mother is better than no mother at all, at least if the surrogate mother can provide some contact comfort for the infant. More recent research has shown that monkey infants fare better with mobile surrogate mothers than when provided with non-mobile surrogate mothers.

Most animals do not have to rely on surrogate mothers. But as we have seen, mothers differ in terms of their mothering behaviors. These differences can influence their offspring's development, including their reactions to novel situations, such as a new environment, a strange animal, or an unfamiliar object, the emergence of independent behavior and the development of species-appropriate social skills. Examples of these influences abound. Rat pups that were rarely licked by their mothers engaged in less exploration and showed greater amounts of distress when exposed to novel situations than did rat pups that were licked more often by their mothers. Maternal differences in pup licking were transmitted from one generation to the next. The more licking that rat pups received from their mothers, the more likely the pups were to lick their own offspring when they became mothers. Perhaps maternal licking by rat mothers comforts rat pups in a similar way to the contact provided by the surrogate mothers given to macaque monkey infants by Harlow. Comforted infants are more likely to feel secure, and so develop the curiosity and exploratory behaviors that characterize their species. Physical contract is a feature of mother–infant interactions for many species, including those that lack arms and hands. Dolphin mothers and their calves often maintain contact as they swim, and it is likely that this contact helps to form the bond between mother and calf.

Another example concerns the dolphin mother described above that restricted her first calf's exploration of his world and his social interactions with other dolphins. As a result, the calf engaged in fewer social activities during his first year of life compared to other calves. The mother's restrictive style hampered the calf's social development and resulted in more frequent solitary activities than social ones once his mother allowed him greater freedom to roam. As a final example, primate infants that have permissive mothers are more likely to explore novel objects and novel environments than do infants that experience more restrictive mothering. These differences persist throughout the primates' "childhood," demonstrating that the consequences of maternal style are sometimes long lasting.

In summary, maternal behavior is very important to the development of infants. Not only does it directly impact the survival of the offspring, but it also affects the development of exploratory, independent, and social behaviors. Mothers with different maternal styles interact very differently with their offspring, thus producing different behaviors in their young. Furthermore, differences between offspring appear to last over time and influence their own parenting behavior. Additional research is needed to better understand the range of maternal and paternal parenting behaviors and their impact on the development of their offspring.

See also Antipredatory Behavior—*Hiding Behavior of Young*
Ungulates
Caregiving—*Incubation*
Caregiving—*Mother–Infant Relations in Chimpanzees*
Caregiving—*Parental Care*
Caregiving—*Parental Investment*

Further Resources

Altmann, J. 1980. *Baboon Mothers and Infants.* Cambridge: Harvard University Press.
Bornstein, M. 2002. *Handbook of Parenting. Vol. 2. Biology and Ecology of Parenting.* Mahwah, N.J.: Lawrence Erlbaum Associates.
Clutton-Block, T. 1991. *The Evolution of Parental Care.* Princeton, NJ: Princeton University Press.
Connor, R., Wells, R., Mann, J., and Read, A. 2000. *The bottlenose dolphin: social relationships in a fission-fusion society.* In: *Cetacean Societies: Field Studies of Dolphins and Whales* (Ed. by J. Mann, R. Connor, P. Tyack, & H. Whitehead), pp. 91–126. Chicago: University of Chicago Press.
Fairbanks, L. A. 1996. *Individual differences in maternal style: causes and consequences for mothers and offspring.* Advances in the Study of Behavior, 25, 579–611.
Goldberg, S. 2000. *Attachment and Development.* New York: Oxford University Press.
Harlow, Harry. 1958. *The nature of love.* American Psychologist, 13, 573–685.

Stan A. Kuczaj & Heather M. Hill

■ Caregiving
Incubation

Incubation behavior refers to the period when the eggs of vertebrates are developing, and most often is used in connection with birds. Embryos are unable to regulate body temperature, and often display a narrow range of temperature tolerance or optimal temperature for development, compared to young, juveniles, or adults of the same species. All birds lay eggs, and most incubate their own eggs until they hatch. For birds, incubation generally refers to a parent bird sitting on the eggs to keep them at the appropriate temperature for development. Eggs are normally deposited in a nest, which might be quite elaborate or merely a depression in the sand, and the parents incubate the eggs only at the nest. One or both parents sit on the eggs to apply the heat necessary for embryonic development. Parents may also shade eggs from intense solar radiation.

Some species incubate eggs until at hatching the young are able to move about on their own and feed themselves (*precocial*), whereas in others hatching occurs while the young are still blind, naked and unable to stand (*altricial*), with gradations in between.

During the weeks of incubation birds are particularly vulnerable to predators because they must sit in one place, and this restricts the amount of time that they can devote to foraging. Selection of a nest site is thus very important because it determines whether the parent and eggs are exposed to predators or inclement weather.

While the term incubation normally refers to the active process whereby parent birds sit on their eggs to maintain a constant and appropriate temperature for the eggs to develop, the term is also used for reptiles. In reptiles (and in one group of birds, see below), incubation is passive in that the female selects an appropriate nest site which will assure that the soil around the nest remains in an optimal state for the eggs to develop. The special case of reptiles will be discussed below in some detail.

For many decades, naturalists merely noted where fish, reptiles or birds laid their eggs and whether there was parental attentiveness or incubation. The assumption was that birds might actively select a particular nest site for their eggs, but that fish and reptiles did not. In the 1940s and 1950s Konrad Lorenz revolutionized the understanding of behavior by defining four ways to examine behavior: 1) biological function (What does incubation do for the animal?); 2) causation (What stimuli result in a particular aspect of incubation?); 3) biological significance (How does incubation contribute to survival?); and 4) evolution (How did various aspects of incubation evolve over time?).

For incubation, the answers to these questions are clear: 1) Incubation maintains the eggs at an appropriate temperature so that development can proceed normally to produce an intact offspring; 2) Birds are stimulated to incubate when they have laid eggs, and are presented with the eggs; 3) Incubation contributes to survival by keeping the embryos at a temperature so that development can proceed and they do not die; and 4) Several different patterns of incubation behavior evolved to assure the appropriate temperature for the developing embryos.

The major aspects of incubation discussed below are incubation patterns in vertebrates, temperature effects and incubation temperature in reptiles, incubation period and temperature, and incubation in birds. Since birds are the primary vertebrates that incubate their eggs, several aspects of their incubation behavior are discussed, including the incubation patterns, gender roles in incubation, incubation bouts, brood parasitism, and incubator birds.

Incubation Patterns in Vertebrates

Vertebrates either bear live young (nearly all mammals) or lay eggs that develop and hatch into young (most other vertebrates). By bearing live young, mammals control the temperature of the developing young because they control their own body temperature. Once eggs are laid, the temperature they are exposed to is either controlled by the parents that sit on the eggs, or by the temperature of the surrounding substrate. While the incubation temperature of most fish, amphibian, and reptile eggs is maintained by the temperature of the surrounding water or soil, alligators and crocodiles construct nests of rotting vegetation. The heat generated from the rotting vegetation is sufficient for the developing embryos.

Birds directly control the incubation temperature of the eggs by either sitting on them to apply heat, or standing over them to shade the eggs when temperatures are exceedingly high. A few species of terns and other shorebirds have been seen on very hot days to fly to the nearby water, wet their abdominal feathers, and return to settle on their eggs, transferring the water to the eggs to cool them. Reptiles do not sit on their eggs to incubate them, but some snakes will stay with their eggs for a few days, presumably to protect them from predators or other females who might dig up the nests while digging their own. While no living reptiles incubate their eggs, some dinosaurs constructed nests in colonies like modern birds and incubated their eggs and cared for their young while they were in the nest. Many fish species show nest guarding behavior while their eggs develop passively in the sun-warmed water.

Blue-footed boobies incubate by covering their eggs with their feet.
Courtesy of Joanna Burger.

Temperature Effects and Incubation in Reptiles

Although incubation in birds has received the most study, recent work shows that the sex of turtles is determined by temperature in the nest. Studies by John Bull, William Gutzke and others in the 1980s showed that lower incubation temperature determined the sex of hatchling turtles. Subsequent studies by Burger and others in the 1990s indicated that even for species without temperature-dependent sex determination, incubation temperature affects the behavior of hatchlings. At low incubation temperatures, young were less able to capture and eat prey, compared to hatchlings from eggs incubated at higher temperatures. Low-incubation hatchlings showed a number of other deficits, including slower recognition of predators, slower movement, and less ability to find shade and avoid the hot sun.

Reptiles can influence the temperature of their developing young by selecting nest sites that are exposed to full sunlight at higher latitudes and selecting shadier locations in more tropical latitudes. Some snakes, such as rattlesnakes and garter snakes, control incubation temperature by bearing live young. Since they are cold-blooded animals and are dependent on sun for temperature regulation, females bask in the sun to increase their own body temperature and that of the developing embryos.

Incubation Duration and Temperature

For an egg, the period of incubation is duration between laying and hatching. For an adult, incubation begins with laying of the first egg and ends when the last egg hatches. Many birds (most ducks, for example) do not begin incubation until they have completed their clutch, and the eggs begin hatching within a few hours of each other. In other species (most gulls and terns and herons), incubation begins sometime after the first egg is laid, and the young hatch asynchronously over a period of several days. For fish and amphibians, the incubation periods range from a few hours to a few days. For reptiles, the incubation period normally ranges from about 45 days to up to 90 days, often depending upon soil temperatures. For example, in snakes, a nest in the open sun during a warm year may hatch in about 55 days, and in the shade in a cool year, eggs may hatch in 80 days. The incubation period is more variable for cold-blooded (*poikilothermic*) species compared to birds (*homeothermic*) because embryonic development is dependent upon external temperatures.

Incubation periods in birds range from about 10 days in small song birds to 80 days for royal albatrosses. Although incubation heat is provided by adult birds, there is still some variation within a species due to erratic incubation during the first few days. Differences in incubation periods among species are due to size of the bird or egg, the length of the bird's life, the stage of the chick's development at hatching (precocial vs altricial), climate, season, and food supply. Most songbirds, with very short incubation periods, are altricial (without feathers, unable to walk or feed on their own) at hatching, while species such as ducks are able to walk about and feed themselves a few hours after hatching (precocial).

Many birds have brood patches to facilitate the transfer of heat from their bodies to the eggs. The *brood patch* consists of one or more areas of the abdominal skin that become denuded of feathers and are richly supplied with blood vessels to facilitate the transfer of heat. Some birds, such as gannets and boobies, do not have a brood patch, but incubation is performed by using their webbed feet, which are highly vascularized.

The embryo has its own physiology, including resisting the flow of heat into the egg from the parent, and promoting heat flow into the egg by controlling its own blood flow. The process of incubation is thus more interactive than previously thought.

The average incubation temperature of a bird's egg is 34° C (about 93° F), although it can vary by a few degrees. There is more latitude below this average than above it. Only a few degree rise in egg temperature will kill a developing embryo, whereas the same decrease in egg temperature merely delays development. For this reason it is more detrimental to keep parent birds off their nests on very hot days than on cold rainy days.

King penguins incubate their one egg by holding it on top of their feet an covering it with their breast feathers—shown here as the lump near the ground.
Courtesy of Joanna Burger.

Incubation in Birds

Incubation behavior has been studied extensively in many species of birds, both those that nest solitarily and those that nest in large groups called "colonies." There are many possible arrangements for incubation in birds, from equal sharing of incubation between the sexes to either male or female taking sole charge. There are several examples of adaptive specialization, such as brood parasitism (most cuckoos and cowbirds) and incubator birds which bury their eggs in a mound of rotting vegetation, or in one case in sand next to hot springs.

The presence of eggs in the nest is an important visual and tactile stimulus for incubation. A nesting bird, returning to its nest after feeding, after disturbance, or after exchanging duties with its mate, will settle on the eggs and adjust its body until the eggs fit against its brood patches. Periodically it will stand and turn the eggs and resettle. Egg turning is essential for normal development in most species, and artificial incubators mechanically rotate eggs to assure the hatching of normal chicks. In many ground nesting birds, if an egg is displaced from the nest, the adult will approach the egg and roll it back into the nest. Niko Tinbergen showed that larger eggs were actually a stronger visual stimulus for egg rolling and incubation than smaller eggs.

In general, in bird species where males and females have the same plumage, both sexes share incubation duties. In many species males and females have different appearances (*sexual dimorphism*) and often do not share incubation. In many species, the male is more brightly colored, and incubation is left entirely to the female, which is usually cryptically colored. This includes ducks, pheasants, and turkeys, as well as some tropical species (tanagers, manikins). In phalaropes the female is more brightly colored than the male, and the male does all of the incubation.

When both parents share in incubation, the incubation bouts may vary from an hour or two in most song birds, to hours or days in albatrosses, to several days for penguins. The length of the time each parents sits on the nest before it is relieved by its mate usually depends on the difficulty its mate has in finding food. Many species of songbirds maintain a nesting territory in which they gather food, and thus fly a few dozen meters at the most to catch insects or fruit. At the other end of the spectrum are seabirds which may have to fly hundreds of kilometers from the nesting island to find food. Birds that nest in colonies have to fly away from the colony to obtain food. Gulls, terns, herons, egrets, cormorants, and ibises normally fly only a few kilometers to catch fish, but albatrosses may fly hundreds of kilometers to foraging grounds, and penguins may have to walk long distances over ice floes and then swim far to obtain food.

There are two special cases of incubation in birds: brood parasitism and incubator birds. *Brood parasitism* occurs when a bird (the parasite) lays its eggs in the nest of another species (the host), allowing the host to incubate and care for its offspring. Some species are obligate brood parasites (most cuckoos, cowbirds, and honeyguides), and never build nests or incubates. Some ducks are facultative brood parasites, building a nest and incubating their eggs, but also laying some eggs in nests of other birds. A parasite has to select a host that is abundant in the same region, and feeds on appropriate foods. Otherwise, the young will not thrive. The parasite also has to lay its egg in the host's nest before the host has completed its clutch, it must have a shorter incubation period than its host, and its egg should resemble that of its host (so it is not rejected by the host). Only then will the parasite's egg be incubated successfully by the host.

Incubator birds, also called mound-builders or megapodes, are restricted to Indonesia, Polynesia, New Guinea, and Australia. They lay their eggs in holes in the ground or in mounds of rotting vegetation and leave them to be incubated by heat from the fermenting vegetation. In some places, the pits are dug in soil through which volcanic steam percolates, incubating the eggs. Some species construct the mound, which can be as large as 35 ft (10.6 m) in diameter and 15 ft (4.5 m) high, and are constructed by several individuals. The eggs are laid in tunnels which may be 3–4 ft (.9 to 1.2 m) long. Some megapodes remain nearby and periodically revisit the mound, check the temperature, and rearrange the vegetation, until it is the correct temperature for incubation, sensed by probing with their bills.

See also Antipredatory Behavior—*Hiding Behavior of Young Ungulates*
Caregiving—*Brood Parasitism among Birds*
Caregiving—*How Animals Care for Their Young*
Caregiving—*Parental Care*
Caregiving—*Parental Investment*

Further Resources

Brooke, M. 1992. *Cambridge Encyclopedia of Ornithology*. Cambridge: Cambridge University Press.
Deeming, D. C. (Ed.). 2002. *Avian Incubation: Behaviour, Environment and Evolution*. New York: Oxford University Press.

Grant, S. G. 1982. *Avian incubation: egg temperature, nest humidity, and behavioral thermoregulation a hot environment*. Ornithological Monographs, 30. American Ornithological Union.

Lack, D. 1968. *Ecological Adaptations for Breeding in Birds*. London: Methuen.

Schreiber, B. A. & J. Burger (Eds.). 2001. *Biology of Marine Birds*. Boca Raton, Florida: CRC Press.

Skutch, A. F. 1987. *Helpers at Birds' Nests*. Iowa City: University of Iowa Press.

Turner, J. S. 1997. *On the thermal capacity of a bird's egg warmed by a brood patch*. Physiological Zoology, 70, 470–480.

Whittow, G. C. 2000. *Sturkie's Avian Physiology*. New York: Academic Press.

Joanna Burger

■ Caregiving
Mother–Infant Relations in Chimpanzees

We have been studying the chimpanzee mind both in the laboratory and the wild. Our main partner in the laboratory study is Ai, a 27-year-old female chimpanzee. For the past 26 years, Ai, whose name means "love" in Japanese, has allowed researchers to study how chimpanzees perceive the world (Matsuzawa 2003). For example, Ai participated in a matching-to-sample experiment on a computer screen. Each color had to be matched with its Chinese–Japanese character, called "kanji." She selected the character for red when a red patch appeared as a sample on the monitor. She is the first chimpanzee who learned to use Arabic numerals to represent numbers (Matsuzawa 1985). Ai recently amazed the world with her numerical memory and cognitive skills. In this memory span test, five numbers appeared on the monitor. When Ai touched the lowest number, the other numbers were masked. She then relied on her memory to touch the hidden numbers in ascending order. Ai's ability to memorize number order and location was comparable to that of an adult human (Kawai & Matsuzawa 2000).

In April 2000, when the cherry blossoms were in full bloom, Ai was presented with an even more challenging task: motherhood. The key question that we wished address in our study of Ai and her baby was the following: How and when does the transmission of knowledge and skills take place? Chimpanzees are humans' closest relatives. Traits such as a complex social structure, cultural traditions and tool use can provide valuable insights into the primate origins of human evolution, cognition, and behavior. This article aims to trace Ai and her baby's first year together.

A group of 15 chimpanzees, three generations of 0- to 37-year-olds, lives at the Primate Research Institute of Kyoto University, Inuyama, Japan. Three steel towers rise from a lush, green section of the Institute. This environmentally enriched compound is home to Ai and her fellow chimpanzees.

Before Delivery

Every day, Ai leaves the outdoor compound and comes to a room to study. It is here that we have been exploring the chimpanzee mind ever since Ai was a year old. One day, Ai was given a chimpanzee doll. Once we realized that Ai had become pregnant, we decided to try to teach her how to hold a baby. In the wild, chimpanzees ordinarily have their first offspring at around age 13 or 14. Ai was in her mid-twenties but had not yet had a successful pregnancy. Children enrich the lives of chimpanzees just as they do humans.' From a research perspective, however, we were extremely interested in discovering how knowledge and skills would pass from one generation to the next. So, we came up with a plan to have her best friend Akira sire Ai's baby through artificial insemination.

Since arriving at the Institute at the age of one, Ai had never closely witnessed another chimpanzee raising offspring. She had little idea of what a baby was, much less a notion of how to care for one. In fact, nearly half of all chimpanzees raised in captivity reject their offspring. We decided to show her video footage of mother chimpanzees in the wild taking care of their newborn. We also showed her a live gibbon baby that happened to be born at our institute around that time, and was being raised by humans. Not only watching but also practicing may be an integral part of learning how to rear a baby. This was the reason why we gave Ai a doll. However,

Ai was given a doll for learning how to rear a baby in April 2000.
Courtesy of Yumi Ozaki.

she was less than eager to hold it. She would handle it—sometimes holding it upside down—but it seemed not to interest her much; we even caught her yawning. The progress was slow. A chimpanzee pregnancy lasts about 8 months. Although Ai continued her nursing lessons daily, she did not have the enthusiasm she usually exhibited for computer tasks. Ai was normally a fast learner but this assignment seemed so different from anything she had ever done before.

Delivery

On April 24, 2 days after the scheduled delivery date, a change was detected in Ai's condition. Staff and students monitored her progress on TV. Ai was in labor. She battled the pain. At 10:50 p.m., Ai finally gave birth to a boy. She clutched her newborn to her breast, as she had been taught. But there was a problem. The baby was not moving. His nose was clogged. Amniotic fluid in the baby's nose and throat prevented him from breathing. Ai began licking and cleaning the baby's face and mouth. Then the baby made a noise. He started to move. Ai had succeeded in clearing the nasal passages well enough to allow her baby to breathe. However, another problem awaited. The baby was in the wrong position and could not feed. TM entered the room and instructed Ai to hold her baby correctly. Twenty hours after birth, the baby at last found the mother's nipple. Ai sat patiently. With the birth and the first feeding, Ai had passed her initial tests as a new mother. The baby was given the name "Ayumu," meaning "to walk," in the hope that he would always take confident steps forward.

Ai gave birth to a boy named Ayumu on April 24, 2000. The photo was taken 9 hours after the birth.
Courtesy of Tetsuno Ochiai.

Introduction to Fellow Chimpanzees

Twenty days had passed. Ai spent most of that time in a room away from her fellow chimpanzees to avoid potential trouble. It was decided that it was finally time to introduce Ayumu to the others. One of the first to be led into the room was Pendesa, who had been like a sister to Ai. Ai and Pendesa hugged each other. It had been a while since they last met, and the reunion was a warm one. Pendesa quickly grew curious about the new arrival, and tried to get a good look at him. Pendesa followed Ai around the room. This upset Ai, who turned her back to her. Even with good friends, the baby came first.

Pendesa realized that she had to be discreet. She looked over at Ai. She was determined to get close to Ayumu. Ai pretended not to notice but still monitored Pendesa's advances. Despite the tension, it was a touching scene. Overwhelmed by curiosity, Pendesa inched closer. Finally, she touched Ayumu gently. But that was as far as she dared to go with Ai watching. Eventually Pendesa retreated. Female chimpanzees in the wild will hold their babies day and night for the first 3 months without letting others approach. This has prevented researchers from studying infant development in any mother-raised chimpanzee baby.

A New Project: Participation Observation

We began a new project based on a triadic relationship of mother, infant, and researcher. In a variant of the participation method, the researcher became involved in the situation in which the mother chimpanzee reared the infant. Thanks to a long-term friendship between the researcher and the mother chimpanzee, the researcher could thus gain access to and interact with the baby even as he was being held by the mother. This is in clear contrast with previous studies in which home-raised chimpanzees under observation were isolated from their biological mothers. We hoped that our project would reveal how a baby chimpanzee, raised by his mother, develops.

However, when TM attempted to touch Ayumu for the first time, Ai gently brushed his hand away. Again, the baby came first. TM pleaded, "Let me have a look. Let me touch him." Ai finally gave in to his requests and allowed him to touch Ayumu. Members of our research group had to be able to videotape Ayumu's face and to touch him if they were to conduct a detailed study. TM decided on a tactic of putting Ai at ease. At 3 weeks after birth, with a simple gestural sign and vocal command, we could ask Ai to lie down and show Ayumu to us.

By the time Ayumu was 2 months old, Ai was completely relaxed. Ayumu stared wide-eyed at the world around him. Ai's cooperation revealed some intriguing facts about chimpanzee development that were previously unknown, such as when a baby chimpanzee learned to recognize his mother's face.

Human babies have been shown to recognize their mother's faces soon after birth. But this was the first time that the same phenomenon could be studied in a mother-reared chimpanzee infant. We showed pictures of various chimpanzee faces to Ayumu and recorded his gaze directed toward them by a tiny handheld videocamera. We found that this baby chimpanzee began to prefer looking at his mother's face over the faces of other conspecifics from about a month after birth. Like humans, baby chimps come to recognize their mother's face visually at a very early stage.

Looking at each other: What does this tell us? When Ai was not participating in experiments, she enjoyed interacting with Ayumu outdoors. Ai often lifted Ayumu high in the air, a game that delights human children as much as it does baby chimps. During such bouts of play, Ai gazed at Ayumu. Intimate moments like these not only act to deepen the bond between mother and infant, but also help the infant to develop communication skills. Again, like humans, chimpanzees make use of various facial expressions to communicate with each other. These constitute an indispensable social ability, learned in infancy during moments like these.

Neonatal Smiling

Another chimpanzee, Chloe, who had participated in experiments similar to Ai's, became a mother on June 19, 2000. Pan, the Institute's youngest adult female, gave birth on August 9, 2000. This resulted in three mother-infant-researcher trios: Ai-Ayumu-Matsuzawa, Chloe-Cleo-Tomonaga, and Pan-Pal-Tanaka. The same experiments carried out with Ayumu were conducted on the new babies by the researchers, each with a strong bond to the baby's mother. It was a wholly unique and precious research opportunity.

The research group had begun round-the-clock observations to monitor the infants' development. A startling discovery occured on the 16th night after the birth of Pal, Pan's baby. Pal smiled with her eyes closed during the night. This was the first evidence of smiling in newborn chimps (Mizuno et al. 2003). As human babies doze off, they often suddenly smile without any explicit stimuli. Pal smiled just like human newborns. This is referred to in scientific circles as "neonatal smiling" but was thought to be unique to humans. The monitoring allowed us to discover that human and chimpanzee babies shared a trait retained over time from a common ancestor. Neonatal smiling happens with the babies' eyes closed, so it is not directed at any particular thing or person. Closer examination of video footage taken of Ayumu just after birth also showed him smiling in his sleep. This landmark discovery suggests that the smile plays an important role right from birth, for humans and chimpanzees alike.

When Ayumu was 40 days old, we observed that Ai tickled him. He was too young to be held in the air, so Ai played with him by laying him on his back and tickling him. Ayumu opened his mouth, and smiled in this face-to-face situation. The neonatal smiling that occurred during sleep gradually disappeared over the first 2 months and was replaced by "social smiling" directed at others when awake. In human babies, smiling is thought to have the function of attracting affectionate attention from the mother.

Social Smiling

At two-and-a-half months old, we could get Ayumu to smile just by wiggling our fingers. After 3 months, Ayumu seemed to smile at everything. To see if Ayumu recognized the faces of those close to him, we showed him a picture of a human or a chimpanzee face. The facial stimuli were greeted with a cheerful response from Ayumu. We also showed Ayumu a mirror. While humans can recognize their own reflection in the mirror by the time they are around two-and-a-half years old, most chimpanzees can only do so much later. To Ayumu, the baby in the mirror was a perfect stranger. But he smiled at it anyway.

This period when Ayumu began smiling at the world around him also corresponded to the time when he developed the ability to sit up on his own. Pendesa, a close friend of Ai's, invited Ayumu to play. She had been fascinated with Ayumu since birth and had worked very hard to become his favorite playmate. Ayumu smiled. When Pendesa brought her face towards Ayumu's, he responded with an enthusiastic smile. Encouraged by his reaction, she continued to make faces. In both chimpanzee and human society, a smile generally communicates friendship. Through interactions like these, infants naturally learn that different facial expressions trigger different reactions in others. The coincident development of smiling and physical growth may lead to greater social activity. Human babies, too, begin to smile frequently at about 3 months old. Smiling draws attention to them and elicits verbal communication. Although smiling is an innate trait, babies must learn to recognize and apply it in the proper context. This is another thing we learned from chimpanzees.

Token Experiment

Late in the summer of 2000, Ayumu was 4 months old and grew more active each day. His presence was causing slight changes to Ai's study habits. Before giving birth to Ayumu, Ai had been learning the concept of tokens. The aim was to see if chimpanzees could comprehend such an abstract human invention as money. The money can be seen as a tool to get various goals. A trial of the token experiment proceeded as follows. First, Ai had to match colors presented to her on a computer screen to the corresponding kanji characters. For each correct answer, she received a coin. At the other end of the room, we had installed a vending machine. When Ai inserted a coin, she could purchase a piece of apple or a blueberry. She came to understand that tokens could be exchanged for other things. She often saved up two or three coins, and then walked over to the vending machine to exchange them for food. It was a pattern she had developed on her own. Ai continued this experiment even after Ayumu was born. But she became distracted when Ayumu started to wander away to play.

It was impossible to concentrate. When Ayumu became tired, he began clinging to Ai for affection. This made it difficult for Ai to move freely between the monitor and the vending machine. Finally one day, Ai did not budge from the monitor for the whole session. She kept solving the questions but ignored the coins delivered. Assured that Ai would not leave him, Ayumu started to play again. Only after solving every single trial did Ai finally gather up the coins and walk to the vending machine. It was a clever solution.

As her son grew older, Ai displayed a flexibility to adapt her behavior accordingly. However, childrearing rarely goes as expected. When Ayumu was 5 months old, he grew more active and began racing about the room. And once he began to fuss, he became a real handful. His demands began to take their toll on Ai. Yet Ayumu was too small to comprehend his mother's mood, and finally Ai burst out in a fit of anger. Nevertheless, rather than to take it out on Ayumu, she chose a nearby student to scream at. Afterwards, Ai returned to her quiet composed self. Raising a child has both its joys and burdens. At her own pace, taking things day by day, Ai was showing us much more than her perception of the world. She was providing insights into good parenting as well.

Winter Snow

It was November 2000. Dr. Jane Goodall, the pioneer of chimpanzee studies in the wild came to see her old friend Ai and her new baby. Dr. Goodall had spent a long time among wild chimpanzees, and observed the arrival of many newborns at Gombe, Tanzania, East Africa. According to her, once chimpanzees give birth to their first baby, they go through a perceptible change, gaining a lot of self-confidence. She thought that Ai may have been going through the same experience.

Dr. Goodall had spent a time with Ai in the same booth several years ago. She looked into Ai's eyes, groomed her a little, and played for a while. Since then, they became friends.

When Dr. Goodall visited Ai this time, she gave the soft panting grunts that chimpanzees utter when they greet each other. Dr. Goodall crouched, watching mother and son in peace. The friendship was renewed.

During Ayumu's first winter, the Institute got more snow than usual. The chimpanzees there had adapted well to the Japanese climate and enjoyed the outdoors all year round. Ayumu turned 9 months old in January 2001. A new experiment had begun related to our main aim of investigating how the transmission of knowledge and skills occurs. This particular study focused on the comprehension of pointing gestures (Okamoto et al. 2002). Ai can comprehend human gestures. The researcher hid a piece of food under one of two cups and then pointed to the one that contained the food. Ai clearly understood the meaning of finger pointing. During the experiment, Ayumu held out his hand. He wanted the cups. He wanted to touch what his mother touched, do what his mother did. Soon, Ayumu's attention was fixed intently on the interaction between Ai and the researcher. His eyes beamed with curiosity.

Ayumu also began to show a strong interest in Ai's computer exercises. But he found it hard to reach the monitor, so we prepared a laptop computer that he could touch freely. Things started simply. If Ayumu touched the picture of a fruit that appeared on the monitor five times consecutively, he received an actual piece of fruit as a reward. The researcher asked Ai to model the behavior. We hoped to observe how Ayumu learned to do what his mother did. When Ai got a piece of apple, Ayumu craved it. Ai shared her food with him, though Ayumu sometimes snagged the prize for himself before Ai had a chance to react. Ai did not seem to mind when Ayumu stole her reward. Soon, Ayumu was absorbed in grabbing the pieces of food and paid no attention to the tasks. Still, Ai diligently continued her exercise. She appeared to be showing Ayumu how things were done.

Ai's behavior followed the pattern of wild chimpanzees using tools around their infants (Matsuzawa et al. 2001). Besides our laboratory studies, we have also been observing a small chimpanzee community at Bossou in Guinea, West Africa. These chimps crack open oil palm nuts with a set of stones to extract the edible kernel inside the hard shell. Infant chimps have yet to learn how to use stones as tools, and it will take some years before they crack their first nut by themselves. Mothers will not prevent their offspring from taking the kernels from their freshly cracked nuts. Instead, like Ai, they indulge the infants. They will even stop what they are doing and wait for the infants to pick up the kernels. Rather than instruct or mold, elders simply model the behavior. Soon, the infants develop an interest in manipulating the stones, initially in a playful manner. They lift stones, step on them, throw and roll them. At one year, they will start to observe carefully how adults use the stones and will begin to try it themselves. For example, practicing infants often put nuts on top of an anvil stone but instead of hitting it with a hammer stone, they slap it with their palm or kick it with their heel. We have seen a youngster placing her chin on an adult's arm as she was hammering: This might provide a feel for the hitting action. However, these young chimpanzees are still not able to combine two stones. They eventually learn to use pairs of stones when they reach the age of at least three and a half years, and many do not get it right until they are 4 to 5 years old or even older. The young chimps learn by watching and doing before finally acquiring the skill. Just like the wild chimpanzees, Ayumu, too, learned the computer skill by watching his mother. In the wild, we cannot observe the exact process of how and when these attempts begin to produce results. A major advantage of this controlled setting is that it allows us to witness the process in close detail.

First Touch to the Computer

On February 16, 2000, at nine-and-a-half months old, Ayumu achieved a breakthrough (Matsuzawa 2002). During the token experiment, while Ai was at the other side of the room, Ayumu stretched out his arm toward the monitor and completed one trial by himself. After Ai had moved away from the monitor, Ayumu touched the start button, calling up a trial. The computer happened to select the kanji character for brown and displayed it on the screen. Ayumu touched the character, at which point he was presented with two choices. He reached out toward the upper square—the correct choice, brown—but could not reach it. He tried again without success. Another angle filmed by a second video camera showed him hoisting himself up on a ledge to reach higher, determined to touch the brown square. It took us by complete surprise. We never dreamed that he would start performing this complex action at only nine-and-a-half months. He had never touched the monitor before, which made this his very first try. And yet he answered correctly. However, we cannot be certain that he chose the brown square with a proper understanding of the kanji character he had seen at the beginning of the trial. In fact, the following sessions revealed that he had made the correct choice only by chance. However, judging from what he did, two things became immediately evident. We knew that he understood the entire course of a trial; that is, that he had to touch three stimuli (the start button, the kanji character, and the choice color square) in succession. Also, he seemed to have a clear intention of touching the brown square, despite the fact that it was further away and harder to reach. Ayumu had been watching his mother working on her computer exercises day after day. In that sense, he had a firm grounding in it. It is likely that the long term exposure from birth was the main reason why he succeeded so suddenly.

Ayumu performs a computer task at 1½ years old.
Courtesy of Akihiro Hirata.

Early Spring

It was March 2001. Ayumu was now 10 months old. Spring had arrived and he was absorbed in playing on a rope hung on the towers in the green compound. The distance that Ayumu wandered from Ai grew longer with each passing day. He was spending more

Acknowledgments: This study was financially supported by the following grants: #07102010 and #12002009 from MEXT, 21COE (A14) and HOPE from JSPS, Japan. Thanks are due to M. Tomonaga, M. Tanaka, M. Myowa-Yamakoshi, S. Hirata, Y. Mizuno, A. Ueno, C. Sousa, N. Nakashima, T. Imura, M. Hayashi and other collaborative researchers of the chimpanzee project 2000 for the study of cognitive development in chimpanzees. We also thank K. Matsubayashi, J. Suzuki, S. Gotoh, K. Kumazaki, N. Maeda, C. Douke and others involved in the chimpanzees' care. Thanks are also due to T. Kim, M. Celli, and D. Biro for assistance with the English text. TM took the role of the main tester of the chimpanzees and MN was responsible for keeping and analyzing long-term video records. A TV program documenting Ayumu's first year of life described in this article was broadcast in Japan by NHK in May, 2001. The English version was co-produced by ANC and MICO, and has been distributed by MICO. A video of the program is available from the authors (for educational purposes only). For further correspondence, please contact the first author at matsuzaw@pri.kyoto-u.ac.jp.

time interacting with other chimpanzees. When Akira, the group leader and Ayumu's father called out in a loud voice—a "pant–hoot"—the other chimps responded. Ayumu too puckered his lips and joined in. Ai carried him on her back to Akira. With a broad smile, Akira beckoned Ayumu to play. Grasping his meaning, Ayumu accepted the invitation. He chased Akira with a playful, smiling face. Communication begins with smiles and gazes. Ayumu will learn a lot more through his contacts with other members of the community.

Ayumu has not made any progressive leaps at the computer since that day in February. But he has been accumulating knowledge daily that will help him jump to the next plateau. Like human children, he moves in fits and starts. Ai has proved to be a patient model for her son. In April, the cherry blossoms were once again in full bloom. Ayumu turned one year old, and he still continues to mature steadily. The knowledge and skills he has acquired from his mother will begin to sink in more rapidly. What will he learn? Will he surpass his mother's abilities? For mother and son, it's a brand new year.

See also Caregiving—*How Animals Care for Their Young*
Caregiving—*Parental Care*
Caregiving—*Parental Care and Helping Behavior*
Caregiving—*Parental Investment*

Further Resources

Kawai, N. & Matsuzawa, T. 2000. *Numerical memory span in a chimpanzee.* Nature, 403, 39–40.

Matsuzawa, T. 1985. *Use of numbers by a chimpanzee.* Nature, 315, 57–59.

Matsuzawa, T. 2002. *Chimpanzee Ai and her son Ayumu: an episode of education by master-apprenticeship.* In: *The Cognitive Animal.* (Ed. by M. Bekoff, C. Allen, G. Burghardt), pp 189–195. The MIT Press Cambridge, Mass.

Matsuzawa, T. 2003. *The Ai project: historical and ecological contexts.* Animal Cognition, 6, 199–211.

Matsuzawa, T., Biro, D., Humle, T., Inoue-Nakamura, N., Tonooka, R., & Yamakoshi, G. 2001. *Emergence of culture in wild chimpanzees: Education by master-apprenticeship.* In: *Primate Origins of Human Cognition and Behavior.* (Ed by T. Matsuzawa). London: Springer-Verlag.

Mizuno, Y., Tanaka, M., Tomonaga, M., Matsuzawa, T. & Takeshita, H. 2003. *Neonatal smiling in chimpanzees.* In: *Cognitive and Behavioral Development in Chimpanzees: A Comparative Approach,* (Ed. by M. Tomonaga, M. Tanaka, & T. Matsuzawa), pp. 56–58. (in Japanese).

Okamoto, S., Tomonaga, M., Ishii, K., Kawai, N., Tanaka, M., & Matsuzawa, T. 2002. *An infant chimpanzee* (Pan troglodytes) *follows human gaze.* Animal Cognition, 5, 107–114.

Tetsuro Matsuzawa & Miho Nakamura

■ Caregiving
Non-Offspring Nursing

Non-offspring nursing occurs when a female mammal nurses the offspring of other females in addition to her own young. This behavior has been reported in at least 60 different species of mammals. It is most common in carnivores and rodents and is more likely to be observed in captivity than in field studies. Non-offspring nursing presents a puzzle to those interested in animal behavior because animals are typically expected to act in ways that maximize their own selfish best interests.

Lactation is one of the most energetically expensive components of mammalian parental care. These costs are exemplified by female phocid seals, which secrete large volumes of lipid-rich milk (50–60% fat) during a period in which the females are also fasting. As a result, young seals absorb up to 6 kg (13 lb) of lipids each day, allowing them to accumulate enough blubber to survive several weeks or months of fasting before they learn to hunt on their own. Many female mammals experience a metabolic rate during lactation that is more than double their metabolic rate in a nonreproductive period. Nursing non-offspring potentially decreases the milk available to one's own offspring and, because of its energetic costs, could decrease the likelihood of having young in the future.

There are two general categories of non-offspring nursing: situations best considered to be milk "theft" and situations more likely to involve "willing" cooperation on the part of the female (communal nursing). An example of milk theft would be northern elephant seals, where some male pups frequently attempt to suckle from other females after they have already been weaned by their own mothers. These pups are sometimes termed "super-weaners" because they gain significantly more weight than those pups that nurse only from their own mothers. To avoid detection, super-weaner pups are more likely to approach females who are sleeping, and whose own pup is already nursing. If a female does detect a pup attempting to nurse, she responds aggressively. In some cases, pups have been killed in these attacks.

As exemplified by the elephant seals, situations described as milk theft appear to be explained as manipulative behavior on the part of the offspring more than cooperative behavior on the part of the female. Milk theft is more likely to occur in mammals that most often produce only a single offspring per litter (monotocous species). In these species, the cost of nursing another's young is particularly high because females are likely adapted to produce only enough milk for one young, and even if their own offspring were to die, continued lactation to support another individual would likely delay the female's ability to conceive another offspring.

In contrast to species in which milk is apparently stolen, some species, such as African lions, appear to "willingly" nurse other cubs in the social group communally. Female lions are reported to nurse non-offspring in about 30% of all nursing bouts. In contrast to the behavior of elephant seal pups described above, lion cubs attempt to nurse from females other than their own mothers whether those females are alert or asleep, and females will lick and nuzzle other offspring nursing from them as they do their own cubs. Females do reject non-offspring more often than they reject their own cubs, but such rejection is not as aggressive as it is in elephant seals.

The apparently cooperative behavior of communal nursing is more likely to occur in mammals that give birth to more than one young, on average (polytocous species). In polytocous species, the costs of sharing milk with another offspring are likely to be lower than that of monotocous species, particularly if the female has lost some of her own offspring. The female must continue to produce milk to support the remaining offspring and milk supply may be less constrained than it is in monotocous species.

In most mammal species, communal nursing is more common in species where females live in small social groups. Such small groups could allow communal nursing to evolve via reciprocity, in which a female preferentially nurses the offspring of females who have previously helped her, because it would be easier to recognize who has and who has not cooperated in small groups. On the other hand, the members of small groups are often closely related, which promotes kin selection, where females share milk preferentially with offspring of her close relatives. In this way, she would be helping to ensure the survival of offspring that are genetically similar to her. In lions, communal nursing is more common when females are in groups of close relatives and less common as the degree of relatedness decreases.

Some authors have suggested that communal nursing may explain why females are likely to raise their young in groups. Although non-offspring nursing may provide some benefits to either offspring (e.g., increased food intake or decreased variation in feeding time) or females (e.g., aiding in the survival of relatives, encouraging reciprocal cooperation), the high cost of milk production makes this an unlikely explanation for social grouping patterns. As indicated above, most species that have been studied extensively have found that females do discriminate their own from other offspring, and only nurse non-offspring under certain conditions. It is more likely that non-offspring nursing evolved as a by-product of raising young in groups rather than as causal factor.

See also Caregiving—*Fostering Behavior*
Caregiving—*Parental Care and Helping Behavior*

Further Resources

Packer, C., Lewis, S. E., & Pusey, A. E. 1992. *A comparative analysis of non-offspring nursing*. Animal Behaviour, 43, 265–281.
Solomon, N. & French, J. (Eds.). 1996. *Cooperative Breeding in Mammals*. Cambridge, Cambridge University Press.

Susan E. Lewis

■ Caregiving
Parental Care

Parental care in sexually reproducing animals begins, essentially, with fertilization of the egg. For some species, care also ends at that point. For others, however, care can be more extensive. Before birth, parental care can include provisioning eggs with particularly high levels of nutrients, creating special coverings or chemical protection for eggs, constructing simple or elaborate nests, carrying eggs or embryos either internally or externally, passive or active defense of eggs or embryos, or acting in ways that improve the physical environment for developing offspring (e.g., increasing water circulation around eggs). Following birth, parental care may involve provisioning food, protecting young, or aiding in survival or foraging by way of social learning. Care can sometimes continue well after the young are nutritionally independent of the parent. Although parental care seems like a "fact of life" for humans and many other vertebrates, there is a great diversity across the animal kingdom in whether parental care occurs and, if so, how much care is provided. Understanding the factors that influence that diversity can help us better understand the situations under which parental care is likely to evolve.

Investing time and energy in parental care, or engaging in care behaviors that could put the parent at higher risk of predation, decrease the likelihood that that parent will survive and reproduce in the future. The magnitude of that decrease will vary depending on the type of care provided by the parent. For example, a sunfish scraping a simple nest along the shoreline of a pond may incur a relatively small cost, whereas a female elephant seal providing milk that is 60% fat to her offspring incurs a high energetic cost. Whether the costs are low or high, parental behavior would only be expected when those costs are balanced and exceeded by benefits. Although one tends to think of the benefits received by the *offspring*, of greater evolutionary importance is how the parent's cost in terms of future reproduction is balanced by how the *parent* benefits in terms of survival of current offspring.

The benefits of parental care fall into four general categories: (1) protection from unfavorable environments, (2) reduced risk of predation or disease, (3) enhanced competitive ability, or (4) provisioning offspring with an otherwise inaccessible resource. A male mallee fowl that constructs a pile of vegetation in which females lay eggs does so to avoid cool environmental conditions that would be detrimental to the developing eggs. As the vegetation in the pile decomposes, it produces heat that effectively incubates the eggs and speeds development. Similarly, many crustaceans living in environments with low levels of dissolved oxygen expend energy to fan their eggs, which increases the amount of water (and oxygen) circulating around the eggs. Predation can also be an important factor favoring parental care. For example, cichlid fish living in environments with high risk of predation are more likely to show care behaviors, such as carrying the eggs and young in the mouth of the parent (mouthbrooding). Animal species that live at high population density in which strong competitive ability outweighs the benefits of high reproductive output are more likely to produce a few, larger offspring and engage more heavily in parental care. Most primates have adopted this strategy of producing a very small number of offspring and directing a great deal of energy toward parental care. Finally, there are some species in which parents possess a critical resource required by offspring for survival. For example, herbivores ranging from termites to prehensile-tailed skinks (a reptile) to sheep have a symbiotic relationship with bacteria and protozoans in their gut that break down the otherwise undigestible cellulose in the plants they eat. These bacteria and protozoans must be transmitted from parent to offspring during a period of parental care.

Parental care is relatively uncommon in invertebrates, but it is found in a variety of crustaceans such as crabs and crayfish, some spiders and other arachnids, and many insect species. Often, such care is fairly rudimentary, such as excavating a safe burrow in which to lay eggs, or carrying eggs for a time until the larvae hatch. Among most social insects (termites, ants, bees), however, parental care is particularly elaborate. In honeybees, for example, "parental" care has evolved to the point where the mother (or queen) is primarily responsible for laying eggs and numerous sterile sisters (or workers) are responsible for nest construction and defense, foraging, and tending and feeding the developing larvae. Thus, most females sacrifice their own opportunities to produce young in favor of assisting in the care and feeding of their brothers and sisters.

Parental behavior is more prevalent in vertebrates than it is in invertebrates, but there is significant variation between the various vertebrate classes in the extent of parental care. Lampreys, which represent some of the most primitive living vertebrates, provide no parental care. They produce thousands of small eggs then die after spawning. Cartilaginous fish (i.e., sharks, skates, and rays) also provide little to no care to free-living offspring (those that have hatched or emerged from the uterus), although most produce rather large eggs that are well provisioned with nutrients. Some sharks and rays invest significantly in the prenatal development of their offspring, however. For example, hammerhead sharks give birth to live young (*viviparity*) that receive intrauterine nourishment from a during a yolk-sac placenta during a gestation period that can last 10 months or more. In other species, such as sand tiger sharks and all rays, eggs hatch but offspring remain within the uterus after a few months development (*ovoviviparity*). Once the nutrients from the yolk sac are absorbed, embryos may receive supplemental nourishment from milk-like secretions from the lining of the uterus. In some cases, they are also nourished by consuming additional eggs, or sometimes embryos, produced by the mother (oophagy or embryophagy).

Parental Behavior in Marsupials

Diana O. Fisher

Birth

Marsupials are born at a very early stage of development. Tiny newborns look like embryos, except that the arms, lungs and nose are much better developed than a nonmarsupial embryo. In kangaroos, the newborn must locate the mother's pouch by smell, and climb up to it without help. Once the baby latches on, the end of the teat swells in its mouth. Newborns of many smaller species of marsupials don't face the problem of reaching the pouch; the mother positions herself so that they virtually drop straight into it. However, many species that eat insects or meat (e.g., pigmy possums, antechinuses, opossums, and Tasmanian devils) give birth to litters. There are always more babies born than the number of teats. This means that the first behavior of newborns is to compete in something like a game of musical chairs, where the slowest ones miss out on a place, and inevitably die.

Life In and Out of the Mother's Pouch

Not all species have an enclosed pouch like a kangaroo. Smaller species with litters have only thickened skin and hairs around the teats, so young begin to drag on the ground after a few weeks. Mothers then leave them in a nest or den, returning frequently to nurse them. The young of meat-eating species, such as Tasmanian devils, spend much time wrestling noisily and chasing each other when their mother is away. Larger possums and koalas do not leave their young behind when they outgrow the pouch, but initially carry them on their backs. Later, young follow behind. Because very young marsupials need to be physically attached to a teat, it used to be thought that care involving the father was unlikely to evolve. Few marsupials have bonds between the sexes that last longer than mating. However, it is now known that some species of Australian possums do live in pairs or extended families, with both parents caring for offspring. An example is the rock ringtail possum. Fathers spend much time holding and defending older young in this territorial, tropical species. Families travel single file with the baby between the parents. If a youngster can't reach a food tree, the mother forms a bridge to the tree with her body. In contrast to other marsupials, young macropods leave the pouch gradually. The mother controls when the joey can enter and leave using her pouch muscles. At first, the joey emerges briefly. It gradually spends longer outside and becomes more adventurous. Young males play-fight, sometimes practicing boxing on their mothers. If danger threatens, the mother calls to her joey to get back in quickly before she flees. Occasionally a kangaroo pursued by dingoes throws her offspring from the pouch. She isn't sacrificing the life of her joey, as is sometimes thought. An ejected joey hides, waiting for its mother to return after she has evaded the predator. When they are too big to be carried, young kangaroos still stay near their mothers until weaning. Joeys of many forest-dwelling wallabies lack this following behavior. Instead, they hide alone in vegetation cover, where they remain inconspicuous until their mothers return to nurse them.

(continued)

Parental Behavior in Marsupials (continued)

See also Caregiving—*How Animals Care for Their Young*
Caregiving—*Parental Care and Helping Behavior*
Caregiving—*Parental Investment*

Further Resources

Fisher, D. O., Blomberg, S. P., & Owens, I. P. F. 2002. *Convergent maternal care strategies in ungulates and macropods*. Evolution 56, 167–176.
Gemmell, R. T., Veitch, C. & Nelson, J. 2002. *Birth in marsupials*. Comparative Biochemistry and Physiology: B-Biochemistry & Molecular Biology, 131, 621–630.
Russell, E. M. 1982. *Patterns of parental care and parental investment in marsupials*. Biological Reviews, 57, 423–486.
Strahan, R. 1995. *The Australian Museum Complete Book of Australian Mammals*. 2nd edn. Sydney: Reed Books.

Bony fish also show significant variation in parental care. About 80% are considered to be broadcast spawners that release eggs and sperm into the water column where fertilization takes place. The adults show no additional care. In those species that do show parental care, it often takes the form of scraping a simple nest or guarding or fanning the eggs. More elaborate care can be seen in mouthbrooding species such as the yellowhead jawfish that guard eggs by holding them in the mouth. Parental care in mouthbrooding species can be provided by females, males, or both parents and can occur either before or after the eggs hatch. Typically, the brooding parent cannot eat during the time they are brooding. Sea horses avoid this cost by depositing up to 300 eggs in a specialized pouch on the male's abdomen. The male then nourishes and protects the developing embryos until they are fully developed. There are also viviparous species such as guppies that give birth to live young following an internal gestation period. In addition to providing protection, some fish care for their young by promoting their postnatal growth and development. This could include fanning the eggs to circulate oxygenated water, guarding against the accumulation of decaying material or fungus, or assisting the offspring as they forage. Parents may lift leaves or agitate the substrate to expose microorganisms for their offspring to consume. In one species (the discus, a cichlid fish) male and female parents are even known to increase mucus production on their bodies to provide food, and possibly antibodies to support the immune system, for the offspring.

As with fish, only a small percentage of amphibians and reptiles actively care for eggs and young. In amphibians, about 10% of species show parental care, which usually involves guarding eggs during development. Anurans (frogs and toads) have more diverse patterns of care than do salamanders and caecilians. These can include guarding tadpoles, transporting tadpoles from a nest to a water source, feeding tadpoles unfertilized eggs, or carrying eggs on the back, in the vocal pouch, or even in the stomach of the parent. There are also a few frogs, salamanders, and caecilians which have evolved viviparity and care for their young internally through the production of secretions from the oviduct or unfertilized eggs. In reptiles, parental care ranges from nest-digging in turtles, to attending or active brooding of eggs in pythons and some other snakes, to protecting and transporting young for extended periods in crocodilians. As with amphibians, ovoviviparity and viviparity have evolved in several groups of reptiles including snakes, legless lizards, and skinks. Some viviparous species have a simple placenta for nourishment of the developing embryos.

In contrast to the other vertebrate groups, well-developed parental care is found in all species of birds and mammals. Most birds build a nest and incubate eggs for some period of time. The only exceptions to this trend are some megapodes (ground-nesting birds from Australasia), that build a nest of vegetation heated by the sun and decomposition of the plant material so the parents do not need to incubate, and brood parasites such as cowbirds and cuckoos that forgo nest-building in favor of depositing their eggs in the nest of another bird. Some species with *precocial young* (those ready to leave the nest almost immediately after hatching) or that feed their young on seasonally abundant food sources such as fruit or seeds show uniparental

Swans lead their babies to the water's edge.
© Sam Kittner/National Geographic Image Collection.

care. However, the juvenile survival rate of many species is greatly enhanced with two parents rather than one, making biparental care the norm. Birds care for offspring through incubation, defense, and social learning. Parents typically provision offspring by providing them with whole or regurgitated food. A few species, such as pigeons, also secrete a milk-like substance from the lining of the esophagus that is fed to offspring.

Mammals fall into three general groups based on their mode of parental care. *Monotremes* such as the duck-billed platypus and the echidna lay soft-shelled eggs. Females have a specialized patch of skin on their abdomen that secretes milk, which the offspring consume by sucking on tufts of hair in this area. *Marsupials* give birth to highly altricial (poorly developed) offspring that are very tiny compared to the size of the adults. These offspring then crawl into a pouch on the female's abdomen, attach to a nipple, and continue to develop for several weeks or months. As the offspring become increasingly independent, they continue to occupy the pouch intermittently for rest, protection, and transportation. *Placental mammals* are those in which offspring develop internally and are nourished by way of a placenta. In addition to the energy invested in offspring during gestation and lactation, parental behavior can include active defense of the young, provisioning of food, carrying offspring, and contributing to social learning.

As is seen in the social insects, parental care in mammals and birds sometimes extends into *alloparental* or cooperative care of the young. This occurs when individuals other than the parents contribute to the successful development of the offspring. Examples of alloparental care include helpers at the nest or den, communal nursing, or eusociality.

Although biparental care does occur, especially in birds, care by only one parent is much more common. Species with internal fertilization tend to have care provided by females whereas those with external fertilization are more likely to have care provided by males. There are three primary hypotheses to explain which parent is most likely to care for the young: paternity certainty, order of gamete release, and association with offspring.

Paternity Certainty: Robert Trivers suggested that in species with external fertilization, males have greater certainty that they are fertilizing the eggs, and thus are more likely to care for the developing offspring. In contrast, in a species with internal fertilization, the female may have mated with other males, so although she is certain to be the mother, the male is uncertain as to whether he is the father. However, theoretical models have demonstrated that paternal care can evolve even in situations with very low reliability of paternity, as long as the care provided by the father increases the survival of the offspring enough to counterbalance the decreased mating opportunities he incurs by staying to care for the young.

Order of Gamete Release: Other authors have suggested that males are less likely to care for young when there is internal fertilization, because they have the opportunity to "desert" the young before the females do, having released sperm before the females release the eggs, and that the opposite is true with external fertilization. This hypothesis is likely to explain patterns of parental care in some groups of animals, such as shorebirds, but it is unlikely to be a general explanation for all parental care patterns. For example, paternal care is more common in frogs, even though males release their sperm into the nest before females release their eggs. Also, in fish species with simultaneous gamete release, males are more likely to provide care even though the hypothesis predicts males and females should be equally likely to care.

Association: In mammals, females are much more likely to provide care, as would be expected given their prolonged association with the developing offspring during gestation and lactation. Some authors have suggested that this simple fact of association also explains the patterns of care between species with internal and external fertilization. With external fertilization, eggs are often laid on a male's territory, such that the defense of the territory becomes a preadaptation for the defense of eggs and young.

As with the above hypotheses, association is unlikely to be sufficient to explain all patterns of parental care. For instance, like the order of gamete release hypothesis, this hypothesis incorrectly predicts that species with simultaneous release of gametes should be equally likely to have maternal and paternal care. Ultimately, which parent provides care is likely to reflect tradeoffs between the increase in offspring survival with the care of one or two parents and the decrease in future breeding opportunities incurred as the result of that care.

The endpoint of the parent–offspring interaction is usually well defined. A digger wasp provides care by digging a burrow and provisioning it with a caterpillar, but once the egg is laid she provides no additional care. Similarly, a garter snake provides care to her developing embryos until they are born, at which point the young disperse and have no additional interaction with their mother. However, in species where care continues after the offspring are born, there is potential for conflict between the parents and the offspring regarding when such care should end. From the parent's perspective, care should end as soon as the costs of care (in terms of future reproductive success) exceed the benefits (in terms of current reproductive success), because parents are related to their current and future offspring equally. However, from the offspring's perspective, care should continue for a longer period of time because they are less closely related to their future siblings than they are to themselves. This creates what has sometimes been called a "weaning conflict."

See also Caregiving—*How Animals Care for Their Young*
Caregiving—*Mother–Infant Relations in Chimpanzees*
Caregiving—*Non-Offspring Nursing*
Caregiving—*Parental Care*
Caregiving—*Parental Behavior in Marsupials*
Caregiving—*Parental Care and Helping Behavior*
Caregiving—*Parental Care by Insects*
Caregiving—*Parental Investment*
Social Organization—*Eusociality*

Further Resources

Alcock, J. 1998. *The adaptive tactics of parents*. In: *Animal Behavior*. 6th edn. (Ed. by J. Alcock), pp. 523–554. New York: Sinauer Associates.

Clutton-Brock, T. H. 1991. *The Evolution of Parental Care*. Princeton, NJ: Princeton University Press.

Rosenblatt, J. S. & Snowdon, C. T. (Eds.). 1996. *Parental Care: Evolution, mechanisms, and adaptive significance. Advances in the Study of Behavior*, 25. New York: Academic Press.

Trumbo, S. T. 1996. *Parental care in invertebrates*. In: *Parental Care: Evolution, Mechanisms, and Adaptive Significance*. (Ed. by J. S. Rosenblatt & C. T. Snowdon), pp.3–51. San Diego: Academic Press.

Susan E. Lewis

■ Caregiving
Parental Care and Helping Behavior

One of the greatest efforts in every species' life is to increase adaptation, and parental care is one of the best ways. It includes those behaviors performed to enhance offspring survival; it is usually directed by parents to their own offspring, but cases are known of adults caring for young of relatives, alien offspring, or even of a different species.

In everyday terms, parental care is considered the action performed toward young, but it actually includes physiological aspects needed to develop sexual cells and embryos as well. Central to this concept is the *Parental Investment* (PI): Since females produce very large and rich eggs, limited in number and at intervals, at mating they have already invested a great amount of energetic resources in offspring. Conversely, males, whose sperm are easily expended because of small size, very large number, and almost continuous production, invest much less energetically. Males are therefore usually less interested in parental care, whereas females stay near the offspring even in danger in order not to waste their great past investment. It is worth noting that the same, but specular, habits can be verified in species where evolution led to complete inversion of parental roles, as in the American jaçana (*Jacana spinosa*) and other birds: The female mates with many males, is greater and does not care for eggs and nestlings, and is aggressive and territorial; whereas males incubate eggs, feed hatchlings, and do not defend the territory.

Anisogamy, the different shape between male and female sex cells, and the related difference in PI are responsible for the evolution of *polygamy*, the reproductive social system where one individual of one sex mates with several others of the opposite sex. Usually, one male mates with two or more females (*polygyny*), as occurs very frequently among mammals, but there are cases of one female mating with several males (*polyandry*), such as the above jaçana. Polygamous species display very unbalanced parental care, which is almost only performed by the sex investing more in the reproductive effort: females in polygyny and males in polyandry. In fact, monoparental care is generally found in polygynic reproductive systems, whereas the biparental one is prevalent in monogamous species. Exclusive paternal care is present in bony fish and in amphibians, but infrequent in other vertebrates.

The environment is an important adaptive pressure for the evolution of parental care. In stable/predictable environments species tend having a reproduction close to *K* selection, that is, with the maximum energetic effort given to offspring. Conversely, if the habitat is unstable and/or with food resources difficult to be found, reproduction goes toward *r* selection, that is, with the maximum of energy used for the production of the maximum number of offspring, but with poor protection afforded. Species undergoing *r* selection produce, therefore, far more numerous offspring, but with low PI. *K* selection instead applies to species bigger in size, producing a small number of offspring, but with high survival probability due to great PI. However, we must bear in mind that both selection types are not a rigid concept and there are many intermediate situations.

Parental care will be obviously more developed in species following the typical *K* selection. Moreover, species with short or long lifespan will tend to follow *r* and *K* selection, respectively. Therefore, *r* selection will be found frequently among invertebrates, whereas *K* selection among vertebrates. Within the latter, however, ectotherm classes (fish as a whole, amphibians and reptiles), smaller and with shorter lifespan, tend generally towards *r* selection, whereas the endotherms (birds and mammals), larger and with longer lifespan, tend toward the *K* one. This sort of classification is however very relative, since both selections can occur within the same class: For instance, passerine birds, with altricial hatchlings, follow *K* selection, whereas waterfowl, with precocious broods, follow *r* selection. Nevertheless, the latter's parental care is still more developed than that of fish or amphibians.

Although being apparently the reverse of parental care, infanticide is not an abnormal behavior, but an evolutionary product. It is relatively widespread and is found in many animal groups, being caused by both environmental and social factors. Infanticide does not offer any advantage to the young being killed, but can advantage the siblings. In fact, in some contexts, as when the litter is too large, the mother herself starts the cannibalistic behavior, killing the less developed offspring, saving the others. The young are killed when still totally dependent on parents, for example, in mammals during lactation, thus during the maximum PI period. In other cases, the young fight each other, while parents do not intervene, a phenomenon known as *fratricide* or *cainism*, from the Bible episode. It occurs frequently in many heron and raptor species, where natural selection led the first-hatched nestling to kill the younger sibling. As a result, the food the parents deliver to the nest can feed sufficiently the reduced number of hatchlings; otherwise, only a few of them would survive to fledge.

Being so important for survival and evolution of animal species, the number and variety of solutions is obviously huge. The number of animals displaying parental care is high as well, although mostly, for the above reasons, among vertebrates. Considering the invertebrates, in fact, we find many species with more or less developed parental care, but restricted to arthropods. The simplest forms involve digging a protected nest for placing the eggs. In more advanced forms, such as ichneumonid wasps, along with the eggs, the mother deposits a food supply, usually a caterpillar, eaten later by the newly emerged larvae. In burying beetles (*Necrophorus*), both parents actively feed each larva with a blackish fluid regurgitated directly into its mouth, very similar to what birds do. This habit is the base of the complex parental care displayed by eusocial insects (ants, bees, termites), the most developed form found among invertebrates. These insect societies have many sterile workers, all queen's children, which defend the group, locate food sources, care for and feed the larvae.

Parental care is virtually absent among agnathans (hagfishes and lampreys), unless simply covering the eggs with sand can be considered parental care. A similar situation occurs in cartilaginous fishes: The most developed care is where internal fertilization led several species to develop viviparity, getting even placentation, that is, embryonic nutrition and oxygenation. Such efficient care is obviously accompanied by a reduction of the number of young produced. Teleosts (modern bony fish) possess instead a greater variety of solutions, in both oviparous and viviparous species. Care spans from male gobies (*Gobiidae*) tending a bunch of eggs attached to rocks and stones and fanning the water to create an oxygenating current, to the construction of true nests with either submerged vegetation (bowfin, *Amia calva*, and three-spined stickleback, *Gasterosteus aculeatus*) or floating bubbles (Siamese fighting fish, *Betta splendens*). Eggs can be protected also by transporting them: Female sea horses (*Hippocampus*) deposit eggs into the male's temporary pouch where fertilization occurs and embryos complete their development, and some African cichlids keep fertilized eggs in the mouth during embryo development (oral incubation); also, after hatching, the young come into the mouth when a potential threat approaches.

Amphibians, primarily the anurans, also display several complex parental care behaviors, which are almost always devoted to protecting eggs from desiccation and environmental harshness, and larvae from predators. Egg guarding and tending is frequent: Eggs are maintained in mud basins built by parents and filled with water; many arboreal anurans (*Hyla, Chiromantis* and others) lay their eggs on a leaf in a nest of mucus produced by the skin or maternal genitalia, the mucous being then "whipped" by the legs until it becomes a soft foam. In *Ichthyophis* (Gymnophiona) and *Amphiuma* (Caudata), the adult encircles the eggs with its body until they hatch within a subterranean den deliberately dug below water level. One further step is carrying the egg cluster on the parent's back, as with the male European *Alytes obstetricans*, using an adhesive substance on the skin. An improvement is accepting the eggs within the skin: The sponge dorsal tissue develops many pouches where the eggs are inserted and then covered until hatching as tadpoles or small frogs (*Pipa*). In *Rhinoderma darwini*, the eggs are stored within the male's vocal sacs. The most bizarre way to transport eggs is shown by the Australian *Rheobatrachus silus*, where incubation is carried out in the female's stomach, which obviously stops feeding for the whole period. Similar examples of protection afforded to tadpoles occur as well.

Although the parental techniques performed by amphibians are very complex, they just help egg or tadpole survival. They can be considered a terrestrial version of parental care displayed already by fish. Surprisingly, reptiles do not display the number or variety of solutions observed in amphibians. The only remarkable examples are the python (*Boidae*) females rolling upon the eggs for several weeks and some viviparous lizards removing membranes from the newborn at parturition. In several crocodile and alligator species, the female, and sometimes the male too, guards the nest with eggs, uncovers the hatching eggs from sand, and then takes newborns one at a time into her mouth, carrying them to water.

Why did reptiles lose many parental care behaviors, already displayed by their relative amphibians? The reason probably lies in the great evolutionary novelty represented by the amniotic egg, the egg with embryo membranes enabling its deposition far from water. When reptiles developed dry-land colonization, in Carboniferous, no serious competitors lived on land; the only danger could come from the amphibians, but they were confined near wet habitats. Such an advantageous evolutionary reproductive structure eliminated many adaptive pressures toward the search for better reproductive success, which in turn reduced the necessity of complex parental care.

Both birds and mammals show a large repertoire of parental care. Many patterns similar to those considered above are evident, but at the same time they produced something very different: the great development of social behavior. This created an enormous qualitative change in parental care. In fact, it is no longer only a behavioral mechanism devoted to allow progeny survival, but also a way to transmit information between generations, even modifying the behavior.

The long duration of care is an important aspect; it is connected with the evolution of endothermy and bigger size. Endothermy needs efficient parental care because offspring energetic requirements increase dramatically soon after birth. In fact, endothermic newborns grow at faster rate than ectothermic contemporaries. Consequently, a similar duration of parental care is connected to greater physiological development of offspring. Suckling, a universal and characteristic behavior among mammals, is not only fundamental for survival of young just after birth, but also a good opportunity to create a sort of symbiosis between mother and child. By staying near their parents after getting independence, juvenile birds and mammals can improve their efficiency in food selection and social behavior learning. Another effect of endothermy in young was the development of a large brain, which then led to the development a greater learning ability immediately after birth, making the passage of information from parents to offspring easier: Imprinting and other learning patterns are evidence of this trend.

Two behavior aspects were developed exclusively by birds and mammals: helping and adoption. Birds in particular (e.g., the Florida scrub jay *Aphelocoma coerulescens*) show nonreproductive subadult individuals helping a pair of adults in parental care. The recipients are usually genetically related to the helpers. It is advantageous for helpers to care for siblings in a life phase when they are not yet able to have a brood of their own, and helpers also obtain the advantage to help survival of individuals bearing a high percentage of their own genes. Similar examples are found in mammals, too, e.g., the black-backed jackal (*Canis mesomelas*). Helping is a fundamental aspect of the odd socio-reproductive system of the African naked mole rat (*Heterocephalus glaber*), showing striking similarities with societies of eusocial insects.

The "spontaneous" parental care displayed by helpers is directly related to the adoption behavior, another example of parental care typical of endotherms. It consists of parents caring for young not their own, even of different species. In gulls, or other marine birds nesting in colonies, adoption is frequent because the close vicinity of nests makes it possible that already mobile nestlings, still dependent on parents, can approach the nest of a neighbor pair. In most cases adoption is, however, an accidental event, especially when occurring between species, a rare but not unknown occurrence. Acceptance of foreign young usually is based on the recognition of "infant signals," identified by Konrad Lorenz in 1943, and borne by all birds' and mammals' young. The adaptive function of infant signals is to be recognized as infant, not as adult, thus stimulating care. The young must use signals to communicate to parents, who must be able to receive them. When signals fade out, parents lose interest in their offspring; weaning is likely based upon a mechanism of this sort. Birds and mammals show a convergence of such visual signals, so that young are recognized as infants even beyond the species boundary. This explains why humans appreciate the young of many mammals and also explains the documented cases of children raised by wild animals.

See also Caregiving—*Helpers in Common Marmosets*
Caregiving—*How Animals Care for Their Young*
Caregiving—*Mother–Infant Relations in Chimpanzees*
Caregiving—*Parental Behavior in Marsupials*
Caregiving—*Parental Care*
Caregiving—*Parental Care by Insects*
Caregiving—*Parental Investment*

Further Resources

Balshine, S., Kempenaers, B. & Székely, T. (Eds.) 2002. *Conflict and cooperation in parental care. Papers of a Theme Issue.* Philosophical Transactions of the Royal Society, Series B, 357(1419), 235–404.

Antipredatory Behavior

A group of meerkats in Botswana keep a close eye out for predators, especially those who fly. They are being vigilant.
Courtesy of Getty Images / Digital Vision.

Frogs will wait for hours for their prey because they are able to blend into the background and thus escape detection. This kind of camouflage is called "crypsis."
© *Anthony F. Chiffolo.*

Art and Animal Behavior

Paintings from Egypt of cows being milked often include a calf standing nearby, suggesting the knowledge that cows milk more easily when their calf is in sight.
© The Art Archive / Musée du Louvre Paris / Dagli Orti.

A representation of resting bison found in a cave in Altamira in northern Spain. This cave art is presumed to have been painted by the Magdalenian people between 16,000-9,000 BC.
© The Art Archive / Dagli Orti.

Behavioral Physiology

A male peacock spreads large iridescent green feathers to get a female's attention. Females prefer males with elaborate tails, and males with elaborate tails have the greatest reproductive success.
Courtesy of Getty Images / Digital Vision.

It's doubtful that any predator could mistake this blue poison-arrow frog in Costa Rica for being anything but what it is, due to its bright color.
© *Tom Brakefield / SuperStock.*

Turtles can bask in the sun to warm themselves because of their ability to blend into the background and thus escape predation.
© *Anthony F. Chiffolo.*

Careers

A professional diver shoots footage of colorful fish for a documentary.
© SuperStock, Inc. / SuperStock.

Caregiving

Lions generally take a communal approach to child rearing, often looking after more than just their own offspring.
Courtesy of Getty Images / Digital Vision.

Caregiving

A male yellowhead jawfish protects fertilized eggs in his mouth.
© Marty Snyderman / Visuals Unlimited.

Cognition

A pufferfish blows itself up so as to appear larger and more formidable to potential predators. Some ethologists refer to this as a form of deception.
© Anthony F. Chiffolo.

Cephalopods

A common octopus (Octopus vulgaris) is almost perfectly camouflaged against its rock rubble environment in about 20 feet of water. This individual would have gone undetected if it had not moved.
Courtesy of James B. Wood.

Communication—Vocal

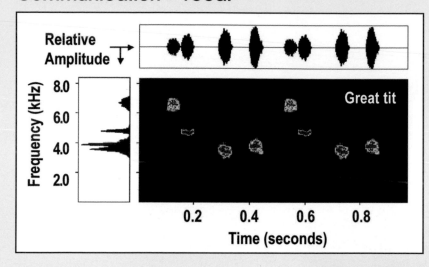

Sonogram of a song of a great tit (Parus major) — at the top is the relative amplitude (usually measured in volts) wave form and, on the left, is the power spectrogram of the song. Two four-note phrases are repeated.

Sonograms of collared dove (Strep-topelia decaocto) coos: a juvenile male (top left), an adult female (top right), and two adult males. The two adult male coos differ in the presence of frequency jumps: one in each of the three coo elements in the bottom right, none in the bottom left. These jumps in frequency are relatively small compared to the harmonic spectral structure of the female coo. However, the sonograms are normalized for amplitude, making the softer female coo, and especially the high frequency harmonics, appear more impressive. The male coos with and without the jumps are much louder than the juvenile and female coos, and this makes the frequency modulations of males perceptually more salient.

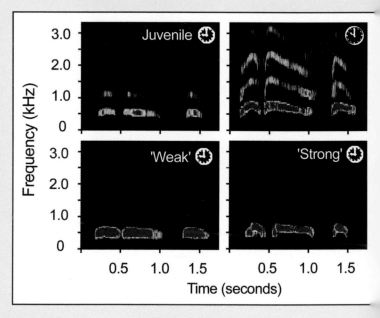

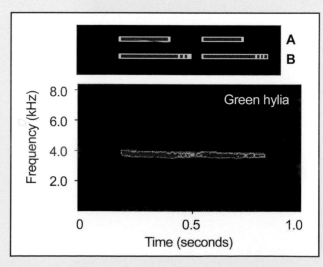

Sonogram of a green hylia song and, on top, a schematic representation of this song when no echoes would be recorded (A) and when both direct and indirect sound would be recorded (B). Similar sound stimuli were used for playback experiments (during which they were transmitted through the forest to the birds, and therefore both short and long songs will have become longer because of accumulated echoes).

All sonograms courtesy of Hans Slabbekoorn.

Communication—Vocal

Whitetail deer can vocalize, especially when they feel threatened, or when they sense a threat to their young.
© Anthony F. Chiffolo.

Coyotes

A coyote and her pup play just outside their den.
© Jack Milchanowski / Visuals Unlimited.

Curiosity

A curious juvenile Japanese macaque examines a photographer's camera.
© Timothy G. Laman / National Geographic / Getty Images.

Weasels are notorious for their curiosity getting them into trouble. Here, a beech marten can't help but to investigate a statue in a garden.
© Manfred Danegger / Photo Researchers, Inc.

Clutton-Brock, T. H. 1991. *The Evolution of Parental Care.* Princeton, NJ: Princeton University Press.

Rosenblatt, J. S., & Snowdon, C. T. (Eds.) 1996. *Advances in the Study of Behaviour. Volume 25: Parental Care Evolution, Mechanisms, and Adaptive Significance.* New York: Academic Press.

Singh, J. A. L., Zingg, R. M. 1966. *Wolf-Children and Feral Man.* North Haven: Archon Books.

Davide Csermely

■ Caregiving
Parental Care by Insects

Most insects do nothing more to provide for their offspring than to lay their eggs in a spot where their hatching young will be able to feed. Some, however, exhibit parental behavior that ranges from covering their eggs with a protective coating to remaining to feed and protect young to forming life-long associations with an extended family. Though not common, parental care has evolved many times in many different groups and has been promoted by very different ecological conditions. It is most developed in some beetles (*Coleoptera*), cockroaches (*Blattaria*), webspinners (*Embioptera*), thrips (*Thysanoptera*), true bugs (*Hemiptera*) and especially the social insects (*Hymenoptera* and *Isoptera*).

Exclusive maternal care is more common (having evolved independently in at least 18 orders) than either paternal care (occurring in only 2 orders) or biparental care (occurring in 4 orders). The most rudimentary care provided by a female is to incorporate toxins into eggs to deter predation or to cover the clutch with a hard shell or wax-like material before abandoning them. This cover may serve as camouflage or protection from parasitic wasps. The most commont form of maternal care is provided by females who guard their clutch of eggs and young nymphs from predators and parasites using chemicals or defensive behaviors. For example the lacebug, *Gargafia solani*, guards her eggs and young until they are mature and, when a predator approaches, she rushes it, fanning her wings.

Other functions of maternal care include the facilitation of feeding and the amelioration of environmental conditions. Female plant-feeding, membracid bugs, *Umbonia crassicornis*, cut slits in the bark so their young can feed. Female salt-marsh beetles, *Bledius spectabalis*, maintain a burrow shaped in such a way to prevent flooding at high tide. They also provision their young with algae and defend them against parasites.

In social insects, the care of offspring is carried to the extreme. There is a wide range of maternal behavior in wasps and bees, from solitary females that provision nests with all the food the developing young will need to the highly social bees and all the ants in which a very large extended family of tens of thousands of individuals performs all the housekeeping, offspring rearing and foraging tasks. Workers in many species even produce sterile eggs to feed to the young larvae. The reproducing individuals, which may be a single queen, often have no other responsibilities.

Maternal care in some insects can take the form of internal development. Cockroaches have a wide range of reproductive modes and maternal investment of this sort. The egg bearing, *Blatella vaga*, produces secretions that feed her newly hatched young. *Diploptera punctata* females undergo a 60-day "pregnancy" during which a highly nutritious milk is secreted from the walls of the brood sac. Young are well developed at birth and are independent of further maternal care.

Exclusive paternal care is found in only about 100 species of insects, and the giant water bug, *Abedus herberti*, is one of the best examples. Females glue their eggs on to the males' wing covers. He stops feeding, aerates the eggs and protects them from predators until the eggs hatch. Males of many species of insects provide indirect paternal care through nutritious offerings to females when they entice her to mate. These may be a "nuptial" gift of a dead insect, such as hangingflies, *Hylobittacus apicaois*; or, as is the case in many crickets and katydids, spermatophores that accompany the sperm and provide females with nutrients that may be passed to her eggs. In this way, males increase the reproductive success of the females.

In insects, biparental care is most elaborate in some beetles and cockroaches and all termites. Males and females in the woodroach, *Cryptocercus punctulatus*, and all termites form life long associations. Woodroaches construct and guard an elaborate burrow system in rotting logs and protect and feed the young until maturity. They rear a single brood that requires about 3 years to mature. In a termite colony, there is usually task specialization based on sex and morphology. In the "higher" termites, sterile workers perform all the offspring care; a king and queen may live 20 or more years and produce a new batch of reproductive males and females every year.

Some biparental care results in serial monogamy. Burying beetles, *Nicrophorus orbicollis*, require a small vertebrate carcass as the resource for their young. This is buried and prepared, and both parents remain with the young to feed and guard them. Young develop quickly on this rich resource. Females usually stay longer than males and each seeks another carcass after they leave. (See the accompanying photos.)

Why do some insects remain to provide care to young rather than seeking another breeding opportunity? And why is this behavior more common in females than males? The costs of providing parental care are the investment in time and energy that could be otherwise spent in future reproduction, exposure to additional danger, and reduction of lifespan. These costs must be outweighed by the benefits of increased offspring survival and viability if parents are to provide care.

Above: Male and female burying beetles on a newly discovered mouse carcass. Right: Burying beetle parent feeding larvae.
Courtesy of
Michelle Pellissier

In herbivorous insects, the most important factor promoting parental care is predation by other arthropods; for insects that rely on unpredictable resources or must construct an elaborate burrow system, the most important factor is competition. A parent must be able to successfully defend its brood against the predator or competitor if parental care is to evolve. Broods guarded by maternal lacewing bugs have a much higher survival rate than those abandoned. Another important factor in the balance of costs and benefits is the probability that the parent will breed again. The herbivorous bug, *Elasmucha putoni*, lays her eggs on a host plant with a short fruiting season and remains to guard her eggs and young nymphs because there is no opportunity to produce another brood and because her protection increases their survival.

Biparental care is favored in insects such as burying beetles that rely on a rich but unpredictable resource like carrion because they also have no guarantees that they will find another during their single season of breeding. Their major competitors for this resource are other burying beetles, and two parents are much more successful defending the carcass and brood than one. Male or female woodroaches are also unlikely to be able to find another mate and breeding opportunity. Although wood is an abundant resource and supplies both food and protection, it is difficult for the young to access and requires gut symbionts to digest. Young grow slowly on this resource and thus the two parents rear a single brood in their lifetime.

Exclusive paternal care is rare in insects because external fertilization is rare. When females mate with multiple males and have internal fertilization, paternity assurance is low for males. In other taxa, exclusive paternal care is associated with male territoriality and external fertilization of eggs in a nest made and guarded by the male. Often he guards the eggs of multiple females at little extra cost and in providing care he not only does not forgo additional matings, but a male carrying eggs may signal his quality as a parent and be more attractive to females. Most of the examples of paternal care in insects fit this behavior pattern. The wing covers of the giant waterbug makes a nest of sorts, and males do carry the eggs of multiple females.

Parental care in insects takes a number of different forms and is often associated with multiple behaviors with multiple functions, like guarding and provisioning. Insects may even show behavioral plasticity and alter the level or duration of their care in response to the loss of a mate or changes in the level of competition. The evolution of parental care can also be accompanied by changes in life history and development; parental species may have defenseless larvae unable to feed on their own, whereas their nonparental relatives have fully independent larvae. This variety is not surprising since parental care has evolved independently many times in insects in response to different selective forces.

See also Caregiving—*How Animals Care for Their Young*
Caregiving—*Parental Behavior in Marsupials*
Caregiving—*Parental Care*
Caregiving—*Parental Care and Helping Behavior*
Caregiving—*Parental Care in Fish*
Caregiving—*Parental Investment*

Further Resources

Choe, J. C. & Crespi, B. J. 1997. *Social Behavior in Insects and Arachnids*. Cambridge, UK: Cambridge University Press.

Clutton-Brock, T. H. 1991. *Parental Care*. Princeton, NJ: Princeton University Press.

Scott, M. P. 1998. *The ecology and behavior of burying beetles*. Annual Review of Entomology, 43, 595–618.

Tallamy, D. W. & Wood, T. K. 1986. *Convergent patterns in subsocial insects*. Annual Review of Entomology, 31, 369–390.

Trumbo, S. T. 1996. *Parental care in invertebrates*. In: *Parental Care: Evolution, Mechanisms, and Adaptive Significance* (Ed. by J. S. Rosenblatt & C. T. Snowdon), pp. 3–51. San Diego, CA: Academic Press.

Wilson, E. O. 1971. *Insect Societies*. Cambridge, MA: Belknap Press.

Zeh, D. W. & Smith, R. L. 1985. *Paternal investment by terrestrial arthropods*. American Zoologist, 25, 785–805.

Michelle Pellissier Scott

■ Caregiving
Parental Care in Fish

Fish are fascinating organisms to study the evolution of parental care. The majority of fish species do not care for their eggs. Fertilization is typically external and the parents leave the spawning site without further care of the embryos. The percentage of fishes that care for their offspring is, however, substantial. In about 20.6% of the teleost families, parental care of some form is present, at least in some species.

If we define parental care as any form of behavior or other biological activities occurring after fertilization and resulting in an increase in the likelihood of offspring survival, we find that different fish species may contribute to the survival of their offspring in very diverse ways.

Diversity of Patterns of Parental Care in Fish

Guarding

One of the simplest forms of parental care is the guarding of the substratum where the eggs have been spawned. This type of parental behavior may be a continuation of territorial defense already initiated before spawning takes place. The simple fact that an adult fish aggressively drives away other animals may, by itself, reduce the risk of predation on the developing embryos. This simple form of nest defense, often provided by the male alone, may be supplemented by behaviors like fanning and rubbing the eggs with the body or fins. These activities promote increased levels of oxygen and help to remove waste products and sediment particles that tend to smother the eggs. In many species, the caregiving adults mouth the eggs and remove dead or infected embryos, although some cannibalism of apparently healthy offspring may also occur.

The spawning site may vary from a patch of bare substratum (like coral or rock) to a hole or crevice, a pit in sand or gravel, or the space under a boulder. The nest site is often made more suitable by cleaning it and removing pebbles or other materials. There are, however, species in which the fish may build quite complex nests, as in sticklebacks and some wrasses.

In some species the parent(s) may also care for their larvae and even juveniles and, in a few cases, may help them to find food, either by allowing the fry to pick mucus from their skin, or by stirring the substratum and suspending small food particles in the water. In extreme cases, the offspring reach sexual maturity while in the natal territory and may help to defend the younger fry while attempting to breed themselves. In these cases, it may be hard to tell to what extent the benefit that the parents gain from these "helpers" is greater than the costs imposed by their parasitic behavior, as they attempt to fertilize part of the eggs and/or spawn their own eggs among those of the larger females. Guarding is found in a very wide spectrum of species, such as bowfins, lungfish, some minnows, many catfish, sunfish, blennies, gobies, damselfish, many cichlids, Siamese fighting fish, and many others.

Internal Gestation

Live bearing evolved independently in many fish lineages, including sharks and rays, coelacanths, and many teleosts, such as surf perches and the guppies, swordtails and mollies (so popular with aquarists), among many others. In this form of parental care, fertilization is internal and males often possess specialized structures, such as modified pelvic or anal fins, that help to introduce sperm into the female genital tract. The nutrition of the young may be provided by large amounts of yolk in the eggs, by secretions provided by the mother, or by structures that are analogous to the mammalian placenta. An interesting example is provided by the living coelacanth *Latimeria chalumnae*. A female about 1.8 m (6 ft) matures several eggs, each the size of an orange (9 cm or 3.5 in in diameter). The embryos develop inside the female, and a single juvenile (one-third the length of its mother) is born after having cannibalized the remaining eggs and embryos. Many sharks provide food for the developing young in a similar way.

External Bearing

Another form of parental care involves transporting the eggs attached to the body of the parent(s), or inside the mouth or the branchial cavity. The ways by which the eggs are carried are quite diverse. In pipefishes and seahorses, the female possesses a long genital papilla that is used to transfer the eggs to the male, through a sort of inverted copulation. The male carries the embryos attached to its ventral surface or in a brood pouch formed by the ventral expansion of the plates that cover the body. In some species, including seahorses, the brood pouch fully protects the eggs and the male seems to provide oxygen and nutrients to the offspring. In these cases, we can speak of a "male pregnancy," analogous in many respects to the pregnancy of females of species with internal gestation.

The stargazers carry their embryos in the space between the pectoral fin and the body. In species of the genus *Kurtus,* the eggs possess long filaments and are suspended from a kind of hook on the head of their father.

Mouth brooding, in which one parent (or sometimes both) picks up the eggs after spawning and carries them in the mouth during development, evolved independently in many teleosts lineages. In some cichlids the embryos hatch and continue to develop inside the oral cavity where they may even begin to feed from particles carried by the water that passes through the parent's mouth.

If we exclude internal gestation that is, by definition, provided by the females, the remaining forms of parental care may, in different species, be provided by males, females, or by both sexes. Contrary to what occurs in many familiar animal groups, like mammals or insects, in fish paternal care is much more common (51 teleost families) than maternal (30 families) or biparental care (23 families). In a few species, the role of both sexes may vary. For instance, in some biparental species (like the *Convict cichlid*) where eggs and fry are jointly defended by a pair, the male sometimes deserts the brood, leaving the female to care alone.

Evolution of Parental Care in Fish: A Challenge to Evolutionary Biologists

Comparative studies show that in fish there are a very large number of examples where male guarding evolved independently from ancestors with no care. There are also known transitions from no care to female care, but these are clearly the minority. From male care, several groups evolved biparental care which, in some cases, originated care by the female alone. These studies also show that guarding of larvae and juveniles frequently evolved as a prolongation of egg guarding, and mouth brooding evolved repeatedly from egg guarding ancestors.

This picture shows that, in fish, not only paternal care is prevalent but it was also the starting point from which several other forms of parental care evolved. We are so used to thinking of parental care as a female trait that we may be surprised by the prevalence of male care in fish. Why should this be so? First of all, our common sense is biased by our familiarity with terrestrial animals. In terrestrial animals, fertilization is internal. In this circumstance, there is a time lag between fertilization and the release of the offspring by the mother. Unless a bond between male and female is present, it would be difficult for males to evolve parental care, even if it was beneficial to them to do so. Thus, we expect that in species with internal fertilization, parental care evolves more commonly in females than in males.

In fish, external fertilization occurs in the large majority of species, providing an ideal opportunity to compare both sexes concerning the likelihood of parental care evolution without the bias imposed by internal fertilization. If providing parental care involves a higher fitness gain for males than for females, its evolution is made easier with external fertilization because the male will be in the vicinity of the eggs when they are spawned. This seems to be the case in many fish.

For a male that defends a nest site, guarding the offspring may involve little additional cost, whereas for the offspring the benefits may be quite substantial. In addition, the effort needed to defend a nest is not a simple function of the number of offspring present, and in many species a male may guard the eggs of several females without much increase in costs. This situation contrasts sharply with what we find, for instance, in birds, which feed their young individually and are much more limited in the number of offspring that they can raise successfully.

Fish typically show indeterminate growth (i.e., they continue to grow after reaching sexual maturity). For females it may be beneficial to limit reproduction in a given year, because if the female survives and grows, she will be able to produce many more eggs in the future when her body and ovaries are larger. In addition, in tropical and temperate conditions, the females of many species are able to spawn repeatedly over a prolonged breeding season. Thus, for females, providing parental care may involve a much stronger limitation in their capacity to spawn again in the future than for males. In conclusion: External fertilization makes the likelihood of parental care evolution more similar for both sexes than internal fertilization; indeterminate growth tends to bias this probability towards males.

In some fish, it has been demonstrated that females prefer to spawn with males whose nests already contain eggs. This female behavior, if common, may have a strong influence in the evolution and maintenance of paternal care because the benefits that males drive from staying in sites where eggs are already present may increase substantially.

Biparental care may have evolved from male care in various ecological contexts, for instance, where nest sites were scarce, predation pressure on offspring was high, or the likelihood of finding new mates was low. In these scenarios, the male would gain more by the assistance of the female than by attracting new mates, whereas the female would gain more by reducing her future reproduction and helping the embryos to survive. If these pressures relaxed in species where biparental care had reached a very high level of efficiency, so that a single parent could care for the entire brood (like in mouth brooding cichlids), male desertion may have led to care by the female alone.

See also Caregiving—*Brood Parasitism in Freshwater Fish*

Further Resources

Baylis, J. R. 1981. *The evolution of parental care in fishes with reference to Darwin's rule of male sexual selection*. Environmental Biology of Fishes, 6, 223–251.
Blumer, L. S. 1979. *Male parental care in the bony fishes*. Quarterly Review of Biology, 54, 149–161.

Blumer, L. S. 1982. *A bibliography and categorisation of bony fishes exhibiting parental care*. Zoological Journal of the Linnean Society, 76, 1–22.

Breder, C. M. & Rosen, D. E. 1966. *Modes of Reproduction in Fishes*. New York: Natural History Press.

Gittleman, J. L. 1981. *The phylogeny of parental care in fishes*. Animal Behaviour, 29, 936–941.

Gross, M. & Sargent, R. 1985. *The evolution of male and female parental care in fishes*. American Zoologist, 25, 807–822.

Gross, M. R. & Shine, R. 1981. *Parental care and mode of fertilization in ectothermic vertebrates*. Evolution, 35, 775–793.

Keenleyside, M. H. A. 1979. *Diversity and Adaptation in Fish Behaviour*. Berlin: Springer-Verlag.

Keenleyside, M. H. A. 1980. *Parental care patterns in fishes*. The American Naturalist, 117, 1019–1022.

Marconato, A. & Bisazza, A. 1986. *Males whose nests contain eggs are preferred by female* Cottus gobio L. (*Pisces, Cottidae*). Animal Behaviour, 34 (5), 1580–1582.

Myers, G. S. 1939. *A possible method of evolution of oral brooding habits in cichlid fishes*. Stanford Ichthyological Bulletin, 1, 85–87.

Perrone, M. & Zaret, T. M. 1979. *Parental care patters of fishes*. American Naturalist, 113, 351–361.

Ridley, M. & Rechten, C. 1981. *Female sticklebacks prefer to spawn with males whose nests contain eggs*. Behaviour, 76, 152–161.

Smith, C. & Wootton, R. J. 1995. *The costs of parental care in teleost fishes*. Reviews in Fish Biology & Fisheries 5, 7–22.

Taborsky, M. & Limberger, D. 1981. *Helpers in fish*. Behavioural Ecology and Sociobiology, 8, 143–145.

Vincent, A. C. J. 1992. *Prospects for sex role reversal in teleost fishes*. Netherlands Journal of Zoology, 42, 392–399.

Wourms, J. P. 1981. *Viviparity: the maternal foetal relationship in fishes*. American Zoologist, 21, 473–575.

Vitor Almada

■ Caregiving
Parental Investment

In many animal species (and in some plants also) parents show some form of care toward the offspring. This ranges from storing food reserves in the egg to increase its survival probability (plants can do this in seeds), feeding the offspring, or guarding both. Any form of care by one or both parents toward the offspring prior to or after birth is known as *parental care*. Parental care may decrease as it is divided among a larger number of offspring. Such is the case of food, because the amount of food a parent can store in its body to produce eggs or live offspring, or the amount of food brought to a nest quickly reaches a maximum that has to be shared among that offspring. This type of parental care is often called *depreciable* (i.e., decreasing) *care*. However, not all types of care have to be shared and in some cases, it may benefit equally a small or a large number of offspring. This, for example, is the case of vigilance in a nest (although when one offspring is born and moves around readily in all directions, watching one individual usually prevents watching any behind the parent). This type of nondecreasing care is known as *nondepreciable care*.

Usually, parental care is not free. The amount of food an animal uses to make eggs or feed the young could be used to feed itself. In some cases, as is the case of milk, this food is actually made from the nutrients circulating in the body of a mammal. Even in the case of vigilance, the time a parent spends looking after the progeny could have been spent looking for food. In most cases, thus, care is associated with an expenditure, therefore called *parental expenditure*. In elephant seals, for example, females loose about 2 kg (4.4 lb) in weight for every one of the 100 kg (220 lb) that its pup gains in the 5 weeks that it suckles its mother. Thus, in weight, the mother's expenditure during lactation is 200 kg (440 lb). This expenditure may represent a different effort for mothers having different weights. For example, for a typical 650 kg (1,430 lb) elephant seal it represents 31% of its body weight, but for a 500 kg (1,100 lb) seal this means 40% of its body weight. That is why expenditure is often measured *relative* to parental weight, energy, time, and so on. It is then called *relative parental investment*.

Because resources (particularly food or body reserves) are limited and expensive, the consequences of such expenditure for parents are often a higher mortality risk (e.g., often animals expose themselves to predators when seeking food), reduced immunity (e.g., reserves are derived from the immune system to produce milk or feed the muscles for food searching), or a reduced ability to reproduce subsequently. The latter is particularly true in females of many species (from insects to mammals) because the ability to produce eggs or to become pregnant depends on body weight, and weight is often lost in reproduction, particularly in food provisioning of the young after birth. That is why current reproduction often lowers subsequent reproduction. In such a case, parental expenditure is termed parental investment. The idea behind this is that, although a parent may increase its expenditure in improving success or numbers of current offspring, there is a cost to pay for such current investment that could have been invested in future reproduction. Parental investment relates to caring for each individual offspring, but it is called *parental effort* when it relates to caring for all the offspring. It is important to note that a greater parental investment does not necessarily mean a larger amount of resources transferred to the offspring. For example, most mammal mothers produce less milk when food is scarce, and young thus grow less, but attempting to maintain some level of milk production is achieved using the body reserves which greatly reduces fertility. Such a smaller milk production in fact means a greater parental investment. Mammals do that because the value of current offspring already born is usually greater than the uncertain possibility of future reproduction.

If the price can be so high, why invest in offspring at all? Investment is at the heart of the nature of sex and starts with the difference between sperm and egg cells. It is believed that during evolution of sex cells, some increased the probability of successful development by adding extra nutrients, thus leading to the evolution of ova or eggs. Although the fusion of such gametes with similar sized cells may have probably increased further the chances of successful development, other cells outcompeted these, increasing their own fertilization chances by becoming small and fast swimming, thus leading to the evolution of sperm. Hence, some form of care and investment is shown in the process of egg production itself. However, the question remains as to why some species give any additional care (vigilance, storing food near the eggs, or feeding the offspring), whereas others do not. There are *phylogenetic* (i.e., historic) reasons for this in some groups of animals. In mammals, for example, milk evolved from sweat because such balanced food made from the mother's body allowed a very fast growth of the young, thus shortening the period in which most animals suffer a greater mortality. However, parental care and associated investment is often shown in all groups ranging from insects to reptiles or birds. Animals in all groups have relied either on reproducing in larger numbers or in care for offspring survival. For example, large schools of fish can spawn simultaneously so that predators can take only a small proportion of all eggs laid. No parental care is needed here. In contrast, where predation risk is high or food is scarce and defendable, it may pay the parents to watch the nest or feed the young. Thus, in insects, for example, parental investment is more common in species living in physically harsher (e.g., desert) or dangerous habitats. Similarly, parental care of the eggs is

Parents Desert Newborns after Break-In! Mother Goes for Food as Siblings Battle to the Death! Father Eats Babies!

Rachel Endicott

Are the above headlines your idea of good parental care? Hopefully not! But in the nonhuman animal world, these and other seemingly strange behaviors are simply part of parents' strategies to raise the greatest number of offspring throughout their lifetimes.

Deserting Newborns

Many bird parents will desert their own eggs if a foreign egg is laid in their nest by birds known as parasites, such as the shiny cowbird, (*Molothrus bonariensis*). These parasitic birds rely on the host parents (the owners of the nest) to feed and care for their young. In these situations the hosts' young are not able to compete with their much larger and more aggressive nestmate and will probably not survive. By deserting their nest, parents reserve their energy for a time when there is a better chance to raise their own nestlings successfully.

Siblicide

Spotted hyenas (*Crocuta crocuta*) live in female dominated hierarchies where cubs are born with sharp teeth and immediately begin battles to establish dominance. Cubs live in small underground dens during their first few months of life and while their mothers are away the cubs may fight with each other. These battles can turn deadly in years with decreased resources; for example, when there is a lack of food. In difficult times, siblicide ultimately benefits the mother, ensuring that all of her food and care goes to the cub most likely to survive to adulthood.

By banding together, female lions reduce the risk to all of the cubs.
Courtesy of Corbis

Eating Offspring

In many species of fish with paternal care, eating his offspring may help a father to improve his reproductive success. If a father's energy reserves are low, his ability to protect and provide for his young is severely diminished. By eating some of his current offspring, a father increases his energy reserves and is better able to raise a greater number of future offspring.

There are of course more palatable examples of parental care:

Communal Rearing

Female lions (often related) live in communal groups called prides where they hunt for, protect and even nurse each other's young. Investing in another's

(continued)

Parents Desert Newborns After Break-In! *(continued)*

offspring decreases the amount of care a mother can provide her own young, and would seem to detract from her lifetime reproductive success. But females benefit in part due to the high rate of infanticide found among lions. When males take over a pride, they usually kill the current immature cubs, allowing themselves to mate more quickly with the females. By living communally, females are better able to fight off invading males, therefore giving their cubs a survival advantage.

Paternal Investment

Male giant waterbugs (*Abedus herberti*) provide high levels of parental care. These fathers carry their eggs on their backs until they hatch. As the eggs begin hatching, the father's feeding instinct is actually inhibited, keeping him from eating his young as they swim away.

Further Resources

In addition to the essays in this encyclopedia, there are many good sources for information on parental care in animals including the following:

Bertram, B. C. R. 1992. *The Ostrich Communal Nesting System*. Princeton, NJ: Princeton University Press.

Clutton-Brock, T. H. 1991. *The Evolution of Parental Care*. Princeton, NJ: Princeton University Press.

Clutton-Brock, T. H. (Ed.) 1988. *Reproductive Success*. Chicago: The University of Chicago Press.

Hosoi, S. Aki & Rothstein, S. 2000. *Nest desertion and cowbird parasitism: Evidence for evolved responses and evolutionary lag*. Animal Behaviour, 59, 823–840.

Kondoh, M. & Okuda, N. 2002. *Mate availability influences filial cannibalism in fish with paternal care*. Animal Behaviour, 63, 227–233.

Manning, A. & Stamp Dawkins, M. 1998. *Animal Behaviour*, 5th edn. (pp. 332–354). Cambridge: Cambridge University Press.

Massoni, V. & Reboreda, J. C. 2001. *Number of close spatial and temporal neighbors decreases the probability of nest failure and shiny cowbird parasitism in colonial yellow-winged blackbirds*. Condor, 103, 521–529.

Riedman, M. L. 1982. *The evolution of alloparental care and adoption in mammals and birds*. The Quarterly Review of Biology, 57, 405–435.

Riedman, M. 1990. *The Pinnipeds*. (pp. 176–312). Berkeley: University of California Press.

Smale, L. Holekamp, K. E. & White, P. A. 1999. *Siblicide revisted in the spotted hyaena: Does it conform to obligate or facultative models?* Animal Behaviour, 58, 545–551.

Taub, D. M. (Ed.) 1984. *Primate Paternalism*. New York: Van Nostrand Reinhold Company.

Verboven, N. & Tinbergen, J. M. 2002. *Nest desertion: A trade-off between current and future reproduction*. Animal Behaviour, 63, 951–958.

more common in freshwater than in saltwater fish, which reflects a similar difference in environmental variability of these habitats.

The difference in investment between sperm and eggs leads to the general rule that females are the sex usually caring for or investing in offspring, whereas the male rarely does it. Thus, a male usually is able to mate with many females and, as a result, many males may end up not mating at all. This has produced intense male–male competition for females in many species, and a strong *sexual selection* within males.

However, all kinds of parental care can be found. In many cases, the effort of caring by any parent does not pay if it does not noticeably raise the chances of offspring survival. Such is the case mentioned above of school fish laying eggs simultaneously or in most insects. When the survival probability of eggs or live offspring left alone is very low compared to its chances if they are cared for, then at least one parent will invest in caring for them. If raising them is so expensive that survival is low even if only one parent cares (e.g., birds that stay for some time in the nest), then there is little point for the male to leave and fertilize another clutch that will suffer the same destiny, and so in these cases, both parents will care. In ducks, hens, and some birds laying large eggs, and in most mammals, the offspring has a reasonable probability of survival with the care of only one parent. In this case the female invests so much in the eggs or gestation (or, in mammals, is the only one able to produce milk), that it is difficult to produce another set of offspring in the current reproductive season, whereas the male can easily mate with several females in the same year. In these cases, it is the female who cares. However, in other cases, such as many fish that spray sperm over laid eggs, any male could easily search for many clutches and fertilize them. Female fish can produce eggs at a lower cost than other groups of animals, but they will lay them only in appropriate nests. Thus, males are the ones building and caring for nests to attract several females and stimulate them for laying. In other species such as the jaçanas, a bird from South America, females invest little in the offspring apart from producing the egg, and it is the male who incubates and cares for the precocious young. In this and other species, called *sex-reversed birds*, roles are reversed and females, larger and more brightly colored, compete for males and lay in several nests in the same reproductive season. This species illustrates clearly one principle about parental investment: The sex investing the most (particularly if the difference is large) is the choosy sex. In most birds and mammals, females choose; in the jaçanas it is the male who chooses.

All the discussion above shows that one sex has greater chances of reproducing (usually the females), whereas the other may achieve a far greater number of offspring if successful, but it is more likely to end up not reproducing at all. *Overall*, both sexes should have the same expectancy of reproducing, as Robert A. Fisher pointed out in the thirties. Taken to the extreme, in the all-male world, the only female is the queen (and vice versa). It is intuitive that parents will react, producing the rarest sex, until the population has equal numbers of males and females. To be more precise, they should tend to spend an equal amount of resources on both sexes. For example, if males cost twice as much to produce as females, it will be twice as expensive to reach the same amount of grand-offspring through males than through females, and parents will tend to have two females per male.

In any case, because males may end up siring more offspring if they are successful, which usually means being the strongest, then it would pay mothers of high body resources or quality to produce males rather than females. Thus, in this and the following cases, parental investment is focused mainly in one sex. This extra care must influence the quality (e.g., strength) of the male when adult. This idea, first proposed by Trivers and Willard (1973), is thus termed the Trivers–Willard Model. In some cases, a greater investment in males is not conducted through a greater probability of having males, but through a greater provisioning of males (e.g., in mammals, giving them more milk). In species with sex reversed roles, as in the jaçana, the theory predicts a greater chance of having females who will take greater care of their male offspring. The overall population does not tend to have a surplus of one sex because the advantage of a particular male is relative to the population.

Calf of Iberian red deer (Cervus elaphus hispanicus) suckling its mother. Milk is one of the important forms of parental care (and investment) in mammals. This subspecies of deer is the first mammal proven to produce a different milk (and also in different quantities) for sons and daughters. They produce more milk and with more protein concentration for males.

Courtesy of Tomás Landete-Castillejos

For example, a 350 lb (158 kg) male deer is no advantage if most males weigh the same.

An extension of this theory is termed *environmental* (in a wide sense) *sex determination*. In some parasitoid wasps that lay a single egg per host, the environment is the size of the caterpillar body that will feed the larva. A greater caterpillar means a greater emerging wasp, and larger wasps benefit more if they are females, because the number of eggs they lay increase with size. In these wasps, thus, female eggs are deposited in large hosts, and male eggs in smaller ones.

As in any investment, animals are sensible to costs. Males, particularly in *polygamous* species, are born larger and require more food than females. In some seals, such as the elephant seal, the difference at birth between sexes is so large that some mothers do not have enough reserves in their body to produce the large male, and below a certain weight they will bear only female pups. In Iberian deer, young growing females reach a trade-off between reproduction and the high cost of males and thus, they tend to have females until they reach adult size.

However, although males are more expensive to raise, in some cases, daughters may be more expensive in the long term. This is because in many mammals, such as most deer, males tend to leave and search for females, whereas daughters stay with their mother, feeding neck by neck and thus competing for food. This may represent a large cost and thus, in some species, high quality females are the only ones that can afford to have daughters. Such competition, and the bias against daughters, may be particularly strong when resources are scarce. This theory is termed *local resource competition (LRC)*. In some cases, LRC biases are enforced through harassment of daughters of low quality mothers, and reduced survival of daughters probably results in these animals giving birth to sons. Overall, LRC biases are more common in primates and some cervids, whereas the Trivers–Willard model (TWM) has been solidly reported in red deer and some other cervids.

Although both theories were postulated focusing on quality, in most cases, they are affected by different characteristics of the mother. Even though quality usually increases with increasing weight and age, in fact, the ability to raise offspring (at least in mammals) increases with weight. Thus, it is likely that TWM bias toward sons is found primarily in heavier mothers. Future competition with mothers is less important as the mother grows older and her life expectancy shrinks. Hence, LRC biases are more likely to be affected by the age of the mother. There is one case in which both theories have been shown to work in the same population. In mouflons, mothers show a LRC bias as they grow older, but within each age class, heavier mothers have a greater probability of having males.

In some cases, biases have been found to change the population sex ratio, although its interpretation is still controversial.

See also Antipredatory Behavior—*Hiding Behavior of Young Ungulates*
     Caregiving—*How Animals Care for Their Young*
     Caregiving—*Incubation*
     Caregiving—*Mother–Infant Relations in Chimpanzees*
     Caregiving—*Parental Care*

Further Resources

Charnov, E. L. 1982. *The Theory of Sex Allocation*. New Brunswick: Princeton University Press.

Clutton-Brock, T. & Godfray, C. 1991. *Parental investment*. In: *Behavioural Ecology: An Evolutionary Approach* (Ed. by J. R. Krebs & N. B. Davies), pp. 234–262. Oxford: Blackwell Scientific.

Fisher, R.A. 1930. *The Genetical Theory of Natural Selection*. Oxford: Clarendon Press.

Hewison, A. J. M. & Gaillard, J. M. 1999. *Successful sons or advantaged daughters? The Trivers-Willard model and sex-biased maternal investment in ungulates*. Trends in Ecology and Evolution, 14, 229–234.

Hiraiwa-Hasegawa, M. 1993. *Skewed birth sex ratios in primates: should high-ranking mothers have daughters or sons?* Trends Ecol. Evol., 8, 395–400.

Trivers, R. L. 1985. *Social Evolution*. Menlo Park, CA: Benjamin-Cummings.

Trivers, R. L. & Willard, D. E. 1973. *Natural selection of parental ability to vary the sex ratio*. Science, 179, 90–92.

Tomás Landete-Castillejos

■ Cats
Domestic Cats

Cats are America's favorite pets. There are approximately 70 million house cats in the United States. However, millions more are homeless (approximately 40 million), many abandoned because of overpopulation and abuse. More than 7 million end up in shelters each year, and millions are euthanised because homes cannot be found for them. It is important that we understand the unique behavior of cats because they contribute to *our* inclusive fitness (survival). They not only contribute to our psychological well being as pets, but they also control vermin (pests) that can destroy our food supply and spread pestilence.

Cats can be feral (wild) or domesticated (socialized to humans). Feral cats are generally raised by feral mothers (who hides her infants from humans) and/or have had little or no human handling from 2–8 weeks of life. This sensitive period for socialization of the cat can be important for cats to become socialized to their own species and alien species (potential predators, e.g., dogs, and potential prey, e.g., rodents). Kittens who have had positive early experiences with alien species will generally act friendly toward them as adults. Domesticated cats have plenty of handling by humans during the sensitive period. However, even though socialized to humans, they generally maintain their independence and individual personality. This programming is thought to be a product of both genetics and early experience, especially during the sensitive period.

Adult cats are relatively solitary, showing little species affinity or social organization, and display territorial behavior (complete with a complex array of markings and other communication signals) that promote avoidance and familiarity. However, this solitary behavior is disrupted when an adult cat encounters a newcomer or a nearly mature adolescent

Wild versus Domestic Behaviors: When Normal Behaviors Lead to Problems

Myrna Milani

Behaviors that increased the probability of their wild ancestors' success unfortunately may undermine that of contemporary domestic animals. Although members of all domestic species may succumb to this paradox, the relatively recent domestication of cats makes them particularly good examples of this phenomenon.

Domestic cats evolved from nocturnal, solitary, and predatory roots that permitted their wild ancestors to succeed in a wide variety of habitats. Small wildcats who hunted alone in limited light at the same time they served as prey for others, acquired skills different from those attained by more social animals, such as wild dogs and wolves, who hunted in groups.

Although small wildcat survival hinges on the ability to catch prey, such does not necessarily hold true for pet cats whose owners view cats as sleek, genteel creatures of comfort who lounge gracefully on their owners' beds. When people who harbor such views awaken to dead mice on their pillows, major breakdowns in the human–feline relationship may occur!

In addition to the finely-tuned predatory skills that enable solitary small wildcats to dispatch prey without assistance, domestic cats inherited two other complementary wild skills that may lead to problems in human households. Successful small wildcats need to know their territories intimately to locate prey easily and escape predators. Enter that famous feline curiosity which enables cats to notice even the smallest changes in their environment. Also, wild felines need some way to mark their space to ward off competitors and other threats. Claw marks and urine sprays form two of the behaviors that evolved to accomplish this.

Predatory behavior combined with feline curiosity may lead to serious health problems for domestic cats, even with no natural prey in sight. Bored house pets attracted by the sound, motion, scent, or texture of toxic spiders, insects or plants may "hunt" and/or consume them. Even humans who abhor the idea of feline predatory behavior often play with cats in ways that trigger the stalk, pounce, pinion, kill, and eat sequence of the predatory response. Cats so stimulated by someone dragging a piece of string may sometimes eat the string. When illness results from these inappropriately targeted displays of normal feline behavior, unknowledgeable humans exclaim, "What stupid animals!" On the other hand, those aware of how these normal feline behaviors may play out in a human household vow never to set up their pets for such avoidable problems.

Similarly, when Fluffy claws the couch or marks furnishings with urine or feces, humans lacking an understanding of normal feline behavior will complain, "What a spiteful, mean cat!" and either punish or get rid of the animal. Compare this to the response of those who understand what cats communicate with these behaviors: "Something's obviously threatening Fluffy. I'm going to find out what it is and what I can do to relieve her fear."

By learning more about normal animal behavior, we can prevent or more effectively treat problem displays in ways that meet our pets' needs as well as our own.

Further Resources

Sunquist, M. & Sunquist, F. 2002. *The domestic cat: History, folklore, ecology, and behavior.* In: *Wildcats of the World*, pp. 99–110. Chicago: University of Chicago Press.

Turner, D. C. & Bateson, P. 2000. *The Domestic Cat: The Biology of Its Behaviour*, 2nd edn. Cambridge: Cambridge University Press.

where dominance must be established. The normal solitary behavior of cats is also disrupted when they are enclosed in a small colony or when a cat of superior strength enters the scene. In these cases aggression and chaos result. However, cats are not antisocial. Feral cats have been observed to have semi-friendly gatherings at night, and females will share infant care and defense against intruders. Domesticated (house) cats living together may also display friendly social relationships.

The kitten, on the other hand, has been described as being highly social in its relationship with both its mother and other kittens. Several phases of kitten socialization have been described. These socialization processes are generally based on the feeding relationships between the mother and her offspring, commensurate with the kitten's sensorimotor maturation. As the mother begins weaning, she spends more time away from the home nest. This encourages the hungry kittens to leave the nest to initiate feedings and to use the mother as a secure base for exploration. Kitten play is also important in the weaning process. If there is only one kitten, the mother takes much longer to wean it. During this time, free-ranging mothers will bring home dead prey and eat it. Later, she leaves dead prey for her young to eat. Finally, she will bring home live prey and allow the kittens to apply their strengthening play patterns (stalking, chasing, pouncing, pinning, slapping, tossing, etc.) to prey catching and killing. She may also take the kittens out on hunting expeditions.

This farm cat sits in the rafters of a barn, watching for rodents.
Courtesy of Corbis

The movements utilized in play and prey-catching behavior are similar and are part of the natural behavioral repertoire that cats apply to a myriad of situations, although the functions may change (e.g., social play may also function to provide opportunities for the acquisition of social communication or metacommunication skills—which will be used in other social situations—and to maintain friendly social relations among members of a litter).

There are many fascinating and intriguing aspects of cat behavior. By understanding how their natural innate behavior patterns are integrated with their human habitats, we can improve the quality of their lives and our own as well.

Further Resources

Bradshaw, J. W. S. 1992. *The Behavior of the Domestic Cat.* Melksham, UK: Redwood Press Ltd.

Fox, M. W. 1979. *Understanding Your Cat.* New York: St. Martin's Press.

Guyot, G. W., Cross, H. A. & Bennett, T. L. 1980. *The domestic cat.* In: *Species Identity & Attachment: A Phylogenetic Evaluation* (Ed. by M. A. Roy), pp. 145–164. New York: Garland STPM Press.

Leyhausen, P. 1979. *Cat Behaviour: The Predatory and Social Behavior of Domestic and Wild Cats.* New York: Garland STPM Press.

Turner, D. C. & Bateson, P. (Eds). 2000. *The Domestic Cat: The Biology of Its Behaviour.* 2nd edn. Cambridge: Cambridge University Press.

Gary W. Guyot

■ Cephalopods
Octopuses, Squid, and Other Mollusks

Cephalopods, which include octopuses, squid, cuttlefish, and nautiluses, are mollusks, but different from the simple, slow-moving snails and clams. Because they have lost the protective shell, octopuses and squid have evolved abilities quite different from those of other mollusks. They have become efficient and fast-moving, intelligent and big-brained, have a short life span, and an excellent skin display system. All of these changes were in response to bony fish, who evolved at the same time as the cephalopods, and are "designers of the skin" of cephalopods.

Efficiency means acute senses, and sensing is the first step leading to behavior. The cephalopod lens-type eye is very much like the vertebrate one, an example of convergent evolution that suggests this is the "one best way" to make an image-forming visual system. This eye can see as well as many vertebrates', though it can't see color. The octopuses and squid have a balance system too, one that is also as good as the vertebrate inner ear at understanding movement in three dimensions. They have a lateral line analogue system that is as good as the fishes' at picking up water movement. Squid can sense as little as millimeters of water movement meters away, important for avoiding predators. These animals have excellent chemical sensitivity, too.

Without the rigid shell, cephalopods don't look as if they have a skeleton to support movement, but they do. Unlike the internal bones of vertebrates or the external skeleton of arthropods, cephalopods have a flexible muscular hydrostat system. Some of the muscles stiffen to act as a skeleton in opposition to other sets which perform actions. So octopuses, for instance, can put their eight flexible arms in loops or circles, make them longer or shorter, twist them or bend all along the length. They can use their suckers to hold like suction cups or fold them sideways to make a pincer grasp. All this takes a lot of nervous system control—three-fifths of an octopus' neurons are outside the brain—and results in a lot of local control of action. An octopus can autotomize or detach an arm, and this isolated arm shows huge local control. It can walk along a wet surface or even pass a small item from sucker to sucker, up to a mouth that isn't there anymore.

Roving Octopuses

Roland C. Anderson

Octopuses are interesting subjects for behavioral studies. They are the most intelligent invertebrate (animal without a backbone). They have recently been reported to learn mazes, unscrew jars to get food, have individual personalities, use tools, and even exhibit play behavior.

One of their behaviors has achieved urban myth status—the ability in captivity to crawl out of a tank to another tank to get food, then crawl back to a home tank, usually to the consternation of their keepers on finding a missing crab or fish and no evidence of a culprit. This behavior was first reported by Henry Lee in 1875 in his book *Octopus—the Devilfish of Fiction and of Fact*. He wrote of his experiences in keeping octopuses at the Brighton Aquarium (near London). He reported the story of an octopus that regularly crawled over the foot-tall barrier of its tank at night into an adjacent tank, where it captured a lumpfish, took it back to its own tank before dawn, and ate it there. This finding was widely reported by the press of the time, and the public was intrigued. A poem was even written about the behavior (see below).

Such behavior by octopuses in captivity is just an extension of their behavior in the wild. They are what the ethologist Eberhard Curio called "refuging predators." They refuge in a den and make forays out to hunt prey, then return to their dens.

But since the Victorian time of Henry Lee, aquarists have learned to not let octopuses stray from their home tanks. Live fish are too valuable to lose indiscriminately for octopus food, and too often octopuses misjudge their ability to travel from tank to tank and end up dead on the floor (in reality, they are probably just trying to get out of confinement). Octopuses are very strong and since they have no bones, they can squeeze through incredibly small openings. At the Seattle Aquarium, a 40-pound (18 kg) octopus moved a tank lid with 66 pounds (30 kg) of rocks on it and escaped. Aquarists nowadays keep octopuses in tanks that are almost hermetically sealed or that have a barrier to their escape, either a high wall or a fence covered by a bristly, uncomfortable coating. The octopuses are not allowed any wandering from home. Still, the escape ability of octopuses has become legendary and the story of the "Straying 'Topus" is still widely believed to be true at any aquarium keeping octopuses.

The Straying 'Topus
BY TOM HOOD, 1873

Have you heard of the Octopus -
'Topus of the feelers eight -
How he left his tank o'po'pus
Lumpfish to disintegrate

To the lumpfish tank as sprightly
As the Brighton coach he'd ride;
For two passengers he nightly
Found convenient room inside.

But it happened Mr. Lawler,
Whom the lumpfish ought to thank,

Caught this very early caller,
"Dropt-in" on his neighbor's tank.

For some weeks the world lumpfishious
Very strangely vanished had;
So the visit was suspicious,
And appearances were bad!

Well for him, this brigand larky
Was not brought before J.P.
(Neither clergy, nor squire-archy)
But to Mr. Henry Lee.

(continued)

Roving Octopuses (continued)

Said he, "Punish on suspicion,
Is a thing I never will -
Catch him in the same position;
Then I'll send him to the Mill!"

Treadmill is a wear-and-tear case,
And Octopus would you see,
Do four men upon a staircase—
Law, how tired the beast would be!

Even though there's a lot of local control of action, cephalopods are known for their large centralized brains, important for behavior variation and memory. The basic structure of molluskan nervous systems is five separate paired ganglia. The cephalopods gathered these systems into a central brain in a ring around the esophagus. There are huge optic lobes below the eyes, important for processing visual information and programming responses. Two separate areas are specialized just for visual and touch-based learning. This learning was extensively studied in the 1950s and 1960s in Naples, Italy, and so we know that the octopus brain's control of learning is similar in many ways to that of vertebrates.

One of the interesting results of this big brain seems to be that cephalopods are very learning-based and have few examples of reflexes—automatic behavior patterns. Much of the work on their learning is based on octopuses, as mobile fast squid are hard to keep in captivity. They can learn to recognize and discriminate visual shapes, to tell objects by touch and texture or even taste. Octopuses can navigate through a maze, and fieldwork tells us this ability is used to help them go out to forage and remember where to return to their sheltering home. They can pick up rocks or shells from the ocean bottom and bring them back to pile in front of their home entrance. This is a form of tool use as defined by Benjamin Beck. Octopuses can even play, an ability we think of as belonging to mammals and birds. Some of them at the Seattle Aquarium watched a floating toy carried to them by the water inflow current and "blew" it back to the inflow opening so it circled the tank over and over, the water equivalent of bouncing a ball.

Much of this flexibility and ability is used in foraging: Octopus and squid will catch almost any marine animal they can find—up to 80 different species of mostly mollusks and crustaceans for the bottom-living octopus, mainly fish for the open-ocean squid, and fish and crustaceans for near-shore cuttlefish. They are specializing generalists, taking many of whatever species is abundant and easy to catch in the local area. Octopuses don't use vision much for prey catching. They mainly hunt by going to likely areas, probing with their arms into the sand for clams, reaching under rocks for crabs and snails, even flaring out the arms and pulling the web down to make a net that catches and holds crabs. Squid and cuttlefish are much more visual, and lie in wait in disguise among floating seaweed, dangle the tentacles as a lure, blow off sand that covers hidden clams or crabs, and chase down and grab small fish.

Even after capturing prey, an octopus may need to use considerable flexibility to get at it. Think of the difficulties of getting inside a protective clam's shell. The octopus can pull the valves apart by arm strength, chip a hole at the valve's edge with its parrot-like beak or drill a hole in the shell and inject a venom that makes the clam's muscles relax so it can be opened easily. All of this depends on the clam's strength, both in muscle and shell. Fragile shells are just broken, medium strength ones pulled apart, and really strong ones chipped or drilled.

The octopus uses its intelligence to figure out what technique to use, too. Weaker clams that are wired shut by an experimenter can't be pulled apart and so are drilled. Drilling location isn't random, either. Within a few tries, an octopus learns to drill over the muscle that keeps clam valves shut or over the heart that keeps the clam alive, not uselessly at the edges into the water-filled mantle cavity. With snails, octopuses learn to drill over the adductor muscle that pulls the snail back into its shell, and this makes pulling it out easy.

Not all cephalopod behaviors are flexible. Near-shore cuttlefish and squid often dig into the sandy bottom to hide during the day. They blow sand or mud out from under themselves with water jets from their funnel, and then throw more sand on top of themselves with cupped arm tips. This is species-typical. One species blows only sand and wiggles its upper body to settle it, another just scoops sand a couple of times from the front, and a third has a front-side-back sequence of sand scooping and throwing. Throwing is strictly paired, with left and right arms scooping and throwing together even if one side isn't getting any sand.

Eggs of the flamboyant cuttlefish (Metasepia pfefferi), with embryos visible, Lembeh Straits, Indonesia.
© Hal Beral/Visuals Unlimited.

The sequence of behavior isn't completely fixed. Put a squid on a gravel bottom and it may try to blow, pick up and place a few "rocks" over itself, turn camouflage colors and wait, or just get up and jet away to a better location.

Camouflage also is a specialty of the cephalopods, who have the best skin color production of any animal. They have yellow, red, and brown pigments in many elastic chromatophore sacs in the skin. Each sac is pulled out by muscles that are controlled by separate nerves. Green or white reflecting iridophores are under these color sacs and are uncovered when all the chromatophores contract. The result is that an octopus, a squid or a cuttlefish can make itself look like anything it wants to and can change in milliseconds. It's amazing that these animals, which have such good color production, have no color vision. It seems that the system evolved to produce camouflage that hides them from fish predators, not for communication with each other. This means that fish are the designers of this skin system.

It's not just the color match that makes cephalopod camouflage so effective. Sometimes the patterns are disruptive—obvious, but looking nothing like an animal. Often the eye is disguised by a bar extending through it, and distracting "eyes" are shown on other parts of the body. Most important, the cephalopods are quick-change artists. As an octopus moves from area to area, it can change from mottled smooth greys on sand to textured brown and greens on algae covered rocks, then to smooth plum purple in the shade. A squid escaping a persistent predator can go to the water surface and imitate floating brown algae, sink down to the bottom and camouflage itself as the gravel or pale and jet away, leaving behind a squid-mimic of a blob of ink. The sequence of action fools a predator which has a "Search Image" of what the squid ought to look like; the speed and unpredictability of the changes confuses it.

This O. macropus, in turtle grass, has raised itself up and is looking at its surroundings. Look closely to see that it has erected some of its papillae. Location: Dry Tortugas. This species is very nocturnal and especially adept at squeezing through small holes in the reef.
Courtesy of James B. Wood.

Probably because they do not have color vision, cephalopods don't seem to use this wonderful skin display system to signal to each other very often. Octopuses have a few patterns that are seen in mating situations, especially one with raised white papillae on a chocolate brown background, used by males who want to mate with a female. Cuttlefish, especially the giant Australian species, have dazzling patterns for courtship. Only open-ocean group-living squid have a wider range of display patterns, and *Sepioteuthis* in the Caribbean may be the champions here.

Despite all this potential intelligence, squid and octopuses don't seem to use it in the range we might expect, maybe because they have fairly simple social lives. Primate high intelligence probably evolved to solve social and not ecological problems, but cephalopods aren't very social. All the coleoid cephalopods are semelparous, with no sex life until the end of their short—usually under 2 years—lifespan. The sexes are separate, and males die after mating. A female octopus lays strings of hundreds or thousands of small eggs and guards and cleans them for weeks or months. After they hatch she dies and the tiny babies usually float off in the plankton, on their own. Cuttlefish and squid lay eggs and die within a few weeks, so they have no maternal care at all. Only the open-ocean squid live in groups and their social organization seems pretty primitive, just a collection of equal units like that of many fish schools. At maturity squid males compete for females, and females test and choose the best, but there seems to be no complex social organization or cooperation in the group. Cephalopods have used their flexibility of action, good sensory ability, and intelligence mostly to eat and avoid being eaten, little of it in relationships with each other, but they still have a fascinating variety of abilities and behaviors.

See also Reproductive Behavior—*Mollusk Mating Behaviors*

Further Resources

Hanlon, R. T. & Messenger, J. B. 1996. *Cephalopod Behaviour*. Cambridge: Cambridge University Press.

Jennifer A. Mather

■ Cognition
Animal Consciousness

Have you ever looked at an animal and wondered what it would be like to be that animal? Then you have thought about animal consciousness. You might have wondered what it would be like to be a tiger or a shark. You probably haven't ever wondered what it would be like to be an earthworm. Perhaps it's like being a rock—in other words, like nothing at all!

Which animals are conscious and which ones aren't? Most people think that there's no black-and-white answer to this question. Perhaps earthworms, or goldfish, have some degree of consciousness—just not the same as ours. But what would that mean? Could it mean that goldfish see, hear, smell, and taste things dimly (or in pale colors)? Or does it just mean that they are aware of fewer things than we are? While the idea of levels of consciousness seems appealing, it's hard to spell out exactly what it means. We know, for example, that honeybees have five different color vision cones, compared to our three, so they can differentiate more colors than we do. It doesn't seem plausible to think that they could be better at making color discriminations based on a dimmer version of consciousness than our own.

Perhaps, then, it is an all-or-nothing question: Some animals (humans at least) are conscious, and others are completely unconscious. They might be very cleverly designed machines, but are no more conscious than the computer on which this article is being typed. If it is all or nothing, then where do we draw the line? Perhaps only animals that can put thoughts into words are conscious. Might humans be the only conscious animals on Earth? If so, then aren't we pretty special? Thus you see how thinking about animal consciousness brings us to central philosophical questions about how we should think about ourselves. Animals are the things in nature most similar to ourselves, so part of trying to find out who we are consists of comparing ourselves to them.

Given how similar we are to other animals, especially chimpanzees and bonobos, our closest relatives among the great apes, it seems incredible that anyone could think that they are just unconscious machines, sharing none of the conscious experiences that we have. How could a whimpering dog that has been trodden upon, not be experiencing pain consciously? Nonetheless, this incredible view has been defended by some philosophers and psychologists. Philosophers and scientists have a long tradition of skepticism—a "show me" attitude that refuses just to go along with what seems to be common sense. There are lots of examples in the history of science where the skeptics were right and common sense wrong. Gallileo's idea that the earth moves, for example, went against the common sense of his day.

The question of what the conscious experiences of other animals might be like is intrinsically interesting in its own right. But it's also a very important question for ethics. Jeremy Bentham, the nineteenth century British philosopher, said that the important question about our treatment of animals is "Can they suffer?" His concept of suffering included conscious experiences of pain and other unpleasant emotions. Scientists' opinions are very important in forming laws and rules about how animals should be treated in research and agriculture. For example, based on the recommendations of a scientific panel, the British Animal Scientific Protection Act draws a line between vertebrates and invertebrates, but makes an exception for the common octopus. Under the British law, octopuses get the same protection from harmful treatment as any mammal, bird, fish, reptile, or amphibian. Do octopuses experience pain consciously? What about other invertebrates that have not been carefully studied? The United States Animal Welfare Act is much less inclusive, covering only mammals and birds. So, what about the fish, reptiles, and amphibians not covered by the U.S. law? Many scientists are skeptical that these questions can be answered on objective, scientific grounds, but this has important consequences for applied ethics, so it is important to understand the strengths and weakness of skeptics' arguments. Before looking at those arguments, however, it will be useful to say a bit more about what is meant by "consciousness."

Meanings of "Consciousness"

The word *consciousness* can have several different meanings. When talking about animal consciousness, only some of these meanings are controversial. There are two very ordinary senses of consciousness that no one disputes their application to many animals. One is the sense in which animals can be awake rather than asleep, or in a coma. Another connects to the ability of animals to sense and respond to features of their environment: It can be said that they are conscious or aware of those features. Consciousness in both these senses is identifiable in a lot of different species of animals and can be studied scientifically. Fish sleep, and earthworms are, in the relevant sense, awake and aware of several things in their environments.

Two other senses of consciousness are controversial when used to talk about non-human animals: conscious experience (also called "phenomenal consciousness") and self-consciousness.

Conscious experience is difficult to define, but one way to approach it is to think about what happens when you see, smell, hear, taste or feel things. Think about looking at a horse, for example. In the presence of a horse, you have more than the abstract knowledge that there is a horse in front of you; the horse *looks* a particular way, it *smells* a particular way, and if you were to lick it, it would *taste* a particular way, too. Because the word "consciousness" is ambiguous, philosophers often refer to this as *phenomenal consciousness* to distinguish it from other senses of consciousness. If you do any of the further reading suggested below, you will also encounter the term *qualia* (which is the plural form of *quale*) which is used to talk about the experiential qualities of phenomenal consciousness: The particular experience of the color of the horse, your particular experience of its smell, and so on, are all qualities of your conscious experience, or qualia for short.

Self-consciousness refers to an organism's capacity to understand itself as an individual who is similar to but distinct from others. Of course, every animal normally has some way of discriminating itself from others—animals typically don't eat themselves, for example. But self-consciousness is a higher cognitive ability involving some sort of self concept. The "some sort of" here is deliberately vague, because it is not at all obvious what that self concept should contain. One idea that has been very important is that a self concept should contain a capacity for thinking about the thoughts (and other mental states) of others. A self concept, in this view, involves the thought that I have certain perceptions, experiences, thoughts, and desires, whereas you may perceive and think about things differently, and have different desires. This is often called having a *theory of mind*.

When people deny that nonhuman animals are conscious, they are usually either denying that they have phenomenal consciousness or that they have self-consciousness. Some people think that these two things are related—that you have to have self-consciousness to have phenomenal consciousness—but many think that phenomenal consciousness is a more primitive capacity that doesn't require the complex ability to think about oneself.

The Experience of Consciousness

A question that philosophers often ask about conscious experiences is whether they are the same for different individuals: Does a horse smell the same to you as it smells to me? Are your color experiences like mine? (Presumably not, if you are color blind.) Could your color experience of the rainbow be a mirror image of mine? (Philosophers call this the "inverted spectrum" problem.) These and related questions seem to present a particular problem for understanding phenomenal consciousness scientifically. Science has methods for dealing with objective, publicly verifiable facts. But it is hard to see how these objective methods could be applied to the subjective experiences of others. I can't have your experiences, so how could I ever know whether your experiences are identical to mine?

So far, these questions apply equally to our ideas about consciousness in other humans as they do to questions about consciousness in nonhuman animals. But nonhuman animals present special challenges. One reason for this comes from differences in sensory equipment. There can be differences with respect to the traditional five senses of humans—honeybee color vision was already mentioned above, and many animals have a much better sense of smell than humans. But many animals also have sensory capacities that are absent in humans. For example, bats and dolphins use echolocation—they send out high-pitched sounds and listen for the reflections—to locate objects around them; and many fish, such as sharks, detect changes in electrical fields. The philosopher Thomas Nagel, in a very influential paper asked, "What it is like to be a bat?" Nagel was not asking his readers to imagine what it would be like if they were to hang around upside down, but rather he was asking what is it like *for the bat*—what are the subjective experiences (phenomenal consciousness) associated with echolocation? Nagel assumes that there is something that it's like to be a bat, but some people think even that should be questioned. And even if you agree with Nagel that bats are conscious in this sense, there are surely some animals about which you would be less sure. Do honeybees have conscious experiences? What about lobsters? Or oysters? Many people think that conscious experience is obviously present in mammals and birds but that it is very unlikely in most invertebrates, such as insects, crustaceans or mollusks (with the possible exception of some cephalopods such as octopus). Reptiles, amphibians, and fish constitute an enormous gray area in many people's minds, which is why many people will eat fish but still call themselves "vegetarians." In summary, phenomenal consciousness raises two main challenges: (1) Can we know which animals beside humans are phenomenally conscious? (The Distribution Question); and (2) Can we know what, if anything, the conscious experiences of animals are like? (The Phenomenological Question).

Nagel thinks that these questions don't arise for humans considering other humans because we are all sufficiently similar to one another that our own experiences can serve as a basis for knowing what it's like to be another human. Perhaps Nagel is wrong about how much we can know about each other. Be we can, at least, have conversations with each other about our experiences, so language is an important part of why we attribute similar experiences to one another. Some animals, such as Kanzi (a bonobo) and Alex (an African grey parrot), have learned to communicate surprisingly effectively with humans using language, but none have reached a level of conversational ability that would allow us to talk with them directly about their experiences. Common sense regards other animals as similar enough to us to make it reasonable to believe that they experience many things as we do, but the skeptic about animal consciousness is able to use the uncertainty that comes from recognizing that other animals are also quite different from us to challenge the claim that we really know anything about animal consciousness at all.

Perhaps the most famous skeptic of all is the seventeenth century French philosopher and scientist, René Descartes. Descartes argued first that mental properties, such as consciousness, don't belong to the physical body, but are properties of an immaterial mind or soul. Secondly he argued that anything animals can do can be done by very complicated arrangements inside physical bodies—in other words, animal behavior could be accomplished entirely by purely physical machines. Descartes thought, however, that the capacity to speak and reason using language could never be accomplished by a machine, so this provided the evidence for humans having both body and mind/soul. (This is an attractive idea for those coming from religious traditions believing in an afterlife for humans.)

Philosophers think it is easier to understand arguments if they are laid out systematically, so here is Descartes' main argument against animal consciousness arranged as a list of premises and conclusions:

1. Attributing consciousness to animals is justified if and only if they act in ways that cannot be explained mechanically.
2. The meaningful use of speech and general reasoning, and only these, cannot be explained mechanically.
3. Animals do not use speech meaningfully or reason generally; therefore,

4. animals do not act in ways that cannot be explained mechanically (follows from 2 and 3); and hence,

5. it is not justified to attribute consciousness to animals (follows from 1 and 4).

This argument is "valid" in logicians' terms—which means that the premises (1 to 3) logically imply the conclusions (4 and 5). So, the only way to challenge the argument is to dispute one or more of the premises. In fact, all three premises have been attacked by philosophers and others from Descartes' own time up to today.

The first premise has been challenged by "materialists" who dispute Descartes' "dualist" view of body and mind as two completely separate kinds of things. According to materialism in the philosophy of mind (not to be confused with the capitalist concept of materialism as a love of material possessions), consciousness and other mental states are to be understood in terms of very complicated brain systems—the human brain, for instance, contains over 100 billion neurons and perhaps as many as 100 trillion connections between them. Human consciousness on this view is a direct product of the enormous amount of machinery in our brains. If that's right, then all the things we do could, in principle, be explained mechanically. The materialist view of things thus leads to a challenge to the second premise above. Modern cognitive neuroscience is increasingly finding out the neural basis for our abilities to talk and reason, and it's no longer widely believed that these things can't ultimately be explained in terms of the brain's machinery.

The third premise has also come under attack. There has been a lot of work recently on language and reasoning in animals. This work is controversial, however, because skeptics maintain that whatever animals are doing isn't really the same as humans. For instance, many linguists say that the abilities of animals to use natural forms of communication, or to use human sign language or artificial symbols on a keyboard, or to understand spoken human words, are really much more limited than human language, and shouldn't really count as language use at all. These issues are discussed in other entries of this encyclopedia. But in any case, Descartes' assumption is no longer safe to make without a sophisticated discussion of the evidence.

These days, not many professional scientists or philosophers accept Descartes' argument, or the dualistic view that it entails. But there are still significant numbers of scientific and philosophical skeptics who accept Descartes' standards of evidence (i.e., they want strong behavioral proof that animals are conscious). Some of these skeptics even want to turn the tables on proponents of animal consciousness by saying it is the proponents who are the real dualists, believing in such intangible things as consciousness. It is also popular among such skeptics to identify language as the thing that separates humans from other animals. Not all the skeptics agree, however. There are some who think that other high-level cognitive abilities are important for conscious experience. One idea is that consciousness emerges from the ability to think about one's own experiences—it is only by turning the mind's eye on itself, so to say, that there can be anything like consciousness attached to sensory experience. One consequence of this view is that even young human children, before the age of four or so, might have no conscious experience. A child's screaming about injections would be designed by nature to get your attention, not to tell you anything about the child's conscious experiences. Likewise for animals, who may yelp and whine, but do so only because it serves the biological purpose of getting others to change their behavior and not because it feels like anything to the animal itself.

Testing the idea of "higher-order" thinking in nonhuman animals—thinking about one's own thoughts—is quite controversial and there are conflicting results from different scientists. Some scientists think they have evidence that certain animals know what others are thinking or seeing, such as when a chimpanzee chooses to go for a piece of food that its rival cannot see rather than one that the rival can see. More skeptical scientists think that there could be other explanations for the chimpanzee's behavior; for instance, that the animal has simply learned to go for food in the right sort of position near a larger object without really understanding that the larger object is blocking the rival's view. Another area where some scientists have thought that higher-order thinking is implicated is in the apparent ability of some animals to recognize themselves in the mirror. The controversy about this interpretation is covered in another entry of this encyclopedia.

Even if it is possible to get evidence for self-consciousness or theory of mind through clever experiments, the connection between these abilities and the more nebulous idea of phenomenal consciousness remains uncertain to many people. Nagel argued that because phenomenal consciousness is private or subjective, it is beyond the reach of objective scientific methods. This claim might be taken in either of two ways. On the one hand it might answer the Distribution Question; that is, by saying no, we can't know that a member of another species (e.g., a bat) has conscious states. On the other hand it might answer the Phenomenological Question; that is, by saying no, we can't know the phenomenological details of the mental states of a member of another species. The difference between believing with justification that a bat is conscious and knowing what it is like to be a bat is important because, at best, the privacy of conscious experience supports a negative conclusion only about what it is like to be a bat. To support a negative conclusion about the former, one must also assume that consciousness has absolutely no measurable effects on behavior (a view called *epiphenomenalism*). But if one rejects epiphenomenalism and maintains that consciousness does have effects on behavior, then there remains hope for detecting those effects through experiment and observation.

Arguments for Animal Consciousness

Most people, if asked why they think familiar animals such as their pets are conscious, would point to similarities between the behavior of those animals and human behavior. Similarity arguments for animal consciousness thus have roots in common sense observations. But they are also reinforced by scientific investigations of behavior and neurology as well as by appealing to evolutionary continuity between humans and other species. Nagel's own confidence in the existence of phenomenally conscious bat experiences is based on nothing more than this kind of reliance on features that mammals share.

Ordinary observers readily judge human and animal behavior to be similar. The reactions of many animals, particularly other mammals, to events that humans would report as painful are easily and automatically recognized by most people as pain responses. High-pitched vocalizations, fear responses, nursing injuries, and learning to avoid things that cause damage are among the responses to noxious stimuli that are all part of the common heritage of mammals. Similar responses also exist to some degree or other in organisms from other taxonomic groups besides mammals.

Neurological similarities between humans and other animals have also been taken to suggest common conscious experiences. All mammals share the same basic brain anatomy, and much is shared with vertebrates more generally. A large amount of scientific research that is of direct relevance to the treatment of human pain, including on drugs that block pain, is conducted on rats and other animals. The validity of this research depends on the similar mechanisms involved, and to many it seems arbitrary to deny conscious pain experiences to injured rats who respond well to the same drugs as humans. Similarly, much of the basic research that is of direct relevance to understanding human visual consciousness has been conducted on the very similar visual systems of monkeys.

Arguments that are based on similarities have an inherent weakness: Critics can always exploit some dissimilarity between animals and humans (of which there are many) to argue that this is the relevant factor the animals are lacking. Mentioning evolutionary continuity between the species may help a bit, but still there is no logical requirement that, just because humans have a trait, our closest relatives must have that trait too. There is no inconsistency with evolutionary continuity to maintain that only humans have the capacity to learn to play chess. Likewise for consciousness. Perhaps a combination of behavioral, physiological and morphological similarities with evolutionary theory amounts to a strong overall case. But in the absence of more specific theoretical grounds for saying that animals are conscious, this combined argument is unlikely to change the minds of those who are skeptical.

One way to get beyond the weaknesses in the similarity arguments is to try to give a theoretical basis for connecting what we can observe about animals to phenomenal consciousness. This would perhaps be possible if we could say just what consciousness is for (by specifying its functions in other words. If consciousness is completely epiphenomenal (i.e., has no measurable effects), then a search for the functions of consciousness is doomed to be futile. In fact, if consciousness is completely epiphenomenal, then it cannot have evolved by natural selection, and it will remain a great biological mystery. On the assumption that phenomenal consciousness is the result of evolution, in human minds at least, and therefore that epiphenomenalism is false, then an attempt to understand the biological functions of consciousness may provide the best chance of identifying its occurrence in different species.

This kind of approach can be traced to Donald Griffin's attempts to force scientists to pay attention to questions about animal consciousness. In a series of books, Griffin (who made his scientific reputation by carefully detailing the physical and physiological characteristics of echolocation by bats) provides examples of communicative and problem-solving behavior by animals, particularly under natural conditions, and argues that these are prime places for scientists to begin their investigations of animal consciousness. Griffin has been strongly criticized for being too vague on the notion of consciousness, and too willing to accept just about any story about clever animals to support his point. Griffin's main positive proposal about the function of consciousness is that consciousness might have the function of compensating for limited neural machinery. This leads to his rather implausible suggestion that consciousness may be more important to honeybees than to humans. But nevertheless, the idea that we must think about what purpose consciousness serves is an important one.

There is another place where common sense may not be a very reliable guide to animal consciousness. The common-sense answer would be that consciousness "tells" the organism about events in the environment, or, in the case of pain and other bodily sensations, about the state of the body so that the body can respond appropriately. But this answer is too general, for quite simple mechanisms that presumably don't involve consciousness nevertheless can provide information to organisms that results in adaptive responses. Your body starts to pull your hand away from a hot stove before you experience anything consciously, because there are neurons going through your spinal cord that "tell" your arm to react without having to go through consciousness. But, the consciousness does come later, and perhaps this helps you to learn from your mistakes. So one promising place to look for the functions of consciousness might be in error correction and learning.

An article such as this perhaps raises more questions than it answers, but the topic would be of little philosophical interest if it were otherwise. And despite the fact that there have been centuries of argument on the question of whether animals have interesting forms of consciousness, modern developments in the neurosciences and much intriguing recent work on animal cognition from a variety of perspectives makes this an exciting time to be working in this area.

See also Behaviorism
> Cognition—*Cognitive Ethology: The Comparative Study of Animal Minds*
> Cognition—*Fairness in Monkeys*
> Cognition—*Mirror Self-Recognition*
> Cognition—*Mirror Self-Recognition and Kinesthetic–Visual Matching*
> Cognition—*Social Cognition in Primates and Other Animals*
> Cognition—*Theory of Mind*
> Emotions—*Emotions and Affective Experiences*
> Empathy
> Play—*Dog Minds and Dog Play*
> Play—*Social Play Behavior and Social Morality*
> Welfare, Well-Being, and Pain—*Behavior and Animal Suffering*

Further Resources

Allen, C. 2002. *Animal Consciousness*. In: *The Stanford Encyclopedia of Philosophy* (Ed. by N. Zalta), Stanford, CA: Stanford University. (http://plato.stanford.edu/archives/win2002/entries/consciousness-animal/)

Allen, C. 2004. *Animal Pain*. In: *Noûs*, 38, 617–643.

Bekoff, M., Allen, C., & Burghardt, G. M. (Eds.) 2002. *The Cognitive Animal*. Cambridge, MA: The MIT Press.

Griffin, D. R. 2001. *Animal Minds: Beyond Cognition to Consciousness*. Chicago: University of Chicago Press.

Nagel, T. 1974. *What is it like to be a bat?*. Philosophical Review, 83, 435–450.

Colin Allen

■ Cognition
Animal Languages, Animal Minds

> *"Animal communication can therefore provide a useful and significant 'window' on animal minds . . . "*
> ETHOLOGIST DONALD GRIFFIN, 1992

> *"If a lion could talk, we could not understand him."*
> PHILOSOPHER LUDWIG WITTGENSTEIN, 1953

What is it like to be a dog? Or a cat? Or any other animal for that matter? What do they think? How do they feel? What does the world seem like to them? Is it anything like it seems to us? These are questions we all have probably asked at one time or another, perhaps especially of our pets. We are fascinated by others' mental lives: We are constantly speculating about what the people around us think and feel, and we are equally fascinated by the minds of animals. In fact, this fascination is a formal preoccupation of some scientists studying animal behavior whose work is designed to gather objective evidence on the character of animal minds. Many forms of behavior might provide useful clues in this regard, but one form of behavior in particular has been supposed to provide an especially clear window into animal minds—communication. In humans, our language, or speech, is the primary way we express ourselves—what we say reflects what we think and feel. Language is, therefore, like a portal to our inner selves. By extension, students of animal behavior have begun to explore animal languages as a possible source of insight into their inner lives.

Of course, deciphering other languages and understanding other minds are not simple matters. In fact, within the scientific community there is disagreement about even the potential accessibility of animal minds. Some, like Wittgenstein, argue that animal minds are not only fundamentally different from our own, but also inherently unknowable. Even if a lion could talk, so the argument goes, we would not be able to understand it. The problem is not so much that the lion's particular language would be foreign to us (which it obviously would be, and this in itself would be a significant hurdle), but the deeper problem is that the lion's world, and its mental experiences of it, are so different from ours that the things it would tell us would be so foreign to our mind that we could not comprehend them. In contrast, others like Griffin admit significant barriers but stress that, despite their diversity of behavior and experience, all animal species are ultimately linked in a massively interconnected, biological web. This biological continuity should include continuity of mental experience which in turn should be detectable through the signs animals use to communicate. Which of these views is right?

Language and Mental Diversity

Our own heavy reliance on speech naturally encourages us to think at first only in terms of *vocal* languages in other animals and to focus narrowly on their function in conveying basic information about events in the world, the way we ourselves often talk about the weather, sports, world politics maybe, and, of course, who's dating whom. Vocal messages are also important in many other species, and in some include communicating similarly basic information about world events—such as the location of food or the presence of a predator. But there are a host of other ways animals communicate and a host of other things they communicate about.

For example, nocturnal fireflies negotiate mating using patterned flashes of light. Male suitors, flying in complete darkness, advertise for females using a sort of Morse code of light pulses, produced by specialized, light-generating cells in their bodies. Interested females return the male's flashes, whereupon the male homes in on the accepting female. Something similar occurs in moths, only the mate-attracting beacons are odorous rather than luminous. Female moths produce and release plumes of a chemical substance, called a pheromone, that is attractive to males. When a wandering male catches the alluring scent, he flies in zigzag formation across the odor trail to determine its direction and then follows its strength gradient back to the female source.

Many other animals also communicate chemically. For example, in ants, a forager roaming in search of food continuously releases pheromone droplets on the ground as it moves. This produces an easy-to-follow chemical trail that the ant can later use to retrace its steps back to the nest, or to a specific location visited earlier, and that other ants in the colony can also use to find food. Other species, including domestic dogs and cats, deposit odorous urine or feces in discrete locations in their range. Volatile compounds within these deposits can convey various messages including the sex and reproductive status of the animal and its possession of the local territory.

In some species, individuals actually spend a great deal of their time communicating with themselves. In very murky rivers and lakes, where vision is next to impossible, some fish navigate and locate food with the help of specialized cells within their bodies, called electrocytes, that can produce an electric current. Using these cells, the fish establish a weak electrical field in the water around them and then use distortions in this field caused by nearby objects with varying resistance properties to "read" the terrain around them. Bats flitting about in the dark, and dolphins swimming in the vast oceans, navigate and forage in a similar way. Both produce short bursts of high-frequency sound waves (labeled, ultrasonic, because they lie well above the limits of human hearing). The sound waves radiate out into the air or water. Many of the sound waves never return, but some are reflected back when they collide with objects in the environment. The animals who sent the signals detect these reflected waves and use the distinctive reflected wave profiles of different objects to respond appropriately. In essence, then, these animals are continuously "talking" to themselves and using the echoes of their own "voice" to monitor the world.

Some animals exploit the opposite, infrasonic extreme of the sound continuum (which lies below the limits of human hearing). Elephants, for instance, sometimes communicate using very low-frequency "rumbles." The rumbles travel exceptionally long distances and serve to coordinate the movements of widely separated individuals of a herd and also to advertise their herd membership to others. Some rumbles are so low-frequency that we humans are more likely to *feel* the rumbles if indeed we detect them at all. However, other species actually make regular use of such felt signals. Kangaroo rats living in the barren desert, for example, drum the ground with their feet creating ground vibrations that broadcast a seismic message of their territorial status to neighbors. Each individual has a unique seismic signature that allows the many residents in an area to identify one another at a distance.

There are a great many other communication systems, but even this very small sampling illustrates the tremendous variety of natural animal languages. It also makes clear that many of these languages, and the worlds to which they are adapted, are entirely foreign to humans. For some species, the world appears as a diverse landscape in which the different peaks and valleys are different chemical odors, varying ultrasonic echoes, or even shifting electrical auras. In fact, many of the communication signals these animals use to probe and respond to their world are ones that we humans cannot even detect naturally. And this fact alone suggests that many animals "see" the world very differently than we do. Does this then mean Wittgenstein was right, that animals' experiences are so different from ours that their minds are effectively closed to us? Perhaps, but not necessarily.

Language and Mental Continuity

In the pitch black of a honeybee hive, bees recently returned from collecting flower nectar perform elaborate "dances" on the surface of the comb. They dance either in circles (the round dance) or trace out small looping patterns that resemble a figure eight (the waggle dance). Other bees crowd around, paying closing attention to the dancer's movements by palpating it with their antennae. Then they leave the nest and fly directly to the nectar-rich flowers from which the dancing bee has just returned. Once again, this mode of communication is peculiar to us—yet another illustration of the great diversity of animal languages. However, there are properties of the bees' dance language that resemble our own. Careful research has shown that the type of dance performed and the detailed steps involved indicate the distance, direction, and quality of food. Bees attending to the dance acquire and act on this information to collect food for the hive efficiently. In an important respect, then, the informational capacity of the dance is roughly analogous to that of human words, which similarly serve to pick out, refer to, or symbolize things in our world. The fact that honeybee communication involves a rudimentary symbolic code suggests that, despite occupying a world markedly different from ours, the bee's mind (however tiny!) may organize some of its experiences in at least crudely similar ways.

In fact, there are a few other examples of simple symbolic communication, the best documented of which concerns the alarm vocalizations of vervet monkeys. Vervets are small, group-living monkeys of Africa. Their size and habits expose them to many predators including leopards, eagles, and large snakes, such as pythons. When one monkey spots such a predator, it gives a loud alarm cry and others in the group immediately take evasive action. Research has shown that the monkeys actually produce three different alarm calls, one for each of these different predator types, and the alarm calls alone cause others to respond in different but appropriate ways. Thus, without having seen the predator themselves, vervets who hear the leopard alarm run for the treetops where leopards cannot go. In contrast, when they hear an eagle alarm, they drop from the treetops where they are vulnerable to a swooping eagle. To the snake alarm, they stand erect and scan the ground around them, looking for the snake. Thus, it seems that the alarm calls of vervet monkeys refer to, or symbolize, these different predators in much the same way as do our words for these same animals.

Several additional parallels to human language come from the elaborate songs of many of the world's birds. Male songbirds sing in spring to attract female mates and to advertise their territories to rival males. Songs are typically composed of multiple different sound elements, called notes, that are grouped together in common themes, called syllables, which in turn are strung together into long, complex melodies. In some cases, the songs of birds in one area differ consistently from those of others of the very same species living some distance away, differences that have been labeled *song dialects* because of their similarity to dialect differences in human languages. The various forms of dialect may arise in similar ways. Neither birds nor humans are born singing or talking. Instead, both acquire, or learn, their language by exposure to adults. Young chicks hear and remember the songs of adults in the areas where they were born. Later, as adults themselves, they rehearse these songs and fine tune their renditions of them to attract female mates of their own. In birds acquiring song, as in humans acquiring language, there is a critical period early in life when they must be exposed to adult song or they will not learn it. However, even with appropriate exposure, slight variations in the songs can accumulate in different areas as a result of minor copying errors, and this may be the source of regional dialects.

A further parallel to human language concerns the fact that syllables can be shuffled about within songs to produce many different melodies. Syllable shuffling is not random, but instead tends to follow specific rules, a phenomenon reminiscent of the rules of human grammar that govern the sequencing of our own speech. There is not yet evidence that the different melodies created by syllable shuffling have different specific meanings to the birds the way grammatical word shuffling of human utterances can produce very different messages (e.g., "dog bites man" versus "man bites dog"). Nevertheless, the shuffling rules represent a sort of *song grammar* that may point to commonalities in the mental organizational principles that are used both to create and to structure variety in communication.

Finally, humans and animals alike use their language for some similar basic social functions, including expression of emotions, mediation of social interactions, easing of social tensions, and servicing of particular relationships. Many monkeys and apes, for example, produce a host of different vocalizations whose tones seem to express different attitudes or emotions, ranging from apprehension and social subordinance, to a desire for social contact, to outright hostility. In baboons and other monkeys, specific calls are also used to calm companions and facilitate close social interaction with them, sometimes even helping to repair previously damaged social relationships. For example, female baboons, keen to handle new babies in the group, cautiously approach the protective mothers and produce a string of low-amplitude grunts. This grunt message seems to reduce mothers' unease, and they often then allow the grunting females to inspect and touch their babies. Similar grunts are sometimes produced after aggressive fights, when former combatants re-encounter one another. After an exchange of grunts, the animals are more likely to interact peaceably, at times even sitting together and grooming one another. In some species, calls are used to monitor and service social relationships at a distance. For instance, when moving and foraging widely, both squirrel and rhesus monkeys produce "contact" calls to stay in touch with particular group members. Close biological relatives and other alliance partners are particularly attentive to each others' calls and are more likely to call in unison. Thus, like humans, animals use their language to coordinate a range of social activities, and some of these uses imply some additional continuity of experience including similar basic emotional states, similar social concerns and apprehensions, and similar underlying attachments to kin and social allies.

A female baboon producing a "contact" call.
Courtesy of Karen Rendall.

It is too early to say definitively whether Wittgenstein or Griffin is truly right. Research on animal minds is only beginning, and we cannot yet say what it is like to be a bat, or a dolphin, a bee, or an elephant. Perhaps, as Wittgenstein argued, it will never be possible to say exactly what it is like to be any of these animals. After all, it is not really possible to say exactly what it is like even to be another human being. For that, you would need to have experienced the world from exactly their perspective which is obviously impossible without actually being them rather than yourself. Putting this special challenge aside, research to date already suggests that some animals have very different perspectives on the world than we do. As Wittgenstein proposed, these different perspectives may well structure their minds very differently from ours. At the same time, though, some of the accumulated research points to dimensions of continuity, revealed, as Griffin proposed, through the signs animals use to communicate. Perhaps the safest conclusion to draw, then, is that the many animals of this planet seem to display both a remarkable diversity and some intriguing continuity of mental experience.

See also Cognition—*Talking Chimpanzees*

Further Resources

Cheney, D. L. & Seyfarth, R. M. 1990. *How Monkeys See the World: Inside the Mind of Another Species*. Chicago: University of Chicago Press.
Griffin, D. R. 1992. *Animal Minds*. Chicago: University of Chicago Press.

Hauser, M. D. 2000. *Wild Minds: What Animals Really Think*. New York: Henry Holt and Company.
Wittgenstein, L. 1953. *Philosophical Investigations*. (Translated by G.E.M. Anscombe), Oxford: Blackwell Scientific Press.

Drew Rendall

Cognition
The Audience Effect in Wolves

Humans appear to modify their behavior depending on who's watching. For example, a teenage boy may do a skateboard trick in front of a girl he is trying to impress. A small child may climb onto the counter to retrieve a cookie from the cookie jar only when parents are not present. This question of an appropriate audience or absence of one may be a fruitful starting place to examine how animals evaluate variables in a dynamic social environment in order to make decisions.

Do other animals change their behavior depending on who's watching? It turns out that golden sebright bantam cockerels do. For example, when food was present, cockerels withheld food calls in the presence of males but emitted food calls in the presence of females. When alone, the cockerels refrained from emitting alarm signals when hawk models were flown overhead but emitted these calls when others were present. These findings are important because we generally characterize these types of calls (signaling behavior) as either a conditioned response or an instinctual response to a stimulus. Both explanations suppose that the animal does not have a choice or cannot modify its response in the presence of a stimulus. This withholding of calls in certain contexts (presence or absence of an audience) indicates that something more complex may be occurring. This audience effect, the change in signaling behavior based on change in social context and not a change in the referent, is one place to start when investigating intentionality and whether or not animals possess a theory of mind.

I was interested if wolves would perform behaviors in order to convey or withhold information. In this case, the referent (stimulus) is relevant biological information and the signaling is a certain behavior. Wolves engage in behaviors that potentially "contain" information for group members. I identified several behaviors that I assigned "information it may contain" and whether or not that wolf would want to convey that information (share) or withhold that information (be secretive).

I examined urine-marking behavior, urine marking food, food caching, cache recovery and pilfering. These behaviors may be useful in conveying information about social rank (frequency of urine marking and urine-marking posture), food ownership (urine marking food), and location of food (caching, cache recovery, and pilfering). I made predictions about whether or not a wolf would want to convey information by performing the behavior in the presence of an

High-ranking wolves tend to urine mark significantly more often in the presence of an audience. Here, a dominant female urinates using the "squat urination posture" (SQU).
© Anthony F. Chiffolo.

audience or withhold that information by performing that behavior in the absence of an audience.

Two captive groups of wolves were studied for four research seasons (1991–1994). As predicted, high-ranking wolves urine marked significantly more often in the presence of conspecifics (members of the same species) than in their absence. The sex and rank of a wolf determined which urination posture was used most frequently in the presence of an audience; the males used the raised leg urination posture, and, in two of three instances, high-ranking females used the squat urination posture. However, in one of the groups, all individuals urine marked more often in the presence of an audience regardless of rank. Urine marking food had no relationship to who actually fed on the food (so in this case, the information about "ownership" was not substantiated); however, consistent with above, high-ranking individuals of one group and all individuals urine marked on food significantly more often in the presence of an audience than in the absence of one. Food caching was performed more often when alone (in the absence of conspecifics) than in the presence of others; however, food cache pilfering (cache recovery by individuals who had not originally made the cache) occurred in the presence of the "cacher" (individual who had originally made the cache) about as often as not. (See the accompanying table.)

These results support in part the hypothesis that wolves modify their behavior depending on the social context. More research needs to be done in order to understand the role of social facilitation and inhibition (especially how the response to an audience varies depending on the rank of the actor and the ranks of the individuals in the audience).

Studies on the audience effect have focused on auditory signals that contain information for conspecifics; however, I think examining how certain behaviors convey information may be more relevant for animals who are less vocal and who exhibit multiple modalities of communication. Logically, social animals learn the effect of their actions on others and can reap the benefits or repercussions of conveying or withholding information.

Why do we care if animals change their behavior when an audience is present? For one, it shows that animals may be evaluating their audience and have more control over their behaviors than originally thought. In addition, the degree to which animals are able to

Audience Effects in Wolves

Behavior	Information	Prediction	Results
Marking	High rank	Convey (for high ranked individuals)	Yes but mixed
Marking posture (RLU* and/or SQU**)	High rank	Convey (for high ranked individuals)	Yes but mixed
Marking food	Ownership	Convey	Information about ownership not supported
Caching food	Food location	Withhold	Yes
Cache recovery (of another's cache)	Pilfering food	Withhold	Equivocal

*raised leg urination
**squatting urination

make decisions about performing behaviors in order to convey or withhold information may be very important in functioning in a social unit. The presence of an audience effect may imply that an animal has a notion of being watched; therefore, this effect may be a precursor to more sophisticated types of cognitive behavior, including imputing a theory of mind (i.e., knowing or assuming that their group mates have knowledge and intentionality or the intention of influencing the behavior of group members or other wolves).

See also Cognition—*Dogs Burying Bones: Unraveling a Rich*
Action Sequence
Cognition—*Food Storing*

Further Resources

Baltz, A. P. & A. B. Clark. 1997. *Extra-pair courtship behavior of male budgerigars and the effect of an audience.* Animal Behaviour, 53, 1017–1024.

Bugnyar, T., Maartje, K. & Kortschal, K. 2001. *Food calling in ravens: Are yells referential signals?* Animal Behaviour, 61, 949–958.

Bugnyar, T., & Kortschal, K. 2001. *Do ravens manipulate the others' attention in order to prevent or achieve social learning opportunities?* Advances in Ethology, 36, 106–107.

Doutrelant, C., McGregor, P. K. & Oliveira, R. F. 2001. *The effect of an audience on intrasexual communication in male Siamese fighting fish,* Betta splendens. Behavioral Ecology, 12, 283–286.

Evans, D. S. & Marler, P. 1994. *Food calling and the audience effects in male chickens,* Gallus gallus: *Their relationship to food availability, courtship, and social facilitation.* Animal Behaviour, 47, 1159–1170.

Marler, P., Karakashian, S., & Gyger, M. 1991. *Do animals have the option of withholding signals when communication is inappropriate? The audience effect.* In *Cognitive Ethology: The Minds of Other Animals* (Ed. by C. Ristau), pp. 187–208. Hillsdale, N. J: Lawrence Erlbaum Associates.

Townsend, S. E. 1996. *The role of social cognition in feeding, marking, and caching in captive wolves,* Canis lupus lycaon *and* Canis lupus baileyi. Ph.D. Thesis, University of Colorado, Boulder.

Susan E. Townsend

■ Cognition
Behavior, Archaeology, and Cognitive Evolution

The human mind evolved in conditions very different from those of the modern world. It follows that if we want to understand a feature of the human mind, we need to understand the conditions under which it evolved. Some features of the human mind, emotions for example, evolved long ago when our primate ancestors lived in tropical trees. Others have evolved more recently. Our ability to conceive of and to organize our actions in space is one of these more recent evolutionary developments. We impose regular geometric shapes on many of the objects we create, and can find ourselves (most of us anyway) in a three-dimensionally conceived universe. Apes do not do these things, and it follows that these abilities evolved in the human past. But how can we study the evolution of something as incorporeal as "mind." Experts in human evolution do have a fossil record of brain size and even gross anatomical shape going back almost 5 million years, but these reveal little about specific abilities.

For the last half of human evolution we have an alternative source of data in the form of an archaeological record. Archaeologists study the traces of human activity—tools, dwellings, garbage, and so on—and these traces tell us something about the minds that organized these activities.

Our ancestors first split from apes in Africa sometime before 5 million years ago. These earliest hominids were bipedal, but were otherwise very ape-like. Two and a half million years ago our ancestors began to make stone tools. These ancestors, probably an early form of the genus *Homo*, knew that fracturing certain kinds of stone produced "flakes" with very sharp edges (see drawing 1 and 2). They also knew that if they removed flakes from alternate sides of a cobble they could produce not just sharp flakes, but a larger "core" tool with a sharp edge (see drawing 3). They did not, however, make tools with an overall shape. They were only interested in making tools with sharp edges which they used for a variety of tasks, including butchering animals they had scavenged. These tools were "ad hoc" in the sense that the hominids made them for a specific task, and then abandoned them on the spot. The tools did not exist in the hominid mind as a distinct category like a "pocketknife." The spatial concepts required to make these tools are simple: "next to," "boundary" (alternate sides of an edge), and "sequence" (one blow after another). All of these are within the abilities of modern apes. Indeed, a few modern apes have actually been taught to flake stone and use the sharp flakes to perform tasks. Cognitively, these early *Homo* were still very ape-like, but in other respects they did not resemble apes. They walked upright on two legs (bipedal), spent a significant amount of time away from trees, competed with carnivores when scavenging carcasses for meat, and carried both tools and meat for distances up to 5 miles. This was a very different kind of ape.

It was not until 1.5 million years ago that our ancestors produced stone tools whose spatial requirements were beyond the abilities of the apes. The hominid responsible, *Homo erectus,* had a larger brain and was much more modern in appearance and behavior. It had developed a way of life that enabled it to expand out of tropical Africa into Asia and southern Europe. This way of life included use of fire and much greater reliance on tools. Unlike earlier hominids, *Homo erectus* made tools with an intentional overall shape. These "handaxes" have an overall bilateral symmetry that reveals something new about

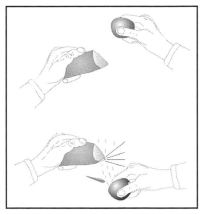

(1) The basic action of flaking stone.
Courtesy of Thomas Wynn.

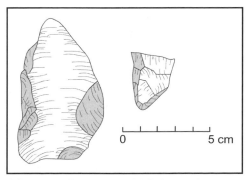

(2) Two-million-year-old flake tools.
Courtesy of Thomas Wynn.

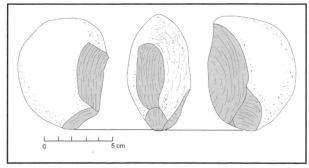

(3) Two-million-year-old core tool.
Courtesy of Thomas Wynn.

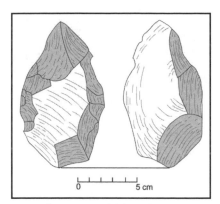

1.4 million year old handaxe.
Courtesy of Thomas Wynn.

the *Homo erectus* mind. When making handaxes, *Homo erectus* had to coordinate two heretofore separate cognitive abilities. The first is the ability to recognize shapes, like symmetry. This is an ability apes certainly have, but they do not coordinate it with the second ability, organizing action in space to produce specific shapes. Not only did *Homo erectus* impose shapes on tools, they also shared an idea of what that shape should be. This standardization of shape indicates *Homo erectus* considered tools to be a separate category of object that existed independent of the tasks they were used for. These seemingly modest developments were a first step toward modern human abilities.

About 400,000 years ago a new set of spatial abilities emerged, associated with the evolution of early *Homo sapiens* out of *Homo erectus*. These ancestors continued to make handaxes, but these tools were now much more regular in shape and many had symmetry in three dimensions—plan, profile, and cross section. This requires not just an ability to coordinate shape and spatial thinking, but the ability to coordinate multiple points of view while making the tool. The symmetry is often very precise—congruent in a geometric sense—indicating that these early *Homo sapiens* injected yet another cognitive ability into their stone working, the ability to conceive of an amount of space (size). The ability of the mind to coordinate shape, size, and point of view is the hallmark of modern spatial cogni-

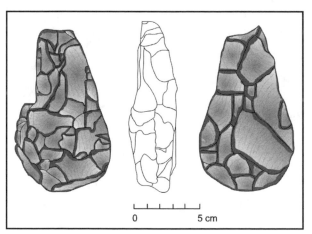

400,000 year old handaxe.
Courtesy of Thomas Wynn.

tion and we can conclude, therefore, that the spatial thinking of these early *Homo sapiens* was modern. This evolution of modern spatial thinking can be linked to other evolutionary developments, especially the evolution of hunting. By 400,000 years ago early *Homo sapiens* had developed effective techniques for hunting large mammals. Archaeologists have even found three complete spears from this time period associated with butchered horse remains. However, other aspects of behavior were not modern. We have no evidence from this time period for art, or decoration, or ritual. Convincing evidence for these does not appear until after 100,000 years ago, and is associated with anatomically modern humans who also had cognitive abilities that enabled contingency planning and innovation—a completely modern mind.

The archaeological record indicates that the human mind is a composite of old and new abilities. But none of these evolved recently enough to be a response to the modern, industrial, information age. We confront the modern world with stone-age minds.

Further Resources

Donald, M. 1991. *Origins of the Modern Mind: Three Stages in the Evolution of Culture and Cognition.* Cambridge, MA: Harvard University Press.

Klein, R., & Edgar, B. 2002. *The Dawn of Human Culture*. New York: Wiley & Sons.
Mithen, S. 1996. *The Prehistory of Mind*. London: Thames and Hudson.
Nowell, A. 2001. *In the Mind's Eye: Multidisciplinary Approaches to the Evolution of Human Cognition*. Ann Arbor: International Monographs in Prehistory.
Wynn, T. 2002. *Archaeology and cognitive evolution*. Behavioral and Brain Sciences, 25(3), 389–438.

Thomas Wynn

Cognition
Cache Robbing

Food caching animals such as squirrels and scrub jays hide food for future consumption, and recover these caches when supplies are less abundant. There is, however, a risk to caching because of the occurrence of cache robbing or *kleptoparasitism* by other individuals. These thieves may be members of the same species, but they need not necessarily be. For example, great tits do not cache food themselves, but they often steal the caches made by marsh tits.

For many species including some tits and chickadees, it is thought that the cache robbers come across the caches by chance. Some mammals such as squirrels may use smell to find the caches. But for birds that have a poor sense of smell, a more accurate method is to rely on memory of where the individual has cached. Furthermore, the ability to locate caches made by others quickly and efficiently may be the important difference between successful cache robbing and potential aggression from the individual that hid the food. So an obvious advantage of remembering where others have cached is that it allows the thief to steal caches efficiently when others have left the scene, thereby eliminating both the costs of caching and of fighting. Members of the crow family such as jays and ravens that are renowned for their intelligence do rely on memory when stealing the caches that other birds have made. Pinyon jays and Mexican jays, for example, are extremely accurate at remembering the location of caches that they saw another individual make.

Food-caching animals also use a number of ways of minimizing the risk of having their caches stolen. For example, some animals such as chickadees and tits may simply stop caching in the presence of potential thieves, whereas others may rely on aggressively defending their caches. The Merriam's kangaroo rat, for example, is a solitary species that usually scatters many caches throughout its territory, but if it sees another rat stealing its caches then it switches from scattering lots of caches to defending a single larder. Several members of the crow family use more complex strategies. Ravens, for example, will not only delay caching if other ravens are in the vicinity and wait until would-be pilferers are distracted or have disappeared before they resume caching, but they will also make fake caches in the presence of observers and preferentially store food behind obstacles so that other ravens cannot see where the caches are being made. And another member of the crow family, the western scrub jay, will return alone to caches it had hidden in the presence of conspecifics, and re-hide them in new places that the observers have not seen. Interestingly, only birds that had been thieves in the past re-hide their caches in new places; naïve cache robbers do not do so, even when they have seen other birds hiding the food. This finding suggests that these thieves relate information about their previous experience as a cache robber to the possibility of future stealing by another individual, and change their caching strategy accordingly suggesting that "it takes a thief to know a thief."

See also Cognition—*Caching Behavior*

Further Resources

Bugnyar, T., & Kotrschal, K. 2002. *Observational learning and the raiding of food caches in ravens*, Corus corvax: *Is it "tactical" deception?* Animal Behaviour, 64, 185–195.
Emery, N. J., & Clayton, N.S. 2001. *Effects of experience and social context on prospective caching strategies by scrub jays.* Nature, 414, 443–446.
Vander Wall, S. B. 1990. *Food Hoarding in Animals.* Chicago: University of Chicago Press.

N. S. Clayton & N. J. Emery

■ Cognition
Caching Behavior

Caching is the act of taking some resource, usually food, that is valuable and plentiful from a situation where it is available to many foragers and moving it to a location where it is much less available or inaccessible to most other animals. Animals cache food when they do not need to eat it for immediate sustenance, and when the food could benefit the cacher at some future time. Caching of food is restricted to birds (e.g., jays, chickadees, nuthatches, shrikes, and some hawks and owls), mammals (e.g., shrews, moles, rodents, and some carnivores), *Hymenoptera* (i.e., bees, ants, and wasps), and a few other groups. Generally, the caching animal takes food that is available to a wide range of foragers and attempts to place it more completely under its own control. The act of caching usually occurs during times of food abundance, whereas retrieval of food generally occurs hours, days, or months later during times of food scarcity.

The distribution of stored food items varies from concentrated to widely scattered. Dense concentrations of stored food items are called *larders*, and the act of preparing a larder is called *larder hoarding*. Individual items or small collections of items stored at widely spaced sites are called *caches*, and the act of preparing these caches is called *scatter hoarding*.

Larders are rich supplies of food that are attractive to other animals and must be defended from potential pilferers. Consequently, animals usually place them in a relatively secure site such as a burrow underground or in a hollow tree. For example, harvester ants carry seeds to their nests and store them in small chambers; red squirrels (*Tamiasciurus hudsonicus*) cut conifer cones and bury them in accumulations of plant litter called middens; honeybees (*Apis mellifera*) collect pollen and nectar from flowers and store them in their hive; and desert rodents sift seeds from the soil and deposit them in chambers near their underground nests.

Scattered caches are not defended. Animals, such as jays, chipmunks, tree squirrels, and chickadees, scatter items throughout their home range and usually hide them well, making them difficult for most other animals to find. These animals are unable to defend larders, either because they are too weak or too busy gathering more food, so they spread the risk of cache loss by making caches small and inconspicuous. These animals usually make caches just under the surface of a substrate. For example, most jays and tree squirrels store individual nuts and acorns 1–5 cm deep in the forest floor, chipmunks bury 1–10 small seeds (e.g., pine seeds) in soil, and chickadees cache small collections of seeds, insects and spiders in the foliage or bark of trees. Most animals don't bother looking for the stored food, but pilfering by other members of their species and other food-storing animals can be an important source of cache loss.

Scattered caches are so well hidden and so numerous that they present a problem of cache recovery by the owners. Individual jays, chipmunks, and chickadees each prepare many thousands of caches each year. They gain an advantage over other animals during cache recovery by

Chipmunks are notorious for their caching behavior.
© Anthony F. Chiffolo.

retaining precise spatial memories of cache sites. The capacity and precision of the spatial memory of some animals is extraordinary. For example, Clark's nutcracker (*Nucifraga columbiana*), a relative of jays and crows, makes 10,000–50,000 caches in soil each autumn scattered over many square miles in high elevation conifer forests. They return to cache sites as much as a year later. Because they cannot smell seeds buried in soil, they must rely on their memories of each cache site. If their memory of the site is more than a few centimeters off of the actual cache location, they would likely fail to find their hidden food. Rodents also have impressive spatial memory, but it appears to be less precise than that of birds because they have a keen olfactory sense, which they use to help relocate cached seeds and to pilfer the caches of other animals.

Animals store food to consume during a future period of food scarcity. Despite this similarity in the adaptive reason for caching, the time between caching and cache recovery varies greatly, from hours to over a year. This variation is caused largely by the qualities of the food being stored, the climate where the animal lives, and the needs of the cacher. Those animals that store food for hours to days are called short-term hoarders, whereas those that store food from weeks to months are called long-term hoarders. Short-term food hoarders include hawks, owls, shrikes and mammalian predators that generally must eat a prey item within several days before it spoils. Northwestern crows (*Corvus caurinus*) forage in the intertidal zone of the western United States and Canada gathering clams and other invertebrates. Because the rising tide enforces a relatively long period when the crows cannot forage, they gather excess food during low tide and cache it above the high tide line. When the tide comes in, they feed from their food stores. In this unique situation, caching and cache recovery are dictated by the tidal cycle. In the winter at high latitudes, chickadees and other birds store food in the morning and then retrieve those items in the afternoon just before going to roost. This behavior ensures that sufficient food can be consumed quickly in the evening to sustain the bird for the long winter night.

Long-term hoarders often store seeds or nuts that are resistant to decay, animal prey under conditions where it cannot spoil, and other types of food resistant to spoilage. Tree squirrels, desert rodents, and chipmunks store seeds and nuts for several months. Foxes bury bird and turtle eggs in the beach sand during the breeding season and then retrieve these items later in the fall and winter when other eggs are no longer available. The short-tailed shrew (*Blarina brevicauda*) bites its prey and injects a venom, which makes the prey comatose. In this state, the prey neither escapes nor decomposes, extending its "shelf life" from days to months. Honeybees represent the ultimate in longer-term food storage. The honey and pollen stored in a hive can sometimes remain in storage for 2 or more years.

Some food-caching animals can have important ecological effects on their environment. Those animals that scatter hoard seeds in soil can act as seed dispersers for plants. A small proportion of cached seeds are never found by animals, and cache depths (1–5 cm) are usually within the range of depths suitable for seedling germination and emergence. The most effective seed-dispersing animals are corvids (jays and nutcrackers) and rodents (squirrels, chipmunks, and a variety of mice and rats). Many of the plants dispersed by these animals have adapted to this mode of dispersal by evolving fruits and seeds that attract seed-caching animals. Most notable in this regard are the nut-producing trees found throughout temperate and tropical regions (e.g., oaks, hickories, walnuts, chestnuts, hazelnuts, almonds, Brazil nuts, and numerous pines). A wide variety of animals (e.g., deer, bears, turkeys, peccaries) that do not store food have come to depend on these seeds and nuts as important foods during the autumn and winter. The coevolutionary relationships between nut-caching animals and the plants they disperse are prime examples of how behavior can have pervasive consequences with ramifications throughout the ecosystem.

All animals do not respond in the same manner to a plentiful supply of food as witnessed by four species of corvids (Clark's nutcracker, pinyon jay *Gymnorhinus cyanocephalus*, western scrub jay *Aphelocoma californica*, and Mexican jay *Aphelocoma ultramarine*) that inhabit the mountains of northern Arizona, and cache pine seeds in autumns when the cone crop is large. These four species inhabit different elevations of the mountains and these altitudes experience very different climatic conditions in fall, winter, and early spring. The highest regions, inhabited by nutcrackers, have cold temperatures, heavy snows, with cloudy, windy weather throughout the winter period. No food is produced during this time, much of the ground is covered with snow, and arthropods, small mammals, and other sources of food are inactive. Thus, inhabitants must either forage for food produced months earlier or rely on food cached earlier. Lower reaches of the mountains have less extreme climatic conditions with less snow, higher temperatures, and more sunny days. Some arthropods, and small mammals are active, and the substrate is partly exposed. Pinyon jays inhabit the higher regions of this intermediate elevational gradient and scrub and Mexican jays the lower reaches. Here conditions are mild, winter temperatures are relatively warm, animals are active year-round, and precipitation is in the form of rain. These four species form a specialization gradient or adaptive continuum based on their intensity to cache, their behavioral and morphological specializations for the harvest, transport, caching of seeds, their dependence on cached seeds to survive the winter and feed their young, and their spatial memory abilities to accurately retrieve their hidden caches. The species forming this adaptive gradient show morphological specializations for seed caching behavior. The nutcracker has a strong, sharp, heavy bill, the pinyon jay has an intermediate bill, and the scrub and Mexican jays have shorter, less stout, blunted shaped bills. These species also possess different types of structures for holding and transporting harvested seeds. The nutcracker has a sublingual pouch, an area under the tongue, that can expand to hold up to 90 pine seeds, and the pinyon jay has an expandable esophagus that can hold up to 40 seeds. These two species carry seeds many miles between harvest areas and caching areas. In contrast, the scrub and Mexican jays can only carry 3–6 seeds in mouth and bill. Also, they do not carry seeds as far as the two former species.

The species most dependent on cached seeds for winter survival and reproductive energy live at the highest elevations and cache the most seeds. For example, nutcrackers may cache up to 50,000 seeds and pinyon jays 26,000 seeds when pine cones are abundant. Scrub and Mexican jays may only store 6,000–8,000 seeds during a good crop year. For these latter species the seed caches may be depleted by the time the spring breeding season approaches. However, nutcrackers and pinyon jays use retrieved seeds for courtship and actually feed some of them to their nestlings (a trait rare among birds). The two species that cache the most seeds are also among the earliest breeding birds in north America. Possibly early breeding is linked to the quantity of seeds cached as they breed earlier in years when the cone crop is heavy and they have created many caches, and later in years when cones are sparse or nonexistent.

Extensive testing in a controlled laboratory setting has revealed that this specialization gradient is also reflected in the special memory abilities of the four species. Nutcrackers and pinyon jays are exceedingly accurate in laboratory tests where the experimenters can control where birds cache, when they can recover, and in which caches they are allowed to recover. Nutcrackers and pinyon jays also perform well in an analogue of the radial arm maze, so their spatial memory abilities goes beyond that of finding food they have hidden.

Food caching is, thus, an important behavioral attribute of some animals. It clearly has survival value in the north temperate and arctic zones during winter, permitting animals to inhabit regions during the season of food scarcity that they otherwise would be unable to occupy. In the tropics and higher latitudes during summer, caching permits animals to sequester resources and better maintain a positive energy balance. This permits animals to forgo foraging or find food more quickly under inclement weather, and it permits animals to reallocate time to other activities (e.g., courtship, territoriality) when those behaviors have potentially greater fitness payoffs. For many animals that store food, caching behavior can be directly linked to increased survival and reproductive success.

See also Cognition—*Cache Robbing*
Cognition—*Dogs Burying Bones: Unraveling a Rich*
Action Sequence

Further Resources

Balda, R. P. & Kamil, A. C. 1998. *The ecology and evolution of spatial memory in corvids of the southwestern USA: the perplexing Pinyon Jay*. In: *Animal Cognition in Nature* (Ed. by R. P. Balda, I. M. Pepperberg & A. C. Kamil), pp. 29–64. London, UK: Academic Press.
Vander Wall, S. B. 1990. *Food Hoarding in Animals*. Chicago: University of Chicago Press.

Russell P. Balda & Stephen B. Vander Wall

Cognition
Categorization Processes in Animals

Animals often adopt adapted behaviors in response to novel stimuli, because they look like familiar stimuli to which the animal knows how to respond. For example, a young monkey that saw adults give alarm calls and then run into the bushes after spotting an eagle would be able to be aware of other raptors (even if they are somehow different from the first one)

Harvard professor B.F. Skinner conducts a psychological experiment with pigeons in which they must match a colored light with a corresponding colored panel in order to receive food, 1950.
Bettmann / Corbis.

and fly toward his mother or into bushes. This behavior expresses an ability to categorize. *Categorization* can be defined as grouping into classes objects sharing one or more common features. This ability simplifies a complex environment, since its entities will be grouped into categories. A common treatment can be applied to the members of each category. Thus, categorization allows, by generalization processes, to predict non-evident properties from evident properties. It is therefore not surprising to find categorical abilities in various animal species, although most of the empirical evidence concerns birds (mostly pigeons) and primates.

Categorization behaviors imply very different levels of cognitive abilities, from basic abilities (like generalizing a response to a certain kind of green to another kind of green), present in many animal species, to more complex abilities (like categorization of relations between relations, found only in humans and apes). The study of categorization processes in animals informs us about their cognitive abilities and about the way they apprehend and organize their world.

Categorization is demonstrated if, after being trained with one set of stimuli, subjects are able to generalize a response to new stimuli according to the category those stimuli belong to. One of the first studies of categorization was carried out by Professors Herrnstein and Loveland in 1964 with pigeons. The birds were trained to peck a switch when they saw a picture including a human, and not to peck when a picture without a human was presented (this is called a go–no go procedure). They received a food reward when their answer was correct, and no food when they were wrong. After this training, the pigeons were able to answer correctly when new pictures were presented. The pictures could be very different—the human figures, too—so the researchers deduced that the criterion used by the pigeons to categorize the pictures into the "peck" or "not peck" classes was the presence or absence of a human and that the birds needed to have a concept of human to solve this task. Since then, many similar experiments were performed with pigeons with many different stimuli: natural stimuli (trees, water, pigeons, fishes, cats, tree leaves), manufactured objects (chairs, cars), and artificial stimuli (geometric shapes, letters or numbers, and even cubic vs. impressionist paintings). Similar abilities were shown in monkeys and apes: They were able to use categories like humans, birds, animals, different monkeys species, colors, alphabetical letters, geometric shapes and so on. Categorization abilities were also shown in other animals, for example in chicken, parrots, or sea lions. Even bees can form simple categories like flowers' colors.

Animals may use a feature analysis to categorize the pictures where all the objects in a category share some perceptual features that the subjects may use as a basis to discriminate between the members and the nonmembers of a category. Some animals may also use a prototype; that is, they compare the new stimuli to a stored representation of a typical member of a class.

However, in those experiments, one cannot be sure that the animals really use concepts to categorize the picture. Since they are categorizing on the basis of a perceptive similarity between the pictures of a given category, they may sometimes use some features of the pictures, without a real understanding of the task. For example, D'Amato and van Sant found in 1988 that monkeys that were apparently able to categorize pictures according to the presence or absence of human figures used, in fact, mostly the presence or absence of reddish patches!

A way of making clear that the subjects do use concepts may be to use functional categories for which there are no common perceptual features. For example, the subjects are trained to sort objects into food vs. nonfood classes, after making sure that there is no perceptual bias, that is, that food objects do not share perceptual features that nonfood objects do not have. Studies of food vs. nonfood categorization were carried out with pigeons, baboons, and chimpanzees. Transfer of this kind of categorization to new objects appears after a very limited training, presumably because it has adaptive importance for the animals: Grouping objects in food and nonfood categories has obvious ecological significance.

Thus, animals may be particularly good at categorizing objects or events that are significant to them. This is also the case with social stimuli like conspecifics. Many animals, including most of the primates, are social; important cognitive problems encountered in natural settings belong to the social area, so those animals may have a special aptitude for solving problems involving social stimuli. Categorization of social stimuli may also give information about the way animals consider their conspecifics and understand social events.

The literature indicates that primates can categorize different pictures of a same individual. The primatologist Verena Dasser showed in 1987 that macaques were able to associate pictures of a familiar conspecific's face with pictures of its body. Experiments about categories of relationships (like kinship or dominance relations) between individuals, that is social concepts, are much less numerous. The concept of mother–child affiliation in monkeys was also tested by Dasser. In her experiment, two female longtail macaques that were shown pictures of their groupmates were able to associate the picture of a mother with the picture of her offspring. Professors Bovet and Washburn (2003) trained three adult male rhesus macaques to categorize pairs of unknown conspecifics presented on a video film according to the dominance status of those videotaped animals. The monkeys were trained to choose the dominant monkey for one category of films (e.g., films showing one monkey chasing another), and then new films were presented (e.g., monkeys fighting). Two of the subjects were able to generalize categorical judgments of dominance to films involving new behaviors. These findings show that monkeys can use abstract social concepts and be aware of the social relationships within their group.

Apes, monkeys and even birds can also use relational concepts like "same" or "different" to categorize relationships between objects. They may for example move a lever to the right when two different stimuli are presented, and to the left when they see two identical objects. But the most abstract kind of categorization is the categorization of relations between relations. In this kind of experiment, carried out by Thompson and Oden, subjects have to match, for example, the pair of letters "AA" to the pair of letters "BB" (because in the two pairs the letters are the same), and the pair of letters "CD" to the pair of letters "EF" (because in the two pairs the letters are different). Among animals, only apes are able to solve such an abstract categorization task.

See also Cognition—*Concept Formation*
Cognition—*Equivalence Relations*
Cognition—*Grey Parrot Cognition and Communication*
Cognition—*Social Cognition in Primates and Other Animals*

Further Resources

Bekoff, M., Allen, C. & Burhgardt, G. M. 2002. *The Cognitive Animal: Empirical and Theoretical Perspectives on Animal Cognition.* Cambridge, MA: MIT Press.

Bovet, D. & Washburn, D. A. 2003. *Rhesus macaques* (Macaca mulatta) *categorize unknown conspecifics according to their dominance relations.* Journal of Comparative Psychology, 117, 400–405.

Cook, R. G. 2001. *Avian visual cognition.* Available: www.pigeon.psy.tufts.edu/avc/

D'Amato, M. R. & Van Sant, P. 1988. *The person concept in monkeys* (Cebus apella). Journal of Experimental Psychology, Animal Behavior Processes, 14, 43–56.

Dasser, V. 1987. *Slides of group members as representations of the real animals* (Macaca Fascicularis). Ethology, 76, 65–73.

Herrnstein, R. J. & Loveland, D. H. 1964. *Complex visual concept in the pigeon.* Science, 146, 549–551.

Thompson, R. K. R. & Oden, D. L. 2000. *Categorical perception and conceptual judgments by nonhuman primates: the paleological monkey and the analogical ape.* Cognitive Science, 24, 363–396.

Tomasello, M. & Call, J. 1997. *Primate Cognition.* Oxford: Oxford University Press.

Zentall, T. R. & Smeets, P. M. 1996. *Stimulus Class Formation in Humans and Animals.* Amsterdam: North-Holland.

Dalila Bovet

■ Cognition
Cognitive Ethology: The Comparative Study of Animal Minds

Cognitive ethology is the comparative, evolutionary, and ecological study of animal minds including thought processes, beliefs, rationality, information processing, and consciousness. Its origins lie in the writings of Charles Darwin and some of his contemporaries and disciples who were interested in the mental continuity between humans and other animals as part of their general argument for evolution from common ancestors. Early Darwinians were interested in understanding variation among individuals and species in terms of natural selection associated with differences in natural history and life history strategies. They believed it important to describe the perspectives of the animals themselves, and they also employed anecdote and anthropomorphism to inform and to motivate rigorous study in conditions as close as possible to the natural environment where selection occurred. Carrying on this tradition, cognitive ethologists are frequently more concerned with the diversity of solutions that living organisms have found for common problems rather than in seeking general mechanisms of learning and memory as revealed by performance in laboratory tasks. Cognitive ethologists emphasize broad taxonomic comparisons among many different species. The emphasis that cognitive ethologists place on individual and species variation, mental complexity, and adaptation in natural environments serves to distinguish cognitive ethology from comparative psychology.

Cognitive ethology is an integrative and comparative approach to the study of animal behavior that is most closely associated with the classical ethologists, Konrad Lorenz and Niko Tinbergen, who shared the 1973 Nobel Prize with Karl von Frisch (discoverer of bee language). In his early work Tinbergen identified four overlapping areas with which ethological investigations should be concerned—namely, evolution (phylogeny), adaptation (function), causation, and development (ontogeny). Cognitive ethology is a significant extension of classical ethology because it explicitly allows for hypotheses about the internal (mental) states of animals, but Tinbergen's framework also proves just as useful for those interested in animal cognition as for those who focus on directly observable patterns of behavior. The methods for answering questions in each of these areas vary, but all begin with careful observation and description of the behavior patterns that are performed by the animals under study. The information afforded by these initial observations allows a researcher to exploit the animal's normal behavioral repertoire to answer questions about the evolution, function, causation, and development of the behavior patterns that are performed in various contexts.

Donald R. Griffin and Modern Cognitive Ethology

The modern era of cognitive ethology and its concentration on the evolution and evolutionary continuity of animal cognition is usually thought to have begun with the appearance of Donald R. Griffin's (1976/1981) book *The Question of Animal Awareness: Evolutionary Continuity of Mental Experience*. Griffin's major concern was to learn more about animal consciousness. He wanted to come to terms with the difficult question of "what is it like to be a particular animal?" But Griffin's concern with animal consciousness represents only one of many important and interesting aspects of animal cognition that have been studied by cognitive ethologists. Because of its broad agenda and wide-ranging goals, many view cognitive ethology as a genuine contributor to cognitive science in general. For those who are anthropocentrically minded, it should be noted that studies of animal cognition have been used to study human autism.

Methods of Study

Ethologists interested in animal minds favor research in conditions that are as close as possible to the natural environments in which natural selection occurred or is occurring. Although research on learning and memory on captive animals can inform the comparative study of animal cognition, cognitive ethologists maintain that field studies are necessary for a complete understanding of the cognitive abilities of animals. Such studies' research may include staged social encounters, playback of recorded vocalizations, the presentation of stimuli in different modalities, observation of predator–prey interactions, observation of foraging behavior, neurobiological techniques, and studies of social and other types of learning. Computer analyses also are useful for those who want to learn what kind of information must be represented in computational models.

There are no large differences between methods used to study animal cognition and those used to study other aspects of animal behavior. Differences lie not so much in what is done and how it is done, but rather how data are explained. The main distinction between cognitive ethology and classical ethology lies not in the types of data collected, but in the understanding of the conceptual resources that are appropriate for explaining those data. Cognitive hypotheses have also resulted in studies that might not otherwise have been considered.

Perhaps one area that will contribute much to the study of animal minds in the future is neurobiology and behavior. Those researchers interested in the cellular or neural bases of behavior and animal cognition and consciousness may use techniques such as positron emission tomography (PET) that are also employed in other endeavors. In general, studies using brain imaging have provided extremely valuable data for humans engaged in various sorts of activities, and, for example, for studies of visual processing in monkeys. But the use of brain imaging techniques is still in its infancy and has not yet been extended to activities going beyond relatively artificial laboratory tasks.

Ethological studies usually start with the observation, description, and categorization of behavior patterns that animals perform. The result of this process is the development of an *ethogram*, or behavioral catalog, of these actions. Ethograms present information about an action's form or morphology—what the animals does and what it looks like—and its code name. For example, in certain situations when a dog crouches on her forelimbs and sticks her hind end in the air, the dog is said to be performing a "play bow," and this action may be designated by the code name "pb."

Descriptions can be based on visual information (what an action looks like), auditory characteristics (sonograms, which are pictures of sounds, bird song), or chemical constituents (output of analyses of glandular deposits, urine, or feces, for example, pheromes). It is essential that great care be given to the development of an ethogram, for it is an inventory that others should be able to replicate without error. Permanent records of observations allow others to cross-check their observations and descriptions against original records. The number of actions and the breadth of categories that are identified in a behavioral study depends on the questions at hand, but generally it is better to pay attention to details and to split rather than to lump actions in the initial stages of a study, and to lump actions when questions of interest have been carefully laid out and finer categories are not needed.

In studies of behavior, it is important to know as much as possible about the sensory world of the animals being studied. Experiments should not be designed that ask animals to do things that they cannot do because they are insensitive to the experimental stimuli or unmotivated by the stimuli. Asking a dog to respond to ultraviolet light is not species-fair, for dogs cannot see UV light. The relationships between normal ecological conditions and differences between the capabilities of animals to acquire, process, and respond to information is the domain of a growing field called *sensory ecology*. A good ethological investigation considers the question, "what is it like to be the animal under study," and pays attention to the senses that the animals use singly or in combination with one another. It is highly unlikely that individuals of any other species sense the world the same way we do, and it is unlikely that even members of the same species sense the world identically all of the time. It is also important to remain alert to the possibility of individual variation.

Some Criticisms of Cognitive Ethology

A balanced view of cognitive ethology requires consideration of critics' points of view. Criticisms of cognitive ethology center on:

1. the notion that animals do not have minds;
2. the idea that (many, most, all) animals are not conscious, or that so little of their behavior is conscious (no matter how broadly defined) that it is a waste of time to

study animal consciousness (cognitive ethology is really a much broader discipline than this suggests, see below);

3. the inaccessibility to rigorous study of animal mental states (they are private) and whatever (if anything) might be contained in them;

4. the assumption that animals do not have any beliefs because the content of their beliefs are not similar to the content of human beliefs;

5. the rigor with which data are collected;

6. the lack of large empirical data bases;

7. the nature of what are called "soft" or weak explanations that rely heavily on theoretical constructs (e.g., minds, mental states) that are offered for the behavioral phenotype under study (they are too anthropomorphic, too folk psychological, or too "as if-fy"— animals act "as if" they have beliefs or desires or other thoughts about something); and

8. the heavy reliance on behavior for generating cognitive explanations.

While most criticism comes from those who ignore the successes of cognitive ethology, those who dismiss it in principle because of strong and radical behavioristic leanings, or those who do not understand basic philosophical principles that inform cognitive ethology, it should be pointed out that more mechanistic approaches to the study of animal cognition are not without their own faults. For example, comparative psychologists often disregard how relevant a study is to the natural existence of the animals under study and pay too much attention to the logical structure of the experiments being performed (the "internal logic" of the experiments) without much regard for the relevance of the experiments to the natural existence of the animals under study (the "external logic" of the experiments).

Case Studies

Case studies illustrating the application of a broadly comparative cognitive ethological approach can be found in recent field research of antipredator behavior in birds and field and laboratory research on social play behavior in various canids (domestic dogs, wolves, and coyotes). Many other examples are found in the references below.

The psychologist Carolyn Ristau studied injury feigning in piping plovers (the broken-wing display) and wanted to know if she could learn more about deceptive injury feigning if she viewed the broken-wing display as an intentional or purposeful behavior ("the plover wants to lead the intruder away from her nest or young") rather than a hard-wired reflexive response to the presence of a particular stimulus, a potentially intruding predator. She studied the direction in which birds moved during the broken-wing display, how they monitored the location of the predator, and the flexibility of the response. Ristau found that birds usually performed the display in the direction that would lead an intruder who was following them further away from the threatened nest or young, and also that birds monitored the intruder's approach and modified their behavior in responses to variations in the intruder's movements. These and other data lead Ristau to conclude that the plover's broken-wing display lent itself to an intentional explanation, that plovers purposely lead intruders away from their nests or young and modified their behavior in order to do so.

In another study of antipredator behavior in birds, Marc Bekoff discovered that western evening grosbeaks modified their vigilance or scanning behavior depending on the way in which individuals were positioned with respect to one another. Grosbeaks and other birds often trade off scanning for potential predators and feeding—essentially (and over-simplified), some birds scan while others feed, and some birds feed when others scan. Thus, it is hypothesized that individuals want to know what others are doing and learn about others' behavior by trying to watch them. This study of grosbeaks showed that when a flock contained four or more birds, there were large changes in scanning and other patterns of behavior that seemed to be related to ways in which grosbeaks attempted to gather information about other flock members. When birds were arranged in a circular array so that they could see one another easily, compared to when they were arranged in a line that made visual monitoring of flock members more difficult, birds who had difficulty seeing one another were more vigilant, changed their head and body positions more often, reacted to changes in group size more slowly, showed less coordination in head movements, and showed more variability in all measures.

The differences in behavior between birds organized in circular arrays when compared to birds organized in linear arrays were best explained by accounting for individuals' attempts to learn, via visual monitoring, about what other flock members were doing. This may say something about if and how birds attempt to represent their flock, or at least certain other individuals, to themselves. It may be that individuals form beliefs about what others are most likely doing and predicate their own behavior on these beliefs. Bekoff argued that cognitive explanations were simpler and less cumbersome than noncognitive rule-of-thumb explanations (e.g., "scan this way if there are this number of birds in this geometric array" or "scan that way if there are that number of birds in that geometric array."). Noncognitive rules-of-thumb did not seem to account for the flexibility in animals' behavior as well or as simply as explanations that appealed to cognitive capacities of the animals under study.

There are other examples that could have been chosen, but these two make the case that chauvinism on either side of the debate as to how to study animal behavior and how to explain animal behavior is unwarranted; a pluralistic approach should result in the best understanding of the nonhumans with whom we share the planet. Sometimes nonhumans (and humans) behave as stimulus–response machines and at other times nonhumans (and humans) behave in ways that are best explained using a rich cognitive vocabulary. Methodological pluralism is needed: species-fair methods need to be tailored to the questions and the animals under consideration, and competing hypotheses and explanations must always be considered.

Those interested in animal cognition should resist temptations to be "speciesistic cognitivists" who make sweeping claims about the cognitive abilities (or lack thereof) for all members of a given species. A concentration on individuals and not on species should form an important part of the agenda for future research in cognitive ethology. There is a lot of individual variation in behavior within species, and sweeping generalizations about what an individual ought to do because she is classified as a member of a given species must be taken with great caution. Furthermore, people often fail to recognize that in many instances sweeping generalizations about the cognitive skills (or lack thereof) of species and not of individuals are based on small data sets from a limited number of individuals representing few species, individuals who may have been exposed to a narrow array of behavioral challenges. The importance of studying animals under field conditions cannot be emphasized too strongly. Field research that includes careful and well-thought-out observation, description, and ethically sound experimentation that does not result in mistreatment of the animals is extremely difficult to duplicate in captivity. While it may be easier to study animals in captivity, they must be provided with the complexity of social and other stimuli to which they are exposed in the field. In some cases this might not be possible.

Cognitive ethology can raise new questions that may be approached from various levels of analysis. For example, detailed descriptive information about subtle behavior patterns and neuroethological data may be important for informing further studies in animal cognition, and may be useful for explaining data that are already available. Such analyses will not make cognitive ethological investigations superfluous, because behavioral evidence is primary over anatomical or physiological data in assessments of cognitive abilities.

To summarize, those positions that should figure largely in cognitive ethological studies include:

1. remaining open to the possibility of surprises in animal cognitive abilities;
2. concentrating on comparative, evolutionary, and ecological questions and sampling many different species including domesticated animals—going beyond primates and avoiding talk of "lower" and "higher" animals, or at least laying out explicit criteria for using these slippery and value-laden terms;
3. naturalizing methods of study by taking the animals' points of view (talking to them in their own languages) and studying them in conditions that are as close to the conditions in which they typically live; often animals do not do what we expect them to do (sometimes prey will approach predators) and knowledge of their natural behavior is important in the development of testable and realistic models of behavior;
4. trying to understand how cognitive skills used by captive animals may function in more natural settings;
5. studying individual differences;
6. using all types of data, ranging from anecdotes to large data sets; and
7. using different types of explanations for the data under scrutiny.

Cognitive ethology is a rapidly growing and exciting cross-disciplinary field of science that is attracting much attention from researchers in numerous and diverse disciplines. Many other researchers look to the results from such cognitive ethological studies to inform their own research.

See also Anthropomorphism
        Behaviorism
        Cognition—*Theory of Mind*
        Cognition—*Social Cognition in Primates and Other*
          Animals
        Comparative Psychology

Further Resources

Allen, C., & Bekoff, M. 1997. *Species of Mind: The Philosophy and Biology of Cognitive Ethology.* Cambridge, MA: MIT Press.

Bekoff, M. 1998. *Cognitive ethology: The comparative study of animal minds.* In: *Blackwell Companion to Cognitive Science.* (Ed. by W. Bechtel & G. Graham), pp. 371–379, Oxford, England: Blackwell Publishers. (from which some of this text has been excerpted)

Bekoff, M. & Allen, C. 1997. *Cognitive ethology: Slayers, skeptics, and proponents*, In: *Anthropomorphism, Anecdote, and Animals: The Emperor's New Clothes?* (Ed. by R. W. Mitchell, N. Thompson & L. Miles) Albany, NY: SUNY Press.

Bekoff, M., Allen, C., & Burghardt, G. M. (Eds.) 2002. *The Cognitive Animal: Empirical and Theoretical Perspectives on Animal Cognition.* Cambridge, MA: MIT Press.

Bekoff, M. & Jamieson, D. (Eds.) 1996. *Readings in Animal Cognition.* Cambridge, MA: MIT Press.

Bshary, R., Wickler, W. & Fricke, H. 2002. *Fish cognition: A primate's eye view.* Animal Cognition, 5, 1–13.

Cheney, D. L. & Seyfarth, R. M. 1990. *How Monkeys See the World: Inside the Mind of Another Species.* Chicago: University of Chicago Press.

Griffin, D. R. 2001. *Animal Minds.* Chicago: University of Chicago Press.

Hauser, M. 2000. *Wild Minds.* NY: Henry Holt.

Marc Bekoff & Colin Allen

■ Cognition
Concept Formation

One prominent educational technique involves showing a series of examples of a particular idea, and from these deriving or pointing out the important commonalties and differences that illustrate a more general principle or truth. As humans, we are experts at detecting the patterns in the world's particulars, and at using the resulting categories, concepts, and rules to guide our behavior. Among the many important benefits of this conceptual ability is that it allows us to engage in behaviors unbounded by our past experience with specific stimuli. We readily apply old principles to new situations, permitting us to generate flexible and adaptive solutions to the novel problems we encounter. A key question in the scientific study of cognition in other types of animals has been whether they similarly classify the world into generalized categories and concepts based on their experience with specific events. If they do, a second important question has focused on analyzing the mechanisms underlying this conceptual ability.

Because we only can indirectly understand what animals are thinking by observing their behavior, making inferences about their internal cognitive states has required careful and controlled experimentation and a precise understanding of how different kinds of stimuli come to control their behavior. *Stimulus control* is said to be present in an animal when it behaves differently in the presence of different stimuli. For instance, when a rat has learned to press a lever for food whenever a tone is played, but not to do so whenever there is silence, the rat's behavior is described as being under auditory stimulus control. Historically, two contrasting views have been offered to describe the nature of this discrimination learning.

The first view centers on the idea that animals memorize the specific connections or *associations* between individual stimuli, responses, and consequences. These types of associative theories have provided a powerful and profitable account of much of the learned behavior that has been studied across widely different groups of animals, ranging from simple invertebrates to complex primates. The widespread distribution of this type of learning across various animals has suggested that associative learning is a general cognitive process that evolved early in the history of life. A second important view has suggested that some animals, including humans, can also learn to use more open-ended and generalized representations, such as categories and concepts, when reacting to the world's events. Thus, these types of animals go beyond simply memorizing specific associations between stimuli and consequences, to abstracting common patterns and rules that allow them to process familiar situations quickly and to react to new stimuli and situations. For example, the ability to represent the common features (e.g., leaves, trunks, roots, and use of chlorophyll) that connect together the perceptually different exemplars of the concept of tree (e.g., oak, aspen, beech, spruce) not only allows one to recognize and categorize further examples of these kinds of trees, but also to recognize new kinds of trees (poplar, birch) in the forest. Clearly the capacity to form concepts is an important and fundamental building block to many advanced forms of thinking.

These two views of stimulus control can be easily contrasted by looking at how your younger brothers and sisters learned the concept of addition. Young children of two or three years of age can often learn from their parents to respond with the correct answers to simple inquiries such as "what is $1+1$ or $2+2$?" A little more probing and testing with some other numbers (e.g., $3+3$; $4+2$) quickly reveals, however, that these children do not understand the concept of addition, They have simply memorized rote answers to a few specific questions. It is only later as kindergarteners, when they become capable of adding any two numbers (within their general number sense), that it is said that they have acquired the general concept of addition. When completely learned somewhat later, this general mathematical principle permits one to combine any two novel numbers that can be imagined. Using the same tactics of probing and testing the limits of any learned discrimination, it is possible to examine whether animals might also be able to learn different kinds of concepts as well.

Conceptual behavior similar to that just described was first recognized in animals using tests involving perceptual categories (Herrnstein & Loveland 1964). Perceptual categories typically involve learning to respond to pictures of common objects, such as cats, chairs, cars and so on. In Herrnstein and Loveland's study, they first trained pigeons to peck for food in an operant testing chamber. In this chamber different kinds of pictures could be projected on a display screen that the pigeons could look at and peck. They trained their pigeons to peck at the display anytime a picture was shown that contained a person or a group of people within it. When they pecked at such pictures, mixed grain was delivered to them from a food hopper. These "people" pictures had various numbers of individuals, appearing in different orientations, sizes, poses, and clothes. The goal of this variety was to make it so that the only common feature joining these pictures together was the open-ended concept of "people." For the purposes of comparison, there was also a second group of pictures that looked similar, except that no human beings were present in them (the figure shows two examples of category).

Representative examples of people (left) and nonpeople (right) pictures from Herrnstein and Loveland's (1964) study. Courtesy of Robert Cook.

Whenever one of these "nonpeople" pictures appeared on the screen, pecking at it never produced food. The pigeons were trained with 40 pictures of each type and soon came to peck quickly and rapidly at the "people" slides and not to peck at the "nonpeople" slides.

This difference in peck rate between the two groups of pictures indicates they were under stimulus control, but the important next step was to determine if this stimulus control was associative or stimulus-specific in nature or was more open-ended and conceptual-like. To examine this question, the pigeons were probed or *transfer tested* with novel pictures of the people and nonpeople categories. Because they had no prior experience with these specific images, they had to rely on their past learning to judge what to do with them. In tests with these novel stimuli the pigeons continued to respond appropriately—pecking at pictures that contained people and not pecking at the pictures that did not. This transfer or generalization to new items within each category forms the operational essence of what we mean by concept formation in animals.

To argue convincingly that an animal has formed a conceptual representation, it has been suggested that five operational criteria need to be demonstrated (Cook 2002).

1. There should be successful discrimination of the categories used to train the discrimination.

2. There should be transfer to novel exemplars of each category (as measured before any differential experience or reinforcement with these new stimuli).

3. The individual items making up each category should be also discriminable from one another.

4. There should be evidence that the transfer items *could* have been discriminated from the training items had the animal been trained to do so.

5. Alternative sources of stimulus control irrelevant to the concept being examined should be ruled out.

Criteria one, two and five are judged the most important of these five. The pigeons in Herrnstein and Loveland's study clearly meet criteria one (discrimination learning during training) and two (novel transfer), but what about number five? Can we be sure that these animals actually formed a concept of "people" in this study, or could something simpler have accounted for their results?

One alternative account might have been that the animals simply memorized the training stimuli rather than learning a generalized concept. To distinguish between such memorization strategy and true concept formation, researchers have studied how animals learn to discriminate "pseudocategories." In these experiments two groups of animals are tested. The "categorization" group receives training with pictures that are divided along their natural categories. Wasserman, Kiedinger & Bhatt (1988), for instance, trained pigeons to discriminate among four different categories (cars, flowers, cats, people). On each trial, the animal would see a picture from one of these categories and then indicate which category it belonged to by pecking at one of four choice locations around the display (each assigned to a different category). In the "pseudocategorization" comparison group, the animals received training with the very same stimuli, but the correct answer for each picture was randomly assigned to one of the four choice locations. Thus, whereas the first group can use categorical identity to learn the discrimination, the second group could only learn to memorize the correct answer to each specific image. It has been consistently found that the categorization group learns faster than the pseudocategorization group. This difference indicates that the animals are doing something more than just memorizing pictures. They are connecting together the common features shared by examples of the same category. Given this, one could (and should) ask whether the animals might be relying on some simpler nonconceptual features to unify the pictures. This sometimes appears to be the case. For example, D'Amato & Van Sant (1988) found that their monkeys might have learned a people/nonpeople discrimination by looking for the simple presence or absence of orange/reddish patches in the pictures, a strategy that led them to misclassify pictures containing objects such as a watermelon or flamingo.

Nevertheless, since the 1960s, abundant evidence from numerous highly controlled studies has strongly suggested that both birds and primates can categorize a wide variety of different objects and concepts based on their common properties and overall similarity (e.g., cats, cars, flowers, people, birds, mammals, people, fish, trees, Picasso and Matisse paintings, female and male human faces, letters, geographic locations). One important further line of research not discussed here has involved using controlled artificial categories that have tried to determine exactly how the animals learn these types of discriminations. (See Huber 2001 for details about the three mechanisms that have been proposed). In general, however, it looks as if both birds and primates can form open-ended perceptual categories that are functionally similar to what humans perceive in the same pictures.

Besides the ability to classify perceptual objects, human behavior is often rule based as well. Humans easily answer questions based on simplifying rules or principles that we have abstracted from the *relations* between objects, and not just the objects themselves. Early on, young children are asked to say whether two things are the same, or specify how two things are different, or whether different things go together (e.g., a glass, plate, and fork). These kinds of higher-order *relational* concepts are important to language, mathematics, analogical reasoning, social relations, and even the fine arts. Recently, increasingly more attention has been devoted to asking if animals can similarly conceptualize the relations among stimuli. The relations that have been tested include: asking whether two things are the same or not (same versus different), whether two dissimilar looking things are functionally equal or related to one another (acquired equivalence), whether two things are identical (matching), whether two or more things are in the correct order (serial pattern learning, use of syntactic rules, transitive inference), and the degree of variability present among a set of items (entropy).

One pivotal relational concept examined in animals has revolved around the twin concepts of same versus different. In a same/different task, the subject is asked to respond "same" when two or more stimuli are identical and "different" when they are not. Cook, Katz & Cavoto (1997), for example, trained pigeons with four general types of highly variable stimulus displays (see the texture, feature, geometric and object examples in the figure; further examples and discussion are available at *http://www.pigeon.psy.tufts.edu/jep/sdclass/sdclass.htm*). Within each of these types, the elements making up the entire stimuli could either be all identical or have some kind of global difference present among them. For instance, the elements of the geometric and texture displays could differ among any of eight different colors or eight different shapes that were randomly selected by the computer (e.g., a blue triangle embedded among a set of red triangles). Likewise, the object displays were created from many pictures of different flowers, birds, fish, and humans, and combining any two of these objects within an array could create a different display. As a result, within and across each of the four types of displays, very large numbers of same and different stimuli could be created (5,056 different and 585 same displays). As before with the picture categorization studies, the considerable variety of these displays was intended to get the pigeons to attend to the only consistent relation present in the displays—the same/different relations between elements.

Cook, Katz & Cavoto found that pigeons could readily learn to discriminate the identity and contrasts between the elements of these displays by making one response (e.g., going

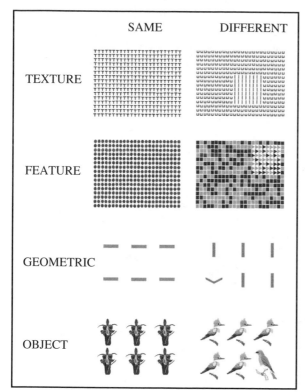

	SAME	DIFFERENT
TEXTURE		
FEATURE		
GEOMETRIC		
OBJECT		

Representative examples of same and different displays from Cook, Katz and Cavoto's 1997 study. Only shape differences are shown in this example.

left), following any type of same display (e.g., going to a left choice), and another response following any type of different display (e.g., going right). Most importantly the birds transferred to novel examples of each display class, suggesting the formation of a generalized relational concept. Further, it was found in subsequent experiments that the majority of these pigeons could also transfer this behavior to a fifth and novel stimulus class (color and gray-scale photographs) that they had never experienced before. This latter transfer is important in showing the relative abstractness of their discrimination, since presumably the more abstract the conceptual representation that has been formed, the greater the range of novel conditions over which it can be extended. Although slightly more difficult to demonstrate in animals than the formation of perceptual categories, increasingly the evidence has suggested that primates and several types of birds can indeed form visually-based relational concepts.

While it appears that primates and birds as classes (at least among the few species studied in detail) can both form both perceptual and relational concepts, Thompson & Oden (2000) have recently suggested that there may be limitations in the capacity of these animals to learn certain types of advanced concepts, in particular those that require judgments about "relations about relations." In this type of same/different judgment, the animals only have to judge whether two things are identical or not (AA versus AB or BA), but further judge whether two sets of stimuli share the same relations or not. If two pairs of stimuli have the same relations within them (both pairs are the same or both pairs are different) the animal would have to say they are the same. On the other hand, if the two pairs of stimuli have different relations within them (one pair is same and the other pair is different) then a "different" response is required. Thompson has argued that only apes can perform these kinds of nested "relations about relations" discriminations, while the conceptual mechanisms underlying the performance of monkeys and birds are not capable of such discriminations. The final word is still out on this general claim, but it nicely shows the logic of how different types and levels of conceptual tasks might be used to illuminate the intellectual differences among animals.

This brief review of conceptual behavior in animals suggests that the basic capacity to recognize and classify the world into categories does exist in birds and primates. Both of these classes of animals recognize the common perceptual similarity of many basic objects and some of the important types of relations that can exist between them. It suggests that this fundamental cognitive capacity is not tied to our use of language, but is more fundamental and evolutionarily older. In general, this type of psychological research comparing different animals helps us learn more about how animals think and process information, and how the underlying cognitive mechanisms are distributed among the members of the animal kingdom. It not only identifies those general information processing principles and mechanisms that are shared by many species, but also isolates those that are unique or specific to individual species or group. This kind of information is critical to our eventually understanding the evolution and function of human and nonhuman animal cognition.

See also Cognition—*Categorization Processes in Animals*
Cognition—*Equivalence Relations*
Cognition—*Grey Parrot Cognition and Communication*
Cognition—*Social Cognition in Primates and Other Animals*

Further Resources

Cook, R. G. 2002. *Same/different learning in pigeons*. In: *The Cognitive Animal*. (Ed. by M. Bekoff, C. Allen, & G. Burghardt), pp. 229–238. Cambridge, MA: MIT Press.

Cook, R. G., Katz, J. S., & Cavoto, B. R. 1997. *Pigeon same-different concept learning with multiple stimulus classes*. Journal of Experimental Psychology: Animal Behavior Processes, 23, 417–433.

D'Amato, M. R. & Van Sant, P. 1988. *The person concept in monkeys* (Cebus apella). Journal of Experimental Psychology: Animal Behavior Processes, 14, 43–55.

Herrnstein, R. J., & Loveland, D. H. 1964. *Complex visual concept in the pigeon*. Science, 146, 549–551.

Huber, L. 2001. *Visual categorization in pigeons*. In: *Avian Visual Cognition* (Ed. by R. G. Cook [On-line]. Available: www.pigeon.psy.tufts.edu/avc/huber/

Thompson, R. K. R. & Oden, D.L. 2000. *Categorical perception and conceptual judgments by nonhuman primates: The paleological monkey and the analogical ape*, Cognitive Science, 24, 363–396

Wasserman, E. A., Kiedinger, R. E., & Bhatt, R. S. 1988. *Conceptual behavior in pigeons: Categories, subcategories, and pseudocategories*. Journal of Experimental Psychology: Animal Behavior Processes, 14, 235–246.

Robert G. Cook

■ Cognition
Deception

Animals use signals to communicate with other members of their species. Signals generally transmit reliable information, such as the sender's fighting ability or its quality as a parent. Communication is usually viewed as a mutually beneficial process because both the sender and receiver benefit from the exchange of honest information. In some contexts, however, the sender of a signal could benefit by giving false information to a receiver, thereby making the receiver behave in an inappropriate manner. Conveying false information is called *deceptive communication*.

Signallers can gain great benefits by sending false information to receivers. In long-tailed dance flies, for example, females send dishonest signals about their reproductive state. Dance fly females cannot catch food for themselves, so they gather together in specific areas and wait for males to approach them. Males bring with them a food item that they give to a female in

A Maculinea alcon caterpillar is fed liquids by a Myrmica ant. A waxy coat of hydrocarbons on the caterpillar's body fools the ants into thinking that the caterpillar is an ant larva. Instead, the caterpillar leisurely feeds upon the ant's brood before hatching into a butterfly.

Darlyne A. Murawski / National Geographic Image Collection.

exchange for a mating. Of course, males do not want to waste time and energy by feeding females that are not ready to mate. Ideally, males should only feed females whose eggs have already developed and are ready for fertilization.

Females house their eggs in their abdomens, so when eggs are ripe the female has a hugely distended abdomen. Males will only feed, and subsequently mate with, females that have distended abdomens. However, females that do not have fertilizable eggs also want to be fed by males. These females therefore swallow air in order to inflate their abdomens. Males cannot tell the difference between females with distended abdomens due to the presence of ripe eggs or due to the presence of swallowed air. So, by swallowing air, females create the illusion that they have ripe eggs and therefore induce males to feed them.

Deceptive communication does not always involve the sending of false information. Some animals are deceptive by *withholding* information that they are expected to provide. An example of this occurs in the penduline tit. In some bird species, it takes only one parent to raise a brood of chicks. This means that one parent can leave, find another mate, and produce another clutch. Both parents want to increase the number of offspring they produce, so they both want to be the first to leave, and they both want their partner to stay and raise the chicks. In most species the male leaves first. This is because he can fly off as soon as he sees that the first egg has been laid (meaning that the female is committed to producing the egg clutch). The female cannot leave until she has laid her last egg. The appearance of eggs is therefore a cue for males to desert. In the penduline tit, however, females have found a way to prevent male desertion. As they lay their first egg, they tuck it under the soft nest lining so that the male cannot see it. They continue to hide each egg as it is laid. Once the entire clutch is laid, the female uncovers the eggs and immediately flies away to look for a new partner. The deceived male is left behind to raise the chicks.

Finding examples of deceptive communication is difficult because deceptive signals are designed to be undetected. This may mean that deception is far more common than we think. For example, it was only through very close observation that Australian zoologists found deception in white-winged chough. This bird lives in areas of very low food availability, and a breeding pair is unable to collect enough food to feed a brood of chicks. They therefore need help from other individuals, usually their older offspring. Birds from previous clutches stay at the nest and help their parents to raise the next clutch by gathering food items and feeding them to the chicks. These "helpers" would not benefit from leaving the nest, finding a mate and raising their own chicks because they would not be able to find enough food to feed their own chicks. They do better to stay with their parents and help raise their younger siblings. Eventually their parents will die, and one of the helpers will then be able to take over the nest and, with the assistance of the helpers who are already there, be able to breed itself.

The individual that gets to take over the nest is usually one that has been a good helper beforehand. It is therefore important for helpers to be seen to be helping. Indeed, helpers who fail to feed chicks are punished. Other group members loudly scold them. What normally happens when a helper feeds chicks is that it flies off and finds a food item, returns to the nest and, in front of the parents and other helpers, feeds the food to the chicks. Some helpers, however, perform "false feedings." Instead of letting go of the food items when they are in the chicks' mouths they pull them back out of their mouths and swallow the items themselves. The parents and other helpers think that the chick has been fed but, in fact, the helper has eaten the food itself. In this way, the false feeder gets the benefits of being seen to help, but does not have to pay the cost of giving away a food item it has collected.

If an individual can benefit so much from transmitting dishonest information, why don't all individuals do it? Deception usually occurs at low frequencies within a population, at most 20–30% of signals are deceptive. If most signals were false, receivers would not benefit by responding to them and they would stop paying attention to that signal. If deceptive signals are relatively rare, however, the receivers will still benefit by responding because they would usually get honest, reliable information. Even though they would occasionally be deceived, receivers would benefit, on average, from responding to the signal.

In some fiddler crabs, approximately 25% of males produce deceptive signals. Male fiddler crabs have one hugely enlarged claw that they use when fighting other males for territories, or when waving to attract females to mate. Sometimes males lose their large claws (either they purposefully drop them when attacked by birds, or it is twisted off when fighting another male). When a male loses his large claw, he can regrow a new one. In most species, the new claw is exactly the same as the old claw. In a few species, however, the new claw is strikingly different. It is the same length as the original claw, but it is much lighter because it has a much thinner exoskeleton; it is much weaker because it has much smaller muscles; and it is a poor weapon because there are no teeth on the inner surface of the "fingers" to interlock with an opponent's claw. Males with regenerated claws tend to lose fights with original-clawed males. However, they do not fight. Instead they bluff that they are strong males by acting aggressively and threatening their opponents. Usually this bluff works and their opponents back off. Furthermore, because the regenerated claws are so light, males can wave them very rapidly. One of the features that females use to select mates is claw wave rate, presumably because it indicates the health of the waver. Males with regenerated claws can wave very rapidly, not because they are particularly vigorous, but because they have such light claws. This means they are very good at attracting females for mating.

In conclusion, most animals communicate with each other using honest signals because both the sender and receiver benefit from sharing reliable information. In some cases, however, senders benefit from transmitting inaccurate information that causes receivers to respond inappropriately. While deceptive communication may be common within the animal kingdom, it only occurs at low frequencies within a species. This is because receivers would stop paying attention to the signal if it usually carried false information.

See also Cognition—*Social Cognition in Primates and Other
  Animals*
  Cognition—*Tactical Deception in Wild Bonnet
  Macaques*
  Cognition—*Theory of Mind*
  Play—*Dog Minds and Dog Play*
  Play—*Social Play Behavior and Social Morality*

Further Resources

Bradbury, J. W. & Vehrencamp S. L. 1998. *Principles of Animal Communication.* Sunderland, MA: Sinauer Associates, Inc. Publishers.

Dugatkin, L. 1999. *Cheating Monkeys and Citizen Bees. The Nature of Cooperation in Animals and Humans.* Cambridge, MA: Harvard University Press.

Pat Backwell

■ Cognition
Dogs Burying Bones: Unraveling a Rich Action Sequence

Dogs often do bury bones and often other objects as well. Most dog owners have seen this. The actions may seem frivolous at first glance, but many fascinating questions follow. These questions can be divided into three broad categories: What, why and how do dogs bury bones? These are among the questions asked by *ethologists* (students of animal behavior) for all action sequences we may be fortunate enough to observe and study. Animal actions are rarely, if ever, frivolous. Often we just need to look harder and open our minds to the mysteries in what we see.

What questions concern descriptions of animal actions. Bone burying (more formally called caching) in dogs is no exception. Watch your own or another dog. Can you identify different parts of the total bone burying action sequence? What starts the sequence, what ends the sequence, what are the movements (coordinated muscular contractions) used in the sequence, how distinctive do you find the individual actions to be, are they arranged in a predictable order, is this order richly variable or simple, fixed and highly predictable (*stereotyped*), where are the bones buried, when are the objects retrieved, and so forth.

The *why questions* concern what ethologists call the *functions* of behavior. Is the animal just bored, playful, or storing food for future use, for example? How would you test your ideas here? This is not always easy to do. *How questions* bridge the whats and the whys: How is it that the animal puts together the sequences we observe, and how is it that the animal accomplishes the task it has set out to accomplish? How questions often lead into questions of causation, and here even little experiments can clarify things. It's not too early to think of experiments (always treating the animals with dignity) that you might do.

Let's step back and look at bone burying and caching more generally. The caching of food items is seen in many animal species, not just domestic dogs. Wolves, coyotes, jackals and foxes cache food, for example. Squirrels cache nuts, mice you might own store food, and even many bird species hide food in various ways. The behavior is common, and thus must serve one or more functions. The behavior also differs in detail across species. Foxes cache food in a simpler (less variable and shorter) overall pattern than do coyotes and wolves, for example. There is a rich range of both similarities and differences one sees in comparing the behavior of different species. This combination of similarities plus differences in the details of behavior form what is called the *comparative method* used by ethologists and other scientists.

A good place to start our observations is noting when and where and how the animal obtains the object (we shall start with a bone) that is buried. Perhaps you gave the dog a bone; perhaps it found one; perhaps the wolf had eaten part of its prey and then buried other parts. If you have a retriever or similar dog, it may fetch and then bury toys as well as bones. In each case the object may be consumed or at least chewed upon as well. What makes it switch to caching? Do puppies do this? Do they do so in a highly skilled manner, or do they display just part of the sequence? And, if they do bury the object, where do they decide to do this? Do they make clear choices? What if they have many bones to cache? Do they put these into one pile (*larder storage*) or do they spread their burying spots around (*scattered storage*)? Already we have many questions to ask.

What actions do the animals use in burying bones (or other objects)? Well, we expect them to dig, but how do they dig? Do they, for example, dig with just one paw or with alternating forepaw movements? Is there a rhythm to these movements? Does it vary? Are there different types of digging movements? Do these movements form a predictable sequence? What happens if the bone is easy to bury (say in loose dirt or mud) or difficult to bury (as in frozen ground or the floor of your home)? It's fun to try to answer these questions. In doing so we find that the richness, of what at first may have appeared to be a frivolous behavior, begins to emerge. Does your dog bury bones in the same way as your neighbor's dog does? Are they the same breed, the same age? Are the circumstances under which you observed them the same? Can you list both the similarities and distinctions that you see? Do you have an idea of what these similarities and differences mean and how you might test them? The list of potentially fascinating questions goes on and on.

Caching behavior in foxes (dog-like carnivores), coyotes and wolves (two species more closely related to dogs, as indicated by the fact that they belong to the dog genus, *Canis*) have been studied in detail. A few of these details may help you in your own observations. First, observers have found it useful to break down the caching sequence into retrieving the bone or other item, carrying the object to a particular site, digging at the ground, dropping the object, and then covering the object. Watch your dog do this. Are there favorite objects or circumstances that elicit caching? Are there favorite sites for burying bones?

Dogs and their relatives first retrieve the food, then carry it while searching for a caching site, then dig with their forepaws, then drop the food, and finally, with a series of tamping and scooping movements with their noses, secure the food into the site. You might be surprised to find that once the object is dropped into its hole, your dog (like foxes, coyotes, and wolves) tamps the object down with its nose, and then in an alternating fashion scoops dirt or other cover over the object, then tamps and scoops, again for a variable period of time. The paws are used for one part of the burying sequence, while the nose is used for another part of the sequence. There are still more details here that

DIG HOLE PLACE FOOD

TAMP SCOOP

The caching behavior sequence in dogs and their relatives.
Courtesy of John Fentress.

you can discover in your own observations. In what ways is the dog you observe similar in its behavior to that reported for foxes, coyotes and wolves?

Sometimes it is useful to take a closer look at just part of an overall behavioral sequence. The repeating and alternating tamps and scoops in the last phases of bone burying are illustrative. To take an example, Simon Gadbois found that sometimes the total tamp–scoop sequence in foxes and wolves was much shorter than at other times, and he wondered if he could predict this difference in tamp–scoop sequence length early in the sequence. He found out something surprising. By taking the *ratio* of initial repeated tamps to initial repeated scoops in the initial tamp–scoop sequence, a good degree of prediction was indeed possible. In this case, as the ratio of numbers in the initial repeated tamps increased with respect to the initial repeated scoops, the length of the overall sequence increased predictably. Note that this is not just the number of initial tamps or the number of initial scoops, but the ratio between these two numbers. An important more general point is that ethologists have often found it useful to look at *combinations* of actions rather than at just single actions to understand their animals more fully. If we go back to our analogy to language, this is perhaps less surprising than it initially appears. To understand someone we are in conversation with, we attend to the combination of individual sounds (phonemes) that make up words, the combination of words that make up sentences, and so on. Both human language and even relatively simple animal actions have layers of organization. Each is important. The combination across these levels provides the whole story.

One observation that applies to each case is that the animals display an *action syntax*. The idea of syntax, borrowed from human language, is that there are sequential rules that link the individual actions together in time (rather like the sequence of sounds you make in speaking to another person). A simple way to amplify this idea is to think of animal actions as consisting of letters, words, and phrases. The simple action properties can be compared to letters that are then coordinated together in a recognizable *unit* of behavior which in turn joins other such units within a broader sequence. This aspect of behavior is often referred to as its *hierarchical* structure. Keeping with the analogy to human speech, sometimes we see modulations in the sequence, where different points of emphasis can be found—rather like we can speak slowly or loudly, quickly or slowly, with a lot of emphasis (inflection) or very little emphasis, and so on. These aspects of speech are called *prosody*. Are the bone burying actions you see also prosodic, or are they more fixed in their form? One more related pair of questions to think about: When you catalog the various action components, are they themselves variable or fixed; are they always clearly distinguishable from one another or do they sometimes blend together (a phenomenon called *co-articulation* in human speech)?

It's, indeed, putting the actions back together that leads one to questions of function or purpose. It's also a hint about simple special observations and experiments that can be done to understand the structure and constraints (and thus origins) of behavior more fully. Konrad Lorenz, one of the founders of modern ethology, noticed that his dog would often go through bone burying actions on the parquet floor of Lorenz's home, finally leaving the bone on the surface of the floor as if the mission had indeed been accomplished! My long-term ethologist friend tells of a simple experiment he did with his malamute dog. As in the case cited by Lorenz, Marc's dog went through bone burying actions on the floor of his house, and then left the scene apparently satisfied that he had done his job. For fun, Marc then went to the bone and gave it to his dog who was now on the other side of the room. In his words, the dog acted surprised, as if saying, "how did you find that bone?!" There is something else to think about here. The actions in these two dogs appeared *hard-wired* to use computer jargon. They persisted even when the mission really was not accomplished. The origins of such fixed actions have long fascinated ethologists and other workers. There are many routes that differ across different cases, always a combination of the animal's genetic makeup and its own individual experiences during development. Simple models of animal *instinct* are no longer satisfactory answers in themselves, but they do point to the importance of the diverse genetic heritages we find in the animal world.

At a different level than has been emphasized here, ethologists and neurobiologists often use action patterns such as bone burying to demonstrate how an animal's nervous system can develop sophisticated "central programs," where the patterns of behavior, once begun, run off with little immediate influence from the environment. A general caution in these cases is that animals may tune into and then shut out particular sources of sensory information as an action pattern progresses. Caching in young foxes and canids may have incomplete elements of movement, and also movements that fail to adjust to their target. Thus both sensory and motor processes become refined as development proceeds. There are clearly rich questions to be asked about the biological control and deeper roots of such actions.

Are these actions and their consequences *important* to the animals? They surely are, but perhaps for many different reasons. When dogs return to their bones, they may eat them. Certainly wild canids and their relatives often consume food items they had previously buried. David Henry has shown how foxes camouflage their scattered stores of cached food by layering dirt, leaves and other objects, thus preserving the food for future use. The tamp–scoop sequencing thus helps pack the food into a secure holding. Through simple experiments one can also show that such hidden burying sites are less often robbed by other animals than are noncamouflaged sites. Fred Harrington has provided data with artificial caches that coyotes and wolves may mark previously revisited sites by marking the now depleted sites with urine. By using olfactory cues to recognize the distinction between the marked and unmarked sites, the animals can in principle increase the efficiency of their subsequent search and retrieval. (Additional recent reviews of caching and other wolf behavior in the field can be found in Mech and Boitani 2003.) Two important summary points from these studies are 1) that comparative observations can enrich what we see in a single species, and 2) that close observations in captive animals can be connected to observations made in the field.

My long-term colleague, Jenny Ryon, has made many fascinating observations on caching in wolves and coyotes (conducted at the Canadian Centre for Wolf Research in Nova Scotia). Among her observations, she reminded me of an instance when pups were born in a group of captive coyotes. The father coyote went into a "caching frenzy." This involved not just food item after food item, but other objects such as blocks of wood and feathers as well. This observation suggests both that caching behavior may vary with circumstance and season and also become so predominant in an animal's repertoire that the actions persist even when their original functions (here, food storage) do not appear to be served. A school teacher friend of mine has told me stories about her retriever which caches favorite toy objects in certain places in her home. The sites are chosen carefully when the still exposed toy is no longer actively used in play, to be retrieved for future use. Caching behavior may well have evolved initially to secure future food stores, but later may have become generalized to other situations, such as toy "hiding." Seeking the evolutionary roots of animal actions remains a fascinating and productive line of research.

Now think again how dogs may be both similar and different from their wild ancestors, and how within the dogs different breeds may also be distinctive while sharing common foundations in their behavior. That is the beauty of the comparative method: noting similarities plus differences, and surprises along with expectations.

I can share another personal slant on the idea of burial site importance. For a period of time I kept a timber wolf (*Canis lupus*) on a farm outside of Cambridge, England. Lupey, as I called him, benefitted greatly from our shared landlady who often offered him scraps of food in his back pen area (or, occasionally in the house). One day I went out as usual to see my normally friendly animal, and was met with a frightening growl. I stepped back, and he wagged his tail. I stepped forward, and he growled again. This sequence was repeated several times before I realized what I had done. I had stepped on his cache of chicken parts! This broke a social rule. Caches are important, and they will often be guarded. The second part of this side story is that Lupey confined his displeasure of me to the cache. When I behaved properly, he was his old friendly self. (I want to make a point in parentheses here: A wild timber wolf is not a domestic dog. In many ways I was fortunate that our relationship worked out so well. I do not recommend that any reader of this book keep a wild animal, especially if the goal is to befriend it. There are too many cases of misfortunes, both for the animal and the person.)

There are many remaining questions about bone burying behavior, its prevalence and form in different species, and its development, that are far beyond the scope of this brief essay. The reader may well think of these questions, and indeed pursue them. To take a single example, how does bone burying develop in the individual animal? Are juvenile actions complete, or do they have to be perfected? We have some observations with young wolves and coyotes that early caching sequences are often incomplete, and that the components are even poorly oriented toward their target of completion. But much more work needs to be done on roles of experience, genetic roots, and so forth. This is true for all animal behavior.

You have learned that bone burying and other action patterns in behavior can often be divided into a number of component parts or properties, but that a fuller appreciation for the richness of animal actions depends also upon looking at the broader contexts within which these individual elements occur: For example, tamps and scoops at the end of a bone burying sequence, and the overall caching sequence occur within the context of other actions and the environment. You have also learned that each observation can help answer questions while at the same time raising other questions. That is the joy of ethological research, as it is with other forms of natural history and science.

Now that you have made your own observations, and developed your own ideas, here is something you might like to try. Make observations with a friend. Write down what you see and what you think your observations mean. Then compare notes. Do you and your friend agree? Where are the similarities and where are the differences? You should find both. This is how science progresses. You might now want to try this again. Or videotape a bone burying sequence and have friends score it. How are their observations and ideas similar and different from each other? Certainly by now you have discovered that bone burying in dogs is not a frivolous action at all. It, as all forms of animal behavior, is rich in its structure, its foundations, and its meaning.

In summary, bone burying in dogs has opened up a host of questions. For some of these questions we have tentative answers. For some issues we have not even discovered what questions to ask. Bone burying is but one example of how rich an initially seeming "frivolous" behavior can be.

See also Cognition—*Audience Effect in Wolves*
Cognition—*Caching Behavior*
Cognition—*Cache Robbing*
Cognition—*Food Storing*

Further Resources

Fentress, J. C. & Gadbois, S. 2001. *The development of action sequences.* In *Handbook of Behavioral Neurobiology, 13: Developmental Psychobiology* (Ed. by E. Blass), pp. 393–431. New York: Kluwer Academic/Plenum Publishers.
(*Extensive review of ethological approaches to development for caching and other natural behavior patterns in animals*)
Harrington, F. H. 1982. *Urine marking at food and caches in captive coyotes.* Canadian Journal of Zoology, 60, 776–782.
(*A nice example of how olfactory marks can be used as a method to enhance the efficiency of subsequent cache retrievals*)
Henry, J. D. 1986. *Red Fox: The Catlike Canid.* Washington DC, London: Smithsonian Institution Press.
(*A lovely summary of red fox behavior and ecology, with useful demonstrations how simple field experiments combined with careful observations can reveal the richness of caching behavior*)
Mech, L. D. & Boitani, L., Eds. 2003. *Wolves: Behavior, Ecology, and Conservation.* Chicago and London: The University of Chicago Press.
(*Excellent recent overview with specialized chapters on wolf behavior, ecology, and conservation based primarily upon field studies*)
Phillips, D. P., Ryon, J., Danilchuk, W., & Fentress, J.C. 1991. *Food-caching in captive coyotes: Stereotypy of action sequence and spatial distribution of cache sites.* Canadian Journal of Psychology, 45, pp. 83–91.
(*This article provides information on how coyotes bury food, including analysis of action sequences and the choice of caching sites.*)

John Fentress

■ Cognition
Domestic Dogs Use Humans as Tools

Dogs live sympatrically with humans across the entire globe, but what has allowed dogs to live together so successfully with humans? Recent research focusing on the problem solving abilities, or cognition, in dogs suggest that part of the answer lies in how dogs are able to utilize humans in solving problems. Dogs excel, relative to other animals, at using human behaviors to learn about the world and modify their own behavior. Many dogs can even be characterized as human tool users because they intentionally ellicit help from humans if they are unable to solve a problem by themselves. Comparative experiments on social problem solving, or social cognition, in wolves, dogs, and foxes demonstrate that dogs acquired these unusual problem solving abilities during domestication. Therefore, this research supports the hypothesis that dogs have been successful in living with humans not only because of changes in their social behavior during domestication, as many have previously suggested, but also because of changes to their social cognition as well.

Making a Puzzle Out of the Making of Man's Best Friend

Dogs were the first species that began living together with humans, and today hold a position in many cultures as a favored companion. In some cases people even raise dogs in lieu of having their own human children. The very earliest archeological evidence of dogs living with humans is from Germany, Israel, and Iraq and indicates that the commensal relationship between humans and dogs began between 12,000 and 14,000 years ago.

For anyone who has owned or interacted with a dog for any amount of time, the fact that dogs were the first domesticated animal to live with humans may not be surprising. Dogs raised by humans can thrive in human company. In fact, many dogs grow to love humans so much that they can suffer from symptoms of anxiety and depression if they are not in visual contact with a human at all times. Moreover, humans enjoy spending excessive amounts of time with and money on dogs. Many people find dogs to be unavoidably attractive playmates, spending hours with dogs on anything from fetch to dress up games. In the United States the obsession with canine companions leads to an annual expenditure on dog food that exceeds 10 billion dollars.

However, although it may seem clear to any dog lover why modern dogs are a favored companion, it is less than obvious how this interspecific mutualism might have originally been fostered. We know from genetic studies that the closest relative of the domestic dog (*Canis familiaris*) is the wolf (*Canis lupus*). By comparing the DNA of wolves and various breeds of dogs, geneticists have been able to determine that all dogs are likely to have originated from a population of Asian wolves that lived somewhere between 40,000 and 15,000 years ago. This population of evolving dogs then underwent explosive expansion, spreading as far as the western United States sometime as early as 9,000 to 10,000 years ago.

Initially scientists argued that, much as people breed dogs today, prehistoric humans must have adopted wolves and, over many generations, intentionally bred them for traits that made them more dog-like. However, today's scientists realize there are problems with this idea. If dogs originated from wolves after generations of intentional breeding by humans, how can this be reconciled with the fact that humans of most cultures, at least in recent history, typically fear and persecute wolves—a persecution that has almost led to their extinction? How could it be that people living tens of thousands of years ago, having no modern weapons, did not find wolves terrifying, but instead found them to be appealing fireside companions? Puzzling yes, but even more so when you realize that adult wolves make horrible human companions. Although beautiful, gregarious, and incredibly intelligent animals, as wide ranging pack hunters, wolves are also free spirited, stubborn, and emotionally reactive. There is a strong dominance hierarchy within wolf groups that is maintained through dominance displays and overt aggression. Although normally gentle and peaceful, adult wolves, even those raised by humans, can easily cause serious injury to any human foolish enough to attempt to force his will on it (i.e., it would be ill advised to attempt training an adult wolf how to sit!). Finally, the origins of our relationship with dogs is further clouded by the fact that, even today, it is far from universal for humans to view dogs as celebrated inhabitants of their communities. Instead, in many cultures dogs are seen as pests that must be tolerated or even eradicated. Therefore, with the knowledge that dogs evolved from a feared predator into a species that many humans shun, it is no longer obvious how dogs evolved from wolves and successfully spread across the entire globe in concert with humans. Thus, if we are to understand why dogs have been so successful as a species, it becomes important to understand what changes occurred during the domestication of dogs from wolves, while also understanding how these changes might have occurred. It is only with such information that we will gain insight into exactly what it is about dogs that has made them so successful at living with humans for so long and in such varied circumstances.

Do Dogs Have Unusual Abilities to Solve Social Problems?

How can we explain the origin of man's best friend if it is no longer clear how or why dogs evolved from wolves and came to live together with humans. In order to uncover clues regarding the origin of dogs and our relationship with them, researchers have gathered information from as many sources as possible, including dog genetics, morphology, physiology, and the archeological record. Additionally, some researchers have studied dog *behavior*. Recently these efforts have focused on examining how the problem-solving behaviors, or cognitive abilities, of dogs might differ from other species of animals, and how these differences might help explain the evolution of our relationship with them.

In 1998 two different comparative psychology laboratories, one in the United States and one in Hungary, published research on how dogs solve social problems. Independently, these two teams of researchers came to the same conclusion. Dogs seem to be skillful at using human social cues, or signals, to learn things about their environments. Both groups of researchers used the same experimental paradigm, or methodology, to study the cognition of dogs. The paradigm was similar in some ways to the famous shell game that many street magicians use to entertain audiences. Thus, while a dog watched from a few feet away, food was hidden by an experimenter in one of two opaque cups that were spread several feet apart. The trick was that, although the dog knew the food was hidden in one of the two cups, there was no way for the dog to see which cup the food was hidden under (the experimenter handled both cups the same way although he only hid food under one cup). However, unlike the street magician, the human experimenter, after hiding the food, tried to help the dog find the food by providing some conspicuous signal to indicate its location. For example, the experimenter might tap on the cup where the food was hidden, point to the correct cup, or simply look toward the correct location. Then the dog was allowed to search for the hidden food. If the dog touched the correct cup first, it received the food reward, whereas if it did not choose the correct cup first, it did not receive the food. Indeed, both groups of researchers found that dogs were excellent at finding the hidden food using any of the signals the experimenters provided. However, a good experimenter should not be convinced quite yet. Dogs are known for their keen sense of smell. Is it possible that the dogs were successful, not because they used the human signals, but instead used olfactory cues (i.e., the scent of the food) to choose the correct cup? To control for this possibility, a test was run that replicated the exact procedure described above, with the exception that the experimenter did not provide any cue to the location of the hidden food. In this controlled situation the same dogs who found food when human signals were available could no longer find the food. Therefore, the failure of dogs to find hidden food in these control conditions demonstrates that dogs find hidden food using human signals.

Perhaps, as someone who interacts with dogs often, you are wondering why this result is interesting. Is it not obvious that dogs use gestures and other social cues provided by humans in all types of situations? Actually, it is not so obvious. The skill dogs show in using human signals is quite impressive in light of work with other members of our own taxonomic group—the primates. Universally, primates are very poor at using signals from humans to find hidden food—even using the exact same experimental methodology just described. Whether it is a capuchin monkey, baboon or even our closest living relative the chimpanzee (sharing in common with us 98.6% of our genetic code), nonhuman primates are totally unsuccessful at finding hidden food if it requires using gestures or signals from humans. This is not to say that primates are not capable of learning how to use human social cues to find food. Primates can learn to use human signals, but only after they have played the game several dozen times. Surprisingly, however, once primates have had enough experience to learn how to use one cue, they do not typically generalize this skill to a new cue or signal. For example, chimpanzees were trained to use a human pointing gesture to find food, but once proficient, they were unable to use the same gesture if the experimenter simply stood several meters away from the two hiding locations. In addition, primates are not skilled at using static cues. When primates do learn to use human signals to find food, they simply rely on the directionality, or motion, involved in making the cue. For example, chimpanzees who are proficient at using a pointing cue are no longer capable if they did not actually see the human extend their arm toward the hiding location. Finally, primates need tremendous experience to learn how to use any type of cue that is novel or arbitrary to them. For example, if instead of using a body part to indicate the location of the food, an experimenter places a wooden block on top of the correct cup, it takes chimpanzees dozens of trials to learn to use the block as a reliable indicator of the food's location.

With the inability of primates to use human signals in mind, the abilities of dogs to use the same type of information skillfully seems quite remarkable. However, how do dogs fair on some of these more difficult tests that were used to uncover how inflexible primates are in using human social cues? Almost without exception, dogs are skillful in the same tasks that primates have previously struggled to understand. First, many adult dogs are able to use various human gestures or signals on their very first try; they do not require dozens of trials to learn how to use each new social cue. Second, dogs do not simply rely on the motion provided by the cue. Dogs can use cues even when they are provided statically. Perhaps most impressively, dogs are not tricked when someone provides a false cue that is in the direction of the food, but is not actually directed at the correct location. Dogs will choose the cup a human is looking at, but will not chose the cup if the human is only looking above it. Finally, dogs are also capable of spontaneously using a range of novel and arbitrary cues to find hidden food. Unlike primates, many dogs, on their first trial, will choose a cup on which a human has placed a wooden block. Taken together, these results are very similar to those obtained with young children, and suggest that dogs have an unusual understanding of human communicative signals and gestures relative to that of other animals.

Both teams of scientists studying problem-solving behaviors in dogs have discovered that dogs have unusual abilities for comprehending communicative gestures and signals provided by humans. Could this unusual ability have played a role in helping dogs succeed in living with humans and spreading across the planet? Indeed, it might seem that skill at using various human social cues could have been extraordinarily advantageous to any animal living in proximity to humans. First, by attending to human gestures and signals, dogs have access to information about the location of things such as danger or food to which they otherwise would not—especially since human vision is far more acute than that of dogs. But perhaps most importantly, a dog that was able to read human social cues flexibly could predict how best to behave across a variety of novel situations in order to avoid making humans angry or perhaps even to find a way to please them. If true, then perhaps the ability to read social cues was an important reason why dogs were able to quickly and successfully spread across the planet with humans.

Did Dogs' Social Cognition Change during Domestication?

Dogs are more skillful than primates at using human gestures and other signals in a number of settings. This suggests the hypothesis that these problem-solving abilities may have played an important role in allowing dogs to survive together with humans, and therefore evolved during domestication. However, while comparisons between dogs and primates suggest this interesting hypothesis, these same comparisons cannot fully test whether it is correct. With the hope of testing whether dog cognition evolved or changed during domestication, both laboratories studying dog cognition have extended their comparisons to include wolves, while also studying how the ability to read human social cues develops in dogs.

If the ability to read human gestures evolved during domestication, then dogs should be more skilled than wolves in reading human social cues. Although a possibility, it is not clear if this is the case. It is also possible that dogs simply inherited their skill at reading social cues from their canid ancestry. Wolves are cooperative hunters

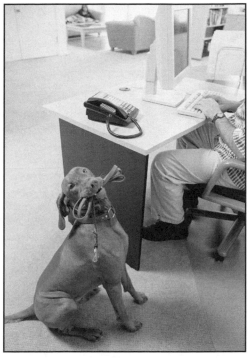

Dogs have subtle, and not so subtle, ways of communicating and learning signals.
© *Lawrence Manning/CORBIS.*

who most probably rely heavily on behavioral cues of group mates and prey alike while working together to bring down their quarry. Therefore, it could be that dogs simply inherited their unusual skill for reading subtle interspecific social cues. If true, this would rule out the possibility that this ability played a central role in facilitating the relationship between dogs and humans.

The same researchers from the United States tested this hypothesis by comparing the ability of seven wolves and seven dogs for their ability to use various gestures in locating hidden food, using the same shell game described previously. The wolves lived at a wildlife sanctuary, but were all raised for several months as puppies by their human caretaker. In addition, as adults, the wolves interacted with people every day—whether it be a familiar caretaker or groups of school children visiting the sanctuary. However, for the tests, to help minimize the chances that the wolves would perform poorly with an unfamiliar experimenter, all of the shell games were performed by their favorite caretaker who had raised them since they were puppies. What happened when the experimenter hid the food from the wolves and then tried to tell them where it was? Regardless of whether the experimenter pointed to, pointed and gazed at, or even tapped on the correct container, the wolves did not find the hidden food using the human cues. Meanwhile, all of the dogs were able to find the hidden food using at least one of the cues that the experimenter provided.

The researchers from Hungary also compared the ability of a different group of wolves and dogs to use human social cues in the shell game. Importantly, the four wolves tested had not only been raised in a human family as puppies, but also had continued to live together with a human family into adulthood. These wolves had the same amount of exposure to humans as the four dogs with whom they were compared (the dogs were raised by the same people as the wolves). Again, even after carefully controlling for exposure to humans, the four wolves were not as skillful as the dogs in using social cues that humans provided for finding hidden food. However, this research team did not stop there. They also conducted a second and perhaps even more impressive study in which they tested how often dogs and wolves would request help from humans in solving a problem they themselves were unable to solve. First, the wolves and dogs were shown that they could easily access a bowl of highly desirable food if they simply removed an obstacle between themselves and the food. Both the wolves and dogs easily solved such a problem. However, the trick was that after the subjects had learned how to obtain the food reliably, the problem was altered so that it was impossible to remove the obstacle. While the wolves worked tirelessly to access the food, continually trying new strategies for the duration of the test, the dogs instead almost immediately gave up and either approached, barked at, or stared at the experimenter—as if requesting help. This finding corroborates previous research showing that dogs will direct humans to hidden objects that they wish to obtain, such as toys or food, while showing that this ability is also not inherited from wolves. Overall, the comparisons of wolves to dogs in the various cognitive tasks used by these two teams of researchers support the hypothesis that dogs' unique abilities to use human social cues evolved during domestication and were not inherited from wolves.

However, there is one more hypothesis that could easily account for the unusual ability of dogs and rule out the possibility that domestication effected the cognitive abilities of dogs. It is possible that dogs acquire their ability to read human gestures and signals during intense exposure to humans as they grow up as puppies in human families. All the dogs tested and compared to primates and wolves were raised in human families. They had interacted with humans for countless hours. Therefore, it is likely that the dogs that participated in the various shell games had been directed to food in their normal interactions with humans on countless occasions before they were ever tested. This would mean that just like primates and wolves, dogs also require dozens of exposures to human signals before they become proficient at utilizing them. If true, the only thing that is unusual about dogs is their intense exposure to humans and not their abilities at reading human social cues. This hypothesis predicts that dogs with little exposure to humans will show less skill at using human cues than dogs with intense exposure to humans. In addition, it also predicts that puppies' skills at using human social cues will improve dramatically as they grow older and have gained more experience with humans. However, neither of these predictions were supported when puppies were tested in the same shell game that had been used with adult dogs. Puppies who were reared in a kennel awaiting adoption by a human family were as skillful at using human social cues as those that had lived with a human family since birth and were attending obedience classes. In addition, when the performance of puppies from different age groups were compared, the youngest puppies performed as well as the oldest puppies.

Two domestic dogs waiting to be fed by their "master."
Courtesy of Corbis.

Therefore, there is no support for the hypothesis that dogs' unusual ability to use social cues is a product of intense exposure to humans during ontogeny.

Taken together, the comparisons of wolves with dogs and between various dog puppies provide no support for either of the two major hypotheses that might have potentially explained the skill dogs show in using human gestures and signals. By default, this suggests that dogs underwent evolution during domestication that effected their social cognition. The evolution not only enhanced the abilities of dogs to attend to interspecific social cues, but also has facilitated their propensity to request help from humans when they encounter an unsolvable problem. In other words, dogs could have as easily earned the nickname "the human tool user" opposed to the title of "man's best friend" if only people had known exactly why dogs are unusual relative to other animals.

What Is Domestication and Can It Cause Cognitive Evolution?

The comparisons between adult dogs, wolves, and puppies of various ages and rearing histories, suggest by default, that the ability that enables dogs to excel at reading human social cues evolved during domestication. However, these same comparisons provide no direct support that domestication might have caused the changes in the social problem-solving abilities observed in dogs. In the hope of testing the domestication hypothesis directly, these comparisons were again extended to include a population of experimentally domesticated foxes for whom all the details of their domestication were known.

First, what is meant by domestication? This might sound like a question deserving of a straightforward answer, but until a Russian geneticist named Dimitri Belyaev began experimentally to domesticate various species of mammals living on a farm in Siberia, it was unclear exactly how animals became domesticates once living in association with humans.

Before the work of Dr. Balyaev, the only certainty was that, without exception, domesticated mammals from cats to cows share a suite of changes from their wild ancestors that seemed to be a result of their association with humans. Domesticates show sweeping changes in physiology, morphology, and behavior. For example, all domesticated mammals have lower levels of stress hormones, show a −20% reduction in brain size, and have a reduction in fear response to novelty. In addition, many domesticates have higher rates of floppy ears, spotted or multicolored coats, smaller teeth and bones, and in some species even curly tails. Although all domesticated mammals seem to have experienced changes in some or all of these traits, it was long a mystery as to how this phenomenon might be explained. Is it that, over history, humans intentionally bred animals that had some combination of these traits? Or alternatively, was it that somehow all of the documented changes were actually correlated with each other: If one changes they all change as an accidental byproduct.

Luckily, the work of the late Dimitri Balyaev has helped in putting this mystery to rest. In 1959, Dr. Balyaev and his colleagues began one of the longest continuing experiments in history with the goal of experimentally domesticating foxes for the purpose of studying the behavioral genetics of domestication. Two separate populations of foxes have been maintained for the entirety of the experiment. Each year the members of the control population have been bred randomly while those of the second population have been selectively bred based on their behavior toward humans. In the selected population, only individuals who were attracted to a human experimenter, as opposed to those being afraid or aggressive, were allowed to breed.

After only 20 generations, the selected fox population began showing all the universal signs of domestication. Not surprisingly, the selected foxes became tame toward humans and even began wagging their tails and barking at the sight of a human. However the selection didn't effect just behavior. Just as with many domesticated mammals, successive generations of selected foxes began having a higher prevalence of curly tails, floppy ears, and reduced tooth and bone size even though none of these traits were selection criteria used by the experimenters. Therefore, Dimitri Balyaev's work with the foxes demonstrated that the suite of changes associated with domestication were all correlated and incidental byproducts of selection against aggressive and fearful behavior. It is only as a result of this work that we now know that domestication is the process by which the least aggressive and fearful animals in a population survive and reproduce at higher levels when living in association with humans.

If domestication is the result of selection against aggressive and fearful behavior, is it then possible that this type of selection during dog evolution is directly responsible for the unusual social problem-solving abilities in dogs? One way to test this hypothesis is to test the experimentally domesticated foxes in the ubiquitous shell game. If the experimentally domesticated foxes are as skillful at using human social cues as domestic dogs are, then this would suggest that selection for domestication can cause enhanced problem-solving abilities in canids. This type of evidence would provide direct support for the hypothesis that dogs' unusual social cognition is a direct result of domestication.

Experiments have now shown that the experimentally domesticated foxes are as skillful as dogs and more skillful than the control foxes at using human gestures and signals when attempting to find hidden food. Therefore, it seems likely that the enhanced social cognitive abilities witnessed in domestic dogs are simply another incidental byproduct of domestication. Given that domestication caused a change in the problem-solving abilities as a result of the transition from wolf to dog, it is likely that reading human social cues played a large role in assuring the success of dogs in living together with humans. Dogs who were not aggressive and fearful toward humans were also more skilled at using their gestures and signals to predict human behavior while modifying their own behavior to the satisfaction of their human groupmates.

Summary of Research on Dog Social Cognition and Domestication

Humans and dogs have an ancient and unusual relationship. Today, in many cultures, dogs enjoy a special position in society as friends or even colleagues. At first glimpse, it seems that this interspecific relationship was all but an inevitable product of human history. However, upon closer inspection, the origin of our commensal relationship has become an evolutionary puzzle. Scientists across a number of disciplines have been working toward understanding how, why, and when dogs became so successful at living with humans. Recently, a new area of research has developed which has focused on studying the ability of dogs to solve social problems. This research has revealed that dogs have an unusual ability to read human gestures and signals in order to solve problems that would otherwise be unsolvable. Comparisons between dogs and chimpanzees dramatically illustrate how gifted dogs really are at comprehending human gestures and signals. This initial finding suggested that dogs may have been successful at spreading across the globe in part because of their unusual abilities to communicate with humans. If true, then the unusual abilities of dogs should have evolved during domestication. Therefore, a number of studies were designed to examine whether dog social cognition was effected by domestication. Comparisons between dogs and wolves demonstrate that the ability of dogs to communicate with humans is not a general ability of canids that has been inherited through common descent. In addition, studies on dog puppies suggest that dogs do not require intense exposure to humans in order to develop their abilities at reading human social cues. Finally, a population of foxes that have been experimentally domesticated share with dogs their unusual ability to read human social cues. Taken together, these comparative and developmental studies show that dog social cognition evolved during domestication and support the hypothesis that these new abilities provided dogs with an important skill in surviving together with humans. As dogs evolved from wolves, those individuals who were less aggressive and fearful while at the same time better at predicting the behavior of humans using human social cues were the most successful at surviving and reproducing. Thus, it maybe that dogs should not only have the title of "man's best friend" but also "the human tool user."

See also Cognition—*Social Cognition in Primates and Other*
    Animals
    Communication—Vocal—*Referential Communication*
    in Prairie Dogs
    Domestication and Behavior
    Domestication and Behavior—*The Border Collie,*
    A Wolf in Sheep's Clothing

Further Resources

Bekoff, M. 2002. *Minding Animals: Awareness, Emotions, and Heart.* Oxford: Oxford University Press.

Coppinger, R. & Coppinger, L. 2001. *Dogs: A Startling New Understanding of Canine Origin, Behavior, and Evolution.* New York: Scribner Press.

Hare, B., Brown, M., Williamson, C. & Tomasello, M. 2002. *The domestication of social cognition in dogs.* Science, 298, 1636–1639.

Mech, L. D. & Boitani, L. (Eds.). 2003. *Wolves: Behavior, Ecology, and Conservation.* Chicago: Chicago University Press.

Miklosi, A., Kubinyi E., Topal, J., Gacsi, M., Viranyi, Z., & Csanyi, V. 2003. *A simple reason for a big difference: Wolves do not look back at humans, but dogs do.* Current Biology, 13, 763–766.

Serpell, J. (Ed.). 1995. *The Domestic Dog: Its Evolution, Behavior and Interactions with People.* Cambridge: Cambridge University Press.

Tomasello, M. & Call, J. 1997. *Primate Cognition.* Oxford: Oxford University Press.

Brian Hare

■ Cognition
Equivalence Relations

Like a computer, an animal's brain can be used as a powerful tool to bring meaning to a jumbled and disorganized array of information. One way the brain simplifies the world is through classification, or the placement of items into categories. True classificatory behavior depends on the learned equivalence or grouping of stimuli. It is well known that many animal species, including chimpanzees, monkeys, dolphins, sea lions, and some birds are capable of organizing their experience along abstract lines. They are able to respond to a certain constant dimension of an object despite variation in other dimensions. This cognitive skill is called concept formation. For example, a sea lion can do very well when required to respond to the shape of an object, such as a triangle or a star, even when the object is altered in size or brightness. In other words, sea lions and a host of other animal species can classify stimuli according to physical attributes along dimensions such as shape. Even though many animals can sort things according to similar physical dimensions, can they classify stimuli that do not share common physical characteristics? Can they classify items and events that are connected by a common location, timing, or purpose? Can they integrate and classify information coming from different sensory channels? The answer to these questions is a very definite "yes" even though classifications according to physical dimensions may be more powerful than those based on more abstract dimensions.

There are many examples from the animal behavior literature on cognition and communication demonstrating that animals can extract meaningful or functional relationships from an assortment of different stimuli. These types of abstract classifications are called *equivalence relations*. To illustrate the importance of equivalence relations in animal behavior, let's go through the following widely different examples and determine what they have in common:

1. Dolphins *echolocate* in murky waters by emitting a train of pulsed sounds or clicks, bouncing the clicks off an object and listening for the returning echoes. In this way they can not only locate underwater objects, but they can also recognize an object echoically and differentiate it from other objects. Following echoic recognition of an object, a dolphin that is allowed to inspect objects visually but not echoically can readily learn that the previously reflected sound cues correspond to the currently reflected light cues from that same object. In other words, despite the dramatic perceptual differences in the stimuli perceived, the dolphin can learn that visual images can substitute for echoic images and vice versa because they refer to the same object. We might say that the echoic stimuli and the visual stimuli have become equivalent in terms of their meaning.

2. Alarm calls that signify particular predators are highly developed in groups of vervet monkeys living in Kenya. Large cats, including leopards, elicit loud "barks" from the monkeys; raptors elicit "coughs"; and snakes, including pythons, elicit "chutters." After recognizing a predator, a vervet gives a predator-specific alarm call, and the other monkeys in the group respond in a predictable way: Leopard calls result in listeners running and climbing into trees, raptor calls result in listeners looking up into the air and then running into thick bushes, and snake calls result in listeners standing up on their hind legs and looking down into the grass. This phenomenon has been investigated experimentally by manipulating the vervet monkeys' vocal communication system by playing sound recordings back to them from hidden loudspeakers. Researchers have concluded that the communication system is language-like in that the monkeys can substitute the predator-specific call for the sight of the predator, and vice versa, and respond appropriately to either cue. The system is analogous to referentiality in human language, where the printed or spoken word functions as a substitute for the object. Just as the spoken words "cup," "cat," and "car" invoke the meaning, or the physical embodiment, of those objects, the barks, coughs, and chutters of vervet monkeys are equivalent to the predators they are associated with.

3. Monkeys and baboons can learn to sort members of the same matriline together because genetically related individuals often share a history of common spatial, temporal, and functional interactions. Thus, when an adult female hears a juvenile scream, she will often

respond by orienting to the juvenile's mother. This behavior implies that the female has learned about the connection between the juvenile and its mother and can respond to one in the absence of the other. There is additional evidence that monkeys can differentiate both their own close companions and those of other individuals. Investigators have found

that female subjects can recognize other individuals by their calls, and that they can classify the callers according to matrilineal kinship. Some monkeys can even be trained to match pictures of mothers in their social group with pictures of their offspring. This type of performance demonstrates that these animals have the capacity to classify individuals on the basis of such abstract properties as "mother–offspring pairs" or "matrilineal kinship." This capacity for abstract classification is clearly illustrated in the case of kin-based redirected aggression in vervet monkeys. If an individual in one matriline threatens an individual in another, then the relatives of those individuals may later fight with one another, acting as if they recognize that the relationship between their relatives is equivalent to their relationship with each other.

4. Male common chimpanzees acquire strong group identities, with all members of their social grouping being treated similarly. In contrast to the vervet monkeys just described, in which individuals learn to sort members of the same as well as other matrilines, group cohesion in chimpanzees frequently occurs among large groups of related males. The members of these groups share communal ranges, and these males live their entire lives within the group where they were born. Bonding between members of these patrilineal alliances is cemented by extreme cooperation and solidarity, which includes events such as mutual grooming, food sharing, the sharing of female consorts, and the exchange of other reinforcers. While interactions between members of an alliance are generally positive and aggression is tempered, the interactions between the members of different alliances can escalate into unusually hostile and violently aggressive attacks by the more numerous and stronger of the two groups. Over a period of many years, such violent attacks are known to have resulted in the complete annihilation of chimpanzee communities consisting of male groups and the females and offspring associated with them. Even inanimate objects associated with a particular group may elicit

The above photographs show an animal performing a matching-to-sample procedure, in which the subject observes the stimulus shown in the center panel (the sample) and then selects corresponding match to the sample from one of the two alternatives on either side. In this experiment, the sample and the correct match do not look physically similar so the subject must make a selection based on its previous experience with the objects shown. For example, to demonstrate the formation of an equivalence relation, a sea lion may be taught two conditional rules; following learning of these rules, the sea lion is tested to determine if other logical relationships emerge without further training. In the series of photos shown, the sea lion is trained to perform the discriminations "given crab . . . select tulip" (top) and "given tulip . . . select radio" (middle). The logical inference "given radio . . . select crab" is tested in the final step (lower). Performance based on the equivalence of the connected stimuli predicts successful performance in the absence of additional training.
Courtesy of Ronald Schusterman.

Fairness in Monkeys

Sarah F. Brosnan

A sense of fairness seems ubiquitous among humans, but does such a thing exist in animals as well? In humans, it is proposed that individuals who have a sense of fairness are likely to be successful in cooperative interactions. Thus, by looking at other highly cooperative species, we can not only learn more about the animals' social behavior, but can learn something about the evolution of co-operation and fairness in humans. Understanding the different stages of the evolutionary development of the sense of fairness allows us to understand more about why this behavior was important and how it could have evolved. The difficulty is that awareness of fairness is proposed to be based on social emotions, like envy, greed, or moral indignation, and it is challenging to uncover animals' emotions.

There are hints that other animals do compare their rewards to those of others. For instance, some animal species live in relatively tolerant societies in which everyone receives some piece of the pie. While not everyone receives the same-sized piece, such a system may lead individuals to expect some level of equity between themselves and others. Individuals who do not get what they think they should receive may react, for instance, by having "temper tantrums." While there is good anecdotal or inferential evidence for reactions to inequitable distributions, in observational studies it is difficult or impossible to get at the underlying causation of the behaviors. Thus, we often turn to experimental work for a more controlled situation. To get around this problem of causation, we ran a controlled experiment on capuchin monkeys to see if they reacted when another monkey got a better deal.

This experiment was intended only to elucidate one aspect of a self-centered sense of fairness; it determined whether the capuchins reacted when another individual got a superior reward. We chose capuchin monkeys because they are a tolerant species known to have high levels of cooperation and food sharing, which made them an ideal species for our study. We paired each monkey with a group mate and watched the monkeys' reactions when their partners got a better reward for doing the same amount of work. In this case, the work was a simple exchange in which the experimenter gave the monkey a granite token, which the monkey could immediately return for a food reward. Food rewards were pieces of cucumber, which capuchins are usually happy to work for, and grapes, which, to a capuchin, are far superior to cucumbers. Compared to their reactions when both received the same reward, subjects were much less likely to be willing to complete the work, a simple exchange, or accept the reward, when their partner got the better deal. Moreover, if the partner didn't have to do the work to get the better reward, but was handed it for "free," the subjects were even more likely to quit participating. Of course, there is always the possibility that subjects were just reacting to the presence of the higher value food, and that what the partner received (free or not) did not affect their reaction. However, in a control test in which the higher-value reward was visible, but not given to another monkey, their reaction to the presence of this high-valued food decreased significantly over time. In tests in which their partner received the grapes, the monkeys increased the frequency of their refusals to participate.

Capuchin monkeys judge both the value of the reward and the effort required when deciding whether they are being treated unfairly. This ability to recognize when one is being treated unfairly is almost certainly one of the stages in the evolution of the complex sense of fairness exhibited by humans. Whether capuchin monkeys are, like humans, using emotion to drive these decisions is unknown. However, it is clear that the sense of fairness has a long evolutionary history in the primate lineage.

Further Resources

Brosnan, S. F. & de Waal, F. B. M. 2003. *Monkeys reject unequal pay*. Nature, 425, 297–299.
de Waal, F. B. M. 1982. *Chimpanzee Politics: Power and Sex Among Apes*. Baltimore: The Johns Hopkins University Press.
Frank, R. H. 1988. *Passions within Reason: The Strategic Role of the Emotions*. New York: W. W. Norton & Company.

aggressive responses similar to those directed at individuals. For example, males may attack and destroy the abandoned sleeping nests of opposing group members that they encounter in their environment. This behavior exhibited by chimpanzees is analogous to humans treating symbols—like statues or icons—as if they are equivalent to the things they denote. Equivalence learning is hypothesized to be the process by which these animals organize information into the functional categories of "friends" and "foes."

We have just looked at a few examples of animal behavior which appear to be the outcome of a generalized learning paradigm called *equivalence*. This paradigm is hypothesized to enable relatively large-brained animals to acquire social and nonsocial skills, and to apply those skills to the solution of a variety of problems posed by their environment. It is believed that animals do this by sorting through the information gathered by the senses so that they can glean regularities and classify relationships between objects, individuals, and events. Although descriptive observations support the idea that animals use equivalence learning, what is the experimental evidence? How does the process enable animals to induce inferences from the familiar to the unfamiliar relations that connect the different members of a given class? What experiments demonstrate that larger classes emerge from a smaller sets of previously learned relationships?

In the laboratory, experiments with trained animals aim to reveal the cognitive skills underlying natural behaviors such as those just described. These experiments generally replace the complex objects, events, and individuals found in natural settings with more simple, well-controlled stimuli. The basis of these experiments is the teaching of predictable relationships or rules about how different stimuli are connected to one another. For example, an animal can learn to select an arbitrary object B when presented with another object encoded A, and then to select object C when presented with object B. This history of associative learning through positive reinforcement results in the learning of two conditional rules: "*given A . . . select B*" and "*given B . . . select C*." Equivalence classification is demonstrated when these learned rules give rise to new, logical inferences that are spontaneously performed by the animal, for example: "*given A . . . select C*" and "*given C . . . select A.*" The photographs on p. 287 depict an experiment that shows the emergence of equivalence relations in a trained California sea lion using different black-and-white patterns such as "crab," "tulip," and "radio." Investigations such as this provide insight into how wild animals classify friends, foes, and a myriad of other related items into meaningful categories. In addition, these captive studies aid in our understanding about how the classificatory behavior of animals relates to the complex linguistic performances and problem solving of humans.

See also Cognition—*Categorization Processes*
    Cognition—*Concept Formation*
    Cognition—*Grey Parrot Cognition*
     and Communication
    Cognition—*Social Cognition in Primates and Other*
     Animals
    Communication—Vocal—*Alarm Calls*
    Dolphins—*Dolphin Behavior and Communication*
    Recognition—*Individual Recognition*
    Recognition—*Kin Recognition*

Further Resources

Cheney, D. L. & Seyfarth, R. M. 1990. *How Monkeys See the World: Inside the Mind of another Species.* Chicago: University of Chicago Press.
Goodall, J. 1986. *The Chimpanzees of Gombe: Patterns of Behavior.* Cambridge, MA: Belknap Press.
Schusterman, R. J., Reichmuth, C. J. & Kastak, D. 2000. *How animals classify friends and foes.* Current Directions in Psychological Science, 9, 1–6.
Schusterman, R. J., Reichmuth, C. J., Kastak, C., & Kastak, D. 2002. *The cognitive sea lion: Meaning and memory in the lab and in nature.* In: *The Cognitive Animal: Empirical and Theoretical Perspectives on Animal Cognition,* (Ed. by M. Bekoff, C. Allen, & G. Burghardt), pp 217–228, Cambridge, MA: MIT Press.
Sidman, M. 1994. *Equivalence Relations and Behavior: A Research Story.* Boston: Author's Cooperative.

Ronald J. Schusterman & Colleen Reichmuth Kastak

■ Cognition
Food Storing

Many kinds of animals store food for later consumption. Some store food temporarily. A leopard (*Panthera pardus*), for example, will carry the carcass of an antelope into a tree to prevent scavengers from consuming it before the cat returns in a few hours to feed. A male northern shrike (*Lanius excubitor*) impales rodents and small birds on thorns or barbed wire to feed to its mate or offspring when the latter are hungry.

Other animals store food in anticipation of shortages weeks or months later. Long-term storage of food can only occur when the food is naturally resistant to decay, like many plant seeds, or the climate is cold or dry enough to slow rotting. Some animals that store food do so as a way to share it with other colony members or dependent offspring (e.g., ants). Other food-storing animals do so because it is impossible to accumulate enough body fat when food is abundant to tide them over to when food is scarce, either because they are too small (e.g., rodents) or because excess weight severely reduces mobility (e.g., birds).

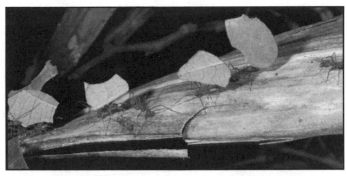

Leafcutter ants bringing cut leaves back to their nest.
George Town/Visuals Unlimited.

There are two general ways that animals create hoards of food for long-term storage. One method is to create a *larder*: Food is collected and brought to one storage location, a kind of animal warehouse. A second way is to *scatter hoard*: A few food items are stored in each of many locations. Food is easier to locate when stored as a larder than as a scatter hoard, and less energy and time is spent recovering food from a larder. However, larders are riskier than scatter hoards: Other animals may discover the larder and pillage its contents, or decay agents such as fungi may attack and ruin the food. Loss of some scatter hoards to competitors or spoilage is not as catastrophic as the loss of a larder, since the other hoards are likely to remain intact.

An example of a scatter hoarding bird.
Courtesy of Brad Sillasen.

Most animals that create larders store the food within their home. Many hymenopteran insects (bees, wasps and ants) store food in their nests, often constructing special chambers for food storage. A single colony of harvester ants (*Pogonomyrmex*) may store more than a liter of seeds, which allows workers to avoid foraging when weather conditions are averse. Chipmunks (*Tamias*) store up to a kilogram of seed within their burrow to consume during the winter when they periodically awake from hibernation.

A few species create extensive larders away from their home. Pika (*Ochotona princeps*) create "hay piles" of drying grasses and forbs near the entrance to their burrows. Beaver (*Castor canadensis*) stick branches and logs into the sediment at the bottom of a pond. In winter, they swim under the ice to the larder, select a branch, and tote it back to their lodge to eat. Acorn woodpeckers (*Melanerpes formicivorus*) painstakingly drill thousands of small holes on the trunks and branches of trees, and in each they place an acorn. These acorn granaries are created and defended communally by several woodpeckers, who also reproduce in a shared nest.

Scatter hoarding is rarer than larder hoarding, occurring principally in some rodents (e.g., squirrels Sciuridae, agoutis Dasyproctidae) and three families of birds: Corvidae (e.g., jays and nutcrackers), Paridae (e.g., chickadees), and *Sittidae* (nuthatches). Scatter-hoarding species are exceptionally interesting for two reasons. First, seed-harvesting scatter hoarders often appear to have coevolved with the plants they exploit. For example, pinyon pines, oaks, and beech trees are dependent upon scatter hoarding by squirrels and corvid birds to disperse the trees' seeds, and have evolved adaptations to make the seeds more attractive to

An example of a larder hoarding bird.
Courtesy of Brad Sillasen.

these animals. Likewise, the birds and rodents have evolved specialized morphologies and behaviors to make them more efficient at harvesting and processing these seeds for storage.

The second reason scatter-hoarding species are interesting is because of what they can reveal about general mechanisms of animal cognition. Animals that store food for long periods, both larder and scatter hoarders, are choosy about what they hoard. Experiments on a variety of food-storing rodents and birds demonstrate that hoarders carefully evaluate both the quality of potential food items to store, including energy value and resistance to rotting, and the costs associated with acquiring and storing the food items, including time, energy, and predation risk. Hoarders balance these considerations to maximize the rate of food hoarding when food is plentiful, or minimize the costs of hoarding when food is scarce or conditions are dangerous. Presently, many scientists are conducting field and laboratory experiments on food-storing animals to learn how they acquire and process multiple kinds of information to make adaptive decisions.

Scatter-hoarding animals must not only store food economically, they must be able to relocate it at a later date. Conspicuous cues like scent marks or trails to the food cache are an invitation to pilferage by other animals, so most animals store at inconspicuous locations, then remember the locations using nearby objects or distant features as landmarks. Scatter hoarders provide some of the most amazing feats of animal memory. Clark's nutcrackers (*Nucifraga columbiana*) and pinyon jays (*Gymnorhinus cyanocephalus*), for example, create over 2,000 seed caches and can recover seeds up to 9 months after storing them. Western scrub jays (*Aphelocoma californica*) appear to remember specific storing episodes, and selectively recover food items based on the type of food item stored and on whether the hoarder was watched by another, potentially pilfering, jay at the time the food was buried. Within two different groups of birds, the Paridae and Corvidae, the size of a particular brain structure, the hippocampus, is associated with the propensity to scatter hoard. The hippocampus is involved in spatial memory; the enlarged hippocampus in scatter-hoarding animals provides some of the best evidence, at present, of a specialization in brain structure associated with specialized behavior.

See also Cognition—*Audience Effect in Wolves*
Cognition—*Caching Behavior*
Cognition—*Cache Robbing*
Cognition—*Dogs Burying Bones: Unraveling a Rich Action Sequence*

Further Resources

Balda, R. P. & Kamil, A. C. 2002. *Spatial and social cognition in corvids: an evolutionary approach*. In: *The Cognitive Animal: Empirical and Theoretical Perspectives on Animal Cognition* (Ed. by M. Bekoff, C. Allen & G. M. Burghardt), pp. 123–128. Cambridge, MA: MIT Press.
Gurnell, J. 1987. *The Natural History of Squirrels*. New York: Facts on File.

Källander, H. & Smith, H. G. 1990. *Food storing in birds: An evolutionary perspective.* Current Ornithology, 7, 147–207.

Langen, T. A. & Gibson, R. M. 1998. *Sampling and information acquisition by western scrub jays,* Aphelocoma californica. Animal Behaviour, 55, 1245–1254.

Shettleworth, S. J. 2002. *Spatial behavior, food storing, and the modular mind.* In: *The Cognitive Animal: Empirical and Theoretical Perspectives on Animal Cognition* (Ed. by M. Bekoff, C. Allen & G. M. Burghardt), pp. 123–128. Cambridge, MA: MIT Press.

Vander Wall, S. B. 1990. *Food Hoarding in Animals.* Chicago: University of Chicago Press.

Tom A. Langen

■ Cognition
Grey Parrot Cognition and Communication

Although psittacids (including grey parrots) are usually regarded as mindless mimics, during the past 26 years I have taught grey parrots meaningful use of English speech (e.g., to label objects, colors, shapes, categories, quantities, absence). Using this code, my oldest subject, Alex, exhibits cognitive capacities comparable to those of marine mammals, apes, and sometimes 4–5-year-old children (Pepperberg 1999). Thus his abilities are inferred not from the standard nonvocal choice tasks involving button pecking or lever pressing common in animal research (i.e., from what are generally called *operant* tasks), but from *vocal responses* to *vocal questions;* that is, he demonstrates intriguing communicative parallels with young children despite his evolutionary distance from humans. I doubt I taught Alex and other parrots these abilities *de novo;* their achievements likely derive from existent cognitive abilities and neurological structures. Given that the walnut-sized psittacine brain is organized very differ-

Even though parrots are often thought of as mindless mimics, Alex has demonstrated a strong ability for complex learning and communication.
Courtesy of Irene Pepperberg.

ently from that of mammals, these results have intriguing implications for the study and evolution of vocal learning, communication, and cognition. Central issues in this research are: the use of interspecies communication as an investigative tool, the capacities this code unveils, and the avian role in the evolution of communication.

Significance of Interspecies Communication

Parrots' vocal plasticity enables us to communicate with them directly (Pepperberg 1999). But why make them acquire our code instead of trying to learn theirs? The answer lies in parrots' above-mentioned existent abilities. I believe that parrots acquire those elements of human communication that can be mapped or adapted to their code. By observing what is or is not acquired, we uncover these elements, and such information allows us to interpret their system. I believe parrots could not learn aspects of reference (e.g., labels that refer to particular colors or to object classes such as "apple") unless their natural code had such referentiality. Although this manner of determining nonhuman referentiality is inferential, direct determination also has a number of difficulties. Moreover, pushing at what might be the limit of avian systems to see what input engenders *exceptional* learning—learning that does not necessarily occur during normal development (e.g., acquiring another species' code; also called allospecific acquisition)—provides detailed understanding of the learning processes. Because richer input is needed for allospecific than conspecific (of one's own species) learning, we can determine how and whether "nurture" modifies "nature" (e.g., alters innate predispositions toward conspecific codes), and thus uncover additional mechanisms for, and the extent of, communicative learning. Again, these mechanisms are likely part of existent cognitive abilities, not something taught *de novo*.

Interspecies communication also has practical applications. It

1. directly states the content of the question—animals needn't determine both a query's aim *and* answer via trial-and-error, as in most operant laboratory paradigms;
2. incorporates research showing that social animals may respond more readily and accurately within ecologically valid social contexts;
3. facilitates data comparisons among species, including humans;
4. allows rigorous testing of the acquired communication code that avoids *expectation* cuing (subjects choose responses from their entire repertoire, not only from relevant subsets, such as buttons representing only "match" and "nonmatch"); and, most importantly,
5. is an open, arbitrary, creative code with enormous signal variety, enabling animals to respond in novel, possibly innovative ways that demonstrate greater competence than paradigms such as the required responses found in operant studies; and
6. thereby allows examination of the nature *and* extent of information animals perceive.

Interspecies communication facilely demonstrates nonhumans' inherent capacities or may even enable complex learning.

How Greys Learn: Parallels with Humans

My greys' learning sometimes parallels the processes observed in human children, suggesting insights into how acquisition of complex communication may have evolved. Referential, contextually applicable (functional), and socially rich input allows parrots, like young children, to acquire communication skills effectively. Reference is an utterance's meaning—the relationship between labels and objects to which they refer—and is exemplified by rewarding birds with the objects that they label. Context/function involves the situation in which an utterance is used, and the effects of its use; initially using labels as requests gives birds a reason to learn the unfamiliar sounds of our English labels. Social interaction accents environmental components, emphasizes common attributes—and possible underlying rules—of diverse actions, and allows continuous adjustment of input to match the parrots' levels of understanding. Interaction engages subjects directly, provides contextual explanations for actions, and demonstrates actions' consequences. In the next section is a description of our primary training technique, followed by experiments to determine which input elements are necessary and sufficient to engender learning.

Model/Rival (M/R) Training

My *model/rival* (M/R) training system uses three-way *social* interactions among two humans and a parrot to demonstrate targeted vocal behavior. The parrot observes two humans handling one or more objects, then watches as humans interact: The trainer presents, and queries the human model about, the item(s) (e.g., "What's here?"; "What color?") and gives praise and the object(s) to reward correct answers referentially. Incorrect responses (like those the bird may make) are punished by scolding and temporarily removing item(s) from sight. Thus the second human is a model for the parrot's responses, its rival for the trainer's attention, and also illustrates the consequences of making errors: The model must try again or talk more clearly if the response was (deliberately) incorrect or garbled, thereby demonstrating corrective feedback. The bird is also queried and rewarded for successive approximations to correct responses; thus training is adjusted to its level.

Unlike other laboratories' M/R procedures, ours also *interchanges* roles of trainer and model, and includes a parrot in interactions to emphasize that one individual is not always the questioner and the other the respondent, and that the procedure can effect environmental change. Role reversal also counteracts an earlier methodological problem: Birds whose trainers always maintained their respective roles responded only to the human acting as questioner. Our birds, however, respond to, interact with, and learn from all humans.

M/R training exclusively uses *intrinsic reinforcers*: To ensure the closest possible correlations of labels or concepts to be learned with their appropriate referent, reward for uttering "X" is X—the object to which the label or concept refers. Earlier unsuccessful programs for teaching birds to communicate with humans used *extrinsic* rewards: one food that neither related to, nor varied with, the label or concept being taught, thereby delaying label and concept acquisition by confounding the label of the targeted exemplar or concept with that of the food. We never use extrinsic rewards. Too, initial use of the label to request the item also demonstrates *functionality*.

Because Alex sometimes fails to focus on targeted objects, we trained "I want X" (i.e., to separate labeling and requesting); his reward is then the right to request something more desirable than what he identifies, which provides flexibility but maintains referentiality. To receive X after identifying Y, Alex must state "I want X," and trainers will not comply until the appropriate task is completed. Thus his labels are true identifiers, not merely emotional requests. Adding "want" provides additional advantages. First, trainers can distinguish incorrect labeling from appeals for other items, particularly during testing, when birds unable to use "want" might not be erring but asking for treats, and scores might decline unrelated to competence. Second, birds may demonstrate low-level intentionality: Alex rarely accepts substitutes when requesting X, and continues his demands.

M/R training successfully demonstrated which elements of input enabled label and concept acquisition, but not what was *necessary and sufficient*. What if some elements were lacking? Answering that question required additional parrots, because Alex might cease learning merely because training had changed, not because of the type of change. With three new naïve greys—Kyaaro, Alo, and Griffin—students and I tested the relative importance of reference, context/function, and social interaction in training.

Eliminating Aspects of Input

We performed seven experiments. First, we compared simultaneous exposure of Alo and Kyaaro to three input conditions:

1. audiotapes of Alex's sessions, which were nonreferential, not contextually applicable, and noninteractive;
2. videotapes of Alex's sessions, which were referential, minimally contextually applicable, and noninteractive; and
3. standard M/R training.

In conditions 1 and 2, birds experienced tapes in social isolation. Condition 1 paralleled early allospecific song acquisition studies; condition 2 involved still-unresolved issues about avian vision and video (e.g., screen flicker and parrots' ability to see in the ultraviolet). We counterbalanced labels across birds, matching training time across sessions.

Second, because experiments by Mabel Rice and her colleagues had shown that interactive coviewers appeared to increase young children's learning from video, a coviewer now provided social approbation for viewing and pointed to the screen with comments like "Look what Alex has!," but did not repeat targeted labels, ask questions, or relate content to other training. Birds' attempts at a label would garner only vocal praise. Social interaction was limited; referentiality and functionality matched earlier videotape sessions.

Third, because extent of coviewer interaction might affect children's learning from video, our coviewer now uttered targeted labels and asked questions.

Fourth, so that lack of reward would not deter video learning, a socially-isolated parrot watched videos while a student in another room monitored its utterances through headphones and could deliver rewards remotely.

Fifth, because birds might habituate to the single videotape used per label (even though each tape depicted many different responses and interactions among Alex and trainers), we used live video from Alex's sessions.

Sixth, because researchers such as D.A. Baldwin had showed that if adult–child duos failed to focus jointly on objects being labeled, the labels were not acquired, a single trainer faced away from the bird (who was within reach of, e.g., a key), talked about the object ("Look, a shiny *key*!," "Do you want *key*?," etc.), but had no visual or physical contact with parrot or object; a bird's attempts at the targeted label would receive only vocal praise, thereby eliminating some functionality and considerable social interaction. Parrots failed to acquire referential use of targeted labels in any non-M/R condition, but succeeded in concurrent M/R sessions.

Finally, we eliminated only some interactive aspects of modeling by having a *single* student label objects, query the bird, jointly attend to objects, and thus interact fully with bird and objects. Griffin did not utter labels in 50 such sessions, but clearly produced labels after two or three subsequent M/R sessions. We suspected latent learning: Griffin apparently stored labels, but did not use them until he observed their use modeled. (Note: Birds that switched to M/R training after 50 video sessions needed about 20 sessions before producing labels.) Although we are replicating video studies with a liquid crystal monitor to see whether cathode ray tube flickering affects learning, results presently emphasize the importance of including reference, demonstrations of contextual use/functionality, and social interaction in training if parrots are to communicate with humans rather than mimic speech.

What Use Does a Wild Parrot Make of Its Abilitiy to Talk?

Irene Pepperberg

A common query, given my data, is "Of what use would these abilities be to wild parrots?" Unfortunately, that question cannot yet fully be answered because little is known about natural grey parrot behavior, particularly with respect to many of the complex abilities they demonstrate in the laboratory. Of course, their need for certain concepts is clear: They must be able to categorize (e.g., types of predators, whether an item is edible or not, whether a flock member is a potential mate or not), to understand same/different (i.e., whether this fruit is different from the one that caused illness yesterday, or the same as that which tasted good last year), to comprehend bigger/smaller (e.g., that this patch of food is better than another), and to recognize absence (e.g., that this nest hole is *not* inhabited by another bird and is therefore up for grabs). In contrast, other concepts such as numerosity or the ability to label (rather than simply recognize) qualities such as color and shape may not seem particularly relevant for the survival of a wild grey. Nevertheless the same evolutionary pressures in nature that, according to researchers like Nicholas Humphrey, may have selected for intelligence in great apes may also have been at work in parrots, allowing all these species to use some general cognitive ability to acquire what appear to be specialized human concepts. Like apes, grey parrots are long-lived and must function within complex ecological and, likely, social environments: They must not only remember great quantities of various types of information for long periods, but must also transfer information across contexts and be able to respond adaptively to change. Thus, like apes, they must forage daily over great distances and keep track of what types of food are in specific places; of when different fruits and nuts ripen in different habitats, cease to exist, or become newly available; and must update this information over the course of 20–40 years. They mate for life, but still may have several partners during their lives because of death or injury to their mates; thus they likely have to learn several of the complex vocal duets that mated pairs of parrots use to identify one another in the dense foliage. If they, like their New World counterparts, have dialects, they must learn new ones if they emigrate from one area to another. We know little about the social structure of the large flocks in which greys live for at least part of the year outside of the mating season, but suspect that these flocks may, like troops of apes, have dominance hierarchies that require birds to remember, and update, the various inter-parrot relationships. In a sense, then, laboratory research on grey parrot abilities may engender greater interest in the natural behavior of these intriguing animals.

Mutual Exclusivity: Studying Subtle Changes in Input

Our parrots' learning may also parallel behavior described in young children as mutual exclusivity (ME). ME refers to children's assumption during early word acquisition that each object has one, and only one, label; at that stage in development, they will deny that a "dog" can also be an "animal." Along with the whole object assumption (that a label refers to an entire object, not some feature), ME supposedly guides initial label acquisition. Thus, given objects X and Y, and knowing that X is a "pinwheel," children will associate Y with the novel label "dax." Researchers such as Jean Liittschwager and Ellen Markman suggest that ME may also help children interpret novel words as feature labels (overcome the whole object assumption), but very young children may find second labels for items initially more difficult to acquire than the first because second labels are viewed as alternatives. Input, however, affects ME: Researchers Gail Gottfried and Stephen Tonks have shown that children who receive inclusivity data (*X* is a kind of *Y*; e.g., color labels taught as additional, not alternative, labels, "Here's a key; it's a green key"), generally accept multiple labels for items and form hierarchical relations; we have demonstrated the same for Alex. Thus, shown a wooden block, Alex answers, "What color?," "What shape?," "What matter?" *and* "What toy?" Parrots given colors or shapes as alternative labels (e.g., "Here's key," and only later, "It's green"), however, have difficulty learning to use these modifiers for previously labeled items. Griffin, thusly trained, initially answered, "What color?" with object labels. Similarly, while learning an object label— *cup*—he answered, "What toy?" with colors and had difficulty acquiring "cup." Thus even small input changes affect label acquisition as much for parrots as for young children.

Combinatory Learning

Based primarily on behavioral data, researchers such as Julie Johnson-Pynn and colleagues argue that a common neural area initially underlies young children's parallel development of communicative and object (manual) combinations, that a homologous area in great apes allows similar, limited, parallel development, and that such data imply a shared evolutionary history for communicative and physical behavior. But Griffin showed comparable limited, parallel combinatorial development of three-item and three-label combinations. Percentages of physical and vocal combinations were roughly equal; despite months of training, vocal three-label combinations emerged only when he more frequently initiated three-object combinations; vocal combinations were generally not those trained; and physical combinations were performed with his beak, not feet. Moreover, unlike Johnson-Pynn et al's *Cebus* monkeys, Griffin was not trained on physical tasks, and we limited training on three-label combinations (*2,5-corner wood/paper*) to see if spontaneous manipulative behavior developed in parallel with vocal complexity.

Although Griffin's behavior—or that of our most advanced subject, Alex (e.g., Pepperberg, 1999)—is equivalent neither to human language nor 2–3-year-old humans' combinatory behavior, we suggest that our greys' behavior patterns match some of nonhuman primates, that parallel combinatory development is not limited to primates, and that a particular mammalian brain structure is not uniquely responsible for such behavior. Responsible areas are likely analogous, arising independently under similar evolutionary pressures and searches for and arguments concerning responsible substrates and common behavior should not be restricted to primates.

Parallel Evolution of Avian and Mammalian Abilities?

Although human and animal behavior are not identical, we must examine many species for information on evolutionary pressures that helped shape existing systems. Such pressures were exerted not only on primates; hence the existence of analogous avian complex communication systems and their bases in analogous brain structures. Moreover, complex communicative systems apparently require, or likely coevolve with, complex cognition: Although communication is functionally social, its complexity is based on the complexity of information communicated, processed, and received; thus contingencies that shape intelligence (social, ecological, etc.) likely shape communication. If intelligence is indeed a correlate of primates' complicated social systems and long lives (i.e., the outcome of selection processes favoring animals that flexibly transfer skills across distinct domains as Paul Rozin suggests, and that, according to Nikolas Humphrey, remember and act upon knowledge of detailed social relations among group members), these patterns might also drive parrot cognition and vocal behavior: Long-lived birds, with complex social systems not unlike those of primates, could use abilities honed for social gains to direct information processing and vocal learning capacities. Add needs for categorical classes (e.g., to distinguish neutral stimuli from predators), abilities both to recognize and remember environmental regularities and adapt to unpredictable environmental changes over extensive lifetimes, and a primarily vocal communication system, then parrots' capacities are not surprising. Whether avian and human abilities evolved convergently (i.e., whether similar adaptive responses independently evolved in association with similar environmental pressures) is unclear, but a common core of skills likely underlies complex cognitive and communicative behavior across species, even if specific skills manifest differently. Questions now remain about such skills: Can birds that understand, for example, same–different, bigger–smaller, and number concepts, show flexible perspective-taking, engage in deceptive conversations, or expand their limited syntactic capacities? Even without such additional demonstrations, we must examine biases about avian abilities: Only by looking for commonalities across species can we develop successful theories about behavioral elements essential to, and the evolutionary pressures that have shaped, complex capacities.

See also Cognition—*Categorization Processes in Animals*
Cognition—*Concept Formation*
Cognition—*Equivalence Relations*
Cognition—*Social Cognition in Primates and Other Animals*

Acknowledgements: Writing this article was supported by the MIT School of Architecture and Planning; research was supported by NSF Grant IBN 96-03803, REU supplements, the John Simon Guggenheim Foundation, the Kenneth A. Scott Charitable Trust, The Pet Care Trust, the University of Arizona Undergraduate Biology Research Program, and *Alex Foundation* donors.

Further Resources

Baldwin, D. A. 1995. *Understanding the link between joint attention and language*. In: *Joint Attention* (Ed. by C. Moore & P. J. Dunham), pp. 131–158. Hillsdale, NJ: Erlbaum.
Griffin, D. R. 2000. *Animal Minds*. Chicago: University of Chicago Press.
Hollich, G. J., Hirsh-Pasek, K., & Golinkoff, R. M. 2000. *Breaking the language barrier*. Monographs of the Society for Research in Child Development, 262, 1–138.
Humphrey, N. K. 1976. *The social function of intellect*. In: *Growing Points in Ethology* (Ed. by P. P. G. Bateson & R. A. Hinde, pp. 303–317. Cambridge, UK: Cambridge University Press.
Pepperberg, I. M. 1999. *The Alex Studies*. Cambridge, MA: Harvard University Press.
Pepperberg, I. M. 2001. *Evolution of Avian Intelligence*. In: *The Evolution of Intelligence* (Ed. by R. Sternberg & J. Kaufman), pp. 315–337. Hillsdale, NJ: Erlbaum Press.

Irene M. Pepperberg

Cognition
Imitation

Imitation has been defined as reproducing an action or behavior after observation, or performing a novel or otherwise improbable action following the exposure to a skillful demonstrator. At present imitation is a hotly debated topic among researchers, and there is a wide disagreement about the definition of imitation, the species that have been shown being able to imitate, and the behavioral mechanisms that support and enable imitative behavior to manifest.

In contrast to the case of animals, scientists agree that humans are the most skillful species with regard to imitative abilities. Shortly after birth infants are able to imitate simple facial gestures produced by their mother or even an unfamiliar experimenter. Child psychologist Andrew Meltzoff observed that babies would respond with tongue protrusion if they see someone to push his tongue through the closed mouth. At a later age they are also able to reproduce simple bodily actions or actions on objects following a demonstration by either an adult or peers; moreover, they seem to remember such demonstrations for more than a day or even much longer. In these types of experiments, the infant sitting in his mother's lap can observe an action on a object by an experimenter; for example, the experimenter touching a piece of building block with his forehead. Actions are chosen on the basis that they are likely novel for the child and that the child has the ability to perform them. After some interval, which can be a few minutes or even 24 hours or more, the infant is offered the same object. Imitation is said to occur if the infant performs the demonstrated action more often then control infants who were not exposed to the demonstration or who have seen a different action. It would seem that infants imitate every "stupid" action that they witness. However, in recent experiments György Gergely has shown that infants are more "fastidious." Strange actions are imitated only when they "assume" some rationality on the part of the demonstrator. After watching an adult turn on a light box with her head by leaning forward when her hands were occupied (she was wrapped in a blanket which she held onto with both hands), infants did not imitate the action. In contrast, they showed imitation when the demonstrator did the same action, but with her hands lying free on the table. Infants might have inferred that the head action must offer some advantage in the particular situation to turn on the light.

Infants can recognize when others are imitating them, and apparently they use imitation not just as a means to learn new ways of interaction with the physical or social world, but also for communication with their young or adult companions. Gestural and action imitation is most frequent before the emergence of language, and its use as a communicative tool declines as the toddlers start to talk. Because of this wide utilization of imitation in human behavior, Meltzoff thought of this ability as the most characteristic human-specific trait, describing us as imitative generalists, *Homo imitans*.

One can suspect that our highly evolved imitative abilities are to be blamed partly for the current chaos in the scientific study of imitation in animals. In contrast to humans, present research suggests that imitative abilities play a very restricted role in the behavioral organization of animals in comparison to humans. Nevertheless, in order to understand the evolutionary origin of the ability, we need to take a closer look at imitation in the non-human species of the animal kingdom.

Early observers of animal behavior often described instances of imitation when they seemed to find some corresponding similarities to human behavior. For example, in the fifties, researchers working on a group of macaques on Koshima Island in Japan often provided the animals with potatoes in order to accustom the monkeys to their presence. On one day these researchers observed that Imo, a juvenile female, picked up the potato from the sand and took it to the sea, where she "washed" it before eating the food. This novel behavior did not only clear the sand from the potato, but also made it more tasteful. Relatively soon, other juvenile monkeys of this troop seemed to have adopted this novel habit. The same group of researchers also observed another skill when rice, which was thrown on the sand, was carried to the sea. This simple technique that separated the edible rice from the unpalatable sand also spread among the individuals of the group. At that time such obser-vations were often interpreted as cases of imitation, partly because researchers thought that this would be the natural explanation for the transmission of such novel behavior in humans.

Subsequent research, however, showed that such simple observational evidence is not enough to assume that imitation is indeed the only possible mechanism at work. Following the report on the "potato-washing" Japanese monkeys, Elisabetha Visalberghi, an Italian ethologist, tried to replicate the phenomenon in the laboratory, but she failed. Although some of the naive monkeys have also acquired the "potato-washing" behavior by observing knowledgeable demonstrators, but after observing the process of the acquisition in detail it was clear that animals were not imitating but were relying on other facilitative processes that resulted in superficially similar behavior at the end.

In order to separate imitation from other types of social learning, researchers needed to refine their definitions. There is consensus that imitation should be restricted to cases where, after a limited period of observation of a skillful demonstrator, the naive observer learns something about the form of the motor behavior. We need to adhere to this con-straining definition because in many cases it turned out that simpler mechanisms could ac-count for matching behavior on the part of the observers. Two alternative possibilities will now be considered in turn, both of which could lead to matching behavior between the demonstrator and the observer, but are not based on imitation.

About 70 years ago, great tits began to show a very ingenious behavior in certain parts of England, and this novel behavior spread rapidly in subsequent years. Some great tits found out how to open the sealed top of milk bottles found at the entrance of family houses. One could assume that observing a skillful bird, a naive observer would perform the similar behavior by the means of imitation, that is, repeating the action that it has seen leading to opening of the bottle. However, there are other possibilities. It could be the case that the skillful individual on the top of the bottle only drew the attention of the observer to the bot-tle, and later the naive bird preferentially approached similar objects and tried to open the lid by the method of trial and error. This process is called *stimulus enhancement* when the mere activity of a companion increases the salience of an otherwise "uninteresting" object or a location (local enhancement), which will be preferentially explored by observers at some later point in time. Stimulus enhancement is also thought to play a role in mate choice copy-ing by some animals. It has been observed that, for example, female guppies preferentially choose a male if he has previously been seen courting another female. The presence of a fe-male and the courting activity might "enhance" the attraction of the male over others.

Another potentially confounding factor is that observers might not imitate the action, but try to achieve a similar change in the environment, which corresponds to the result of the demonstrator's action. For example, if pushing a lid to the side allows access to food, the observer might try to remove the lid by whatever means it can. It will not necessarily aim to copy or reproduce the action performed by the demonstrator; instead, it will more likely try to achieve a similar change in the environment. If the animal has only a limited set of motor actions available for that task, and the change to be achieved is also constrained, similarity between demonstrator and observer behavior will be inevitable even without any attempt of imitation on the part of the naive subject. To distinguish this form of social learning from im-itation, primatologist Michael Tomasello introduced the term of *emulation*.

According to the present view, only a critical experimental approach can distinguish imitation from other alternative explanations, but it has proved to be a very difficult task to provide appropriate experimental evidence. Further, it seems that whether or not experi-ments support imitative abilities depends on the particular theoretical formulation used by researchers. At present there are many categorizations along which simpler and more complex forms of imitation are distinguished. Moreover, it has turned out that researchers need also understand the organization of the motor action and its mental representations before they can work out a model for imitation. Richard Byrne has distinguished three levels of imitation. At the first level a relatively familiar motor action is imitated in a novel context (*contextual imitation*); at the second level simple novel actions and/or their sequences are imitated (*production imitation*); and finally, at the third level the structure of a complex action is imitated (*program-level imitation*).

After several experimental designs to test for simple imitative abilities, many researchers claim that the so-called "two-action test" is the most appropriate method for revealing such abilities. The basic outline of the procedure is as follows. The experimenter trains demonstrator animals to perform one or another possible action on a simple manipulandum such as a lever. For example, quails have been trained either to step or peck at a lever in order to receive a piece of food. Importantly, both actions lead to the same movement of the lever (it was pushed downward); the only difference was the form of motor action displayed by the demonstrator. It has been found that naive quails preferred to show the behavioral action that corresponded to the behavior of their demonstrator. This means that after watching a demonstrator peck at the lever, the observer showed enhanced frequencies of lever-pecking behavior some minutes later. This result supports the view that quails and some other animals tested with this procedure have the ability to reproduce a behavior after witnessing an action displayed by a social companion. Further experiments have shown that quails imitate only if the demonstrator gets food as a result of performing the action. This means that the observers neglect "unsuccessful" displays. Although the results of such experiments seem to provide clear evidence for imitation in general, they can be criticized by the fact that the to-be-imitated actions (pecking, stepping, etc.) are very familiar to the observers; in other words, they are already part of the action repertoire of the individual before the experiment. Considering imitation as a form of learning then, it is not obvious whether the observers have acquired any kind of new "knowledge." One could assume that watching a companion display a simple behavioral action facilitates the use of a corresponding action in the observer, as in the case of barking, when dogs readily join in a barking chorus after hearing another dog emit such vocalization.

One way to solve this problem would be to adhere more strongly to the definition of imitation used at the beginning of this essay, requiring novel or "improbable" actions to be replicated. Many researchers have recognized that it is very difficult to verify what is novel for an individual of a given species, because one would need to ensure that the demonstrated action had been neither performed nor seen by the observer before. Because experiments are often performed with adult animals, it is unlikely that they are naive with regard to the demonstrated behavior. Therefore, novelty of any motor action must be thought of as a relative concept that depends on many factors. Although the pecking or the stepping action utilized in the two-action test might have been familiar to the observers, it is likely that they did not produce them in the particular circumstance earlier; that is, they did not peck at a lever placed in a special experimental box. This means that, in order to be successful, the observer needed not only to perform a pecking movement, but also orient it according to the actual requirements of the environment. However, this line of argument suggests that the particular form of the pecking was not necessarily learned by observation, but the actual motor command was controlled by the environment as perceived by the observer. So again in this case, imitation behavior is doubtful (if one assumes that imitation should lead to the acquisition of novel motor behavior), even if, by the facilitatory effect of the demonstrator's action, the observer has learned something about the usefulness of a possible action and its particular form in a given environment (*contextual imitation*).

This foregoing analysis suggests that theoretically it is very difficult to prove imitation at the level of simple actions, and, as shown previously, simple mechanisms might be equally effective. However, the ability to imitate could be advantageous if the demonstrator performs novel sequences of actions. Even if the individual actions are familiar to the observer, the sequence could be novel, and therefore a novel motor action sequence could be learned as a result of imitation (*production imitation*). To test this possibility, primatologist Andrew Whiten introduced a novel task to some chimpanzees. They observed how a human demonstrator removed the latches of an "artificial fruit" in order to get access to some edible food placed inside. The artificial fruit was a closed plastic box, which could be opened only if the animal or human removed a series of bolts and pins that locked the lid. Naive chimpanzees were shown one possible sequence of actions, and then they were allowed to open the "fruit" themselves. The results of these experiments showed that chimpanzees are able to learn the sequences of actions on the basis of observation because observers followed the action sequence shown by their particular demonstrators even though the fruit could be opened in other ways.

This laboratory evidence seems to provide some support for the assumptions that complex action sequences observed in apes living in their natural habitat might be also acquired by imitation. After observing the elaborate food processing behavior of the mountain gorilla, it has been suggested that young animals might learn this skill by imitation. The technique used to process the leaves of the nettle was described by Richard Byrne as "amassing a bundle of leaf blades without the stems of petrioles that host the strongest stings, and fold the bundle so that only a single leaf underside is exposed when the parcel is popped into the month." He also supposed that naive observers of skillful adults would not necessarily observe and learn all the minute motor movements of the action, but learn the basic structure of the action, described as a "program," that has to be known in order to process the plant. This kind of "program-level" imitation was also used to account for various "human-like" performances in the case of apes that were living in close contact with humans. A wide range of such imitative behaviors have been reported by Anne Russon and Birute Galdikas in the rehabilitant orangutans living in Tanjung Putting National Park (Indonesia); for example, orangutans were observed "making fire" or "weeding." One animal, Supinah, was observed washing clothes by dipping them into the water, rubbing them with soap, and subsequently with a brush, and finally wringing the wet clothes. Supinah did not imitate a particular action sequence, but more likely imitated the structure of the washing behavior that consists not only of a sequence of actions, but also of subgoals that must be achieved before the actor can proceed to the next task. For example, clothes need to be made wet before they can be rubbed with soap, so the action (dipping in the water) is not what matters, but the condition of the clothes. Up to now the only problem with this approach was that experimental evidence was lacking. Recently, the "artificial fruit" test was changed in a way to accommodate the possibility of "program-level" imitation. Three-year-old human children seemed to learn the structure of the task preferentially, and ignored the actual sequence, suggesting that this type of imitation learning is present in humans. Only further research can tell whether this ability is also present in our primate relatives.

Another line of research was interested in whether some animals could understand the concept of imitation. In the so called "do as I do" task, the animal has to learn that it should imitate any action of the demonstrator after a single observation on some kind of command (Do this!). The first such experiments were run with a chimpanzee, Viki. At the beginning

Viki was shown a certain set of actions that had to be repeated by her in order to get a reward. This training was terminated when the chimpanzee reliably imitated all demonstrated action. Researchers supposed that if during the training phase the chimpanzee acquired the concept of imitation then it should also imitate actions demonstrated for the first time. The results were encouraging because the chimpanzee was able to imitate many novel actions after a single observation. Recently, this experiment has been repeated with two other chimpanzees, and the outcome was the same. Both animals imitated more then half of the demonstrated novel actions, such as lip smacking, clapping hands, grabbing the left thumb by the right hand, pointing with the index finger to the nose, touching the ear, and so on. Interestingly, similar experiments failed with macaques, but not with dolphins which can imitate both humans and other dolphins in the "do as I do" test, as it has been reported by Louis Herman. For example, dolphins were shown a somersault under water either by a human demonstrator or by a dolphin companion, which was instructed by trainers giving visual signs (unobservable for the naive dolphin). Interestingly, the observer dolphins performed the same action in response to both demonstrations in spite of the fact that humans and dolphins used different body parts in acting. When the human demonstrator put an object in a basket by using his hands, the dolphin copied the action by using his rostrum. Simple human bodily gestures were also imitated. If for example, the human leaned forward and bent her head and upper body downward, as a response the dolphin lowered its own head, hunching forward at the water surface. These results not only suggest that dolphins have a concept of imitation, but also that they understand something about the functional significance of the action; that is, they do not aim for an accurate reproduction of the demonstration (which is impossible because of the anatomical differences between dolphins and humans), but they try to imitate the significant invariant aspect of the action.

Imitation has also interested researchers because it seems to be (still) a puzzle how an organism is able to plan a motor action based on observing another animal. The problem is that the demonstrated behavior is observed from a different perspective, and often the learner can not see its own actions during the performance, for example, in the case of facial imitation. Some have tried to explain imitation by assuming processes that resemble simple instrumental learning. In this explanation, observers learn to use corresponding actions after observing the demonstrator because matching behavior will lead only or mostly to reward. Others assume that at least in humans there is a built-in ability to imitate others. It has been supposed that infants actively compare the visual information about the observed demonstrator behavior with their own movement in space. There are also cognitive theories that relate imitative skills to perspective taking; that is, the observer has the ability to put himself in the perspective of the other in order to perform the matching action.

The main problem with most of these theories is that their viability is very difficult to test for. The recent discovery of the so-called mirror neurons, however, gives some hope for providing a mechanism for at least some forms of imitations. Giacomo Rizzolatti and others have observed that some population of neurons in the cortex of the monkey responds equally to an action performed or seen to be executed by the other. The same neurons were firing when the monkey had seen a companion (either human or another monkey) reaching for a piece of nut or when he was allowed to perform the same action. This discovery suggests that in the brain there is a mechanism that segments observed behavior into simple actions, and at the same time these actions are mapped onto the individual's own behavioral repertoire. Even if there are likely no mirror neurons for all types of actions, this neural mechanism is a candidate for explaining simple forms of contextual imitation.

Unfortunately, there are a few issues that present research on imitation has not dealt with to a great extent. In the case of animals it is not clear what the function of imitation is and how it contributes to the survival of the individual. One could argue that, if it is so difficult to find cases for imitation in comparison to alternative ways of learning, then this ability might play only a restricted role in the lives of animals. Further, even if some animals are capable of learning by imitation, there is little knowledge about whether a motor pattern acquired this way will actually "survive" and become incorporated into the individual's behavioral repertoire or would be counteracted by an alternative individually-experienced behavioral pattern. Interestingly, researchers describing one of the most complex action patterns of nut opening in chimpanzees seemed to find little evidence for imitation, mainly because it takes several years (6–7) to learn this skill.

At present there also seems to be a human–animal (ape–non-ape?) dichotomy with regard to the social function of imitation. In birds and some mammals, imitation seems to be under the strong control of reinforcement. Dolphins and apes seem to be less dependent on it, and humans are often pictured as "imitating for its own sake." In our species, imitation is not only important as a means to learn about a motor behavior, but also has an equally important function as a communicative social "tool." Therefore it remains to be seen whether the human imitative ability is a homologue of an animal trait or if it is specific invention of ours.

See also Learning—*Evolution of Learning Mechanisms*
　　　　　Learning—*Insight*
　　　　　Learning—*Social Learning*
　　　　　Learning—*Social Learning and Intelligence in Primates*
　　　　　Self-Medication

Further Resources

Byrne, R. W. 2002. *Imitation of novel complex actions: What does the evidence from animals mean.* Advances of the Study of Behavior, 31, 77–105.
Dautenhahn, K. & Nehaniv, C. 2002. *Imitation in Animals and Artifacts.* Cambridge, MA: MIT Press.
Galef, B. G. & Heyes, C. M. 1996. *Social Learning in Animals: The Roots of Culture.* New York: Academic Press.
Miklósi, Á. 1999. *The ethological analysis of imitation.* Biological Review, 74, 47–374.
Whiten, A. & Ham, R. 1992. *On the nature and evolution of imitation in the animal kingdom: Reappraisal of a century of research.* Advances in the Study of Behavior, 21, 239–283.
Zentall, T. & Galef, B. G. 1985. *Social Learning: Psychological and Biological Perspectives.* Hillsdale, NJ: Erlbaum.

Ádam Miklósi

■ Cognition
Limited Attention and Animal Behavior

Consider listening to a friend at a party: There are numerous conversations carried out simultaneously, which you can readily hear, yet you listen to only one person, tuning out the other sounds. Or imagine driving on a major highway trying to read the bumper sticker on the car just in front of you. Even though you can usually attend to the surrounding

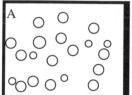 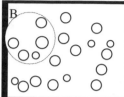 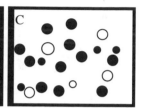

The effect of unfocused versus focused attention on the ease of detecting a cryptic target. All three panels contain the same target and background items at identical spatial configurations; the target in all three panels is the same circle, which is slightly larger than all other circles. Providing additional information about the target allows the reader to focus attention on either a selected area, the dotted circle in 1B, or the subset of white circles in 1C, making it easier and faster to detect the target than in the case of unfocused attention (1A). The target, which is at the middle right side in the circle in 1B, appears in that same location also in 1A and in 1C.
Courtesy of Reuven Dukas.

landscape and traffic, focusing on the bumper sticker results in neglecting the rest of the visual field. These two examples illustrate the possible effect of limited attention on animal behavior.

Limited attention means that the brain has a restricted rate of information processing. This implies that animals must often attend to only a subset of the relevant information that may affect their survival and reproduction. Neurobiological research involving either electrophysiological recordings of single neurons in monkeys or brain imaging of large neuronal populations in humans reveals that, when subjects face a difficult detection task involving a few distinct stimuli, focusing attention on a given stimulus is correlated with enhanced response and sharpened selectivity of the neurons that process that stimulus and a diminished activity of the neurons processing other stimuli. Simultaneous behavioral tests reveal that the probability of detecting the given stimulus is increased while the probability of detecting other stimuli is decreased.

Search Image

Limited attention may determine animals' feeding behavior and, consequently, the visual appearance of animals and plants. Many animals must search for *cryptic prey*, prey that is visually similar to its surrounding background and hence difficult to detect. Some animals are known to show search image, meaning that they focus their search on one cryptic food type while bypassing familiar, equally profitable distinct food types. For example, chickadees searching for camouflaged light green caterpillars may not notice the cryptic dark green caterpillars on the same foliage. Controlled laboratory experiments indicate that limited attention underlies animals' tendency to show search image. When blue jays (*Cyanocitta cristata*) had to search simultaneously for two cryptic target types, their overall rate of target detection was 25% lower than when they had to search for only a single target type at any given time. Dividing attention between the two difficult search tasks reduced foraging performance compared to focusing full attention on a single task. This suggests that, under many circumstances, animals searching for cryptic food items should focus attention on one food type while ignoring others. Additional experiments with blue jays and pigeons indicate that subjects focusing on searching for one target type are more likely to detect that type than other target types.

Because predators must focus their limited attention on the visual attributes characterizing a certain prey species, individuals of that species carrying genes that make them appear somewhat different from the common individuals might be overlooked by the predators and hence incur higher survival rates. Such a selective advantage of rare individuals can help maintain *polymorphism*, the existence of two or more distinct classes of individuals within a population. A genetically based polymorphism may result in *speciation*, defined as the generation of two or more new species from a single ancestral species. The enormous variation in leaf shapes of closely related co-occurring passion fruit plant species (*Passiflora spp*) and the extreme diversity of wing patterns within and between moth species may have partially resulted from the focused searching behavior of their butterfly herbivores and avian predators, respectively.

Search Rate

At any given time, animals can either attend to a wide angle of the visual field and coarsely perceive their surroundings, or focus on a narrow angle and extract fine details. The focus of attention can rapidly shift in width and spatial position, but there is always a negative correlation between the angle and level of detail per degree of the visual field. Animals searching for conspicuous targets can employ a wide attentional angle, which allows them to search a large area per unit time. But animals searching for cryptic targets must use a narrow attentional angle covering only a small area per unit time. Consequently, whereas a search for conspicuous items allows for a high search rate, an effective search for cryptic targets requires a low search rate. Thus limited attention may determine animal movement patterns.

Balancing Foraging and Predator Avoidance

Limited attention may also affect animal survival in nature. Most animals must simultaneously search for food and attempt to avoid ending up as their predator's meal. The amount of information processing required for that dual task of feeding and antipredatory behavior may exceed the animal's attentional capacity. In a controlled laboratory study, when blue jays focused on detecting cryptic targets at the center of the visual field, they were three times less likely to detect peripheral targets than when required to detect conspicuous targets at the center of the visual field. Even though the two experimental treatments (easy and difficult central detection) involved identical visual fields, identical conspicuousness of the peripheral targets, and identical frequencies of target appearance within the visual field, the difficult central detection treatment required more attention devoted to the center of the visual field, resulting in a much reduced frequency of detecting the peripheral targets than during the easy central detection treatment. This experiment suggests that, under natural settings, foragers engaged in a difficult food-searching task may be less likely to notice peripheral objects such as approaching predators. That conclusion is in agreement with data indicating increased frequency of predation when animals are engaged in foraging, playing, or courtship behavior compared to being unoccupied by such attention-demanding tasks.

In summary, limited attention requires animals to focus on only a small subset of the available information. This constraint may determine animals' diet choice and their rate of movement through the environment. The need to focus on a limited amount of information at any given time means that animals engaged in one important task sometimes do not notice peripheral events, such as a stealthily approaching predator.

See also Feeding Behavior—*Grizzly Foraging*
     Feeding Behavior—*Scrounging*
     Feeding Behavior—*Social Foraging*
     Social Evolution—*Optimization and Evolutionary*
       Game Theory

Further Resources

Chun, M. M. & Marois, R. 2002. *The dark side of visual attention.* Current Opinion in Neurobiology, 12, 184–189.

Dukas, R. 1998. *Cognitive Ecology: The Evolutionary Ecology of Information Processing and Decision Making.* Chicago: University of Chicago Press.

Dukas, R. 2002. *Behavioral and ecological consequences of limited attention.* Philosophical Transactions of the Royal Society of London. B., 357, 1539–1548.

Kastner, S. & Ungerleider, L. G. 2000. *Mechanisms of visual attention in the human cortex.* Annual Review of Neuroscience, 23, 315–341.

Reuven Dukas

■ Cognition
Mirror Self-Recognition

Mirror self-recognition (MSR) has been a controversial topic in the field of comparative psychology since it was first reported by Gordon Gallup in 1970. He provided behavioral evidence that chimpanzees were able to understand the nature of their mirror image, meaning that they recognized themselves in the mirror. Consider the phenomenon. A mirror image is a two-dimensional representation of the world, just like a picture or a photograph. However, unlike a still image, it is dynamic and mimics the behavior of the viewer. A chimpanzee who demonstrates MSR must understand that the mirror image is an actual representation in both time and space, and also, that it is not simply another chimpanzee looking back at her. It is, in fact, herself. But what does this imply? Are the mental abilities that allow MSR associated with other, more sophisticated, cognitive skills? To answer that question, a more detailed consideration of MSR is necessary.

To date, most individuals from all of the great apes have shown evidence of MSR through self-directed behaviors, the mark test, or both.
Courtesy of Jessie Cohen, National Zoo.

In Gallup's study, individual chimpanzees were each presented with a full-length mirror placed in front of their enclosure beyond their reach. Initially, the chimpanzees were agitated and responded as if their reflection was an unfamiliar individual. Within two or three days of exposure, they gradually stopped their social responses and began to use the mirror in novel ways. For example, they would make faces as they looked at the mirror, blow spit bubbles, or watch themselves chew food. Further, the chimpanzees also inspected areas of their bodies that they were normally unable to see, such as their teeth. Importantly, they used the mirror to guide their fingers during these investigations. Gallup termed these behaviors "self-directed," and interpreted them to mean that the chimpanzees recognized themselves in the mirror. To validate his interpretation, Gallup devised a test for mirror self-recognition. This test involved sedating the chimpanzees and applying an odorless red dye on one eyebrow and the opposite ear. When the chimpanzee awoke, she would have no idea that the dye was present unless she looked in the mirror. If she specifically touched the dyed areas, using the mirror to guide her fingers, this would be strong evidence for MSR. All of the mirror-experienced chimpanzees in Gallup's study passed this "mark test," thus supporting his interpretation of self-directed behaviors. Both measures are now used in studies of MSR.

Since Gallup's ground-breaking study, a number of scientists have investigated MSR in a variety of species, including dolphins, elephants, monkeys, and apes. Most of the research has centered on primates, which will be the focus of this essay. To date, individuals from all of the great apes (chimpanzees, bonobos, gorillas, orangutans, and humans) have shown evidence of MSR through self-directed behaviors, the mark test, or both. Among all of the other species of primates, only gibbons have demonstrated behaviors that are suggestive of MSR. While numerous studies have attempted to document MSR in other primate species, such as macaque monkeys, none have provided clear evidence of mirror-guided self-directed behaviors or passing the mark test. This difference across species is particularly interesting since individuals who have failed to show MSR clearly understand the reflective properties of a mirror and can use mirrors to obtain information about their environment. For example, macaque monkeys can use a mirror reflection to guide their hand in finding foods or objects that are otherwise hidden. Pygmy marmosets can learn to use a mirror to locate and monitor the position of rivals. When they threaten those rivals, they turn toward the rival's location directly, rather than threatening the mirror image. Yet, in these cases, the monkeys give no indication that they recognize themselves in the mirror.

In those species who demonstrate MSR, it does not appear immediately in young individuals, but must develop. In humans, the development of MSR parallels the development of other representational skills, such as symbolic play. In carefully controlled studies using a version of the mark test (one that does not require sedation), human infants do not show MSR until 15–18 months of age, and some as late as 24 months. The developmental course of MSR has not been well studied in other great apes, but there is some evidence that chimpanzees achieve this ability as early as $2\frac{1}{2}$ years of age.

There are numerous examples of individual nonhuman great apes that clearly and convincingly demonstrate MSR. However, despite prolonged mirror exposure and ample opportunity to interact with their reflection, there are also individual great apes that have not demonstrated any mirror-mediated, self-directed behaviors or the ability to pass the mark test. This variation among individuals has been documented in all of the great apes. Some of these exceptions are most likely related to deprived rearing conditions characterized by lack of appropriate social contact and/or little or no mental stimulation. In other cases, well-socialized apes that have grown up in enriched environments also fail to demonstrate MSR. In these cases, the explanation for this apparent lack of ability remains unclear. The suggested possibilities range from simple lack of interest in the mirror to a cognitive deficit that interferes with comprehension of their reflection. Do these differences in performance mean that there are, for example, two separate kinds of chimpanzees—those who are capable of demonstrating MSR and those who aren't? The answer is no.

The discovery of MSR in chimpanzees revealed the presence of a cognitive complexity that was previously assumed to be exclusive human territory.
Courtesy of Jessie Cohen, National Zoo.

Organisms appear to show three classes of behavior as they gain experience with their own reflection. These classes are social, contingent, and self-directed. Social responses to the mirror vary, ranging from greeting or play behaviors to threats or a fear response. Contingent behaviors are those which appear to test the relationship between an individual's movements and those of the mirror image. An example would be waving the arm or hand while watching that action in the mirror. Self-directed behaviors are distinct since they involve the use of the mirror to guide a behavior that is focused on the body. An example would be an orangutan using the mirror to groom his face. In this situation, an important distinction is that the ape moves her hand away from the mirror, and toward her own body while observing her reflection. Her actions indicate a clear understanding of the nature of the mirror image. Generally, when first exposed to a mirror, individuals initially show social responses. These behaviors wane as contingent and self-directed behaviors emerge. This progression may occur gradually over hours or even days. In experimental situations, the mark test is used to validate interpretations about self-directed behaviors.

Even though all of these behaviors are involved in MSR, each is not necessarily supported by the same mental abilities. Social responses to the mirror indicate that an individual is "reading" the image, although there is no comprehension of what it actually represents. These social behaviors decrease, most likely because the "responses" produced by the mirror image are not reciprocal, as would be expected from an actual social partner. However, these actions do provide information that begins to establish a correspondence between the mirror image and the individual's own movements. Although contingency behaviors emerge as social behaviors decline, a specific relationship has not been demonstrated experimentally. However, the rise in contingency behaviors suggests that the individual has perceived some relationship or correspondence between her own body movements and position in space with that of the mirror image. We interpret contingency behaviors as explicit testing of this correspondence. Self-directed behaviors are probably the outcome of an individual's understanding of the mirror image based on information that is acquired during contingency testing. So, let us return to the initial question that was posed. Are the mental abilities that allow MSR associated with other, more sophisticated, cognitive skills?

Most researchers that have investigated MSR agree that a relationship exists between mirror self-recognition and the phenomenon termed *theory of mind* (TOM). Theory of mind refers to the ability of one individual to understand their own knowledge and perceptions, and attribute that ability to others as well. Individuals with TOM use this understanding to project thoughts and intentions to others. That is, based on their own thoughts and experiences, they have formed a theory of how minds work. Having TOM allows for a range of abilities that are otherwise impossible. Intentional deception is one clear example. Unlike deceptive behaviors that are primarily controlled genetically, commonly referred to as "instinctive," TOM-based deception relies on more complex mental abilities most commonly associated either with distraction from, or concealment of, information. A now classic example involves a group of captive chimpanzees living on a large island.

Watching from their indoor enclosure, the apes witnessed their caretaker carrying a box of fruits onto the island. Well out of sight, he buried the fruits, leaving just a bit exposed so that they could be found. As he returned with the empty box, the chimpanzees eagerly anticipated their release onto the island. Once outside, the apes dashed about searching for the fruit. However, none were able to find the correct spot, or so it appeared. Later in the day, as the group began their afternoon nap, one individual waited until all of the other individuals were resting, and unnoticed, headed off directly to where the fruits were buried. Apparently having noticed their location earlier, but concealing that information, he enjoyed all of them

Individual Variation and MSR: The Response of Two Orangutans upon Seeing Their Reflection

Robert W. Shumaker and Karyl B. Swartz

The study of MSR in nonhumans has a relatively short history, and the results have generated lively debates for scientists working in this field. These deliberations are normal and healthy, serving to refine ideas, interpretations, and methodologies. A fundamental aspect of this process is a shared understanding among colleagues of the basics that are associated with MSR. The specifics of a valid "mark test" makes a fine example. In this procedure, the subject must be completely unaware that a mark has been applied. Before the mirror is presented, the subject's behavior is carefully observed and recorded to ensure that no touches are directed at the mark, providing evidence that the subject is truly naive. Once the mirror is exposed, the subject passes the test when they specifically attend to the mark by actively using their reflection. Interpreting the results of a properly conducted mark test should be straight forward, but the responses of some individuals can make this difficult. A study of MSR conducted by Daniel Shillito with orangutans provides good case studies for discussion.

Shillito and his colleagues worked with a group of captive-born orangutans who had never been formally tested on their responses to mirrors. After careful planning and preparation, including all of the proper control procedures, mark tests were begun. Two individuals provided notable results.

Junior, an adult male orangutan, proved to be an exceptionally interesting subject. He showed no awareness of the mark until the mirror was presented. Once his reflection was visible, he quickly noticed the mark and touched it with his finger as he looked directly into the mirror. After touching the mark, he smelled and inspected his finger, although the dried paint had left no residue. Without a doubt, he clearly passed the test by using his reflection to guide his fingers toward the odorless paint that was streaked across his brow. In every way, he provided a convincing response. Even without the presence of a mark, Junior demonstrated a particularly strong interest in his reflection. He sat in front of the mirror for long periods, inspecting areas of his body that were otherwise invisible to him. He would open his mouth into a wide yawn and examine each of his teeth. In addition, he observed his genitals, carefully touched the area around his eyes, and even gently exposed the inside of his eyelids as he gazed intently into the mirror. On at least one occasion that might be considered self adornment, he draped a burlap bag over himself like a shawl and looked at his reflection. One interpretation is that he was admiring his appearance.

Bonnie, an adult female orangutan, behaved quite differently when exposed to a mirror. Although attentive and interested, she showed no clear indication that she recognized her reflection. When given the mark test, she provided a unique response. As with Junior, she was completely unaware that a mark had been painted on her forehead. Upon presentation of the mirror, she looked directly at her reflection. She made no attempt to touch the mark, but then quickly walked away from the mirror. She moved directly across her enclosure to a water spigot, turned it on, and vigorously rubbed her forehead under the stream. The mark was completely washed away. Officially, Bonnie failed the mark test. Would you agree with this interpretation?

himself and then returned to the group. In this case, the deceiver clearly had information that he withheld from the group. By not acting on what he knew, until he was satisfied that the other group members were sufficiently occupied away from the fruit, he eliminated any potential competition. His ability to understand that he knew something that the rest of the group did not, and to deceive them by concealing that information, is best explained by theory of mind.

Although there is clear overlap in the mental abilities that support MSR and TOM, this does not mean that the two are equivalent. An individual who shows MSR has the mental capacity to make use of the information provided by his reflection to reveal knowledge of himself as a separate entity. The mirror is the methodological bridge between this self concept and behavior. The mirror provides an opportunity for the individual to demonstrate knowledge that is otherwise private. However, this level of understanding does not sufficiently explain the range of behaviors that is associated with TOM. In our view, showing self-directed behaviors or successfully passing the mark test is not sufficient evidence to support the notion that TOM is present. Even though the cognitive skills that are necessary for MSR are also essential for TOM, at least three additional capacities are required as well. Specifically, they are the ability to consider one's own thoughts, the ability to understand that those thoughts are distinct from those of other individuals, and the ability to use that knowledge to predict or manipulate another individual's behavior.

The cognitive ability that allows MSR is directly related to the other, more sophisticated, mental skills necessary for TOM. It is the first in a suite of interdependent cognitive capacities that lead to the expression of TOM.

The discovery of MSR in chimpanzees revealed the presence of a cognitive complexity that was previously assumed to be exclusive human territory. That phenomenon has now been extended to all of the other great apes as well. Current research into TOM is proceeding in a similar manner, providing evidence that the mental skills that are present in the nonhuman great apes exceed previously accepted limits. As our understanding of the minds of the great apes continues to progress, the boundaries that separate humans from the other great apes continue to be blurred. As Charles Darwin suggested, the mental differences between humans and the other great apes are in degree, not in kind.

See also Cognition—*Animal Consciousness*
Cognition—*Mirror Self-Recognition and Kinesthetic–*
Visual Matching

Further Resources

Bekoff, M., Allen, C., & Burghardt, G. M. 2002. *The Cognitive Animal*. Cambridge, MA: MIT Press.
Gallup, G. G. Jr. 1970. *Chimpanzee self-recognition*. Science, 167, 86–87.
Mitchell, R. W., Miles, H. L. & Thompson, N. (Eds.). 1997. *Anthropomorphism, Anecdotes, and Animals*. Albany, NY: SUNY Press.
Parker, S. T., Mitchell, R. W., & Miles H. L. (Eds.) 1999. *The Mentalities of Gorillas and Orangutans: Comparative Perspectives*. Cambridge, MA: Cambridge University Press.
Tomasello, M. & Call, J. 1997. *Primate Cognition*. New York: Oxford University Press.
de Waal, F. B. M. 1998. *Chimpanzee Politics: Power and Sex among Apes*. Baltimore: Johns Hopkins Press.

Robert W. Shumaker & Karyl B. Swartz

The ideas expressed in this essay were developed in part during research funded by NIH grant GM08225.

■ Cognition
Mirror Self-Recognition and Kinesthetic–Visual Matching

Recognizing yourself in a mirror or on video seems to be a simple process. After all, when you look in a mirror, you see someone who looks just like you. Recognition seems more difficult when you think about how you knew what you looked like before you ever recognized yourself in a mirror. Without a visual representation of yourself to begin with, how did you ever figure out that the image in the mirror is an image of yourself? This question becomes even more complicated when you realize that language is unlikely to be a significant part of the explanation: Human children begin to recognize themselves at around 18 months of age, a time when they have a limited ability for language; and great apes (chimpanzees, bonobos, gorillas and orangutans) who do not use language (and even ones who do), as well as other animals such as dolphins, killer whales, and lesser apes, can recognize themselves in mirrors. One suggestive explanation for self-recognition, associated with comparative psychologist Gordon Gallup, is that you have a self-concept (or a sense of self, or self-awareness) which allows you to figure out what you look like, but the "self" in each of these expressions of self that allows you or me to recognize ourselves remains unexplained. Being aware that I exist, or that I have (am) a self, seems necessary at some level for self-recognition, but how I might use such awarenesses to know what I will look like in a mirror is hard to understand. A more straightforward explanation of mirror self-recognition, initially proposed by developmental and comparative psychologist Paul Guillaume in the 1920s, is that you can recognize a match between what your body feels like (which is variously called kinesthesis, somasthesis, or proprioception) and what you (or someone acting like you) look like. To get a sense of kinesthetic–visual matching, think of the game "Simon Says," in which children are supposed to recreate (or not) activities of another person. How can they do this? The likely answer is that when they see someone else perform an action, they know what that feels like (and would look like) using their own body: For example, someone's arm moving to the top of his or her head looks like what your own arm moving to the top of your head feels like. This kinesthetic–visual matching, once you become aware of it, seems effortless and uncomplicated, but it isn't: Monkeys, dogs, and other animals do not appear capable of self-recognizing or recreating another's actions, and thus appear not to have the ability for kinesthetic–visual matching.

Guillaume believed that mature kinesthetic–visual matching developed from the same process that allows imitation of others to arise: a gradual building of awareness of similarities between an infant's own actions and those of others based on recurrent experiences in which the infant and others simultaneously perform the same actions (usually toward the identical object or a similar one), which then become associated with each other. The infant in these situations sees both its own and other's movements, and feels its movements at the same time it sees its movements, all of which create associations between visual and kinesthetic aspects of similar actions. More recent researchers argue that a significant part of the basis for imitation of others is already present in a limited ability for kinesthetic–visual matching at birth or soon after. In my view, a more generalized ability for kinesthetic–visual matching is necessary for self-recognition and the more elaborate imitation-based activities of 18-month-old children, including generalized imitation of others, pretending to be another, and recognition of being imitated. So far evidence of such a generalized kinesthetic–visual matching ability in apes and dolphins is present only in imitation of others and self-recognition, and indeed these are observed in only some members of each species tested. Evidence of a generalized capacity for imitation of others can be demonstrated when animals can be trained to replicate almost any behavior by someone else or when animals spontaneously show bodily recreation of a variety of behaviors not normally shown by these animals. Evidence of self-recognition is usually demonstrated either by self-exploration—looking in the mirror at parts of its body the animal normally cannot see, such as parts of the face, inside the mouth, or the back—or by wiping something off the face that doesn't belong there, such as a spot of facepaint or a sticky paper dot placed there without the animal's knowledge, after seeing one's image in the mirror. Surprisingly, animals who show one piece of evidence of self-recognition do not always show the other; the more difficult activity, and the one usually used for evidence of self-recognition in children, seems to be wiping something from one's face that doesn't belong. The greater frequency of self-exploration suggests that animals may more easily know general properties of their bodies (such as that their mouth has an inside or where their nose is) than specific properties (such as that their face should not have a red dot over their eyebrow). The fact that recognizing a photograph of oneself or delayed video images of oneself occurs later than self-recognition in mirrors for human children suggests that the image of yourself based solely on visual processes is a later development in the understanding of self, presumably derived from acquaintance with your mirror image.

Even more elaborate understandings of the self, such as evaluations of yourself from someone else's perspective, most likely build on the understandings of self derived from kinesthetic–visual matching, but also seem to require language or other symbolic systems. Sign-using apes show some evidence of an evaluative self, including using mirrors to transform their visual image or calling their own actions "good" or "bad," but human children develop their evaluations of themselves much further. How complicated the evaluative self can be in nonhuman species, especially those taught to understand signs, is an open question. But the understanding of self present in kinesthetic–visual matching is clearly one we humans share with at least some other animals.

See also Cognition—*Mirror Self-Recognition*
Cognition—*Animal Consciousness*

Further Resources

Mitchell, R. W. 1997. *A comparison of the self-awareness and kinesthetic-visual matching theories of self-recognition: Autistic children and others*. Annals of the New York Academy of Sciences, 818, 39–62.

Mitchell, R. W. 2002. *Subjectivity and self-recognition in animals*. In: *Handbook of Self and Identity* (Ed. by M. R. Leary & J. Tangney), pp. 567–593. New York: Guilford Press.

Nadel, J., & Butterworth, G. (Eds.) 1999. *Imitation in Infancy*. Cambridge, UK: Cambridge University Press.

Parker, S. T., Mitchell, R. W., & Boccia, M. L. (Eds.) 1994. *Self-awareness in Animals and Humans*. New York: Cambridge University Press.

Swartz, K. B., Sarauw, D., & Evans, S. 1999. *Comparative aspects of mirror self-recognition in great apes*. In: *The Mentalities of Gorillas and Orangutans* (Ed. by S. T. Parker, R. W. Mitchell & H. L. Miles), pp. 283–294. Cambridge, UK: Cambridge University Press.

Robert W. Mitchell

■ Cognition
Social Cognition in Primates and Other Animals

Early studies of learning and intelligence in nonhuman primates involved laboratory experiments that were specifically designed to test abilities related to some component of human cognition—for example, concept formation or language learning. Although valuable for their precision and control, these experiments suffered from a number of drawbacks. Subjects were usually tested with arbitrary stimuli, such as different-shaped blocks, tones, or human words, that had little biological validity to the species being tested. Moreover, subjects were usually singly-housed animals that had few if any contact with other individuals. As a result, little attention was paid to the animals' natural communication or social behavior.

It was not until the pioneering work by Japanese scientists on Japanese macaques (*Macaca fuscata*) in the 1950s that primate studies began to focus on the behavior of known individuals living in natural social groups, and on the cognitive capacities displayed by primates in the absence of human training. These and the many studies that followed revealed the complexity of primate social behavior. They suggested that animals not only recognize others as individuals, but also behave differently toward one another depending on kinship, sex, dominance rank, reproductive condition, and their previous history of interaction.

For example, in such Old World monkey species as macaques (*Macaca* spp.), vervets (*Cercopithecus aethiops*) and baboons (*Papio cynocephalus spp.*), groups are composed of up to 100 individuals. Females remain in their natal groups throughout their lives, whereas males emigrate to neighboring groups at sexual maturity. Females form dominance hierarchies according to matrilineal rank, and offspring "inherit" their mothers' ranks. The stable core of vervet and baboon groups, therefore, is a hierarchy of matrilineal families. Although females maintain close bonds with their matrilineal relatives throughout their lives, they also interact with non-kin and often compete to form bonds with the members of high-ranking families. In contrast to female rank, male dominance rank is based primarily on age and fighting ability. In some species, unrelated males form alliances to compete with higher-ranking individuals.

Social groups of monkeys, then, are composed of individuals of varying degrees of genetic relatedness and dominance ranks. Relationships are simultaneously cooperative and competitive, and alliances are common. As a result, there seems to be strong selective pressure for individuals to recognize not only their own close kin, dominance rank, allies, and rivals, but also the kin, ranks, allies, and rivals of others. A number of observational and experimental studies have now documented this form of "third party" recognition in several monkey species. What remains unclear is the extent to which primate social cognition might differ from that of other animals.

Several authors have argued that the selective pressures imposed by complex social interactions have favored the evolution of large brains in primates, including humans. Two sorts of evidence support this "social intelligence" hypothesis. First, in primates the size of the neocortex is positively correlated with group size. If one accepts the view that increasing group size inevitably leads to increasing social complexity, large brains may have evolved in response to the demands of social life.

Second, whereas throughout the animal kingdom brain size increases with increasing body size, brain size to body weight ratios also differ from one taxonomic group to another. Primates have larger brain-to-body-weight ratios than other mammals. Structural differences in the brain are also apparent. Within primates, the proportion of the brain devoted to the neocortex increases progressively from monkeys to apes to humans. Within the neocortex, ape (and especially human) brains have a particularly enlarged prefrontal cortex, an area known to be involved in many forms of abstract thought, rule learning, and decision making.

Paralleling these anatomical correlates, primates do indeed differ from many other species in at least one measure of socially complex behavior, patterns of alliances. Alliances occur during aggressive interactions, when two animals cooperate to threaten or chase a third. In most species of animals, allies are usually close relatives. Monkeys and apes, however, also regularly form alliances with nonrelatives. When soliciting alliances with non-kin, monkeys and apes consistently select partners whose relative rank is higher than their opponent's. To select partners strategically, according to their relative power, an individual must know not only its own relative rank but also the rank relations that exist among others. Such "third party" knowledge can be obtained only by observing interactions in which one is not involved and making the appropriate deductions. Moreover, as group size increases the need for such knowledge places increasing demands on individuals, because larger groups produce an explosive growth in the number of triadic relations. If primates are, in fact, unique in forming strategic alliances, and if strategic alliances require knowledge of the relations that exist among others, then the social competition found in large groups offers one explanation for primates' unusually large brains.

Early evidence that monkeys recognize other individuals' social relationships emerged as part of a relatively simple playback experiment designed to document individual vocal recognition in vervet monkeys living in Amboseli National Park, Kenya. Field observations revealed that mothers often ran to support their juvenile offspring when the juveniles screamed during aggressive interactions. Mothers appeared to recognize the calls of their offspring. To test this hypothesis, we designed a playback experiment in which we played the distress scream of a juvenile to a group of three adult females, one of whom was the juvenile's mother. As expected, mothers consistently looked toward the loudspeaker for longer durations than did control females. Even before she had responded, however, a significant number of control females looked at the mother. In so doing, they behaved as if they recognized not only the identity of a signaler unrelated to themselves, but also associated that individual with a specific adult female.

In an attempt to replicate these results, we carried out similar experiments on baboons in the Moremi Game Reserve, Okavango Delta, Botswana. In these trials, two unrelated female subjects were played a sequence of calls that mimicked a fight between their close relatives. The females' immediate responses to the playback were videotaped, and both subjects were followed for 15 minutes after the playback to determine whether their behavior was affected by the calls they had heard. In separate trials, the same two subjects also heard two control sequences of calls. The first control sequence mimicked a fight involving the dominant subject's relative and an individual unrelated to either female; the second mimicked a fight involving two individuals who were both unrelated to either female.

After hearing the test sequence, a significant number of subjects looked toward the other female, suggesting that they not only recognized the calls of unrelated individuals, but also associated these individuals with their kin (or close associates). Females' responses following the test sequence differed significantly from their responses following control sequences. Following the first control sequence, when only the dominant subject's relative appeared to be involved in the fight, only the subordinate subject looked at her partner. Following the second control sequence, when neither of the subjects' relatives was involved, neither subject looked at the other. Finally, following a significant proportion of test sequences, the dominant subject approached and supplanted the subordinate (a mild form of aggression). In contrast, when the two subjects approached each other following the two control sequences, the dominant rarely supplanted the subordinate.

Taken together, these experiments suggest that baboons and vervet monkeys recognize the individual identities of group members unrelated to themselves, and that they recognize the social relationships that exist among these animals. Monkeys appear to view their social groups not just in terms of the individuals that comprise them but also in terms of a network of social relationships in which certain individuals are linked with several others. Moreover, their behavior is influenced not only by their own recent interactions with others but also by the interactions of their close associates with other individuals' close associates.

Dominance relations offer another opportunity to test whether animals gain information about other individuals' relationships by observing their social interactions. Like matrilineal kinship, linear, transitive dominance relations are a pervasive feature of social behavior in groups of monkeys. A linear, transitive rank order might emerge because individuals simply recognize who is dominant or subordinate to themselves. In this case, a linear hierarchy would occur as an incidental outcome of paired interactions and there would be no evidence to suggest that animals observe others' interactions. Alternatively, a linear hierarchy might emerge because individuals genuinely recognize the transitive dominance relations that exist among others: A middle-ranking individual, for example, might know that A is dominant to B and B is dominant to C, and therefore conclude that A must be dominant to C. Because a female's dominance rank in many species of monkeys is determined by the rank of her matriline, knowledge of another female's rank cannot be obtained by attending to absolute attributes such as age or size. Instead, it requires that animals monitor other individuals' interactions.

Several observations suggest that monkeys do recognize the rank relations that exist among others females in their group. For example, dominant female baboons often give a quiet, tonal "grunt" when they approach more subordinate females with infants and attempt to handle the infant. Grunts seem to facilitate social interactions by appeasing anxious mothers and signaling friendly intent, because an approach accompanied by a grunt is significantly more likely to lead to subsequent friendly interaction than is an approach without a grunt. Occasionally, however, a mother will utter a submissive call, or "fear bark," as a dominant female approaches. Fear barks are an unambiguous indicator of subordination; they are never given to lower-ranking females.

To test whether baboons recognize the rank relations of others, we and Joan Silk designed a playback experiment in which adult female subjects were played an anomalous call sequence in which a low-ranking female apparently grunted to a high-ranking female and the higher-ranking female apparently responded with fear barks. As a control, the same subjects heard the same sequence of grunts and fear barks made causally consistent by the inclusion of additional grunts from a third female who was dominant to both of the other signalers. For example, if the anomalous, inconsistent sequence was composed of female 6's grunts followed by female 2's fear barks, the corresponding consistent sequence might begin with female 1's grunts, followed by female 6's grunts and ending with female 2's fear barks. Some subjects were higher-ranking than the signalers, others were lower-ranking. Regardless of their own relative ranks, subjects responded significantly more strongly to the inconsistent sequences, suggesting that they recognized not only the identities of different signalers but also the rank relations that exist among others in their group.

The ability to rank other group members is perhaps not surprising, given the evidence that captive monkeys and apes can be taught to rank objects according to an arbitrary sequential order, the amount of food contained within a container, their size, or the number of objects contained within an array. What distinguishes the social example, however, is the fact that, even in the absence of human training, female monkeys seem able to construct a rank hierarchy and then place themselves at the appropriate location within it.

Humans routinely classify others according to both their individual attributes, such as social status or wealth, and membership in higher-order groups, such as families or castes. Although nonhuman primates recognize other individuals' dominance ranks and kin relations, it is unclear whether they classify others according to both criteria simultaneously. Humans make such higher-order classifications easily, and as a result recognize that not all superficially similar interactions have equal significance. For example, in Shakespeare's *Romeo and Juliet*, we discount Mercutio's teasing of Romeo as trivial because both Mercutio and Romeo are allied with the house of Montague. When Mercutio aims his taunts at Tybalt, however, we regard his behavior as more ominous, because Tybalt is a Capulet. Our responses are guided in part by our tendency to organize social relationships into a hierarchical structure—in this case, familial affiliation—that is governed by a functional set of rules: Quarrels between families are potentially much more destructive than quarrels within families.

To test for the existence of similar classification in baboons, Thore Bergman, Jacinta Beehner, and we once again used a playback experiment in which adult females heard a sequence of calls mimicking a fight between two other females. The experiments used a within-subject design. On separate days, the same subject heard one of three different call sequences: (1) an anomalous sequence mimicking a *within-family* rank reversal (e.g. B_3 threat-grunts + B_1 screams); (2) an anomalous sequence mimicking a *between-family* rank reversal (e.g. C_1 threat-grunts + B_3 screams); and (3) a no-reversal control sequence consistent with the female dominance hierarchy. In some control sequences the signalers were related to each other (e.g. B_1 threat-grunts and B_3 screams); in others, they were unrelated (e.g. B_3 threat-grunts + C_1 screams).

Having already found that subjects respond more strongly (by looking toward the speaker for longer durations) to playback sequences that were inconsistent with the current dominance hierarchy than to those consistent with it, we predicted that subjects would respond more strongly to both of the rank reversal sequences than to the control sequences. We further predicted that, if baboons were simultaneously sensitive to both rank and kin relations, they should respond more strongly to sequences that simulated a between-family rank reversal than to sequences that mimicked a within-family rank reversal. In species such as baboons, both within- and between-family rank reversals are rare. When within-family rank reversals do occur, however, they typically involve only two individuals and have little effect on social relationships outside the matriline. In contrast, occasional between-family rank reversals represent major social upheavals in which all the members of two or even more matrilines may lose or gain rank. They therefore have the potential to influence the rank relations of many individuals.

One possible confound arose because members of the same matriline occupy adjacent ranks, while members of different matrilines are often more widely separated in rank. As a result, subjects might respond more strongly to a between-family rank reversal sequence simply because the reversal involved individuals of more disparate ranks. We controlled for rank distance by ensuring that a proportion of the within-family rank reversals involved signalers from large matrilines who were separated by as many as seven ranks. Similarly, a proportion of the between-family rank reversals involved signalers who were adjacent in rank; that is the lowest-ranking female in one matriline and the highest-ranking female in the next.

Subjects responded significantly more strongly to sequences that mimicked a between-family rank reversal than to both within-family rank reversal sequences and no-reversal control sequences. In contrast, although subjects on average responded for a longer duration to within-family rank reversal sequences than to control sequences, this difference was not significant. Among control (non-reversal) sequences, subjects responded equally strongly to those that mimicked a within-family dispute as to those that mimicked a between-family dispute.

Subjects' responses to apparent rank reversals were unrelated to the rank distance separating the two signalers. In both within- and between-family rank reversal sequences, subjects responded as strongly to apparent reversals involving closely-ranked opponents as to those involving more distantly-ranked opponents. Similarly, there was no evidence that between-family rank reversals elicited stronger responses because members of different matrilines were rarely close to one another, or interacted at lower rates than non-kin (Bergman et al. 2003).

Results suggest that monkeys classify others simultaneously according to both their individual attributes and their membership in higher-order groups, and that they do so in the absence of human training. Baboons appear to understand that their group's female dominance hierarchy can be subdivided into matrilines. As a result, they may recognize that, although predictable rank relationships are maintained both within and between matrilines, the latter are qualitatively different from the former.

More speculatively, results may be relevant to theories concerned with the evolution of language—long considered the most fundamental feature distinguishing human from animal cognition. In language, humans deduce the meaning of sentences by arranging words into nested, hierarchical groups. In this "tree structure," words are defined according to the functional role they play in noun phrases, verb phrases, and so on. At present, it is not known whether the formation of such rule-governed, hierarchical groups is unique to language or might originally have evolved to serve other, non-linguistic purposes, for example in the domains of number, spatial memory, or social relations. Our results suggest that baboons organize their companions into a hierarchical, rule-governed structure based simultaneously on kinship and rank. Like words in a sentence, individual baboons are recognized both as individuals and according to the functional roles they play in society as holders of a particular dominance rank and members of a specific kin group. The selective pressures imposed by life in complex groups may therefore have favored cognitive skills that constitute an evolutionary precursor to at least some components of human language.

Do primates differ from other animals in their ability to infer third party social relationships? We can identify at least three competing hypotheses. The first argues that primates are in fact more intelligent than non-primates. This intelligence is reflected not only in tests of captive animals but also in primates' superior ability to keep track of complex social relationships. The difference between primates and nonprimates is qualitative and fundamental, and will be corroborated by future research.

The second hypothesis maintains that selection has favored the ability to recognize other individuals' relationships in all species that live in large, complex social groups. According to this hypothesis, monkeys only appear to have a greater capacity to recognize third party social relationships because they have received more attention than nonprimates living in similarly large groups. Once this imbalance in research has been redressed, differences between primates and other animals will disappear, to be replaced by a difference that depends primarily on group size and composition. There is some neuroanatomical support for this hypothesis: In carnivores, as in primates, there is a positive correlation between neocortex size and group size.

A third hypothesis claims that neither phylogeny nor group size and composition have influenced animals' ability to gain information about other individuals' social relationships.

It argues, in effect, that there are no species differences in "social intelligence." Monkeys, for example, appear to excel only in their ability to recognize allies' and opponents' relative ranks because their large social groups allow them to display this knowledge. In contrast, studies of species that live in small social groups have to date focused primarily on observers' abilities to assess the dominance of only two individuals. Once monogamous and even solitary species have been given the opportunity to reveal what they know about the social relationships of many different individuals, they will be shown to possess a level of "social intelligence" that is no different from that found among animals living in large social groups.

At present, it is difficult to test among these hypotheses, although some relevant information is beginning to emerge. For example, data from dolphins (*Tursiops truncates*) and hyenas (*Crocuta crocuta*) suggest that nonhuman primates are not the only mammals in which individuals acquire information about many different individuals' social relationships. When competing over access to females, male dolphins form dyadic and triadic alliances with selected other males, and allies with the greatest degree of partner fidelity are most successful in acquiring access to females. The greater success of high-fidelity alliances raises the possibility that males in newly formed alliances, or in alliances that have been less stable in the past, recognize the strong bonds that exist among others and are more likely to retreat when they encounter rivals with a long history of cooperation.

Like many species of Old World monkeys, hyenas live in social groups comprised of matrilines in which offspring inherit their mothers' dominance ranks. In a 1999 study, Kay Holekamp and her colleagues played recordings of cubs' "whoop" calls to mothers and other breeding females. As with vervet monkeys and baboons, hyena females responded more strongly to the calls of their offspring and close relatives than to the calls of unrelated cubs. In contrast to vervets and baboons, however, unrelated females did not look at the cubs' mothers. One explanation for these negative results is that hyenas are unable to recognize third-party relationships, despite living in social groups that are superficially similar to those of many primates. Alternatively, hyenas may simply be uninterested in the calls of unrelated cubs.

In fact, hyenas' patterns of alliance formation suggest that they do monitor other individuals' interactions and extrapolate information about other animals' relative ranks from their observations. During competitive interactions over meat, hyenas often solicit alliance support from other, uninvolved individuals. When choosing to join ongoing skirmishes, hyenas that are dominant to both of the contestants almost always support the more dominant of the two individuals. Similarly, when the ally is intermediate in rank between the two opponents, it inevitably supports the dominant individual. These data provide the first evidence for a nonprimate species that alliance partners may be chosen on the basis of both the allies' and the opponents' relative ranks. They are consistent with the hypothesis that hyenas are able to infer transitive rank relations among other group members.

Recent research on "eavesdropping" by birds and fish indicates that even animals living in small social groups are capable of monitoring other individuals' social interactions to acquire detailed information about their relative dominance or attractiveness as a mate. Often, this information is of necessity restricted to a few other individuals. Among territorial bird species living in small family groups, for instance, questions about the ability to track social relationships among many other individuals are largely moot, because the opportunity to monitor interactions among all possible neighbors rarely arises. Eavesdropping on the competitive singing duets of strangers, for example, allows territorial songbirds to extract information about the two contestants' relative dominance. Whether these birds would also be capable of recognizing a dominance hierarchy involving numerous individuals remains unclear. Although many species of songbirds form flocks during the winter, little is known about the social interactions that take place within such flocks, or the degree to which flock members recognize other individuals' relative ranks.

One recent experiment attempted to test the prediction that socially-living birds will display enhanced abilities to make transitive inferences by comparing the performance of highly social pinyon jays (*Gymnorhinus cyanocephalus*) with relatively nonsocial western scrub jays (*Aphelocoma californica*). Using operant procedures, subjects were required to order a set of arbitrary stimuli by inference from a series of dyadic comparisons. Although subjects of both species learned the sequence order, pinyon jays did so more rapidly and accurately than did scrub jays. Although not conclusive, these results lend support to the hypothesis that social complexity may be correlated with superior performance in tasks involving the ranking of multiple stimuli.

Similarly, very little is as yet known about the ability of nonprimate mammals or birds to recognize other individuals' social relationships. Colonial white-fronted bee-eaters *Merops bullockoides* offer one example of an avian society in which there would appear to be strong selective pressure for the recognition of other individuals' kin groups. Observational evidence suggests that bee-eaters may recognize other individuals and kin groups and associate these groups with specific feeding territories, although this has not yet been tested experimentally.

Clearly, more data are needed from both natural and laboratory studies before we can make any definitive conclusions about cognitive differences between primates and other animals, or between species living in large as opposed to small groups. It remains entirely possible that apparent species differences between primates and other animals in the recognition of third party social relationships are due more to differences in the social context in which eavesdropping occurs than to any cognitive differences in the ability to monitor social interactions. Given the opportunity to evaluate the social relationships of many different individuals, species living in small family groups and even primarily solitary species may well be shown to have similar abilities to those living in large social groups. It is to be hoped that future research will attempt to investigate the extent to which gregarious species in taxa other than primates are capable of recognizing the close associates and allies of other group members, and to determine the neural correlates of this ability.

See also Cognition—*Animal Consciousness*
Cognition—*Cognitive Ethology: The Comparative Study of Animal Minds*
Cognition—*Equivalence Relations*
Cognition—*Fairness in Monkeys*
Cognition—*Mirror Self-Recognition*
Cognition—*Mirror Self-Recognition and Kinesthetic–Visual Matching*
Cognition—*Theory of Mind*
Empathy
Play—*Dog Minds and Dog Play*
Play—*Social Play Behavior and Social Morality*

Further Resources

Bergman, T. J., Beehner, J. C., Cheney, D. L. & Seyfarth, R. M. 2003. *Hierarchical classification by rank and kinship in baboons.* Science, 302, 1234–1236.
Cheney, D. L. & Seyfarth, R. M. 1990. *How Monkeys See the World: Inside the Mind of another Species.* Chicago: University of Chicago Press.

Harcourt, A. H. & de Waal, F. (Eds.) 1992. *Coalitions and Alliances in Humans and Other Animals.* Oxford: Oxford University Press.

Humphrey, N. K. 1976. *The social function of intellect.* In: *Growing Points in Ethology* (Ed. by P. P. G. Bateson & R. A. Hinde), pp. 303–321. Cambridge: Cambridge University Press.

Macgregor, P. (Ed.) 2004. Animal *Communication Networks.* Cambridge: Cambridge University Press.

de Waal, F. & Tyack, P. (Eds.). 2001. *Animal Social Complexity: Intelligence, Culture, and Individualized Societies.* Cambridge, MA: Harvard University Press.

Robert M. Seyfarth & Dorothy L. Cheney

■ Cognition
Tactical Deception in Wild Bonnet Macaques

Tactical deception involves the display by an actor of certain behaviors, drawn from its normal repertoire, in situations where they are likely to be misinterpreted by the audience; this usually leads to some tangible benefit for the actor with or without some corresponding cost to the audience. All such manipulative acts are thus functional, and most cases of primate deception can be included in this category.

A total of 128 events of deceptive social interactions was observed in three troops of bonnet macaques over a period of 4 years. Individuals in all the troops exhibited comparable levels of deception, ranging from an average of 0.0025 to 0.0060 acts/hour/individual, but differed widely with regard to the social situations—competition for food, mates and grooming partners, as well as aggressive interactions—during which tactical deception was displayed.

Bonnet macaques also displayed a remarkable individual variation in the performance of deceptive acts. Some individuals exhibited a very high frequency of these behaviors, at levels significantly greater than that shown by others. Such deceptive abilities also appeared to be independent of age and dominance ranks of the actors. Moreover, there were striking differences in the distribution of deceptive acts across the 15 categories of deception commonly displayed. The simple types of deception shown included concealment by hiding behind a physical barrier or away from the troop, by inhibiting interest in an object, and by ignoring; distraction by calling, by threat, and by close-range behavior; creating a neutral or a positive image; and deflection to third party. Some deceptive behaviors were complex, involving a rapid sequence of simple deceptive acts; these included concealment by hiding behind a physical barrier and by inhibiting interest in an object; concealment by inhibiting interest in object and distraction by close-range behavior; distraction by leading and by calling; distraction by threat and by close-range behavior; concealment by hiding behind a physical barrier, distraction by close-range behavior and by threat.

Human-like deception requires that the actor who signals information creates a false belief in the audience. The signaler thus needs to recognize that the audience's mind can be in a state of knowledge that is different from one's own and that it is possible to alter and, hence, control others' mental states without necessarily changing one's own. Are the deceptive acts displayed by nonhuman primates truly intentional in this sense, attributable to a theory of mind? Does the actor actually attempt to alter the mental states of another individual through its deception? Or, has experience simply taught the deceiver the use of certain behavioral strategies that, in particular situations, lead to predictable responses from the audience and thus allow the actor to achieve a desired goal?

The fact that certain individuals are more adept at deception and that the ability to deceive is independent of other individual attributes, including age, indicate that many of these acts could involve mentalism on the part of the actor rather than simple behavior analysis (since the latter would usually imply that rates of deception would increase with age and/or experience). Individuals who deceived at relatively higher levels also did so in many more different categories. Such individuals, thus, may have indeed been better cheaters with perhaps greater insights into the power of manipulative behavior than other individuals. This was particularly illuminated by a young subadult male who displayed 9 of the 16 acts of deception performed by the eight males of that troop; remarkably, these nine acts belonged to nine different categories of deception!

Several events of deception by the macaques involved acts of physical concealment in which the actor either simply hid from the target behind some physical object or performed a behavior surreptitiously behind a barrier, occasionally leaning out to inspect the target individual. This kind of visual perspective-taking, estimating what would be visible from another individual's point of view, has also been documented in other primates, notably chimpanzees and baboons. This ability to recognize and utilize the geometric perspective of another individual has earlier been equated to being able to represent correctly another individual's mental representation in one's own mind.

Most study individuals did not invariably use deceptive strategies in apparently identical situations, a result not expected if these acts were being performed in response to certain behavioral contingencies alone. Another form of volitional control of deception involved the display of very different categories of deceptive acts at enhanced levels by certain adult males following changes in the social environment—when they emigrated out of one troop and joined a neighboring one. A major difference that these individuals faced in the two situations was that of their dominance ranks, which fell drastically once they had joined the new troop. The perception of their specific positions in the rank hierarchy in the respective troops as well as the changing demands of the new social milieu may have thus triggered a completely different repertoire of deception in these two males.

If, indeed, some acts of tactical deception displayed by bonnet macaques truly involves mentalism, such manipulation is likely to be intentional, involving some recognition of the mental states of the audience. This would mean, in simple terms, that an individual performs a deceptive act in order to change the belief system of the audience; it can then take advantage of the false belief that it has generated to achieve a particular personal goal.

See also Cognition—*Deception*
       Social Evolution—*Social Evolution in Bonnet*
          Macaques
       Social Organization—*Social Knowledge in Wild*
          Bonnet Macaques
       Tools—*Tool Manufacture by a Wild Bonnet Macaque*

Further Resources

Byrne, R. W. & Whiten, A. (Eds.) 1988. *Machiavellian Intelligence: Social Expertise and the Evolution of Intellect in Monkeys, Apes, and Humans.* Oxford: Oxford University Press.

Byrne, R. W. & Whiten, A. 1990. *Tactical deception in primates: The 1990 database.* Primate Report, 27, 1–101.

Tomasello, M. & Call, J. 1997. *Primate Cognition.* New York: Oxford University Press.

Whiten, A. (Ed.) 1991. *Natural Theories of Mind: Evolution, Development and Simulation of Everyday Mindreading*. Oxford: Basil Blackwell.

Whiten, A. & Byrne, R. W. 1988. *Tactical deception in primates*. Behavioral and Brain Sciences, 11, 233–273.

Anindya Sinha

■ Cognition
Talking Chimpanzees

Cross-Fostering

Although chimpanzees have great difficulty using their voice to speak, they can freely move their hands, meaning that a gestural language of hand signs is well suited to their abilities. Psychologists R. Allen and Beatrix Gardner recognized this in their sign language studies with young chimpanzees. In 1966, the Gardners brought a 10-month-old chimpanzee named Washoe to the University of Nevada at Reno where they began their cross-fostering study. In 1972 the Gardners began a second cross-fostering project with four other infant chimpanzees. Moja, Pili, Tatu, and Dar were born in American laboratories, and each arrived in Reno within a few days of their birth. Moja arrived in November 1972, Tatu arrived in January 1976, and Dar in August 1976. Pili arrived in November 1973, and he died of leukemia in October 1975. Cross-fostering a chimpanzee is not at all like having a pet dog or cat. Even though pet owners love their pets, they do not treat them like children. In the Gardners' cross-fostering research they treated the infant chimpanzees just as if they were human children. They changed their diapers, played games and read them stories when they went to bed at night (Gardner & Gardner 1971).

The Gardners and their students on the cross-fostering project used only American Sign Language (ASL) to talk with the chimpanzees throughout the day. They would talk about everyday events and special things that might interest the chimpanzees such as "that chair" or "see pretty bird." They would ask the chimpanzees questions and answer questions. When the chimpanzees would ask for a toy or a game, they always tried to comply. The Gardners stated that: "Washoe, Moja, Pili, Tatu, and Dar signed to friends and strangers. They signed to each other and to themselves, to dogs and to cats, toys, tools, even to trees" (R. Gardner & Gardner 1989, p. 24). The chimpanzees' use of signs to talk with each other and their human companions was a very important part of their lives. Just as with human children, the Gardners found that as the chimpanzees grew so did the size of their vocabularies, the number of signed utterances they used each day, the variety and length of their signed phrases, and the complexity of their phrases.

Cultural Transmission: A Baby Chimpanzee Learns His Language from His Mother

Today the chimpanzees live at the Chimpanzee and Human Communication Institute (CHCI) where we have continued to explore how the chimpanzees acquire signs and use them to communicate with humans and each other. After the Gardners' research with Washoe, Moja, Peli, Dar and Tatu, scientists knew that chimpanzees could learn sign language from humans. But a question remained as to whether or not they could pass their language onto their children like we humans do. The question was answered in 1979 when Washoe adopted a

10-month-old son, Loulis. To find out whether Loulis would learn signs from Washoe and other signing chimpanzees without human intervention, we humans did not sign when Loulis was present except for seven specific signs, *who, what, where, which, want, sign,* and *name.* Instead we used vocal English to communicate with him and each other in his presence. Loulis began to pick up and use signs from Washoe after being with her only 7 days. At 15 months of age he combined signs into two- and three-sign utterances, and at 73 months of age he had acquired 51 signs from Washoe and his other signing chimpanzee companions. (Fouts & Mills 1997). Loulis' vocabulary and his use of phrases grew over time. He began with simple one-sign phrases such as "hurry" and shortly started using two-sign phrases such as "gimme drink" and "you chase." As he continued to grow up he began to use three-sign phrases such as "hurry you tickle." Loulis' phrase development was very similar to the cross-fostered chimpanzees raised by humans. Like human children, the development of phrases grew gradually in Loulis and the cross-fostered chimpanzees (B. Gardner & Gardner 1989). Loulis' acquisition of phrases is particularly impressive since it occurred in the absence of human signing, and his only models for how to use signs were the other signing chimpanzees.

Washoe, the Gardners' first foster chimpanzee.

Courtesy of the Chimpanzee and Human Communication Institute.

Remote Video Taping

In June 1984, the signing restriction around Loulis ended, and we turned our attention to Loulis' use of signs through the use of remote videotaping (RVT). RVT is a technique used to record the behaviors of the chimpanzees with no humans present. In this method video cameras were mounted just outside of chimpanzees' enclosures and were used to record the chimpanzees inside the enclosures. The cameras were connected to television monitors and a VCR in another room completely away from the chimpanzees. Only one camera recorded at a time and the VCR operator could control which camera recorded.

Deborah Fouts (1994) made 45 hours of RVT recordings to examine Loulis' interactions with the other chimpanzees at CHCI, Washoe, Moja, Tatu, and Dar. Loulis initiated 451 interactions, both signed and non-signed, with the other chimpanzees. Forty percent (181) of those interactions were directed to his male peer, Dar. Loulis used 206 signs in his interactions and 114 of those were directed toward Dar. She also reported 115 private signs that Loulis made in the absence of interactive behaviors such as looking toward another individual. He was talking to himself just as we might talk to ourselves when we are alone.

Later we looked for more instances of private signing by the other chimpanzees in the 45 hours of RVT. We found 90 instances of private signing. We classified these into categories of private speech that humans use. We later recorded 56 more hours of RVT and found 368 instances of private signing (Bodamer, Fouts, Fouts & Jensvold, 1994). In both samples one of the most common categories of signing was *referential* (59% in the 56-hour sample). In this category the chimpanzee signed about something present in the room; for example, naming the pictures in a magazine. The *informative* category, an utterance that refers to an object or event that is not present, accounted for 12% in the 56-hour sample, and 14% in the 45-hour sample. An example of this category was when Washoe signed "Debbi" to herself when Debbi was not present.

One category of private signing was imaginative signing. *Imaginative signing* is when the chimpanzees use their signs to play "pretend." We found six instances of imaginary play. There were four instances of *animation* in which the chimpanzee treated an object as if it was alive. For example Dar signed "Peekaboo" to a stuffed bear. There were four instances of *substitution* in which the chimpanzee pretended as if one object were another. For example, Moja wore a shoe and signed "shoe." She then removed the shoe, put a purse on her foot, and zipped it up and called it a "shoe" (Jensvold & Fouts 1993).

Conversational Context

RVT provides a way to discover what the chimpanzees do without any humans around to influence them. At other times we are interested in controlling variables and measuring the chimpanzees' responses within the context of their typical daily signing interactions with their human caregivers. In the Gardner experiments and in our experiments the chimpanzees were free to leave the testing situation and were free to respond to their world with any behavior. The following studies that we describe were all conducted in naturally occurring signing interactions between the chimpanzees and their human caregivers while still using careful scientific controls.

At the original CHCI facility the chimpanzees had access to connected enclosures in four different rooms. One of the enclosures was across the hall from a human workroom. When a human friend was in the workroom, the chimpanzees often came to the nearby enclosure to request objects or activities. They often made noises if the human was not looking at them. Bodamer and Gardner (2002) systematically studied these interactions. The human friend would sit in the workroom with his back toward the chimpanzees' enclosure. When the human was not facing the chimpanzees, the chimpanzees made noises, such as bronx cheers, and rarely signed. When the chimpanzee made a noise, the human would turn and face the chimpanzee immediately or after a 30-second delay. The few times the chimpanzees signed they used signs that made noise, such as "dirty" where the back of the hand hits the bottom of the jaw making a loud teeth-clacking sound. With force this sign is very noisy. In the delay condition the noises became louder and faster. Once the interlocutor faced the chimpanzees, they signed and stopped making sounds. Using a naturally occurring situation, this experiment showed that the chimpanzees initiate interactions and sign spontaneously, and know when a person cannot see them and that they must make a sound to attract their attention to see their signing. Using signs to talk to someone who can't see you is not very effective. The chimpanzees were smart enough to make sure the person was watching them before they signed to them. This means that they understand the basic requirements for a conversation.

In another test of conversational skill, the human friend pretended not to understand the chimpanzees during a conversation. Whatever the chimpanzee signed the friend asked "What?," or "Huh?," or "Who wants that?" (Jensvold & Gardner 2000). The chimpanzees frequently added more information, making their original statement clearer. The chimpanzees showed that they can help a person who doesn't understand by adding more information. This shows that these chimpanzees are skilled conversationalists.

Using rigorous methodologies that allow the chimpanzees to demonstrate their behaviors in natural situations of daily life, sign language studies of chimpanzees show remarkable similarities between human and chimpanzee behaviors. These chimpanzees acquired the signs of ASL from humans and other chimpanzees. The chimpanzees used signs when conversing with each other, even when no humans were present. They used the signs to sign to themselves and in imaginary play about things that were present as well as not present. They initiated interactions with humans and appropriately adjusted their conversations to changes in the human's signs and questions. Sign language studies fill some of the gaps between humans and the rest of nature that were created in the minds of philosophers and are maintained by human arrogance.

See also Cognition—*Animal Languages, Animal Minds*
Cognition—*Animal Consciousness*
Cognition—*Theory of Mind*
Gorillas—*Gorillas/Koko*

Further Resources

Bodamer, M. D., Fouts, D. H., Fouts, R. S. & Jensvold, M. L. A. 1994. *Functional analysis of chimpanzee* (Pan troglodytes) *private signing.* Human Evolution, 9, 281–296.

Bodamer, M. D. & Gardner, R. A. 2002. *How cross-fostered chimpanzees* (Pan troglodytes) *initiate conversations.* Journal of Comparative Psychology, 116, 12–26.

Fouts, D. H. 1994. *The use of remote video recordings to study the use of American Sign Language by chimpanzees when no humans are present.* In: *The Ethological Roots of Culture* (Ed. by R. A. Gardner, B. T. Gardner, B. Chiarelli & F. X. Plooij), pp. 271–284. Netherlands: Kluwer Academic.

Fouts, R. & Mills, S. T. 1997. *Next of Kin.* New York: William Morrow Publishers.

Gardner, B. T. & Gardner, R. A. 1971. *Two-way communication with an infant chimpanzee.* In: *Behavior of Nonhuman Primates.* (Ed. by A. Schrier & F. Stollnitz), vol. 4, pp. 117–184. New York: Academic Press.

Gardner, R. A., & Gardner, B. T. 1989. *A cross-fostering laboratory.* In: *Teaching Sign Language to Chimpanzees.* (Ed. by R. A. Gardner, B. T. Gardner & T. Van Cantfort), pp. 1–28. Albany, NY: SUNY Press.

Jensvold, M. L. A., & Fouts, R. S. 1993. *Imaginary play in chimpanzees* (Pan troglodytes). Human Evolution, 8, 217–227.

Jensvold, M. L. A., & Gardner, R. A. 2000. *Interactive use of sign language by cross-fostered chimpanzees* (Pan troglodytes). Journal of Comparative Psychology, 114, 335–346.

Roger Fouts, Mary Lee Jensvold, & Deborah Fouts

■ Cognition
Theory of Mind

Theory of mind (TOM) refers to the ability to reason about the mental states—the beliefs, desires, and intentions—of other individuals. Having a TOM is considered a hallmark of human cognition. Consider, for example, the following scenario: A boy opens a cookie jar, looks inside, frowns, and walks away. Most people interpret this scene as follows: The boy *wanted* a cookie, *thought* there was a cookie inside the cookie jar, but *realized* that he was wrong. Thinking about the actions of others in such mental states terms allows our species to predict not only another individual's behavior, but their unobservable thoughts as well.

Because a TOM is so important in adult human cognition, many have become interested in both the development and evolution of our TOM abilities. Much of the developmental work on TOM uses a classic task known as a "false belief" experiment. The logic behind false belief experiments is that children should use their TOM to think about another person's belief even when that belief is different from their own beliefs. In one version of a false belief task, an experimenter asks what a child thinks is inside a box of Smarties candy (Smarties is the British version of M&M's). Most children reply that they think Smarties are inside the box. The experimenter then opens the box to reveal something unexpected (e.g., pencils) inside the box. The experimenter then closes up the box and asks the child what another person will think is inside the box. Children older than four years of age correctly reply that another person will probably have a false belief about what's inside the box—the others will think that Smarties are in the box. Younger children, however, perform differently. They mistakenly think that others will have the same belief about the box that they do, namely that pencils are inside. Results from studies like these have demonstrated that children between three and five years of age undergo important developmental shifts in their ability to understand the beliefs of others.

Psychologists have also investigated whether other animals share our human TOM abilities. Such comparative work has focused mostly on nonhuman primates, particularly chimpanzees. A number of now classic experiments suggested that primates know very little about the minds of others. Chimpanzees, for example, watched as an experimenter stared at or pointed at one of two possible food locations. Chimpanzees chose randomly in these studies, ignoring the experimenter's intent to communicate knowledge of the food's location. Similarly, chimpanzees ignore information about what experimenters can and cannot see when choosing whether to beg for food. Recently, however, researchers have begun using more ecologically competitive tasks to ask what primates know about the mind, allowing subjects to compete against others for hidden pieces of food. These recent studies suggest that chimpanzees use information about what competitors do and do not know when vying for contested pieces of food. In contrast to previous work, these new studies hint that at least one nonhuman animal may possess some TOM abilities.

See also Cognition—*Animal Consciousness*
Cognition—*Deception*
Cognition—*Social Cognition in Primates and Other Animals*
Cognition—*Tactical Deception in Wild Bonnet Macaques*
Cognition—*Theory of Mind*
Play—*Dog Minds and Dog Play*
Play—*Social Play Behavior and Social Morality*

Further Resources

Hare, B., Call, J. & Tomasello, M. 2001. *Do chimpanzees know what conspecifics know?* Animal Behaviour, 61, 139–151.
Perner, J. 1991. *Understanding the Representational Mind.* Cambridge, MA: MIT Press.
Povinelli, D. J. & Eddy, T. J. 1996. *What young chimpanzees know about seeing.* Monographs of the Society for Research in Child Development, 61, 1–152.
Tomasello, M., Call, J. & Hare, B. 2003. *Chimpanzees understand psychological states—the question is which ones and to what extent.* Trends in Cognitive Sciences, 7, 153–156.
Wellman, H. 1990. *The Child's Theory of Mind.* Cambridge, MA: MIT Press.

Laurie R. Santos

■ Communication—Auditory
Acoustic Communication in Extreme Environments

In the animal kingdom, social interactions depend on the exchange of information. Various signals such as acoustic, visual, mechanical, electrical, and chemical are used to carry information between individuals. Only some groups of arthropods and vertebrates emit acoustic signals to communicate. Sound is an efficient and essential means of communication, propagating quickly over long distances even in obstructed or dark environments, without leaving trails. Sound-based communication relies on an emitter transmitting an acoustic signal carrying information that is collected by a receiver. As with any signaling process, sound communication may suffer from environmental constraints during propagation through air, water or solid substrates. Acoustic waves are susceptible to various modifications due, for example, to absorption, reverberation and backscattering by obstacles, or by excessive background noise. Thus the received signal is likely to be different than that first emitted. This represents a major constraint for communication. In some cases this constraint is so strong that various animal species have evolved specific adaptations providing an efficient transfer of information between individuals.

To illustrate specific adaptations, two extreme situations for sound communication can be considered. The first situation occurs for songbirds in a forest environment with numerous obstacles such as trees, shrubs and plants attenuating long-range communication. The second situation may occur in an environment where the background noise in colonies of birds and mammals becomes so loud that it represents a major constraint for normal acoustic communication. Thus, one can ask the question, "how does acoustic communication in extreme environments remain efficient?"

Long-Range Communication in Forest Songbirds

The Problem: Long-Range Propagation-Induced Modifications

For birds of the forest, loud advertising songs are an important means of maintaining communication. The songs function as a means of establishing territorial boundaries, repelling competitors, and attracting mates. Birdsongs are designed to propagate over distances varying from 30 meters (100 feet) to more than 1 kilometer (3500 feet) depending the bird species. They represent an energy-saving substitute for movement across and within specific territories, and when visual behavior is impaired by various obstacles. However birdsong is also subject to attenuation, or weakening, and modification during transmission. This is particularly true in thick forests, especially in the tropics where there is dense vegetation such as abundant tree trunks, intertwined branches, and excessive leave growth. Such forests constitute an extreme environment for acoustic transmission, substantially modifying long-range signal-to-noise ratios by absorption, echoes by reverberation, high-frequency filtering, and distortion of amplitude and frequency patterns. However, forest songbirds perform efficient long-range acoustic communication during territorial competition and mate attraction. How can this be achieved?

The Solution: Adaptive Strategies

Emitting and receiving behaviors. One can often observe that birds use perches in trees to sing. It has been demonstrated using propagation experiments in the field, with a sound emitting and recording system, that the choice of the best perch for song transmission

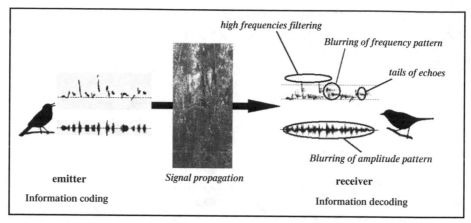

Influence of the forest environment on the structure of propagated acoustic signals. During transmission through dense vegetation, the signal-to-noise ratio decreases, the amplitude and frequency patterns are blurred, high-pitched waves are absorbed, and reverberations add echoes to notes, then filling silences.
Courtesy of Nicolas Mathevon.

depends on the constraints of the forest. In other words, the heterogeneity of the vegetation can be assessed by the bird and the signaling behavior modified accordingly. Generally the highest perch location selected by the bird usually corresponds to the place where the surrounding vegetation is the least dense. Sound experiments have also shown that improved sound reception can be obtained by increasing the height of the perch. This suggests that selecting the highest perch results in a greater benefit to songbirds in terms of improved communication conditions, particularly when they act as receivers of songs rather than the senders of song. Thus, depending on its perching behavior, a forest songbird has the possibility to improve the range and reliability of acoustic communication.

Another behavioral characteristic of forest songbirds, influencing the efficiency of communication, is the choice of the right time of day to begin singing. The song activity of male birds often follows a circadian rhythm with a peak of emission at dawn resulting in the so-called "dawn chorus." It has been argued that one of the possible reasons for dawn singing is that sound transmission conditions are much better at this time of the day. Indeed, diurnal microclimatic variations may induce sound degradation by atmospheric turbulence, absorption, and an increased level of abiotic background noise. However, recent acoustic experiments have shown that sound signals are not able to propagate more efficiently over longer distances at dawn compared with other times of the day. Moreover the "dawn chorus" is characterized by a high background noise with many birds singing together. Therefore dawn may be the right hour for private communication because signals will only reach the closest receivers, such as a mate, thus avoiding potential eavesdroppers and, in particular, surrounding competitors.

Coding–decoding processes. An important aspect of acoustic communication concerns the encoding of the transmitted sound and the decoding by the receiver. Field experiments have shown that the energy of high-pitched sounds (>4 kHz) can be more easily absorbed by the environment than low-pitched sounds. Furthermore, low-frequency modulation is less sensitive to propagation-induced attenuation than high-frequency modulation. It is well known that the forest birdsongs are usually lower pitched and are frequency-modulated at a lower level than the songs of birds living in open habitats. However, this is not true for all forest birds. Indeed, sound emission is constrained by several factors such as, for example, the anatomy and physiology of the bird. A tiny bird like the winter wren *Troglodytes troglodytes* is unable to emit a low-pitched sound due to its small size. However it lives in the denser parts of the forest where high-pitched sounds propagate badly over long distances. Bird species like the wren compensate their apparently inadequate signals by a wider tolerance during the decoding process, with the receiver able to recognize the species-specific identity of the sender even if the song has been highly degraded during propagation. Indeed, the wren does not pay much attention to echoes, changes in amplitude modulation or high-frequency filtering, but relies on frequency modulation to decode the relevant information.

Propagation-induced degradation of sound can also be a source of information for the receiver. Since the degree of degradation is relatively proportional to the transmission distance, the receiver should be able to assess the position of the singer. Numerous experiments support the idea that birds use sound degradation to range the acoustic source (see Naguib & Wiley 2001, for a review of these ranging processes).

Individual Vocal Recognition in Birds and Mammals Colonies

The Problem: High Risks of Signal Masking and of Confusion between Individuals

Some animals, like numerous seabirds, such as penguins, and pinnipeds, such as fur seals, gather together on the seashore to breed and rear their offspring within large colonies. When an adult comes back to the colony from a foraging trip in the ocean, it has to find its mate and/or its offspring within the colony despite the great number of individuals. These encounters are essentially mediated by voice. In a colonial environment the noise generated by individuals combines to form an almost continuous background. Especially in bird colonies such as king penguins, *Aptenodytes patagonicus*, the signals are partly absorbed by the mass of bodies. Such extreme "jamming" of the acoustic signals represents a serious constraint on family recognition. Thus the risk of confusion between individuals is high due to their overcrowding. For instance, when the subantarctic fur seal female *Arctocephalus tropicalis* comes back from hunting at sea, she may have to relocate her own pup among several hundred similar-looking animals in the rookery. Nevertheless, both the fur seal and the king penguin are able to locate their young within minutes of arriving back on shore. How is this possible?

(Left) Colony of king penguins Aptenodytes patagonicus. *The size of this colony is very impressive: more than 1 million birds pairs! Here, behavioral processes implied in accurate vocal recognition between mates and between parents and chick must be especially efficient. (Right) A king penguin calling (spectrogram above). Notice the characteristic stretching position of the bird that avoids strong signal absorption by the bodies of the other individuals.*
Courtesy of Nicolas Mathevon.

The Solution: Adaptive Strategies

Discriminative abilities. An experimental study has shown that subantarctic fur seal pups acquire the ability to recognize their mother's voice when they are just 2–5 days old, and that the mother delays her first departure to sea accordingly (Charrier et al 2001). This learning process is very efficient. When the mother comes back two or three weeks later, the mother–pup pair needs only a few minutes to meet each other within the crowded colony.

In some cases, this strong learning is reinforced by astonishing abilities to extract biologically pertinent signals from the background noise. For instance, king penguin chicks perform an accurate recognition of their parents' voices even if the level of these signals is −6 decibels (and, for some chicks, −9dB) under the level of the background noise. By comparison, in the Adélie penguin *Pygoscelis adeliae* which forms less dense colonies with a lower background noise level, chicks becomes unable to identify the call of their parents as soon as its level goes under the noise (i.e., <0 dB).

Emission and reception behavior. Colonial animals can maximize the emission and the reception of signals by displaying specific postural behaviors. For instance, a king penguin waiting for its mate raises its head, increasing the opportunity to hear the signal of its partner. During signal emission, these same birds stretch their neck and point their bill to the sky, allowing the sound signal to avoid absorption by the bodies of other individuals.

Coding-decoding processes. To be efficient in a colonial context, animal vocalizations supporting the recognition process must contain highly individualized features, characterizing the vocal signature of each individual. These cues must be resistant to propagation through a noisy environment. This is the case in all species of seabirds and colonial pinnipeds that have been studied so far. For instance, both "female attraction call" and "pup attraction call" emitted respectively by subantarctic fur seal pups and females fulfill this condition.

The redundancy of information is also a crucial cue. High redundancy in a signal improves the probability of a message to be received in a noisy environment. In colonial birds and mammals, redundancy in sound signals is, for instance, supported by the presence of numerous harmonics (i.e., sounds with a complex frequency spectrum, composed by a fundamental frequency and several harmonics). Fur seal mothers need only 2–3 harmonics to recognize their pup's voice, whereas the pup's call is composed of 8–12 harmonics. Moreover, calling individuals repeat their call several times. This temporal redundancy is likely to enhance signal detection by the receiver. A study has shown that king penguins

Subantarctic fur seal Arctocephalus tropicalis *pup and mother calling each other. Both pup's and mother's call show an individual signature allowing them to vocally identify each other among several hundred similar-looking individuals of the colony.*

Courtesy of Nicolas Mathevon.

increase their rate of call repetition with the speed of the wind—an important factor impairing the reception of sound signals—and consequently the background noise level. (See Aubin & Jouventin 2002 for a review.)

Colonial birds and fur seals use a multiparametric analysis to perform individual recognition, securing the recognition process. The frequency composition (*spectrum*) of the sound and the characteristics of the frequency modulations are the two main categories of parameters which are used by colonial animals for individual recognition of their mate or their young. In the two biggest species of penguins, the king penguin and the emperor penguin *Aptenodytes forsteri*, the complexity of the vocalizations is enhanced by the presence of "two voices," a phenomenon present in some species of birds, characterized by the presence of two harmonics series generating beats. Both these species rely on these "two voices" to perform individual vocal recognition, and this increases considerably the potential of the acoustic signals for coding individuality.

In a forest environment, the difficulty for the emitter is to control and to improve the "active-space" of its signals. To do that, the singer has to code information in acoustic parameters resistant to propagation and to adopt an adapted emission behavior. From the receiver's point of view, the problem is to optimize the quantity of gained information. The solution is then to be tolerant to sound degradation during the decoding process and to optimize the reception behavior.

In a colonial environment, the difficulty for the signaler is to address a particular receiver. Once more, the solution here is to use resistant acoustic parameters for information coding and to emit signals with an important redundancy of information to secure the transmission. From the receiver's point of view, the problem is to discriminate the relevant signal against the background noise. Besides an adapted reception behavior, the receiver has to show high discriminative abilities and to be very tolerant to sound masking during the decoding process.

In both cases, in spite of extreme acoustic conditions, the use of adaptive strategies allows animals to communicate efficiently. This demonstrates that the influence of the environment plays an important role in the complex process of the evolution of acoustic communication signals and behaviors.

See also Communication—Vocal—*Alarm Calls*
Communication—Vocal—*Choruses in Frogs*
Communication—Vocal—*Referential Communication in Prairie Dogs*
Communications—Vocal—*Vocalizations of Northern Grasshopper Mice*

Further Resources

Aubin, T. & Jouventin, P. 1998. *Cocktail-party effect in king penguin colonies.* Proceedings of the Royal Society of London B, 265, 1665–1673.

Aubin, T. & Jouventin, P. 2002. *How to vocally identify a kin in a crowd ? The penguin model.* Advances in the Study of Behavior, 31, 243–277.

Bradbury, J. W. & Vehrencamp, S. L. 1998. *Principles of Animal Communication.* Sunderland, MA: Sinauer Associates, Inc.

Catchpole, C. K. & Slater, P. J. B. 1995. *Bird Song, Biological Themes and Variations.* Cambridge, UK: Cambridge University Press.

Charrier, I., Mathevon, N. & Jouventin, P. 2001. *Mother's voice recognition by seals pups.* Nature, 412, 873.

Charrier, I., Mathevon, N. & Jouventin, P. 2003. *Vocal signature recognition of mothers by fur seal pups.* Animal Behaviour, 65, 543–550.

Dabelsteen, T. & Mathevon, N. 2002. *Why do songbirds sing intensively at dawn? A test of the acoustic transmission hypothesis.* Acta Ethologica, 4, 65–72.

Hauser, M. D. 1996. *The Evolution of Communication.* Cambridge, MA: MIT Press.

Holland, J., Dabelsteen, T., Pedersen, S. B., Larsen, O. N. 1998. *Degradation of wren* Troglodytes troglodytes *song: Implications for information transfer and ranging.* Journal of the Acoustical Society of America, 103, 2154–2166.

Mathevon, N., Aubin, T. & Dabelsteen, T. 1996. *Song degradation during propagation: Importance of song post for the wren* Troglodytes troglodytes. *Ethology*, 102, 397–412.

Naguib, M. & Wiley, R. H. 2001. *Estimating the distance to a source of a sound: Mechanisms and adaptations for long-range communication.* Animal Behaviour, 62, 825–837.

Nicolas Mathevon

■ Communication—Auditory
Audition

Communication is essential to the survival of every individual. Without the ability to convey information to *conspecifics* (members of your own species), and at times *heterospecifics* (members of other species), one would be unable to engage in basic behavioral necessities, such as finding a mate or escaping from predators. Different species have evolved elaborate systems of communication that facilitate behavioral interactions across a range of contexts. In acoustic communication, the subject of this essay, it is the ability of animals to generate vocalizations and to respond appropriately to vocal sounds from their own species (and sometimes other species) that is under consideration. Data on acoustic communication are available in a range of taxonomic groups, such as insects, anurans, reptiles, birds, and mammals. The breadth of data available on acoustic communication in nonhuman animals is far too extensive to review here. We will discuss the role of acoustic communication in two general contexts crucial to an individual's survival: avoiding predators and competition for finding a mate. While the examples discussed here focus on specific examples, many of the principles that underlie these communicative systems can be generalized to other species.

Avoiding Predators

One of the most persistent problems facing animals is the task of avoiding potential predators. They have to be constantly on the lookout for both themselves and close relatives. As a result of this constant pressure from predators, many species have evolved alarm calls. These are calls whose sole function is to alert others to the presence of something dangerous. The alarm calls of some species of mammals and birds appear to convey information primarily about predator types. The acoustic structure of many species' alarm calls varies depending on predator type; that is, how the alarm call sounds can indicate which predator is nearby. Aerial predators elicit acoustically different calls than terrestrial predators, and such calls evoke specific escape responses by listeners. Alarm calls are so useful that some animals actually listen in on the alarm calls of other species to assess danger levels. In other animals it seems as though alarm calls convey information about the level of response urgency—indicating, for example, how close the predator is. These calls seem to

Noisy Herring

Ben Wilson

Imagine the scene; it's 2 a.m. in a deserted marine laboratory on the Canadian west coast. The only illumination comes from the blinking red light on my tape recorder. Thirty herring are lazily swimming about in a tank six feet away. The lab is silent, pitch dark. My finger hovers over the play button on the tape recorder. I'm about to expose the fish to the hunting sounds of an echolocating dolphin and measure how they respond. I count down the seconds—5, 4, 3, 2—and then a moment before I hit play, the room fills with a loud raspberry sound. Six seconds it goes on and then silence. I sit, frozen. The tape player never left the pause position, it wasn't my seat creaking, the tank pumps and unnecessary electronics are off. Even the phone is unplugged. I wait for the practical jokers to giggle or expose themselves. They don't. And then it comes again. A series of clicks—rapid, quick-fire at first—becoming more drawn out, like a fingernail moving over the teeth of a comb. I reach out in the dark to touch the source of the sound. My hand bangs into my desk speaker, the one that's attached to the monitoring microphone dangling in the tank. The sounds are coming from inside the tank! Something must have fallen into the tank or it's leaking or . . . I dash for the light switch. Blinking under the lights, I see that nothing has changed; the fish look a bit startled but are otherwise normal. I sit there all night with the lights on, but hear nothing more.

Fish, in the popular consciousness, are not considered particularly noisy creatures. How wrong this is. In fact, probably as many fish as amphibians make noises. Think of all the croaks, peeps and whistles that tropical frogs issue. But what about herring? It turned out that the scientific literature contained nothing but rudimentary descriptions of some rather unspectacular herring sounds. So, the inevitable happened—I began a series of increasingly weird experiments.

It turned out that the herring made all kinds of pulsed noises, but the most common were the fast repetitive tick (FRT) sounds that I'd heard on that first night. Feeding the fish or withholding food made no difference to the number of sounds they produced. Nor did scaring them with a dip net or with the scent of sharks. However, the time of day did. The FRTs were nocturnally produced, mostly between dusk and midnight. The number of herring in the tank also proved important. Low-density fish populations produced few sounds, but at higher densities the *per capita* rate went up disproportionately. In other words, the herring made these sounds in the company of their conspecifics. So much for the *what*, but *how* were they made?

Fish have evolved the ability to produce sounds in many different ways—vibrating swimbladders, grating jaws, clicking spines, slapping fins, and so on. But there were no clues on how herring might do it. To my ear the FRTs were like the noise you might make by blowing out of pursed lips. The air-filled swimbladder in herring, rather unusually, has two openings—one leading to the stomach and the other exiting the body, via a thin tube, near the anus. Perhaps the sounds resulted from shunting air through one of these openings. There was nothing for it but to take a dead fish and give it a good squeeze underwater. Sure enough, with sufficient finger pressure on either side of the body cavity, the deceased

(continued)

Noisy Herring (continued)

would produce a short burst of sound pulses and stream of bubbles. But what about live fish? Scottish colleagues came up with a solution—specialized cameras that could film herring in apparent darkness. Finally it was possible to watch the fish as they made the sounds. The moment we heard each FRT sound, one fish would produce a stream of bubbles from its anal region. Other experiments showed that this was likely air from the swimbladder rather than digestive gas from the gut. So we have a mechanism—ejecting air from the anal duct—one that, at the time, had never been described in any fish before. Bizarrely, given the many decades of in-depth herring research, at precisely the same moment, a group in Sweden was coming to similar conclusions!

So we have the *what*, and a *how*, but we don't have the *why*. These sounds, of course, could result purely from the necessary processes of the fish managing the air in their swimbladders. But the social mediation hints at a communicative role. If so, the information transfer is between whom and for what reason? Perhaps the fish use it to maintain contact in darkness or maybe detect prey or signal to predators. It just isn't known. But, one has to wonder, if the sounds are useful to herring, how does the din of modern ship engines, wind farms, military sonar, seismic blasting and all the other human related noise that pollutes our coastal waters affect remaining herring populations? At least, now that we know that herring make and perhaps listen to these mysterious sounds, we can make efforts to find out.

Further Resources

Wilson, B., Batty, R. S., & Dill, L. M. 2003. *Pacific and Atlantic herring produce burst pulse sounds*. Biology Letters, 10, 1098.

---

vary with the type of hunting and speed of an approaching animal rather than the exact predator type.

Monkey Alarm Calls: Attending within and between Species

One of the most persistent problems primates encounter is the task of avoiding potential predators. Most primate species serve as prey to at least one kind of predator, and many species are hunted by a number of different predator types. For example, red colobus monkeys (*Procolobus badius*) are eaten by eagles, leopards, chimpanzees, and human poachers. Predation can be the number one cause of death in some species, bringing about more deaths than disease, injuries, and other causes. Vervet monkeys (*Cercopithecus aethiops*) are a prime example. In vervet monkeys, predation alone accounts for over 70% of all deaths. With mortality rates as high as these, it is likely that strong selection acted on this species to evolve a system for avoiding predation. In the auditory domain, these tactics can include the detection of predatory-specific acoustic signals as well as the production and recognition of alarm calls.

In response to hearing or seeing a predator, many primates produce alarm calls. More impressively, a number of primate species distinguish between different predators, producing acoustically different alarm calls for different classes of predators. The best known antipredator tactic is the alarm-calling behavior of vervet monkeys. Vervet monkeys in the Amboseli National Park in Kenya are preyed upon by at least three different types of predators: eagles, leopards, and snakes. Because each of these predators hunts in a

Vervet monkeys are the best of example of a species that produces distinct alarm calls for different types of predators. The graphs shown on the right are called "spectrograms"; they are "pictures" of the sounds. The x-axis is time and the y-axis is frequency. The darkness of the bands indicates how loud that particular component is. These particular spectrograms show the structure of the vervet's "eagle," "leopard," and "snake" alarm calls.

Vervet photo by Asif Ghazanfar; sounds courtesy of Marc Hauser.

different way, no single general-purpose antipredator tactic would be effective against all. When faced with an eagle, for example, a vervet monkey's safest response is to hide as close to the ground as possible. However, when faced with a leopard, the ground is the most dangerous place to be; instead, vervets must immediately move as high off the ground as possible, and clamber up the nearest tree. Vervets must distinguish between these different predators and react in a manner that is appropriate for the way each predator attacks. One way vervets manage to categorize different predators is through the use of three predator-specific alarm calls—one for aerial predators, one for leopards, and one for snakes. In experiments where one plays back previously recorded alarm calls through a hidden speaker (when there is actually no predator in the area), vervet individuals always reacted to the calls in a predator-appropriate way. When subjects heard playbacks of leopard calls, they ran into the trees, but when they heard eagle alarm calls, they hid under bushes on the ground. These results suggest that vervets naturally distinguish both between the different predator classes and the vocalizations associated with those classes.

The alarm calls of one's own species are not the only kind of acoustic information relevant for predator detection. Many primates live in the same area with other alarm calling animal species who are preyed upon by the same types of predators. As you might imagine, the ability to detect, learn and respond appropriately to the alarm signals of other species would be extremely useful. There is now much evidence to suggest that primates do just this; many primate species can learn and respond adaptively to the alarm calls of other species that live in the same area. In southern India, for example, bonnet macaque (*Macaca radiata*) groups are found in association with Nilgiri langurs (*Trachypithecus johnii*), another species of monkey, Hanuman langurs (*Semnopithecus entellus*), and sambar deer (*Cervus unicolor*), but the frequency of such associations varies between groups of bonnet macaques.

All four species fall prey to leopards and produce alarm calls upon detecting a leopard. Scientists compared the responses of bonnet macaques to playbacks of their own alarm calls with playbacks of alarm calls produced by the other three species. They found that there were no differences in the time it took to run for safety following their own versus other species' alarm calls. That is, the bonnet macaques treated the alarm calls of the two langur species and the deer with as much urgency as they would their own alarm calls. On a group-by-group basis, however, the responses to alarm calls of species to which bonnet macaques were not frequently exposed were significantly different from the responses to their conspecific alarm calls. This suggests that sufficient experience is necessary for bonnet macaques to learn the alarm calls of other species.

Meerkat Alarm Calls for Predators and Urgency

Meerkats (*Suricata suricatta*) are a species of cooperatively-breeding mongoose that live in groups of 3–33 individuals. The typical group consists of adults, juveniles and dependent young. They live in open semidesert areas and are most active during the daylight hours when they forage in the sand for insects and small vertebrate animals. While digging, meerkats must constantly be on the lookout for predators. As a result, they must frequently interrupt their digging to stand upright and take a look around. Of course, the more they have to look up for potential predators, the less time they can spend finding food. The group can decrease their vigilance (the time they spend looking out for predators) and increase their foraging time (the time they spend looking for food) if members take turns acting as sentinels for the entire group. Indeed, meerkats will sometimes take turns acting as lookouts by standing upright (usually at an elevated position, like on a termite mound) and scanning the horizon for potential predators. Members often announce that they are acting as a sentinel by making short, soft calls, and they call continuously for the time that they are guarding to let group members know that they are still on watch duty. Upon hearing these sentinel calls, other members lower their own vigilance levels and can spend more time foraging.

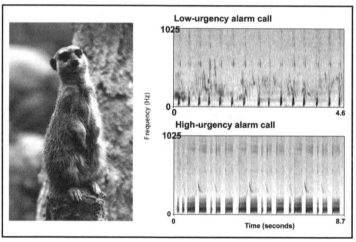

Meerkat standing sentinel. This meerkat is acting as a lookout while the rest of his group is foraging beneath. If he sees something dangerous, he will start producing the appropriate alarm call. The spectrograms show alarm calls to a terrestrial predator in two different contexts: low-urgency (top panel) and high-urgency (bottom panel).

Photo by Asif Ghazanfar. Meerkat calls are courtesy of Dr. Marta Manser.

If they see an approaching predator, sentinels and foraging individuals can emit several different types of alarm calls. Different alarm calls are produced according to the class of predators. For mammalian predators (primarily, jackals) there is one type of alarm call, and for aerial predators (like eagles and goshawks) there is another acoustically-distinct alarm call type. A final, third type of alarm call is less specific. It is produced upon detecting snakes, the feces or hair of predators and/or unfamiliar meerkats. Detection of all of these things elicits this third alarm call type, and group members approach the caller and either mob the snake to scare it away or investigate further (if it's hair, feces, etc). How do we know this? Both field observations and playback experiments have provided compelling evidence. Playback experiments of the different alarm calls elicited predator-type specific escape responses from the meerkats. So, even in the absence of the actual predator, if an aerial predator alarm call is played back, meerkats respond by freezing and crouching, scanning the sky and/or running as fast as possible to the nearest bolthole (holes that they make specifically for escaping dangers); there is little time to waste against the stoop of a raptor. In contrast, upon hearing a terrestrial predator alarm call, they have a qualitatively different response: The group gathers together at the same safe place and decide whether or not to leave the area together. Given that foraging is of vital importance to the meerkats, it is in their best interest not always to panic and run for cover if a predator is sighted. It would be more efficient to assess how dangerous the situation is, for example, by determining how much of a threat a predator poses. How might they do this?

Remarkably, the alarm calls of meerkats convey information not only about the predator type, but also the level of urgency (particularly, how close the predator is to the group). The acoustic features of the calls that encode information about the level of urgency are different from those that vary between types of predators. The features that indicate the level of urgency are the same for the different predator call types. For example, as the level of urgency increases, the calls change from clear sounding to noisy and harsh. Furthermore, urgency is also reflected in the rate of calling and the duration of a bout of calling. For aerial alarm calls, call bout duration and call rate decreases with increasing levels of urgency. For terrestrial alarm calls, call duration increases, but call rate decreases with increasing levels of urgency. Again these findings are strongly supported by playback experiments.

Competition for Finding a Mate

Although avoiding predators and locating food are essential tasks for survival, an animal's ultimate success in life is determined by its ability to produce offspring. This requires finding high-quality mates. In a wide variety of mating systems, mate choice is based on an assessment of auditory signals presumed to correlate with fitness.

Many animals spend a considerable amount of time and energy in elaborate displays which involve repeated vocalizations. These displays generally serve two functions: to intimidate rivals and to attract mates. Vocalizations—their acoustic structure and how frequently they are produced—can allow listeners to assess the health of the vocalizer. This is important for competitors because it allows them to assess whether a physical confrontation is worthwhile, and it is important for potential mates because it allows them to assess whether the caller is "fit" enough to choose as a partner.

Red Deer Roars

In the early fall, red deer (*Cervus elaphus*) females (hinds) cluster together in particular areas of their home ranges and are then joined by male red deer (stags). These stags have spent the many previous months in bachelor groups, preparing themselves for the mating season. Older, bigger stags, usually over 5 years old, collect and defend groups of hinds (harems) and try to protect their harems from other stags. The older, experienced stags are good at defending their harems, while younger stags tend not to be. Younger stags often hang around another stag's harem and try to abduct hinds on the sly. Defending a harem is not easy and harem-holders spend very little time feeding because they spend most the day collecting and herding the hinds and chasing off other stags.

During the 4–5 week period when they are defending harems, red deer stags roar loudly and repeatedly. Within an individual, these roars are produced at a rate that can vary between periods of low and high roaring activity. The high period generally corresponds to times where there is male–male competition. Roaring repeatedly is costly to the male for two reasons:

1. It uses up precious energy resources. In fact, over the course of the breeding season a stag's ability to roar and his body weight decrease considerably; and
2. Roaring increases the chances that a predator will hear him.

It has therefore been suggested that roaring rates indicate the health of the caller because only a very fit individual can produce calls at high rates and afford to risk detection by a predator. So how do other stags and female red deer (hinds) respond to the roars of another stag?

In many large animals, changes in fighting ability within breeding seasons or across the lifetime of individuals are related to changes in body condition but not to obvious changes in body size. In situations where a conflict of interests is likely to lead to a fight, opponents might be expected to assess each other on traits that are related to variation in body condition. This appears to be the case for red deer stags: Roaring can serve to scare off potential rivals and avoid fights because roars allow the listener to assess the fighting ability of the roarer. Indeed, there is a strong correlation between the maximum and average roaring rates of an individual and his fighting ability. Competing stags usually engage in roaring contests only in situations where it is not clear from body size differences who is likely to win a physical contest. Observations and playback experiments showed that stags answered each others' roars and that their roaring rate was related to that of their opponent. That is, each tried to out roar their opponent.

Hinds also listen in on the roaring abilities of stags. Although they are discouraged by the harem-holding stags, hinds can readily enter and leave harems. Knowing that there is a more fit male nearby could compel them to leave their current group. And like rival male stags, females use the rate of roaring to assess the quality of males. Indeed, experiments where roars were played back through a hidden speaker demonstrated that females are more attracted to higher roaring rates than lower ones. However, the situation is not so simple that hinds will leave a harem as soon as it hears a more robust male. Familiarity with a male may also be a component. Once a hind enters a harem, it is increasingly exposed to the roars of the harem-holding male and, due to his proximity, the roars are louder and more frequent than that of rival stags. Importantly, the degree to which a stag is familiar can also be a reliable indicator of male quality because familiarity relates to the ability to retain the female within its harem. It has been shown, using playbacks, that after controlling for roar rate, loudness, and roar length, hinds can discriminate between different stags by voice alone. However, it remains to be tested whether hinds actually choose stags that are more familiar to them.

Primate Copulation Calls

In many monkey species, individuals (males, females or both sexes) produce copulation calls. These calls are produced immediately before, immediately after, or during copulation and serve as auditory cues for reproductive status. In rhesus monkeys (*Macaca mulatta*), only males produce copulation calls and it is always *during* copulation. The evidence suggests that such calls are useful for assessing male quality (how healthy a male is) by females. Three pieces of evidence suggest that this is the case. First, the number of calling males decreases with increased competition for sexually-active males. Competition levels between males will be high if there are only a few females around. Second, males, independent of their status in the group, who produced copulation calls received more aggression from conspecifics. Finally, copulation calling males have a greater mating success (in terms of number of copulations) than silent males. In essence, males who call are more fit because they can withstand the aggression of other males following their copulation calls.

Both rhesus monkeys (top) and Barbary macaques (bottom) produce copulation calls. However, in rhesus monkeys, only males produce these calls, whereas in Barbary macaques, only females do.
Photos by Asif Ghazanfar.

In yet another species of macaque, the Barbary macaque (*Macaca sylvanus*), only females produce copulation calls. Once again, playback experiments were used in attempt to figure out what effect these calls may have on male listeners. Here's what was found: Following playbacks of their copulation calls, Barbary macaque females were mated sooner. Playbacks to pairs of males revealed that only the higher-ranking of the two would approach the sound source while the other male stayed behind. These results suggest that these female Barbary macaque copulation calls are used by females to attract mates and, importantly, to attract the best mate (the higher-ranking males). This is an indirect mechanism of female choice. But males can also be choosy. For example, it would be wise for a male Barbary macaque looking for a suitable mate to select, and fight for, a mate who is at the peak stage of fertility. Indeed, by listening to the copulation calls of females, Barbary macaque males can actually distinguish the reproductive states of the females. Playbacks of female copulation calls produced when she is most likely to ovulate elicited stronger responses from males than calls produced during periods when she is unlikely to be ovulating. The dominant frequency and/or the duration of the copulation call are two acoustic cues males may use to discern the reproductive states of females. Taken together, the data from both rhesus and Barbary macaques suggest that these primates can and do use copulation calls to influence their reproductive success.

Frog Advertisement Calls

Mating in many frog species occurs in a system known as *lekking*. A *lek* occurs when a group of males congregate around a specific location and attract females there for mating. For frogs, leks occur at night around pools of water, such as ponds and lakes. Here males produce species-typical vocalizations, known as advertisement calls, known to attract females. Male frogs do not provide females with food or aid in parental care. In fact, the only thing that males give to females is the DNA in their sperm. Therefore, females must use information encoded in the acoustic structure of the males' advertisement calls to choose their mate. Selection, therefore, must act on the structure of the call to provide females with cues about the caller's quality. While a range of data show this pattern in different frog species, exactly what features are used for mate choice, and why, seems to vary between species.

The tungara frog (*Physalaemus pustulosus*) inhabits the neotropical rainforests of Central America. The male advertisement call of this species consists of two acoustically distinct components: one long frequency-modulated "whine" and 0–5 short noisy "chucks." An elegant series of experiments have shown that females prefer males that produce calls with chucks over those calls that lack them. However, producing advertisement calls is quite costly to males because one of this species' primary predators, bats, have evolved the ability to use the chuck component of the call to detect the location of frogs. As a result, producing a lot of chucks will attract females and increase the likelihood of mating, but it will also attract bats and increase the likelihood of being eaten. Although this represents a paradox for the male, it explains why females use this component of the call to choose their mate. In essence, the male is communicating his quality as a mate to the female. In other words, a male that is able to produce many chucks demonstrates that it has a calling strategy that allows him to emit chucks and avoid predators, a behavior that is likely encoded within his genes. As such, a female mating with that male will pass those genes on to her offspring and increase their survivorship and the likelihood that they reproduce successfully and pass their genes on.

The Value of Acoustic Signals

Many species evolved systems of communication that facilitate mediating behavioral interactions with conspecifics. Although each species evolved unique systems of communication for this purpose, many species possess acoustic signals that are used in similar contexts. The widespread occurrence of species producing acoustic signals in certain contexts, such as to warn group members about the presence of a predator and for males to announce their quality to females, suggest that selection acted on different species to evolve functionally similar signals. As reviewed above, two types of acoustic signals that are observed in the vocal repertoires of a range of species are predator alarm calls and mating calls. Given the prevalence of acoustic signals in these contexts across nonhuman animals and insects, an important question is why these species evolved acoustic signals in these contexts rather than other types of signals. The answer to this may be that acoustic communication affords individuals some advantages over communication in other modalities. For example, unlike olfactory signals that may persist over long periods of time, acoustic signals are temporally discrete; the signal is produced and received almost simultaneously, with only those individuals within a certain range of the caller able to detect the signal. Like acoustic signals, visual signals are also temporally discrete, but they require that the signal producer and receiver be in close visual contact. In the case of acoustic signals, signaler and receiver could be separated by considerable distances and out of sight of one another. Some primate vocalizations can be heard hundreds of feet away, whereas the songs of whales are detectable at over 1,000 kilometers (620 mi). Ultimately, species will communicate with conspecifics using signals in different modalities for different purposes. However, the unique properties of acoustic signals make them particularly useful in the contexts discussed in this essay. Continued research on acoustic communication is likely to further elucidate the functional significance of acoustic signals both within a species and across taxonomic groups.

See also Communication—Auditory—*Acoustic Communication
in Extreme Environments*

Further Resources

Bradbury, J. W. & Vehrencamp, S. L. 1998. *Principles of Animal Communication*. Sunderland, MA: Sinauer.

Ghazanfar, A. A. (Ed.). 2002. *Primate Audition: Ethology & Neurobiology*. Boca Raton, FL: CRC Press.

Hauser, M. D. 1996. *The Evolution of Communication*. Cambridge, MA: MIT Press.

Hauser, M. D. & Konishi, M. (Eds.). 1999. *The Design of Animal Communication*. Cambridge, MA: MIT Press.

Seyfarth, R. M. & Cheney, D. L. 2003. *Signalers and receivers in animal communication*. Annual Review of Psychology, 54, 145–173.

Asif A. Ghazanfar & Cory T. Miller

■ Communication—Auditory
Bat Sonar

What is it like to be a bat? Imagine yourself flying around in the dark, feeding in summer by catching small flying insects on the wing. Early in the winter you fly into a totally dark cave to find a spot with just the right temperature and humidity where you can hibernate all winter when no insect food is available. You are smaller than a mouse, and your wings are thin very flexible membranes stretched between elongated finger bones, your legs and your tail. When you want to land, you maneuver into a stall with your head down just below something you can grasp with your hind claws; and you can do this all in less than a second. How do you know what is out there in the dark, and how do you avoid bumping into things? You can't see anything in a deep cave, and even if a scientist has covered your eyes, you can still dodge small wires as well as ever. You do this by *echolocation*, making sounds and locating things by the echoes they return. Sometimes this is called bat radar, but bats use sound waves and not radio waves, so their echolocation is a form of sonar.

A typical North American bat, the little brown bat (*Myotis lucifugus*), sends out very brief pulses of sound at frequencies above the range that humans can hear. Each pulse lasts only 0.001 to 0.005 second or 1 to 5 milliseconds, and in these few milliseconds the frequency falls by about an octave, from 80,000 to 40,000 cycles per second.

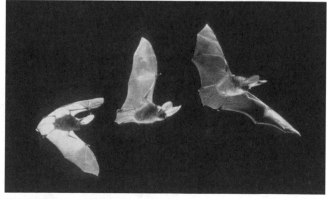

Bat in flight, stroboscopic view.
© Carolina Biological / Visuals Unlimited.

In scientific terminology, cycles per second is called Hertz, and thousands of cycles per second, kiloHertz, abbreviated kHz. These brief pulses of ultrasonic sound are well above the 20 Hz to 20 kHz range of human hearing; and they are repeated at repetition rates from about 10 to 30 per second when the bat is "cruising" with no special orientation problems. But when it has a difficult task such as dodging a fine wire this repetition rate may double.

To understand bat echolocation, it helps to think in terms of the space occupied by these pulses of sound and the time scale at which bats operate. To appreciate their properties, it is much more convenient to use the metric system. Bats determine distance to the source of an echo by the time between emitting a sound and hearing an echo. These sound waves are spherical surfaces expanding from the bat's mouth at about 34 cm per millisecond (1,130 feet per second), so that a 2 millisecond pulse extends 68 cm (27 in) from beginning to end. The wavelength at the initial 80 kHz is 4.3 mm (.17 in) and the 40 kHz at the end is 8.6 mm (.34 in).

Scientists first learned about bat echolocation by impairing various senses. Blinding had no effect, but tightly plugging a bat's ears caused it to collide with obstacles. Bats can dodge even small wires in total darkness by hearing their echoes. Later scientists learned that they also use echolocation to locate, pursue, and catch flying insects. They do this very rapidly and accurately. In a typical insect pursuit, a bat detects an insect when one or two meters away, turns or dives to intercept it, comes close enough to surround the insect with the wing membranes, eats it and flies on to continue hunting, all in less that one second. In a typical insect catch, the bat increases its pulse repetition rate from about 20 pulses per second to a short burst at 200 per second. Electronic bat detectors translate these inaudible ultrasonic sounds into brief clicks that we can hear, and in the 200 per second burst the clicks fuse into a feeding buzz.

A cloud of bats fills the twilight sky over Sarawak.
© Klum/National Geographic Image Collection.

There are many other kinds of bats, most living in tropical or semitropical regions. The large flying foxes of Africa, South Asia, Australia, and many of the Pacific islands eat fruit or nectar and have large eyes. They do not fly into totally dark caves or use echolocation. Many of the bats that do use echolocation gather other kinds of food than flying insects. Some take insects or other invertebrates from the ground or vegetation. They rely heavily on passive listening to sounds made by their prey and use echolocation for general orientation rather than prey capture. Other bats trawl with specialized hind claws for insects and small fishes at the surface of water. And many echolocating bats in the New World tropics feed on fruit or nectar and use echolocation mainly to avoid colliding with obstacles.

Dolphins and toothed whales, such as the sperm whale, use echolocation underwater for general orientation and to locate fish on which they feed. Dolphins are especially expert at detecting and discriminating between small objects by sonar. But echolocation is not limited to bats and dolphins. We can all do it, although we can't hope to be nearly as proficient as bats or dolphins whose ears and brains are highly specialized for extracting detailed information from echoes. Blind people make extensive use of echolocation to detect the presence of objects before touching them. This is often called "facial vision," because the blind person feels there is something close ahead but often does not know that this feeling depends on echolocation. But if his ears are tightly plugged or if he can't make any sounds, he loses most of this ability. Unfortunately, human echolocation is severely limited compared to bat or dolphin sonar. These specialized echolocators can discriminate important echoes from other sounds, but our brains are far less effective at this. Sighted people can learn with extensive practice to detect objects by echolocation when their eyes are covered, and under very quiet conditions any of us can detect large objects in this way, although it takes considerable practice.

So what is it actually like to be a bat? Some scientists and philosophers argue that we can never find out what a bat is aware of because bats depend so much on echolocation. But blind people, and any of us with enough practice, can also detect large objects by echolocation. Bats are shouting at their environment, and paying attention to changes in what it sounds like; and these changes tell them about solid objects and even about tiny flying insects. Thus we can learn at least a little about what it is like to be a bat. And perhaps it is not all that different from being ourselves when we concentrate on one special way of finding our way in the dark.

See also Bats—*The Behavior of a Mysterious Mammal*

Further Resources

Fenton, M. B. 1992. *Bats*. New York: Facts on File.
Kovacs, D. 2001. *Noises in the Night, the Habits of Bats*. Austin, TX: Steck-Vaughn.
Richardson, P. 2002. *Bats*. Washington, DC: Smithsonian Institution Press.

Donald R. Griffin

■ Communication—Auditory
Long Distance Calling, the Elephant's Way

Imagine you are trying to say something to a friend, such as "Where are you?," but the friend is two miles away. Even if you yell at the top of your voice your friend will not hear you. Now, if you and your friend are huge elephants, and you use a very deep, low voice— so low that it cannot be heard by people—your elephant friend two miles away could probably hear you. Elephants can communicate over such long distances because they can make very, very low sounds which produce waves in the air that are different than the sound waves produced when we speak or yell in a higher-pitched voice. The low-pitched sound waves produced by elephants are called rumbles and are referred to as infrasonic because they are below the hearing range for humans and other mammals such as dogs and cats. Elephants produce these low-pitched sounds because their vocal cords, and the parts of the throat involved with making the sounds, are so huge and the vibrations so slow. It's like the sounds produced by the heaviest, longest strings of a bass in an orchestra.

Elephants also have special adaptations in their ears so that they can hear these low-pitched sounds. The important thing about low-pitched sounds is that they are not nearly as broken up or blocked by trees, rocks, hills and other obstacles like normal human vocalizations. Low-pitched sounds or rumbles, therefore, can travel long distances.

Communication over long distances, such as two miles, is important to elephants because they have large families and travel long distances, and parts of the families can temporarily break up and be miles apart. Their low-pitched vocalizations allow them to keep track of each other and to make simple communications, such as: "Where are you?," "I'm in trouble," and "Come here." Scientists working with elephant rumbles believe that the low-pitched sounds can be recognized individually by other members of the family. In other words, one elephant receiving rumble signals from another two miles away will know which elephant is calling and whether the other is in trouble or asking the other one to come closer. As far as we know, elephants are unique in the world in their ability to produce these low-pitched sounds that travel such long distances.

Scientists have recently learned that elephants have another means of special communication for long distances. The amount of energy that it takes to produce the low-pitched sounds is so great that an elephant's whole body vibrates as it produces these sounds. If you were near an elephant you would not be able to see the vibrations, but if you touched the elephant on the side, you would be able to feel the vibrations move through the body. These vibrations also cause vibrations in the earth which travel away from the elephant and can reach distances of a couple of miles. Elephants feel these vibrations through their feet and trunk and may be able to recognize the vibrations as being produced by another elephant a long way off.

Visual depiction of sonic (top) and seismic (bottom) vibrations between two elephants. The seismic vibrations arrive later.
Courtesy of Lynette and Benjamin Hart.

You can get an idea of what may be going on in the elephant's world if you imagine standing on the ground near a railroad crossing as a train is going by. Trains are big and make many vibrations in the ground. If it were possible to hold your ears so you could not hear the train approaching, you would still be able to feel the train approaching through the vibrations in the ground. We believe this is similar to what elephants feel when they are sensing vibrations made by another elephant a mile or two away. The vibrations in the earth are referred to as *seismic vibrations* to differentiate them from *sonic vibrations* that travel through the air.

As elephants produce low-pitched infrasonic vibrations and at the same time produce seismic vibrations; however, the two vibrations travel at different speeds, but they feel or sound somewhat similar at the receiving end. The elephant is using two channels of communication, but the signals arrive at slightly different times. Naturally, the further apart the animals are the greater the difference in travel time will be between the two types of vibrations. This means that the elephant on the receiving end can not only get a message from a family member, identify the family member by the nature or signature of the sound vocalization, but by comparing the infrasonic sounds it hears with the ear with the vibrations it feels in the feet, get some idea of how far away the sender of the message is.

Such means of communication are particularly important for elephants that live in a jungle habitat, like Asian elephants that are found in India and Thailand, where it is difficult to see other family members at a distance, and where the trees and other vegetation tend to block ordinary vocalizations. Rumbles and seismic vibrations are also important to elephants living in the African savannah, but perhaps the different species of elephants, living in two different types of environmental habitat, have different needs and uses for these special infrasonic sounds and seismic vibrations.

Elephants live in large, complex families that split up from time to time and must often communicate with each other over long distances. This requires a large, complex brain. In fact, elephants have one of the most complex brains of all the mammals living on the surface of the earth and the largest brain of all land mammals. This large brain enables the elephant to manage family relationships and complex communication.

See also Communication—Tactile—*Vibrational Communication*

Further Resources

Arnason, B. T., Hart, L. A. & O'Connell-Rodwell, C. E. 2002. *The properties of geophysical fields and their effects on elephants and other animals.* Journal of Comparative Psychology, 116, 123–132.

Bradbury, J. W. & Vehrencamp, S. L. 1998. *Principles of Animal Communication.* Sunderland, MA: Sinauer Associates.

McComb, K., Reby, D., Baker, L., Moss, C. & Sayialel, S. 2003. *Long-distance communication of acoustic cues to social identity in African elephants.* Animal Behaviour, 65, 317–329.

O'Connell-Rodwell, C. E., Arnason, B. T. & Hart, L. A. 2000. *Seismic properties of Asian elephant* (Elephas maximus) *vocalizations and locomotion.* Journal of the Acoustical Society of America, 108, 3066–3072.

Payne, K., Langbauer, Jr., W. R. & Thomas, E. 1986. *Infrasonic calls of the Asian elephant* (Elephas maximus). *Behavioral Ecology and Sociobiology,* 18, 297–301.

Lynette Hart, Benjamin Hart, Caitlin O'Connell-Rodwell, & Byron Arnason

■ Communication—Auditory
Ultrasound in Small Rodents

The most commonly discussed attributes of sound are volume and pitch. The *pitch* or frequency of a sound is measured in the number of cycles per second. The different notes played on a piano are an example of differences in the pitch of a sound. The key farthest to the left on the keyboard is at 27.5 cycles per second or 27.5 Hertz (Hz), whereas the key farthest to the right on the keyboard is at 4,186 Hz. The range of our hearing is from about 20 Hz to 20,000 Hz. Sounds below 20 Hz are called *infrasounds* and those above 20,000 Hz are called *ultrasounds*. This is, however, a very human-centered approach to describing the world of sound. Many other animals have auditory worlds different from ours. African elephants and Asian elephants hear what are for us infrasounds, and a large number of animals hear in the ultrasonic range. The ultrasonic world is filled with familiar species like crickets, grasshoppers, moths, bats, our pet dogs, and small rodents like house mice, deer mice, hamsters, and gerbils. Many species that can hear ultrasounds can also make ultrasounds, and these sounds can be crucial for their normal existence.

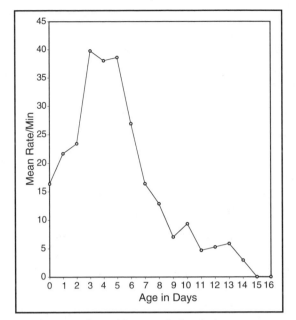

Developmental trend in rate of ultrasonic vocalization for young mice tested at 20° C.

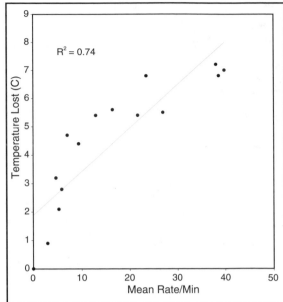

Temperature lost following 3 minutes at 20° C.

Newborn small rodents make ultrasounds for about one or two weeks after they are born. These ultrasounds are most often associated with the young cooling off. For about one or two weeks after they are born, small rodents cannot regulate their own body temperature, and so their internal body temperature is at ambient temperature, that is, the temperature of the world around them. Under normal circumstances, the mother will keep the young warm with her body inside a nest that acts like insulation. But if the young are left alone for a long time or if they tumble out of the nest, they will cool down and begin to make ultrasounds. These ultrasounds can alert and guide the mother to the pup.

A newborn mouse or other small rodent that is out of the nest will be retrieved by its mother usually within about a minute. If a pup is placed in full view, the mother will retrieve it and bring it back to the nest, but she will also find the pup if it is out of view. How does she do this? This simple experiment demonstrates the answer. If you put the pup in a clear glass tumbler placed on its side with the opening facing away, the mother will not find the pup. However, if you face the opening toward her, with the glass no longer blocking the sound, she will find the pup and bring it back to the nest. She finds the displaced pup by homing in on the ultrasounds coming from the chilled pup. With the proper equipment, humans can hear these ultrasounds. A special microphone and the appropriate circuitry can translate the ultrasounds into our audible range where they sound like countable chirps.

The temperature inside a nest filled with 5-day-old house mouse (*Mus musculus*) pups and the nursing female will be about 35° to 38° C (95 to 100.4° F) At this temperature, the pups will not make any ultrasonic vocalizations. If you take a single pup out of the nest and keep it at 36° C (96.8° F), it still does not vocalize. But, drop its temperature to 20° C

(68° F) by placing it in a cold chamber, and it will make ultrasounds. The rate of ultrasonic vocalization per minute will vary as a function of age. The first figure shows the developmental trend in the rate of ultrasonic vocalization for young tested at 20° C (68° F). These data are from 3 litters of *M. musculus* totaling 17 young that were born in my laboratory. Slightly higher rates are associated with colder temperatures and lower rates are associated with warmer temperatures. If the temperature is at or near 2° C (about 36° F), then the pup will become comatose and stop vocalizing.

If you take a mouse pup from a 36° C (96.8° F) nest and place it in a 20° C (68° F) chamber for 3 minutes, it will lose body heat as measured by a temperature probe. Like ultrasonic vocalizations these data also show a developmental trend. The second figure shows the temperature lost following 3 minutes at 20° C. Both measures can be graphed together as in the third figure. This scatter diagram with its fitted linear trend line shows that the rate of ultrasonic vocalization and the development

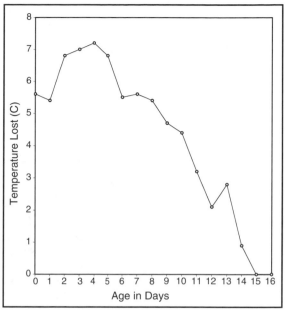

Combination of previous two graphs—showing the association of rate of ultrasonic vocalization and the development of temperature regulation.

of temperature regulation are associated. The correlation between the two measures is $r = +.86$. Squaring the correlation gives $r^2 = .74$. Seventy-four percent of the variability in the average rate of ultrasonic vocalization is accounted for (caused by) the variability in the development of temperature regulation. It is common to find r^2 values between .70 and .85 for a large variety of rodent species indicating that ultrasound production is linked to the development of temperature regulation.

An internal body temperature of 35° to 38° C (95° to 100.4° F) is necessary for the normal progress of physical and behavioral development. The newborn pup has a circular problem to solve. To develop physiological temperature regulation, and thereby maintain a constant internal temperature, the pup needs to maintain an internal body temperature of 35° to 38° C. The rodent pup solves this problem by using a behavioral mechanism that is developmentally linked with the inability to regulate its body temperature. As the pup cools down for whatever reason, it makes ultrasounds. The nursing female locates the pup, provides warmth, and the nest she built helps maintain the necessary temperature. This complex behavioral system continues until physiological temperature regulation has developed.

If the same pups are tested day after day as done here, then the developmental method employed is a longitudinal one. If different pups are tested each day, then the method used is called cross-sectional. Using both longitudinal and cross-sectional methods allows one to discover if any experiential factors, such as handling or repeated exposure to low temperature, affect the developmental profile. When the appropriate comparisons are made, the conclusion is that the developmental pattern in the development of ultrasound production as seen in the first figure is independent of experience.

See also Caregiving—*Parental Care*
   Caregiving—*Parental Care and Helping Behavior*

Further Resources

The original discovery that young rodents emit ultrasounds when they cool down was made by Zippelius & Schleidt (1956). The first studies of the developmental changes in ultrasonic vocalization in rodents were done by Noirot (1966) using albino mice and by Hart & King (1966) using deer mice. A review of much of the early work on the development of ultrasonic vocalization in rodents can be found in DeGhett (1978). An excellent summary of the uses of ultrasounds in all animals is in Sales & Pye (1974). The nature of and uses of both infrasound and ultrasound is examined in detail by Pye & Langbauer (1998). Downer (1988) provides a broad view of all of the sensory systems of animals that includes a very readable chapter on infrasound through ultrasound.

DeGhett, V. J. 1978. *The ontogeny of ultrasound production in rodents.* In: *The Development of Behavior: Comparative and Evolutionary Aspects* (Ed. by G. M. Burghardt & M. Bekoff), pp. 343–365. New York: Garland Press.

Downer, J. 1988. *Supersense: Perception in the Animal World.* New York: Holt & Co.

Hart, F. H. & King, J. A. 1966. *Distress vocalization in two subspecies of* Peromyscus maniculatus. Journal of Mammalogy, 47, 287–293.

Noirot, E. 1966. *Ultrasounds in young rodents. I. Changes with age in albino mice.* Animal Behaviour, 14, 459–462.

Pye, J. D. & Langbauer, W. R., Jr. 1998. *Ultrasound and infrasound.* In: *Animal Acoustic Communication: Sound Analysis and Research Methods.* (Ed. by S. L. Hopp, M. J. Owren & C. S. Evans), pp. 221–250. New York: Springer-Verlag.

Sales, G. & Pye, D. 1974. *Ultrasonic Communication by Animals.* London: Chapman & Hall.

Zippelius, H.-M. & Schleidt, W. M. 1956. *Ultraschall-Laute bei jungen Mäusen.* Naturwissenschaften, 43, 502.

V. J. DeGhett

■ Communication
Electrocommunication

Electric Signals

Many animals live in a sensory world very different from ours. However, although we lack the appropriate sense organs to perceive stimuli from this world, we can overcome some of our own limitations by making use of modern technological developments. This has enabled us to discover "sixth senses" in many animals. Remarkable examples are a number of cartilaginous and bony fish. They can sense electric signals originating from physical or biological sources, even if these signals are of such low intensity that we can only detect them by using extremely sensitive and highly sophisticated equipment. Many of these fish have even gone one stage further and developed specialized organs that produce electric discharges. As will be shown below, these discharges are carriers of a specific behavioral repertoire—the electric behavior.

Strongly and Weakly Electric Fish

Electric fish are divided into two groups—strongly electric and weakly electric—on the basis of the discharge voltage. Strongly electric fish generate voltages between several tens of volts and approximately 600 volts. (For comparison, the voltage of the electricity supplied in the United States is 110 volts, and in most European countries 220 volts.) Representatives are the marine electric ray (*Torpedo*) and the freshwater electric catfish (*Malapterurus*). The discharges are used to stun prey, and some of these strongly electric fish can be dangerous to man.

By contrast, the signals produced by weakly electric fish are far too low in voltage to be detected, unless amplified. In 1951, Hans Lissmann of the University of Cambridge in the U.K., and one of the pioneers of electric fish research, was the first to succeed in making the electric organ discharges of weakly electric fish visible on an oscilloscope and audible by means of an amplifier and loudspeakers. Given that the voltage usually does not exceed one volt (measured through external electrodes placed near the head and tail of the fish), it is not surprising that these signals remained unknown for so long.

The well-known electric eel (*Electrophorus electricus*) incorporates features found in both strongly and weakly electric fish. This South American freshwater species has three different electric organs. The main organ and one half of the Hunter's organ produce powerful electric shocks, whereas the other half of the Hunter's organ, as well as the Sachs' organ, generate only weak electric discharges.

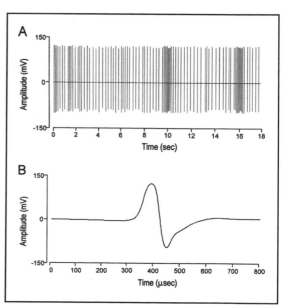

Wave-type versus Pulse-type Fish

Weakly electric fish have been found among both freshwater and marine species. The best-examined group are the freshwater weakly electric fish, which comprise two taxonomic orders of teleosts: the Gymnotiformes, which live in South and Central America, and the Mormyriformes, which are native to Africa. All mormyriforms, except one species (*Gymnarchus niloticus*), produce electric pulses that are followed by relatively long intervals of "silence." The pulses are very short, lasting from 250 microseconds (1 microsecond = 1/1,000,000 seconds) to

*Electric organ discharge of the elephant nose (Gnathonemus petersii). The presentation at a compressed time scale (**A**), showing the individual discharge pulses as vertical lines, reveals the rather irregular discharge rhythm. The expanded time scale (**B**) resolves the details of the waveform of an individual pulse.*

Electric Fish

Günther K. H. Zupanc, Daniela G. Meissner, and Jonathan R. Banks

A number of fishes have, in the course of evolution, developed specific electric organs by which they produce electric discharges. These fish species include the African order Mormyriformes; the South American order Gmnotiformes; the electric catfish (family Malapteruridae); the stargazer (*Astroscopus*), a perciform fish; the electric rays (suborder Torpedinoidei); and the skates (suborder Rajoidei). The voltages of these discharges, measured through electrodes placed close to the head and tail vary between less than 1 volt and several hundred volts. The electric fish use the electric discharges produced for stunning prey, orientation in the closer environment (so-called active electrolocation), species recognition, and intraspecific communication, including sex discrimination. The ability to communicate through electric signals is commonly referred to as *electrocommunication*.

Somewhat independent of the existence of electric organs, a large number of aquatic vertebrates have developed sense organs specialized in the perception of weak electric fields. This list includes both electric fish and species that are not capable of producing electric signals themselves, such as the aquatic forms of apodan and urodelian amphibians, the Australian egg-laying mammal platypus, sharks, and nonelectric catfish. In electric fish, the various types of electroreceptors are used to sense electric signals from both biological and nonbiological sources, including—and most importantly—the electric field generated though the animal's own electric organs. In nonelectric animals, the electroreceptors mediate a passive reception of the fields emitted by other organisms or originating from geophysical and electrochemical sources. This passive mode of reception guides localization of prey by enabling the predator to detect muscle and nerve potentials generated by the prey animal. Furthermore, this mode is possibly used for long-distance navigation. Such a function, suspected to exist, for example, in sharks, could be mediated by detection and analysis of currents induced by the animal's own swimming through the earth's magnetic field.

8 milliseconds (1 millisecond = 1/1,000 seconds). Because of their discharge pattern, fish of this group are referred to as pulse-type fish. The first figure shows a recording of a pulse sequence, as well as an individual pulse of such a fish.

In contrast, most gymnotiforms, as well as the mormyriform species *Gymnarchus niloticus*, produce discharges that resemble continuous waves when displayed on an oscilloscope screen. This continuous form of the signal is due to the fact that each discharge cycle is immediately followed by the next one. Because of the wave-like appearance of the signals on the oscilloscope screen, the fish that make up this group are referred to as wave-type fish. The signals produced by two of such fish, the knifefish *Eigenmannia* sp. and the brown ghost *Apteronotus leptorhynchus* are shown in the figure on page 355.

Electric Organs

The electric discharges are produced by specialized electric organs consisting of individual columns of electric elements called electrocytes. These columns often extend over nearly the entire length of the trunk, whereas in other species the electric organ is more restricted to the tail region. In most electric fish species, the electrocytes are modified muscle cells, and the electric organs are, therefore, called *myogenic*. However, in the gymnotiform genus *Apteronotus*, the electrocytes are derived from axons of motor neurons of the spinal cord. Hence, the electric organ is classified as *neurogenic*.

The electric organ discharge is generated by synchronous activity of many electrocytes. Individual electrocytes produce—similar to normal muscle or nerve cells—potentials of approximately 100 millivolts (1 millivolt = 1/1,000 volts). The number of electrocytes arranged in series determines the voltage of the electric discharge. For example, the electric eel may have 6,000 electrocytes arranged in series. When discharged synchronously, the potentials of these cells combine to generate a sum potential of 600 volts. This sum potential is commonly referred to as the electric organ discharge.

Both mormyriforms and gymnotiforms have larval and adult electric organs. The larval organ appears a few days after fertilization of the egg. As the adult organ gradually differentiates some weeks later, the larval organ begins to degenerate. In several species, including *Pollimyrus isidori*, the electric discharges produced by the larval organ are remarkably different from the discharges generated by the adult organ. In *Pollimyrus isidori*, in which the male guards the nest with the eggs and larvae, the larval discharges are thought to assist the parental fish in recognizing and locating the young fish.

Electroreceptors

In 1958, Hans Lissmann showed in another pioneering behavioral study that electric fish not only produce electric fields, but also can sense their self-generated electric signals. This stimulated an intensive search for electroreceptors organs, and within a few years after Lissmann's discovery several research groups reported the identification of such sense organs. As the work of these and other scientists has shown, the electroreceptor organs are derived from the lateral line system, which functions as a mechanosensory organ in fish and the aquatic forms of amphibians. Such studies have also revealed that the ability to sense electric fields is, to a certain degree, independent of the ability to produce electric discharges. Sharks, for instance, are electroreceptive, but lack the ability to produce electric signals.

In general, electroreceptor organs consist of up to several hundreds of receptor cells, joined by support cells. These receptor organs are embedded in the skin and distributed over the body surface, with particularly high concentrations in the head region. Electroreceptors in electric fish measure the voltage of the electric field across the skin, thereby functioning as tiny voltmeters. Based on their morphological appearance and physiological properties, two types of electroreceptor organs can be distinguished: ampullary and tuberous. The ampullary organs are most sensitive to electric fields of low frequency, whereas the tuberous organs exhibit maximum sensitivity to frequencies within the range of the fish's own discharge frequency. Subtypes of the latter category of electroreceptors enable the animal to analyze different physical properties of electric signals produced by themselves or other animals, or originating from physical sources. Thus, electroreceptive animals can acquire a comprehensive electric "image" of the world around themselves by means of their electroreceptor system.

Electrolocation

The discharges generated by electric fish create a three-dimensional electric field surrounding their body. Objects in the closer vicinity of the fish distort this electric field in a specific way, and these distortions are projected onto the skin surface in the form of an "electric image." Objects of higher conductivity than the water (e.g., a metal rod or a water plant) "attract" the electrical current lines, resulting on average in a higher density of current entering the skin of the fish near the object and, thus, in an increase in the voltage drop across the skin. This causes an increase in activity of the electroreceptors, and this information is "reported" via sensory nerves to the brain. Objects of lower conductivity than that of the water (e.g., a plastic rod or a stone) have the opposite effects. Such objects "repel" the electrical current lines of the fish's electric field, resulting on average in a lower density of current entering the skin opposite the object, and a decrease in the voltage drop across the skin. This causes, finally, a decrease in activity of the electroreceptors. On the other hand, objects with the same conductivity as the surrounding water are electrically invisible to the fish. This ability to detect, localize, and analyze objects by monitoring the self-generated electric field is referred to as *active electrolocation*. Experiments have shown that this capability is restricted to the detection of objects within a few centimeters around the fish. Active electrolocation is thought to have evolved as an adaptation to life in turbid, murky water (a considerable number of weakly electric fish live in such habitats) and in the dark (many weakly electric fish are active at night).

The electrolocation capability of weakly electric fish was first demonstrated by Hans Lissmann in the late 1950s using an elegant behavioral approach. Lissmann trained the weakly electric fish *Gymnarchus niloticus* to distinguish between objects of different conductivity, for example two cylindrical porous pots filled with solutions differing in salinity. In the course of training, an approach to one porous pot was rewarded with food. Approach to the other was followed by mild punishment consisting of chasing away the fish. The fish's ability to discriminate the objects based on differences in conductivity was tested by analyzing the differences in approach when the porous pots were presented, without rewarding the fish for a "correct" response or punishing it for an "incorrect" response. The positive results of these and other experiments demonstrated that the fish are, indeed, capable of electrolocation, and that they can distinguish objects solely based on their electric properties—even when the differences in conductivity are minor.

Species Recognition and Intraspecific Communication

Comparison of the electric organ discharges of sympatric species, that is, species living in the same habitat, has suggested another potential function of the electric behavior. Even when investigators examined a large number of sympatric species, in most cases they could readily distinguish the discharges based on one or several of the following parameters:

- discharge type (pulse-type versus wave-type)
- waveform of the individual discharge pulse in both wave-type and pulse-type fish
- number of pulses produced per time unit

The following figure illustrates the differences between two wave-type species, the knifefish (*Eigenmannia* sp.) and the brown ghost (*Apteronotus leptorhynchus*). As these recordings demonstrate, the knifefish produces fewer discharges than the brown ghost; that is, the discharge frequency of the knifefish is lower than the frequency of the brown ghost. Furthermore, the waveforms differ quite markedly between the two species. The knifefish emits discharges resembling simple sine waves, whereas the discharges of the brown ghost are clearly more complex.

In wave-type species, similar differences are often also observed between the two sexes of the same species. For example, in the knifefish, sexually mature males discharge at lower frequencies than gravid females, with little overlap at intermediate frequencies. In the brown ghost, this sexual dimorphism is reversed—in the latter species, females discharge at lower frequencies than males. In the knifefish, an additional type of sexual dimorphism exists regarding the waveform. Female discharges are more sine-wave-like than male discharges.

The existence of such differences does not prove that the fish actually use them for species recognition or sex discrimination. However, conditioning experiments have shown that the fish can discriminate electric signals differing in frequency and waveform. In these experiments, two electric signals differing, for example, only in frequency are played back via stimulation electrodes to a fish kept in an isolation tank. Then, during the training stage of the experiment, the fish are trained to discriminate the two signals by rewarding the fish's approach to the electrodes with food when it is stimulated with one of the two frequencies. Conversely, if the fish approaches the electrodes when stimulated with the other frequency, they either do not receive any food reward, or they are mildly punished, for instance by being chased away from the electrodes. This association of the different frequencies with reward and non-reward (or punishment) is repeated several times.

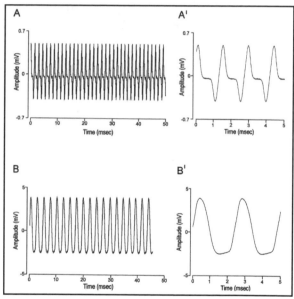

Oscilloscope traces of the electric organ discharges of the brown ghost, Apteronotus leptorhynchus *(A) and the knifefish,* Eigenmannia *sp. (B). Each of the two species produces signals which appear as continuous waves on the oscilloscope screen. In the examples given, the brown ghost generates approximately 34 electric signal cycles within 50 milliseconds, equivalent to a discharge frequency of 680 Hz. Conversely, the knifefish generates approximately 21 electric signal cycles within 50 milliseconds, thus discharging at a significantly lower frequency of 420 Hz. Part of these recordings of the brown ghost and the knifefish are shown at expanded time scales (A' and B', respectively) to show the differences in waveform between the two species.*

The elephant nose shown in various behavioral situations. **A.** At rest, the fish likes to stay in caves or, in this case, in a clay pipe. In such a situation, the fish typically emits only a few pulses, separated by rather irregular intervals. **B.** If a second fish is introduced into the tank, both fish stay initially at a short distance from each other. **C.** They then frequently adopt a head-to-tail stance alongside each other, with their "chin" appendage (sometimes incorrectly called a "nose") projecting rigidly forward. **D.** If the intruder fails to move off, the territory-holder attacks and rams the intruder. **E.** Finally, the intruder is beaten by the territory holder. The defeated fish turns light brown. Now, both fish have curled in their chin appendages. The social interactions shown in **B-E** are accompanied by specific patterns of electric organ discharges which clearly differ from the resting discharge.

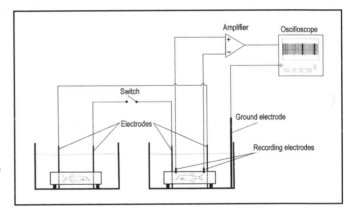

Experimental setup to demonstrate electrical interaction between two elephant nose fish (Gnathonemus petersii). A single fish is kept in each tank. The two aquariums are electrically connected via wires attached to electrodes, which are placed near the fish. This connection can be controlled by an electrical switch. In addition, the electric organ discharges of one of the two fish are monitored through recording electrodes, which also allow the researcher to follow the signals produced by the other fish as long as the electrical connection between the two tanks is kept active. The latter discharges appear as pulses of significantly lower amplitude compared to the pulses of the first fish.

In the next stage of the experiment, the two signals are played back using the identical procedure as during the training stage, but without any reward or punishment; and the fish's response is recorded.

Such experiments have shown that knifefish can discriminate between electric sine waves differing in frequency by as little as 0.5 Hz at stimulus frequencies around 500 Hz, thus corresponding to roughly 0.1% frequency difference. This frequency discrimination ability is similar to the one of humans, and markedly better than the typical values found in

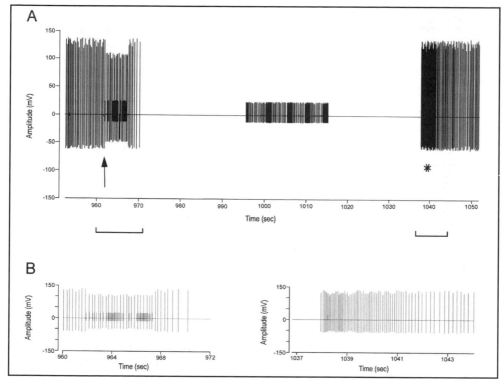

Modulation of the discharge rhythm of the elephant nose (Gnathonemus petersii) *by electric organ discharges of a member of the same species. For the recordings, a setup as shown in the previous figure was used. The discharge pulses of the first fish are represented by the large vertical lines in **A**, the discharges of the second (neighboring) fish by the smaller vertical lines. A few seconds after electrically connecting the two tanks, and thus stimulating the first fish with the discharges of its neighbor (indicated by arrow), the first fish slows down its discharge rhythm and, another few seconds later, even completely ceases discharging. During this period of silence, the neighboring fish produces a salvo of discharge pulses, to which the first fish does, however, not respond immediately. Some twenty seconds after this discharge salvo, the first fish starts discharging again (*), displaying an extremely high pulse repetition rate within the first few seconds. Two parts of the recording of **A**, as defined by the brackets below the time scale, are shown at expanded time scales in **B**.*

other vertebrates. Similarly, knifefish can distinguish between electric signals based on waveform only, for example male and females waveforms. Although these experiments still do not conclusively prove that the fish use these parameters for species recognition or sex discrimination, they show that they have the sensory capability to perform such tasks.

Another important, and from an ethological point of view certainly the most interesting feature of the electric system, is the ability of weakly electric fish to produce different electric behaviors depending on the social context, and to interact through these signals with members of the same species. This capability, together with the sex discrimination ability based on differences in the electric discharges, is commonly referred to as *electrocommunication*. Studies have shown that the range over which electric signals can be exchanged is limited to a maximum of approximately 1 meter. This restriction is a consequence of the physical properties of the electric field generated by electric fish. This field resembles, in first approximation, the electric field of an electric dipole. As is known from electrostatic theory, the field strength of the latter declines with the inverse cube of distance. Thus, if a fish doubles the distance from its neighbor, the amplitude of the electric signal received declines to one eighth. This decrease in signal intensity is much more pronounced than that of acoustic signals, which decline only with the inverse square of distance.

The mode through which information is encoded for electrocommunication differs between pulse-type fish and wave-type fish. In pulse-type fish, the individual pulses are extremely stereotyped and change very little over shorter periods of time—even if these periods involve days or weeks. This makes the pulse waveform ideally suited to convey "long-term" information, such as sex or species identity of the fish. "Short-term" information, on the other hand, is encoded by the number of pulses produced over a certain time period (the so-called pulse repetition rate), and by the discharge rhythm, commonly described by the sequence of intervals between individual pulses.

Observations have shown that the latter two discharge parameters change specifically with different behavioral situations. The photograph sequence shows the elephant nose in different behavioral situations occurring in the context of aggressive encounters. Such encounters, which largely take place through electric interactions, can usually be mimicked by a simple experimental setup illustrated here. Two elephant nose fish are kept isolated in aquariums, each equipped with one tube to provide a hiding place for the fish, and a pair of electrodes linked to the electrodes of the other tank via wires. This electric connection can be activated or interrupted by a switch. In addition, the electric organ discharges of one of the two fish are monitored through recording electrodes.

At rest, that is as long as the connection between the two tanks is not activated, the fish typically emits less than 10 electric pulses per second, often displayed at a rather irregular rhythm (see the final figure). As soon as the electric circuit is closed, the fish can perceive the discharges of its neighbor (visible on the oscilloscope screen as pulses of markedly lower amplitude). Within a few seconds, the fish alters its discharge pattern, first reducing the number of pulses produced, and finally completely ceasing the emission of pulses. After a minute or so, the fish starts discharging again, but now at a much higher pulse-repetition rate compared to the rate produced at rest. At this stage of the encounter, the fish sometimes generate more than 100 pulses per second. Moreover, the intervals between the individual pulses become now rather irregular.

Are these different discharge pattern used as signals; that is, does one fish transmit information to the other fish? The answer to this question can be obtained by employing artificial electric fish: so-called fish dummies or fish models. They consist of plastic rods with electrodes inside them. Through these electrodes, the isolated fish is stimulated with previously recorded discharges or synthesized electric signals. The behavioral response to this artificial fish is then observed. Such experiments have shown that the fish's response does, indeed, depend upon which discharge patterns are played. For example, aggressive discharge signals usually elicit aggressive behavior, while resting signals do not, generally, evoke attacks directed toward the model.

Waveform species employ a different mechanism to generate signals used or social communication. Normally, the frequency and waveform of these discharge are extremely stable in an individual fish. For example, a brown ghost may discharge at 700 Hz, changing its frequency throughout the day by no more than 1 or 2 Hz. Over 10 minutes, the frequency changes are so small that their detection requires precision instruments. As such, the electric organ discharge of these fish are some of the most stable phenomena known in biology.

Despite this normally observed constancy, changes (modulations) in both frequency and amplitude do occur, especially during social encounters. These changes are very brief, typically lasting only a few milliseconds. One type of these modulations is called "chirps" because the signals resemble chirps of crickets when made audible through a loudspeaker connected to the amplifying equipment. In some species, frequency modulations may be followed by a complete cessation of the electric organ discharge.

In some species, such as the knifefish *Eigenmannia* sp., chirping occurs only during the breeding season. During the night, when the fish spawn, a male may chirp almost continuously up to 80 times a minute. Playback experiments have revealed more about the function of chirping in this species. When recorded chirps of a male were played back to an isolated, sexually mature female via a fish model, the female could be induced to lay eggs! It is likely that the continuous male chirps bring the female into a physiological state in which she is ready to spawn. This is an especially intriguing example of the function of social signals.

Further Resources

Bullock, T. H. & Heiligenberg, W. (Eds.) 1986. *Electroreception*. New York: John Wiley & Sons.

Moller, P. 1995. *Electric Fishes: History and Behavior*. London: Chapman & Hall.

Turner, R. W., Maler, L. & Burrows, M. (Eds.) 1999. *Electroreception and Electrocommunication*. The Journal of Experimental Biology, 202, Issue 10.

Zupanc, G. K. H. 1988. *Fish and their Behavior*. 2nd edn. Melle: Tetra-Press.

Zupanc, G. K. H. 2004. *Behavioral Neurobiology: An Integrative Approach*. Oxford/New York: Oxford University Press.

Günther K. H. Zupanc, Daniela Meissner & Jonathan R. Banks

■ Communication
Honeybee Dance Language

Following Karl von Frisch's finding that the dance movements of bees conveyed the distance and direction to food sources to other worker bees (*Decoding the Language of the Bee; Nobel Lecture*), a rash of exciting discoveries about the behavior and sensory capabilities of honeybees followed. For example, he found color vision in honeybees that included the perception of ultraviolet light, wavelengths not used by the human eye. Further explorations included studies of how the bees used the sun as a compass, how they seemed to be able to use the compass even when the sun was behind clouds, and how the dances differed among species of honeybees. Frisch's conclusions, though, have not been without controversy. Some biologists have felt that they could not replicate his findings, and have argued that bees use an odor-based system of navigation to take them to food sources that other bees in the colony have discovered.

As Frisch had concluded, the angle on the comb of the straight-run of the dance indicates the direction of the food source. A bee that runs straight up on the comb is indicating to her followers that the food source is in the direction of the sun; a bee who runs straight down shows a direction away from the sun. Intermediate angles (between up and down) on the comb communicate angles to the right or left of the sun. This seems like a perfect solution to the problem of communicating direction; bees can use the sun as their reference point on a compass and fly in the appropriate direction. The only problem with this, and

Insects like this honeybee have been found to convey sophisticated information to one another about food resources and enemies.
© *Anthony F. Chiffolo.*

this is a huge problem, is that the sun appears to move in the sky. In fact, the sun's apparent movement is so rapid that from the time a bee leaves the colony to when she arrives at a food source, it will have changed positions by several degrees. Thus the compass communicates misinformation, rather than information.

How can the dance language system work, then? From a human perspective, this seems like a difficult problem to solve. In reality, for you to use a sun compass as accurately as the bees do, you would need a watch to calibrate the sun's position with the time of day. Knowing this relationship, you could then compensate for the changing position of the sun in the sky as you travel.

At the time Frisch developed his dance language hypothesis, it was inconceivable that an insect could accurately tell time without a mechanical aid; certainly humans could not accomplish this feat. Nevertheless, this was exactly what Frisch and his students found. As time progresses, dancing bees modify the angle of their dance to compensate for the sun's movement. Further, as recruits fly to the flowers, they compensate so that their flight path is not distorted by the sun's apparent movement. Now we know that many animals have a well-developed time sense, and that compass navigation using the sun, moon, or stars is a common attribute of a broad range of animals. Investigations into the dances of honeybees opened the way for a general appreciation of internal clocks in animals.

If you have a clock, you need to be able to set that clock. But it is not enough to set your own clock; for communication to work, the bees' clocks must be synchronized. Individual bees use the daily light–dark cycle of their surroundings to set their clocks, but social interactions among the bees result in their synchronization. The only bees that need to be held to a daily activity cycle are the foragers—the bees that work outside the colony. Younger bees, which work only inside the colony, do not show a daily rhythm in their activity patterns.

Once the problem of apparent movement is solved, the sun seems like an excellent orientation cue. What happens, though, if the sky is partly cloudy? In particular, what if the sun is covered by clouds? To the human eye, the sky away from the sun appears uniformly blue, as featureless as the open ocean. Again, it would seem that the compass should fail. In most environments, cloudy days occur frequently, and bees probably can't afford to forage only when the sun is visible. The sky, though, is not as featureless as it seems to our eyes. The vibration of light waves can be oriented in uniform directions when light passes through a substance like the earth's atmosphere; physicists call light which is vibrating in a uniform manner "polarized." In fact, there is a pattern of polarization of light in the earth's atmosphere that you can visualize by popping the lenses out of polarizing sunglasses. Hold one lens stationary so you can view part of the blue sky through that lens, and then put the other lens in front of the first lens. Slowly rotate the second lens. You'll see that the effect of the polarization of the lenses is that as you rotate one lens the amount of light coming through changes. When the polarization of the light by the two lenses is in the same direction, the maximum amount of light comes through; when the polarization is in different directions, less light comes through.

Now, take one lens and aim it at a part of the blue sky (never directly at the sun!). As you rotate the lens, you'll see the same effect of changing light intensity. If you keep the lens at a constant angle, you'll see that different parts of the sky have different directions of polarization. In fact, there is a pattern of polarization in the sky, and with practice you can learn to predict the sun's location by knowing the polarization of a patch of sky.

Humans can't see the polarization of the sky, but honeybees can. As a result, their sun compass functions well, even when the sun is obscured; they can navigate by extrapolating the position of the sun from a small patch of blue sky. In each *ommatidium*—the units that make up the compound eye of insects—there are stacks of fingerlike projections, microvilli. Within these microvilli, the light sensitive pigment molecules are arranged in lines, so that they respond to light coming with specific planes of polarization. This means that any pattern of polarization in the bee's view will cause its eye to respond in a unique way, giving the brain information about polarization as well as color and form. Polarized light perception is now known in many animals.

One important outcome of Frisch's discovery of the dance language of the honeybee has been the realization that humans lack many of the sensory capacities enjoyed by animals. Our knowledge of the ability of animals to sense and use ultraviolet light, polarized light, and magnetism, as well as the accurate use of an internal time sense, all stemmed from studies of dance language of the honeybee. While not all animals possess all of these capabilities, scientists who study animal behavior must be continuously aware of the possibility that the animal under study is able to sense much more about its environment than we can.

Bee researchers now realize that bees use communication tools in addition to the waggle dances. One interesting signal is the tremble dance, which is performed by bees that return from good food resources, only to find themselves delayed in handing their food over to another bee. While Frisch and others had recognized the tremble dance, only recently was Thomas Seeley able to assign a meaning to this dance. The tremble dance seems to play a key role in regulating foraging effort by a bee colony, helping it to adjust the amount of incoming food to its capacity to handle that food.

Another fascinating discovery is that honeybee swarms use the dance language to communicate about possible new homes. When a swarm (a queen and a few thousand worker bees) leave their old nest, they find a branch and settle there, making a conspicuous hanging ball of bees. Scouts leave the swarm and search for new places to nest. Not only do scout bees dance to indicate possible nesting sites; the bees use the dance communication to reach a consensus about which of the possible new homes is the best choice.

Some scientists have criticized Frisch's basic assertion, that bee dances actually communicate distance and direction to their fellow workers. Adrian Wenner, with his students and collaborators, argued that Frisch and subsequent investigators never rigorously tested the dance language hypothesis by eliminating possible alternative explanations. Wenner's experiments point to the use of an odor map by bees; he suggests that the dance movements are irrelevant and that instead recruits are cued by odors on the surface of the dancer.

In response to these criticisms, James Gould designed an elaborate experiment that attempted to separate the dance from all other possible cues. He was able to have dancers who perceived the time of day differently from the bees following the dance. In other words the dancers directed the bees to one location, based on their time perception, while the dance followers interpreted a different direction, following a differing time perception. His result clearly showed that bees used the dance information.

The critics found fault with Gould's experiment, and in fact, no scientific experiment will ever be perfect. The most significant mistake a scientist can make is to collect data trying to prove a point, rather than testing a hypothesis with an open mind about the result. Have Frisch and subsequent workers whose data supports the dance hypothesis made this mistake? Have Wenner and his followers made the same mistake in an effort to disprove the dance hypothesis? Despite existing controversies, the overwhelming majority of current evidence supports the dance language hypothesis. However, bees are clever, surprising animals and there is no doubt that they can use odors in sophisticated ways. At least for some skeptics, final resolution of this issue awaits further experimentation.

See also Communication—Vocal—*Referential Communication in Prairie Dogs*

Further Resources

Frisch, K. von. 1967. *The Dance Language and Orientation of Bees*. (Translated by L. E. Chadwick), Cambridge, MA: Belknap Press of Harvard University Press.
Gould, J. L. 1975. *Honey bee recruitment: the dance-language controversy*. Science, 189 (4204), 685–693.
Seeley, T. D. 1995. *The Wisdom of the Hive: The Social Physiology of Honey Bee Colonies*. Cambridge, MA: Harvard University Press.
Wenner, A. M. & Wells, P. H. 1990. *Anatomy of a Controversy: The Question of a "Language" among Bees*. New York: Columbia University Press.

Michael D. Breed

■ Communication
Modal Action Patterns

The bucking of a horse, the begging response of newly hatched birds, the grooming behavior of a dog, the head-bobbing of a lizard, and the fighting response of stickleback fish are all examples of behavior patterns. Each of these behavior patterns is a response to a particular stimulus or a combination of stimuli. As we have all experienced, many of the sights, smells and sounds in the environment stimulate us and cause us to respond in a certain way. For example, walking into a restaurant filled with delicious smells of cooked food may cause us to salivate. Collectively, the sights, smells and sounds that influence us are known as stimuli, and the reactions they elicit are responses that include physiological changes, reflexes, and behavior patterns that reflect "natural" and readily observable units of behavior.

Ethologists, those who study animal behavior in a naturalistic environment, are particularly interested in behavior patterns, which are considered to be the central unit in the study of animal behavior. Historically, ethologists have used the terms, instinctive movements, species-typical behaviors, motor coordinations, behavior patterns, consummatory acts, and fixed action patterns, to describe the units of behavior. Of these terms, "fixed action pattern (FAP)," coined by the famous ethologist, Niko Tinbergen, is particularly well-known. *Fixed action patterns* are stereotypical and species-typical behaviors that are elicited by specific signals from the environment. Once elicited, a FAP will often continue until completion.

In the late 1960s George Barlow, a well-known ichthyologist and ethologist, recognized the need to standardize behavioral terms used by ethologists. Barlow realized that although FAP was commonly used to describe behavior patterns, many behaviorists would agree that the units of behavior are not necessarily "fixed." In fact, most organisms have the ability to modify their behaviors through experience to varying degrees. To eradicate this misconception of the fixity of behavior, Barlow proposed that the term, *modal action pattern* (MAP), be used in place of FAP.

According to Barlow, a unit of behavior may be categorized as an MAP if it has the following characteristics:

1. the behavior pattern is a recognizable movement pattern that can be easily observed and quantified,

2. the behavior pattern is the simplest unit of behavior that one cannot easily subdivide into smaller units, although some of its components may occur independently or in other MAPs, and

3. the behavior pattern can be observed in similar forms throughout an interbreeding population.

Some of the most interesting MAPs can be observed when animals are feeding. The feeding repertoire, particularly the phases of immobilization and consumption once a suitable prey item is captured, contain some excellent examples of MAPs. For example, many snakes are known to constrict their prey before consuming them. Constriction is a MAP in which prey is immobilized by two or more points on a snake's body. This behavior pattern readily follows the MAP characters outlined by Barlow. Constriction behavior is easily observed in captivity (one would be very fortunate to happen upon a snake feeding event in the wild) and aspects of constricting behavior, such as the overall duration can be

Example of constriction, a MAP in which prey is immobilized by two or more points on a snake's body. Courtesy of Rita Mehta.

quantified. Secondly, constriction is a single unit of behavior within the snake feeding repertoire. It would be very difficult to reduce constriction behavior into smaller parts and, if reduction were possible, the components may not be very useful. Lastly, constriction can be seen in similar forms throughout an interbreeding population of snakes. Although some populations may exhibit more variability than others, constriction is still a readily recognizable behavior pattern.

In the late 1970s, Harry Greene and Gordon Burghardt found constriction behavior to be prevalent in many snake species. However, with careful comparative analysis, it was discovered that the constricting behavior of boas and pythons was different from the constricting behavior observed in the distantly related rat snakes and king snakes. Boas and pythons constrict their prey with tight winding coils, whereas rat snakes and king snakes constrict their prey in a more irregular manner. Boas and pythons exhibited a highly stereotypical constricting pattern, whereas rat snakes and king snakes had more variable constricting patterns.

Also, rat snakes and king snakes could alter their MAPs in response to the size of the prey item. Rat snakes did not constrict small prey items that could easily be swallowed, whereas

Boas and pythons constrict their prey with tight winding coils, whereas rat snakes and king snakes constrict their prey in a more regular manner.
Courtesy of Rita Mehta.

large prey items would be automatically constricted. Similarly, dead prey were not always constricted by rat snakes but were frequently constricted by boas and pythons. The ability to change prey handling techniques to match the difficulty of the prey item is advantageous because constriction appears energy intensive. Efficient use of time is important and snakes that constrict prey items take a much longer time to feed than those that simply seize and swallow their prey. This is significant because feeding is a time when a snake may be blind to its own predators.

Variation in MAPs exhibited by different species can be an important tool for the ethologist to use in animal classification. As early as 1910, Oscar Heinroth, a pioneering ethologist and ornithologist, described MAPs that were indicative of all geese and MAPs that were distinct for the different genera of swans. This type of behavioral categorization continues today, and modern day ethologists, organismal biologists, and behavioral ecologists are interested in the evolution of certain MAPs and the persistence of MAPs within a particular group of animals. Although boas, pythons, rat snakes, and king snakes are related, rat snakes and king snakes are more closely related to each other than to boas and pythons. Boas and pythons are each other's closest relatives and share the same type of constricting MAP. Therefore, constriction behavior may be used as an additional line of evidence to illustrate relationships among snake groups.

See also Play—*Social Play Behavior and Social Morality*

Further Resources

Barlow, G. 1977. *Modal action patterns.* In: *How Animals Communicate* (Ed. by T. A. Sebeok), pp. 98–134. Bloomington, IN: Indiana University Press.

Burghardt, G. M. 1973. *Instinct and innate behavior: toward an ethological psychology.* In: *The Study of Behavior: Learning, Motivation, Emotion, and Instinct* (Ed. by J. A. Nevin & G. S. Reynolds), pp. 322–400. Glenview, IL: Scott, Foresman.

Dewsbury, D. A. 1978. *What is (was?) the "fixed action pattern?"* Animal Behaviour, 26, 310–311.

Greene, H. W. & Burghardt, G. M. 1978. *Behavior and phylogeny: Constriction in ancient and modern snakes.* Science, 200, 74–77.

Grier, J. W. & Burk, T. 1992. *Biology of Animal Behavior* St. Louis: Mosby-Year Book, Inc.

Rita Mehta

■ Communication—Olfaction
Chemical Communication

Chemical signals are the most ubiquitous mode of animal communication. Insects provide the most famous examples of *pheromones*, chemicals produced by exocrine glands for communication with others of the same species. However, at least some members of many major animal groups, such as mollusks, annelids, fish, amphibia, reptiles and mammals,

Dog Scents and "Yellow Snow"

Marc Bekoff

Dogs spend a lot of time with their well-endowed nostrils stubbornly vacuuming the ground or pinned blissfully to the hind end of other dogs. They have about 25 times the area of olfactory (odor) epithelium (which carry receptor cells) in their nose and have many thousands more cells in the large olfactory region of their brain (mean area of 7000 mm^2) than humans (500 mm^2). Dogs can differentiate dilutions of 1 part per billion, distinguish t-shirts worn by identical twins, follow odor trails, and are 10,000 times more sensitive than humans to certain odors. They can also detect mines and the feces of other animals.

When dogs wiggle their noses and inhale (suction) and exhale (snort), they concentrate odors, pool them into mixtures and expel others. Like their wild relatives (wolves, coyotes, foxes, and jackals) dogs gather much information from the symphony of odors left behind in their nostrils. Urine provides critical information about who was around, their reproductive condition, and perhaps their mood. Dogs expel millions of gallons of urine wherever they please (more than 1.5 million gallons along with 25 tons of feces per year in New York City alone) and use it well.

Odors are powerful stimulants. It is said that the famous psychoanalyst Sigmund Freud used soup smells to stimulate clients to recall past traumas. Although my companion dog, Jethro, enjoyed visiting his veterinarian, he showed fear if he went into an examination room where the previous canine client was afraid. Fear is conveyed via a pungent odor released by the previous dog's anal glands.

Now, what about sniffing other dog's urine? I conducted a study of Jethro's sniffing and urination patterns. To learn about the role of urine in eliciting sniffing and urinating, I moved urine-saturated snow ("yellow snow") from place to place during five winters to compare Jethro's responses to his own and other dog's urine. Immediately after Jethro or other known males or females urinated on snow, I scooped up a small clump of the yellow snow in gloves and moved it to different locations.

Moving yellow snow was a useful and novel method for discovering that Jethro spent less time sniffing his own urine than that of other males or females. Other researchers have also noted that males dogs (and coyotes and wolves) spend more time sniffing the urine from other males compared to their own urine. Dogs also usually spend more time sniffing urine from females who are ready to mate ("in heat") compared to urine from males or reproductively inactive females.

The differences in Jethro's response to the displaced urine from other males or from females are worth noting, especially when considering "scent marking" behavior. "Scent marking" is differentiated from "merely urinating" by a number of criteria that include sniffing before urinating followed by directing the stream of urine at urine that is already known to be present or at another target. When Jethro arrived at displaced urine he infrequently urinated over or sniffed and then immediately urinated over ("scent marked") his own urine, but he sniffed and then immediately scent marked the displaced urine significantly more often when it was from other males than when it was from females. While domestic dogs are usually not very territorial (despite myths to the contrary), their wild relatives are often highly territorial, and they show similar patterns of scent marking behavior in territorial defense.

(continued)

Dog Scents and "Yellow Snow" (continued)

This brief foray into the olfactory world of dogs gives some idea of what the dog's nose tells the dog's brain. You can easily repeat this simple experiment (and risk being called weird). The hidden tales of yellow snow are quite revealing about the artistry of how dogs make sense of scents.

See also Communication—Olfaction—*Mammalian Olfactory Communication*
Communication—Olfaction—*Chemical Communication*

Further Resources

Bekoff, M. 2001. *Observations of scent-marking and discriminating self from others by a domestic dog* (Canis familiaris): *Tales of displaced yellow snow.* Behavioural Processes, 55, 75–79.
Dunbar, I. 1978. *Olfactory preferences in dogs: the response of male and female beagles to conspecific urine.* Biology of Behavior, 3, 273–286.
Wells, M. C. & Bekoff, M. 1981. *An observational study of scent-marking in coyotes.* Animal Behaviour, 29, 332–350.

use chemical communication. Birds are the only important exception; as far as we know, birds lack chemical communication.

Why use chemical communication? Odors can reach their target in the dark, carry around objects, and move over distances too great for sound. In small animals, for which production of loud sounds is particularly difficult, odor provides a unique opportunity for long-distance communication. Carried by currents of air or water, an animal's odor can be more effective than visual, audible, or electrical signals. A less volatile odor signal can be left behind on an object, marking a territorial boundary or a trail, and remain effective in communication long after the animal has gone. Many chemical signals can be used at will, released when needed, and stopped when production might be disadvantageous or costly. These features combine to favor the evolution of chemical communication.

Most often, chemical communication is used to attract or choose among potential mates. Odors, carried by wind or water, signal the presence of a receptive animal, usually the female, and attract potential mates from a wide area. Typically, a female produces a species-specific blend of chemicals, which, when carried in water or air, attract males to her location. The complexity of the mixture, which scientists call a multi-component pheromone, helps the male to be sure he is finding a female of his own species; even closely related species of insect produce different sex pheromone blends. Chemical signals may also convey information used in choosing a mate, such as the dominance status of an animal, its reproductive potential, or its genetic makeup. The genetic information can be quite important because it allows potential mates to avoid inbreeding or to seek mates that will yield higher quality offspring.

Male calling of females is less common, but not unknown. Travel to a pheromone site is generally riskier, in terms of exposure to predators, than staying in one spot and producing pheromone. Evolutionary forces usually favor males accepting this risk, while females are more likely to avoid such hazards. In fact, females may use the ability of males to survive risks as a test of male fitness as a mate. In the few species in which males call females, the male usually possesses a valuable resource, such as a territory with food or shelter, which, if the female joins the male, will enable the female to be more successful in producing off-spring. Males in possession of territories, such as certain species of cockroaches, often use pheromones to call females for mating.

When a male and female have found each other, chemicals may provide additional information used in actually deciding whether or not to mate. In many rodents, such as house mice, odors are more similar among closely related animals (sibs and half-sibs) than among unrelated individuals. These signals, which are produced in the urine, allow genetically similar animals, such as brothers and sisters, to avoid mating. Mice which differ only in their major histocompatibility genes (MHC) produce different urinary odors; this link between MHC and odors used in discriminating relatives from nonrelatives is present in many animals, including anemones, fish, and amphibia, as well as mammals.

Within social groups, chemicals can identify group membership, signal dominance status, communicate the location of food resources, and even regulate the physiology of other group members. Most species of social insects, bees, ants, wasps, and termites, use surface chemicals to determine whether an animal belongs in the nest or not. Even within a species, surface chemicals vary among colonies. In honeybees, the ability to determine whether a bee belongs or not is used to prevent a colony from being robbed by other colonies.

Trail pheromones are used by ants, termites, and stingless bees to mark pathways between colonies and food sources. Trail pheromone communication helps to coordinate movements of masses of worker social insects and is a key to the foraging success of many species. Trails work well in the dark, in complex habitats such as tropical rainforest, and under circumstances in which one forager needs to communicate with other foragers without direct contact. In some species of ant and termite, trail pheromone communication works so well that the foragers have lost their eyes in the course of evolution.

In honeybees, queens produce an array of chemicals which regulate worker behavior and physiology. While queens do not "direct" the activity in the hive, the presence of the queen pheromones causes workers to form a court around the queen, prevents worker egg-laying, and suppresses the rearing of replacement queens. When the queen is removed, workers start rearing new queens.

Many social insects use alarm pheromones to coordinate defensive responses. Honeybee alarm pheromone, isopentyl acetate, has a banana-like smell to humans and is released when a bee stings. The alarm pheromone alerts other bees, which are stimulated to fly and are then attracted by the motion of the predator. In addition to honeybees, alarm pheromones are common in ants.

An interesting example of simple chemical communication in a vertebrate social group are fish alarm pheromones, which are produced in skin cells of ostariophysian fish (a large group of fish that includes many familiar aquarium species, such as minnows and tetras). Injured fathead minnows release a pheromone, hypoxanthine-3-N-oxide, into the water, alerting nearby minnows of the presence of a predator. Similarly, tetras avoid predators that have previously ingested the alarm pheromone; this helps them to stay away from predators that have eaten tetras. Alarm pheromone production in schooling fish augments other methods of identifying predators, such as visual recognition.

The interpretation of mucus trails laid by mollusks, such as slugs or snails, is more controversial. The trails could serve as pathways to feeding areas or as routes to mates. However, the mucus trail itself may serve as a trap for algae, so snails follow trails in order to collect food, rather than using them for communication.

The main limitations of chemical communication come from the dilution of chemicals as they radiate in air or water from their source and from the tendency of odor plumes to break up into disconnected ribbons. Natural selection has compensated for dilution of pheromones in air or water by giving animals efficient structures for filtering rare molecules from their surroundings and highly sensitive sensory organs. The elaborately branched antennae of many moths and beetles allow them to strain molecules from the surrounding air, and at least in moths, only a few pheromone molecules are required to stimulate a signal from the antenna to the brain. As a rule, animals' sensory systems are much more responsive to pheromones than to "general odorants" found in the environment.

To find the source of the pheromone, a searching animal must be able compensate for its movement in and out of contact with the odor. Moths accomplish this by flying upwind when in the odor plume, and crosswind when out of it. Flying across the wind maximizes their chances of finding an odor-carrying stream of air.

Some chemical signals are not pheromones. Generally, we think of pheromones as chemicals whose production has evolved for communication. Other chemicals, like metabolic waste products, surface secretions designed for waterproofing, or food odors that have passed through an animal, may be used as sources of information, even though evolution has not shaped them for this purpose. As a result, not all odor communication is intentional. An animal may send a signal that betrays itself, attracting predators, causing others to judge it an unsuitable mate, or causing it to be excluded from a social group which it is attempting to enter.

Chemical signals, whether pheromones or unintentionally produced odors, may attract trouble. Predators or parasites may use their prey species' pheromones as a pathway to food. Insects put themselves at risk for predation when they use sex pheromones to advertise their presence; the insect world abounds with examples of pheromones which turn their producer into a target. A species of digger wasp, which preys on other insects, is attracted to the sex pheromone of a stink bug, which it then preys upon. Outside the insects, slug-eating snakes use their prey's mucus trail to lead them to their next meal. Odors like this, used by one species to locate another, are termed *kairomones*.

Do humans communicate using chemical signals? The debate on this question is controversial, but there is good evidence that odors synchronize the menstrual cycles of women who live together, such as in a sorority. Infants can recognize and respond to the axillary (armpit) odors of their mother, and adult humans can use odor to recognize a t-shirt worn by their spouse or partner. The ultimate human pheromone, of course, would be a mate attractant, but so far no such chemical has been conclusively identified.

See also Communication—Olfaction—*Mammalian Olfactory Communication*
Frisch, Karl von (1886–1982)

Further Resources

Agosta, W. C. 1992. *Chemical Communication: The Language of Pheromones*. San Francisco: W. H. Freeman.
Bell W. J. & Cardé, R. T. (Eds.) 1995. *Chemical Ecology of Insects*. Sunderland, MA: Sinauer Associates, Inc.
Cardé, R. T. & Minks A. K. (Eds.) 1997. *Insect Pheromone Research: New Directions*. New York: Chapman & Hall.
Doty, R. L. & Müller-Schwarze, D. (Eds.) 1992. *Chemical Signals in Vertebrates 6*. New York: Plenum Press.

Vamder Meer, R. M., Breed, M. D., Winston, M. L., & Espelie, K. E. (Eds.) 1998. *Pheromone Communication in Social Insects*. Boulder: Westview Press.

Michael D. Breed

■ Communication—Olfaction
Mammalian Olfactory Communication

Animals have various means of communication, relying heavily on visual, vocal, and auditory systems to send and receive messages; however, smell is the central sensory modality by which most animals communicate with their environment. Chemical signals are vital to all living creatures and, among mammals, play an especially prominent role in the lives of rodents, rabbits, elephants, ungulates (hoofed animals), carnivores, and primates. Odor signals have the quality of being independent of daylight and persistent over time (even when the signaler is no longer present). Such reliable signals supply a wealth of information, and their detection allows animals to identify food, select mates, recognize their young, and maintain social bonds. Odors also facilitate navigating the environment, demarcating territories, locating predators, and sensing or warding off danger. Animals even communicate their emotional states, such as alarm or fear, through changes in body odor.

From an evolutionary standpoint the sense of smell is primal—one of the most ancient of senses. All eukaryotic organisms have evolved a mechanism to recognize olfactory stimuli in the environment, to translate even complex sensory features into neural code, and to transmit this information to the brain, where it is then processed to create an internal representation of the external world. The olfactory sense is thus able to distinguish among a seemingly limitless number of chemical compounds present at very low concentration. This mechanism (the sense of smell) depends on sensory receptors in the nasal cavity that respond to chemicals possessing certain molecular properties. *Odorants* are typically volatile, airborne compounds that are soluble in water, evaporate easily, have some ability to dissolve in fat, and are of limited molecular weight. Such chemicals are carried by inhaled air to smell detectors (or chemoreceptors) found in the olfactory epithelium located high in the nasal cavity. Odor molecules become dissolved in the lipid-rich mucous layer of the nasal cavity and stimulate these receptors, causing them to send impulses or signals to the brain for processing. The brain's olfactory center, the olfactory bulb, provides a spatial map that identifies which of the numerous receptors has been activated. The quality of an odor therefore may be encoded by different spatial patterns of activity in the olfactory bulb.

Involatile compounds can also contain olfactory signals, but may require physical contact, such as licking, to enter the olfactory pathway. For instance, males of many species frequently sniff and lick the female's vulva, vaginal secretions, or urine, to obtain information about her reproductive state. On licking an odor source, mammals can detect chemicals not only by taste and olfaction, but also by a separate neural pathway involving chemoreceptors of the *vomeronasal organ*. This organ, which is present in most terrestrial species (excluding higher primates and certain bats), includes a pair of elongated, fluid-filled sacs that are lined with chemoreceptors and connect to the roof of the mouth (or palate). When mammals close their nostrils and curl their upper lip in a characteristic grimace, called the *flehmen response*, they transfer tasted odors directly to the vomeronasal organ.

An adult male ringtailed lemur (Lemur catta) marking a sapling with secretions from his wrist gland. On closer inspection you can also see that the bark has been scarred numerous times by the wrist spurs of males. The wrist (or antebrachial) spur is located adjacent to the wrist gland, and through the action illustrated in this photo, the male leaves a composite signal with both visual and olfactory components.
Courtesy of David Haring.

An animal's reliance on olfactory cues is therefore revealed in certain anatomical features, including the complexity of the nasal chamber, the extent of the olfactory epithelium, the development of the vomeronasal organ, and the relative size of the olfactory bulb (which is proportionate to the acuteness of the organism's sense of smell). For instance, among primates, the greater olfactory capabilities of the more "primitive" prosimians (e.g., the lemurs of Madagascar), as compared to monkeys, apes, and humans, are reflected in their relatively long snout and in the presence of a moist fleshy pad (or rhinarium) at the end of the nose. These characteristic features give lemurs a somewhat dog-like appearance. By contrast, the greater emphasis on vision in "higher" primates is accompanied by a reduction in the size of olfactory structures in the brain and of the entire olfactory apparatus (i.e., a decrease in the size of the snout) as well as evolutionary changes in the skull, eyes, and brain, that are associated with visual development.

Dependence on olfactory cues is also related to the animal's ecological niche—a relationship again well illustrated within the primate order. For instance, whether an animal is active by night (nocturnal) or by day (diurnal) and lives in the trees (arboreal) or on the ground (terrestrial) influences whether it requires greater olfactory versus visual competence. As primate lineages evolved, with animals increasing their daytime activity and moving out of the trees, species became less reliant on olfaction and more invested in vision. As mentioned, prosimians have elaborate means of olfactory communication, reflecting their predominantly arboreal and nocturnal lifestyle. New World monkeys are similarly arboreal, but have settled into a diurnal niche. Because visual cues are less effective in dense vegetation, but more perceptible during daylight hours, these monkeys have compromised on sensory specialization, relying instead on both olfaction and vision. By contrast, the diurnal Old World monkeys and apes (including humans) are also often terrestrial, occupying a more open habitat, and have sacrificed olfactory acuity to specialize on visual capacities.

The influence of behavioral ecology on sensory development is similarly evident upon examining the interactions between species (*interspecific communication*), including predator–prey relations, where both the hunter and the hunted possess well-developed olfactory skills. For example, many African predators, such as lions and spotted hyenas, are active at dawn and dusk (*crepuscular*), largely to avoid physical exertion in the heat of the day, and consequently use odor cues to locate prey. Likewise, nocturnal scavengers, such as jackals and brown hyenas, exploit odors associated with a kill to gain a free meal. Because the competition between different predator species is fierce, hyenas also rely on odor cues to warn of lion presence, in order to avoid a potentially lethal encounter. In captivity, spotted hyenas have demonstrated this ability to discriminate or classify odors through differential responses to various categories of odorants. On the flip side of the coin, prey animals are also endowed with an acute sense of smell to avoid being featured on the evening menu; hence the trademark stealth of lions approaching their migratory feast from down wind.

In addition to possessing a remarkable ability to detect odors in the environment, many mammals produce their own characteristic scents. These function primarily as messages between members of the same species. Such chemical signals can be derived from feces, urine, saliva, genital discharge, and specialized skin glands located in different regions of the body. Mammals produce a vast array of glandular secretions with enormous quantitative and qualitative variation across taxonomic groups. Some glands occur quite commonly across species. For instance, many species of deer make use of various combinations of the following odor sources: forehead glands, preorbital sacs (that collect secretions from the orbit of the eye), tarsal organs (situated on the inside of the hock), metatarsal organs (on the outer surface of the hind leg), interdigital glands (on the fore and hind feet), and caudal glands (on the tail). Likewise, many primates possess specialized glandular fields in the gular (throat), sternal (chest), abdominal, palmar, circumgenital, and perianal regions. Conversely, some species are distinguished by possession of

An adult female ringtailed lemur (with her twin offspring) engaged in anogenital marking. She presses her hind quarters up against a tree and then wiggles her bottom gently from side to side. Courtesy of David Haring.

highly specialized structures. For instance, ringtailed lemurs have brachial (upper arm) and antebrachial (forearm) organs that serve as odor reservoirs, and antebrachial spurs for marking branches or their own tails, which are waived at other males during ritualized "stink fights." Similarly, elephants have unique temporal glands and produce secretions that are most notable in males during *musth* (or rut)—a physiological and behavioral state associated with heightened sexual and aggressive activity.

Not surprisingly, there is also great variation in the methods animals use to transfer their scent to the surroundings, called *scent marking*. For example, primates engage in such awkward behavior as "chinning," "chest rubbing," "mouth wiping," and "anogenital rubbing"—all variant forms of causing friction between the scent producing organ and a substrate (which at times may be another animal). Thus, "urine washing," which is also common among primates, involves depositing and rubbing urine on the hands or feet, and is either reserved for oneself or applied to a mate.

Brown hyenas, striped hyenas, spotted hyenas, and aardwolves (the four surviving members

This "pasting scene" illustrates olfactory communication in spotted hyenas (Crocuta crocuta). The adult animal that is walking away from the camera has just deposited a paste mark on a blade of grass (note the raised tail and the bulging anal sac). Meanwhile, the youngster in the foreground is intently sniffing the "pasted" blade of grass while another adult is patiently waiting its turn to take a sniff. Used by permission of Paula White.

of the family *Hyaenidae*) also have an elaborate scent-marking repertoire involving long-lasting, composite signals. They urinate and defecate in conspicuous latrines located throughout their territory, frequently depositing odor from interdigital glands by additionally scratching the soil vigorously enough to leave scented furrows. Like most carnivores, hyenas possess a pair of large supra-anal sacs that produce a thick, durable, glandular secretion or "paste" that is extruded by squatting and everting the anal pouch over the item being marked. Pasted scent marks, often deposited along territorial borders, are visible from a distance of several meters and can remain potent for a month or longer. Regarding similarly derived odors in other taxa, the award for potency, however, goes to the mustelids (e.g., badger, otter, mink, ferret, weasels, martins, etc.), exemplified by the noxious consequences of a defensive skunk encounter. More benign uses of anal glands are illustrated by our canine companions, for whom a well-placed cold nose constitutes acceptable social etiquette.

Mulitiplicity of form suggests multiplicity of function, not only between species but also within a species. Thus, solitary species may use scent marks to keep intruders at bay, whereas social animals may use olfactory cues to synchronize or facilitate group behavior and maintain contact or solidarity with their offspring and other group members. For example, mammals that live in family groups, such as the beaver (a large member of the rodent family), possess a clan-specific odor (a "castor print" of sorts) that helps maintain cohesion. Inhibiting or disrupting the olfactory system, therefore, can have severe consequences on social behavior. Female prosimians that have had their olfactory bulb removed, for instance, have lower infant survival and behave like social isolates even within a social group, suggesting that odors are crucial for promoting and maintain social bonds.

Migratory species, such as reindeer, may use scent marks to lay or follow tracks, whereas sedentary species may use scent marks to define their territorial boundaries and advertise ownership. A "land-owning" beaver, for example, deposits odor from castor sacs as territorial proclamation. This animal's readiness to explore fresh marks is well known by trappers, who exploit such curiosity to increase capture success. In comparable fashion to the beaver, social carnivores often perform goal-directed border patrols, when animals cruise the boundaries of their territory to freshen waning scent marks and scan the horizon for trespassers. Patrolling involves both exploratory and preventative behavior, serving to assess neighboring territories as well as to minimize the risk of costly aggressive encounters by fortifying existing boundaries. For scent marking to function in territorial advertisement, marks should be individually identifiable. Indeed numerous studies of various mammalian species have shown that unique odor cues are contained in scent gland secretions. Moreover, as reflected by the differential responding of animals when presented with various classes of conspecific odors, these scent "signatures" convey information detectable to the recipient about the sender's species, subspecies, individual familiarity, kinship, sex, dominance status, and even identity.

Additionally, whereas males may scent mark to ward off potential competitors, females may scent mark to advertise their sexual status and attract suitors. For scent marking to provide the male with information on the reproductive condition of the female, so too must odors reflect sensitive changes to internal hormonal state. And indeed, not only do frequencies of scent marking change across an individual's reproductive cycle, scent marks also produce measurable changes in the behavior or physiology of those animals that are exposed to the odor. Thus, urine from sexually active females can increase reproductive hormones (i.e., testosterone levels) in socially isolated males, thereby enhancing the males' reproductive desire. Conversely, scent marking can also be inhibitory, as the same female's urine can suppress the reproductive cycle of subordinate females, serving as an olfactory method of birth control. Analogously, urine from dominant males can decrease the testosterone concentrations and increase the stress hormones (i.e., cortisol concentrations) in isolated males.

As scent marks require a certain "freshness" to maintain efficacy, with more recently deposited odors having stronger stimulus value than those of older sources, scent marking becomes quite competitive between individuals, influencing rate of occurrence and actual placement of scent marks. For instance, rates of scent marking tends to be rank-related, in that the most dominant animals (males or females) scent mark more frequently than do lower ranking animals. Competitive scent marking is also expressed in "counter-" or "over-marking," whereby one animal deposits its scent directly on top of a previous animal's mark. Female rodents, such as hamsters and voles, preferentially remember or value the top scent, suggesting that over-marking is a form of competitive olfactory communication by which the "over-marker" gains the upper hand.

An adult spotted hyena in the act of defecating. Spotted hyenas frequently deposit feces in well-used latrines. Latrines are distributed throughout their territory, but are especially concentrated in core areas where other animals would be most likely to encounter them. Because of the high calcium in their diet (from eating bones), hyena scat is resistant to degradation and can remain visible for over a year. Consequently, feces also constitute a composite signal.
Used by permission of Paula White.

Odor cues are thus functional in all aspects of mammal life, with undoubtedly numerous messages being sent and received in the environment that, at present, we are still unable to discern. Mounting evidence suggests that odors carry similar, though potentially more subtle, messages for humans, and play a far greater role in influencing our behavior and interactions with others than was traditionally recognized. Further research into this fascinating area is guaranteed to reveal important new insights into the sensory world that surrounds us.

See also Communication—Olfaction—*Chemical Communication*
Communication—Olfaction—*Dog Scents and "Yellow Snow"*

Further Resources

Albone, E. S. 1984. *Mammalian Semiochemistry: The Investigation of Chemical Signals between Mammals.* New York: Wiley.

Belcher, A. M., Smith, A. B., Jurs, P. C., Lavine, B. & Epple, G. 1986. *Analysis of chemical signals in a primate species* (Saguinus fuscicolis): *Use of behavioral, chemical, and pattern recognition methods.* Journal of Chemical Ecology, 12, 513–531.

Brown, R. E. & Macdonald, D. W. (Eds.) 1985. *Social Odors in Mammals* (Vol. 1–2). Oxford, U.K.: Clarendon Press.

Doty, R. L. 1976. *Mammalian Olfaction, Reproductive Processes and Behavior.* New York: Academic Press.

Drea, C. M., Vignieri, S. N., Kim, H. S., Weldele, M. L. & Glickman, S. E. 2002. *Responses to olfactory stimuli in spotted hyenas* (Crocuta crocuta): *II. Discrimination of conspecific scent.* Journal of Comparative Psychology, 116, 342–349.

Ewer, R. F. 1973. *The Carnivores.* New York: Cornell University Press.

Halpin, Z. T. 1986. *Individual odors among mammals: Origins and functions.* Advances in the Study of Behavior, 16, 39–70.

Ralls, K. 1971. *Mammalian scent marking.* Science, 171, 443–449.

Schilling, A. 1979. *Olfactory communication in prosimians.* In: *The Study of Prosimian Behavior* (Ed. by G. A. Doyle & R. D. Martin), pp. 461–542. New York: Academic Press.

C. M. Drea

■ Communication—Tactile
Communication in Subterranean Animals

Living Underground: Advantages and Disadvantages

All animals have different adaptations and ways of life related to the places where they live, feed and reproduce. Each environment has different *ecotopes* (the full range of adaptations of a species to the environmental and biotic variables affecting it) that one or many species can occupy. Some of these ecotopes are easier to see, and others are far from our immediate direct experience. The subterranean way of life is one of the latter, and protection against predators is one of the reasons why many animals exploit it. *Fossorial animals* (that is, those adapted to digging, such as kangaroo rats, armadillos) find some protection in their constructions, but only the animals that are strictly subterranean (i.e., moles, pocket gophers) are almost always protected against attacks from other species that are not subterranean themselves. By living underground, these species also gain other advantages as the control over some microclimatic conditions inside the burrows, like temperature, humidity or oxygen and CO_2 levels.

Nevertheless, the advantages of living underground are counterbalanced by some disadvantages. Living, feeding, and reproducing underground could constrain subterranean species to adapt their behavior in many ways. Each animal has to build his path and refuge through the soil, thus they should have morphological and behavioral adaptations allowing them to dig and to build burrows, except for interstitial animals (animals that are about the same size of the particles constituting the environment they live in). Many of them have to obtain oxygen, constraining them to dig in permeable soils allowing gas exchange or to build connections to the surface to somehow control this exchange. Some species could go to the surface to feed with the consequent loss of security, but others are adapted to find their (sometimes very restricted) food sources underground. Reproducing underground is also difficult for any species, except for those self-sufficient *hermaphroditic animals* (i.e., those animals that can reproduce within themselves). *Bi-sexuated species*, which have both males and females, should possess behaviors allowing them to get together and mate, and this is an important problem especially for those species in which individuals are solitary throughout the majority of the year. If the solution to the problem of moving through soil and building a refuge is usually the development of digging skills and the adaptation of some organs, like paws, to do it, and if the solution to the feeding problem is go to the surface and lose security or to adapt to eat subterranean animals or plant parts, what is the solution to the social and reproductive problems? The main solution to this problem is communication.

Solving the Problem of Living Underground: Subterranean Rodents as an Example

The animals known as "subterranean rodents" are a group of species pertaining to different rodent families dispersed over almost all continents (the Americas, southern Africa, Asia and Middle East). These species are not directly related to each other, but the animals are very alike because of the convergent adaptations they have developed as a way to adapt to the same environment (e.g., pocket gophers in North and Central America, "mole–rats" in Africa, "bamboo rats" in Asia and "tuco-tucos" in South America). Among these families we can find different group organization systems, from solitary animals to complex societies where the members present a characteristic division of labor (*eusociality*) similar to that found in social insects like ants or bees. The similarities between these groups of animals are striking and can be found in morphology, physiology, ecology and behavior. These features allow us to study them in a comparative way, considering them as a "coherent" group of organisms that share the same kind of adaptations. Nevertheless, they are "equivalents" but not equal.

Subterranean rodents face the same kind of problems of many subterranean organisms: They need to communicate with other individuals of the same species and, at the same time, cope with the constraints imposed on communication by the environment they live in. Thus, subterranean rodents cannot use some communication channels like vision because of the lack of light inside the burrows and the impossibility of transmitting visual signals through soil from one burrow system to another.

Chemical, acoustic and tactile channels of communication are severely restricted in their use, and the animals must exploit "unusual" channels like the seismic one. The different characteristics of the communication channels employed also determine differences in channel and signal usage between solitary and so-

A mole pushes its upper body up through the earth into the sunlight.
© *Mark E. Gibson / Corbis.*

cial species. As a result of this quick examination of the communication possibilities of subterranean rodents, we have to conclude that we may find different communicative strategies for different species, based on the particular combination of environmental and social characteristics each species shows.

Regardless of the social systems that different species have, the integrity of the territories—in the subterranean rodents' case corresponding to the total extension of the burrow system—should be maintained, and, to do so, territorial advertising or warning signals are needed. These signals should be detected by possible intruders at considerable distances from the burrow in order to prevent tunnel invasions and unexpected encounters that could end in direct aggression. This behavior, even if many species of subterranean rodents show middle-to-high levels of aggression, is usually the last resource, and animals try to avoid fights because of the danger injuries represent for their physical integrity and overall health.

Territorial signals, directed to other members of the same species, need to travel between burrows to attain their target. If we consider the different communication channels available, we can discard visual signals because of their impossibility of transmission in this situation. A similar problem can be detected for chemical signals because, to be active over mid-to-long distances, they should be released in the open and transported by the wind. Even if some chemical marks could be deposited inside and at the entrances of the burrows, the danger of a direct encounter with an invading animal is still high. On the other hand, tactile signals are by definition close-contact signals (animals should touch each other to exchange information) and thus unsuitable for transmitting an aggressive or warning message. We are left, then, with only two alternatives: acoustic and/or seismic communication, and these two modalities are employed by subterranean rodents.

Seismic signals, substratum vibrations produced by the animal tapping over the soil with some body part (paws, head, teeth), are the most suitable for this type of communication because the surface vibration waves provoked in a solid could travel considerably farther than any other type of signal. These kinds of signals are used by many subterranean rodent species, such as *Spalax ehrenbegi* (blind mole-rat) from Israel, which is a head thumper, beating its head against the burrow ceiling to produce signals, or *Georychus capensis* (Cape mole-rat), an African mole-rat that uses its hindlegs to signal.

Acoustic signals, usually vocal sounds produced by individuals and propagating through air, can be used to advertise territories. Even if the filter effect that the soil has on sounds is important, especially over high frequency sounds, some low frequency sounds can still reach neighboring burrows. Some South American rodents of the genus *Ctenomys*, like *C. pearsoni*, *C. mendocinus* and *C. talarum* use their sounds as warning/territorial signals (South American tuco-tucos). For both acoustic and seismic signals, the best frequencies for signal propagation are constrained by soil characteristics and the way the animal produces the vibrations, and these constraints determine that the variability in the signals—and then, the potential for encoding different kinds of information into them—depends on the rhythmic pattern with which they are produced.

Any of these long-distance signals can also be used to encode many kinds of information about the signaler: species, sex, individuality, motivation (internal willingness to perform a determinate type of behavior), age and others. These characteristics allow the animal to use these signals also as identifiers and attractants to other individuals during the reproductive period, especially in solitary animals. In *Georychus capensis*, thumping is sexually dimorphic, and males and females alternate in performing the signals. In *Ctenomys pearsoni* and other relatives, the rate of delivery of territorial vocal signals is higher in males than in females, whereas in *C. talarum* only males emit this kind of signal. For solitary individuals, it is crucial to reach neighboring burrows with their signals and allow other animals to identify the sex of the signaler and its position in space, to attract them into close contact and mate. In social animals, this is usually not necessary because males and females live together, but sometimes mating should be directed to members of other colonies as a way to avoid inbreeding.

In solitary species, when future mates are together or when pups are still dependent on their mother, and during the majority of the lives of social animals, all of the above mentioned communication channels are, theoretically, available. Visual communication is still almost useless in dark burrows and for animals which, in many cases, have reduced eyesight or are blind. Acoustic signals continue to be limited to low-frequency sounds along tunnels except when individuals are close to each other. Low-frequency sounds are the only means capable of traveling medium distances inside burrows because high-frequency (and, as a consequence, short wavelength) sounds are very straight in propagation and then suffer from rebounds and scattering against the burrow walls and corners. Seismic signals could be used inside burrows, but their best performance is over long distances. As a consequence of these characteristics, inside their burrows subterranean rodents use low-to-medium frequency sounds, chemicals (mainly odors in urine or feces, and sometimes products of some specialized glands) and tactile signals to communicate.

As a result of these differences in performance, the use of any kind of signal and communication channel is intimately related to the "meaning" each signal try to convey. Consequently, to warn putative intruders and limit their territories, subterranean rodents will use seismic or low-frequency sound signals, capable of attaining their targets at considerable distances from the emitter. Threatening signals are usually emitted as mid-frequency sounds or the use of chemicals. On the other hand, sexual arousal, courtship signals, sexual receptivity, pair bonding, and mother–pup bonding, and adult or pup contact signals can be emitted using soft high-frequency sounds, some chemicals, and tactile signals. Looking closely at this "signal distribution" related to motivation and meanings, we can see that there exists a continuum going from long- to medium- and short-distance signals that is somehow "translated" into the use of a parallel continuum of seismic-acoustic-chemical-tactile signals by the animals. Thus, signals are designed to get to their targets in the most efficient way, taking into account the situation they are signaling, and avoiding excessive energy expenditures and probably also avoiding reaching non-intended receivers.

Besides those design adaptations imposed by the environmental constraints, solitary and social subterranean rodents seem to use almost the same kinds of signals, but social species usually have a greater number of signals in their repertoire. This difference supposedly arose from the permanent interaction individuals have with each other inside a society, leading them to develop their repertoires to match a greater number of behavioral or meaning "categories" needed in complex social situations.

Another source of variation among signals and repertoires derives from the different motivational sources producing them. Some signal design features could be shaped by different motivational and evolutionary origins. As an example of this, vocal signals expressing aggression or warning tend to have a guttural quality and to be uttered in low frequencies (like dog growls), at least in some vertebrate groups. Sometimes, low-frequency signals are also related to body size because the bigger the body, the bigger its resonance chamber, and the more easily are the lower sounds produced (the same way drums of different sizes produce different sounds). On the other hand, nonaggressive or submissive sound signals tend to be emitted in higher frequencies and to have a whistled quality, just opposite of the characteristics of the aggressive signals (this is known, from Darwin, as the "antithesis principle"). These kinds of design rules allow listeners to readily differentiate among aggressive and nonaggressive signals.

All of these rules, designed during the evolution of the different species and groups, and acting during their development, have produced the actual repertoires of these animals, allowing them to cope with some of the problems of living and reproducing in the subterranean ecotope.

See also Burrowing Behavior
        Communication—Tactile—*Vibrational Communication*

Further Resources

Burda, H., Bruns, V. & Müller, M. 1990. *Sensory adaptations in subterranean mammals*. In: *Evolution of Subterranean Mammals at the Organismal and Molecular Levels* (Ed. by E. Nevo & O. Reig), pp. 269–293. New York: Wiley-Liss.

Francescoli, G. 2000. *Sensory capabilities and communication in subterranean rodents*. In: *Life Underground: The Biology of Subterranean Rodents* (Ed. by E. A. Lacey, J. L. Patton & G. N. Cameron), pp. 111–144. Chicago: The University of Chicago Press.

Heth, G., Frankenberg, E., Pratt, H. & Nevo, E. 1991. *Seismic communication in the blind subterranean mole-rat: Patterns of head thumping and their detection in the* Spalax ehrenbergi *superspecies in Israel.* Journal of Zoology, London, 224, 633–638.

Mason, M. J. & Narins, P. M. 2001. *Seismic signal use by fossorial mammals*. American Zoologist, 41, 1171–1184.

Gabriel Francescoli

■ Communication—Tactile
Vibrational Communication

When the first vertebrate animals appeared on land, these amphibians had ears that could "hear" vibrations from the soil on which they lay, but they could not "hear" sound passing through the air, as we do. Animals were thus able to communicate through signaling with vibrations before they could call to one another with airborne sound. We know that animals communicate through a variety of mechanisms, using their senses for seeing, hearing, feeling, smelling, and tasting their environments. We also know that a large number of animals, from insects and spiders to mammals, produce vibrations that can be detected and interpreted by other members of their species.

Vibrations are like sounds, but they travel at frequencies below the level that can be detected by human hearing. Humans can "feel" vibrations that are detected by receptors in our skin and joints, and we are all familiar with pleasant vibrations from a favorite band, as well as unpleasant ones from construction work. Still, humans may not use vibration as a communication channel to signal to other humans. For any event to be a communication signal, it must be a specific message sent through a medium that can be decoded by a receiver. If a vibration modifies the behavior of members of the same species as the sender, then it is a signal. Otherwise, it is merely a "noise."

Philip Brownell was able to show in the late 1970s that scorpions can estimate both distance and direction of prey from vibrations the prey create in the substrate. Until that time "everyone knew" that small animals were incapable of gathering any biologically useful information from vibrations in their environments! Brownell's work, although not suggesting that scorpions signal each other with vibrations, was so compelling that it did open the door to others to investigate previously baffling behaviors in the context of vibrational communication. Advances in computers and technology have since allowed scientists to discover a broad array of contexts in which messages are communicated among individuals via vibrations.

Why would it be better to drum or thump on the soil or a plant stem to get attention than just to make a sound, or flash like a firefly, or release a scent into the air? One common thread that seems to bind together some animals that communicate through vibrations is that they live in noisy environments where sounds in the air might not travel without distortion, especially in desert or grassland habitats. They may be trying to communicate in windy places or in stormy seasons. Banner-tailed kangaroo rats (*Dipodomys spectabilis*) foot-drum inside their burrows on windy nights to communicate with their close neighbors through vibrations. On calm nights, they drum outside their burrows to communicate over longer distances with sounds carried through the air. Jan Randall has found that these signals are individually distinct and can be used to advertise territories, as well as to communicate a variety of other information.

Vibrations might also be more effective than other types of signals when an animal is hiding from a predator. Rex Cocroft has found that tiny thornbugs (*Umbonia crassicornis*) communicate within family groups as they all perch on the same stem. Mothers and young send vibrations that can be decoded by family members on that stem, but flying wasps or other predators cannot "eavesdrop" as they are searching for a meal. The young signal together to their mother when a wasp appears, and continue to signal as one of their siblings is being attacked. The mother sends out a vibrational signal when the wasp has left that calms the brood. It is interesting that the mother does not respond to a single offspring, but she does come to the rescue if the whole brood signals her.

Prairie mole cricket (*Gryllotalpa major*) males respond to vibrations sent through the soil by other males singing nearby, but they do not respond to the airborne component of these songs. These males will change their own chirping calls to alternate in rhythm with the chirps in an artificial vibration signal, but they will ignore a chirping call sent only through the air. This represents an interesting bimodal communication mechanism because the males are calling to flying females to attract them for mating, but the vibrations they produce during these songs provide information to their closest male neighbors. Neighbors appear to use this information as they spatially aggregate their singing burrows in a lek mating arena.

Most of what we know about communication through vibration has been learned since Brownell's work in the late 1970s: We have found that animals use vibrations for mate location and recognition, as well as to advertise mate quality; to warn others that a predator is nearby, or to solicit help from a protective parent; to recruit nest mates to help gather food and to warn them of danger. Our study of this very old communication channel may help us to better understand mysterious behaviors of even the most familiar animals, such as bison and elephants, who appear to be able to maintain contact with family members over long distances without the use of a telephone.

See also Communication—Auditory—*Long Distance Calling,
the Elephant's Way*
Communication—Tactile—*Communication in
Subterranean Animals*

Further Resources

Bradbury, J. W. & S. L. Vehrencamp. 1998. *Principles of Animal Communication*. Sunderland, MA: Sinauer Associates.
Brownell, P. H. 1984. *Prey detection by the sand scorpion*. Scientific American, 251, 86–97.
Cocroft, R. 1999. *Thornbug to thornbug: The inside story of insect song*. Natural History, 108, 52–57.
Hill, P. S. M. 2001. *Vibration and animal communication: A review*. American Zoologist, 41, 1135–1143.
Lewis, E. R. & P. M. Narins. 1985. *Do frogs communicate with seismic signals?* Science, 227, 187–189.
Randall, J. A. 2001. *Evolution and function of drumming as communication in mammals*. American Zoologist, 41, 1143–1156.

Peggy Hill

■ Communication—Visual
Fish Display Behavior

Members of the family Anatibantidae are small tropical fish native to Asia and Africa that are common in aquariums. The family contains climbing perch of Asia (*Anabas testudineus*), dwarf gouramis (*Colisa lalia*), and kissing gouramis (*Helostoma temmincki*), but perhaps the best known are Siamese fighting fish (*Betta splendens*), and paradise fish (*Macropodus opercularis*), both popular in home aquariums.

Anabantid fish are fascinating for several reasons. They have a special respiratory organ, known as the labyrinth, in the gill cavity which enables them to breathe air. These fish can live in oxygen-depleted water by rising to the surface and gulping air, and climbing perch can crawl across land and breathe air for a considerable period. In their native habitats, many of the anabantid fishes live in stagnant ponds or rice paddies of Asia where they are important predators of mosquito larvae. Behaviorally, males of most species of anabantid fish are territorial around bubble nests which they construct on the surface of water. They attract females with colorful display patterns; they show intricate mating patterns, and remarkable paternal behavior in caring for eggs and young. If eggs or newly hatched fry drop from the nest, males retrieve the falling eggs or fry in their mouths, and spit them back into the bubble nest. This retrieval continues for 3–5 days after hatching. The females are driven from the nest territory during this time.

Mutual lateral display of Chinese paradise fish. The display is characterized by darkening colors, spreading fins, and body vibrations. This display typically occurs when two strangers meet, and is most intense between males.
Courtesy of Charles Southwick.

The territorial displays of male bettas (the generic name is also used as a common name), and paradise fish are fascinating to observe and easily evoked in aquariums by introducing a different male into their territorial space. In bettas, this response is so striking that male bettas must normally be kept in isolation in jars or tanks. Paradise fish males are also best kept separate in tanks from other males of the same species. When a new male is introduced into the tank of a male betta, the territorial display involves an aggressive frontal approach. The resident male approaches the interloper with flared opercula (gill covers), spread fins, and enhanced red or blue colors. The stranger usually responds in a similar way. The fish circle around each other, approaching again with flared opercula and spread fins. After several such displays, the resident male then attacks the fins of the stranger, often biting pieces of fin. This becomes a real fight, hence the popular name, Siamese fighting fish. If the fish are not separated, the fins of the loser will be shredded. In time, they will grow back if not injured too badly. In Bangkok and the villages of Thailand, many shops keep domesticated bettas on their shelves in individual jars for patrons to pay for placing them together and betting on their fights.

Paradise fish share some of the same life cycle features—male territoriality, intricate mating, male bubble nests, paternal care, and aggressive displays in confronting strangers—but their displays are fundamentally different than the bettas'. Instead of frontal displays and fin biting attacks, male paradise fish have a head-to-tail orientation known as the "mutual lateral display." In this mutual lateral display, the opposing fish align themselves in a head-to-tail fashion, spreading their fins as broadly as possible, and wrapping the tail fin around the head of the opponent, so it is most directly in front of the opponent's eye. This display is accompanied by a vibration of the entire body, a behavior which must send pressure waves toward the opponent. Also, the colors of the fish are darkened. The fish spontaneously relax these displays, circle, and then return repeatedly for more mutual lateral displays. This typically occurs for several minutes, sometimes 15 or 20, and then attacks may occur; but unlike bettas where the attacks are directed toward the fins, the attacks of paradise fish are directed toward the mouth and jaws. Sometimes these attacks result in a "jaw lock" in which the fish then show a rhythmic slow motion struggle, often quite dramatic, until one fish breaks away and retreats.

In nature these aggressive territorial conflicts result in the retreat and departure of one combatant—in aquariums, it is essential to separate the fish before one becomes seriously injured. The display patterns of anabantid fish illustrate many aspects of vertebrate social behavior—communication through several modalities, colorful display, territoriality, complex mating patterns and parental care, and aggression and dominance—all of which can be readily observed in laboratory aquariums.

Further Resources

Forselius, S. 1957. *Studies of anabantid fishes*. Zoological Beitrag, 37, 93–595.

Innes, W. T. 1956. *Exotic Aquarium Fishes*. Philadelphia: Innes Publishing Company, pp. 391–417.

Southwick, C. H. 1968. *Display patterns in anabantid fish. No. 817*. In: *Animal Behavior in Laboratory and Field* (Ed. by A. W. Stokes), pp. 99–103. San Francisco and London: W.H. Freeman and Co.

Southwick, C. H. & Ward, R. W. 1968. *Aggressive display in the paradish fish*, Turtox News, 46(2), 57–62.

Charles Southwick

■ Communication—Vocal
Alarm Calls

When alarmed by predators, individuals of many species emit loud vocalizations known as alarm calls. Calls may be directed to other members of the caller's species to warn them about the presence of a predator, or to create pandemonium during which time the caller may escape. Calls may also be directed to the predator and may function to discourage pursuit. If alarm calls create pandemonium or discourage pursuit, the caller, by calling, increases its own chance of survival; such behavior requires no complex explanation. However, when calls are directed toward other members of the caller's species, the very act of signaling may also alert the predator to the caller's presence. The explanation of why animals emit potentially costly alarm calls to help others was initially an evolutionary paradox.

The solution lies in considering the fitness obtained by helping relatives survive. By calling, yellow-bellied marmot females warn their vulnerable offspring and presumably increase their survival. Thus calling, a form of maternal care, increases the caller reproductive success. Even more complex is the calling behavior of black-tailed prairie dogs. These highly social rodents are sensitive to the relationships of nearby individuals and call more when they have other relatives, even those that are not offspring, within earshot.

The structure and function of signals are interrelated. For instance, we expect signals that are directed to a predator to be "obvious." *Mobbing calls* are a specific type of alarm call that animals produce to rally assistance and drive out typically low-risk predators. Many species produce mobbing calls. For instance, many nesting birds will scold and "mob" a jay or crow that comes too close to its nest, and nesting crows will emit mobbing calls when a raccoon comes too close to their nests. Mobbing calls are loud, broad-bandwidth, and rapidly repeated vocalizations. These characteristics make them easy to localize by both potential helpers, as well as the predator. In contrast, alarm calls by birds that are elicited by aerial predators which are hunting them are difficult to localize because they have a relatively narrow bandwidth and fade in and out. Being near a hunting raptor is very risky, and while animals may warn others, they do so in a way that reduces their own conspicuousness while simultaneously warning others.

Alarm calls may communicate different types of information. The calls of ground squirrels and marmots communicate the relative risk a caller experiences when it calls, whereas the calls of chickens and vervet monkeys communicate the species or type of predator. Alarm calls from suricates, a social mongoose, communicate both relative risk and predator type. To understand the meaning of alarm calls, it is important to study the situations under which individuals call and how they respond to calls being broadcast from hidden speakers. In vervet monkeys, snakes elicit "chutters," leopards elicit "barks," and raptors elicit "coughs." And, when these monkeys heard these vocalizations broadcast through hidden speakers, they responded as though there were a snake, leopard, or raptor in the area. Snake calls caused vervets to stand on their toes and look around for snakes (an appropriate response on the savannah). Leopard calls caused the monkeys to run to trees and move out to peripheral branches where leopards could not reach them. Raptor calls sent vervets, caught in the open on the ground, into the central branches on trees—a good place to avoid raptors. If vervets were already in trees, raptor calls signaled a nearby raptor, and monkeys hearing them descended to the ground. These behaviors provided key evidence that these calls function to communicate predator type rather than escape strategy.

See also Antipredatory Behavior—*Predator–Prey*
        Communication
        Cognition—*Equivalence Relations*
        Communication—Vocal—*Referential Communication*
        in Prairie Dogs

Further Resources

Blumstein, D. T., Steinmetz, J., Armitage, K. B. & Daniel, J. C. 1997. *Alarm calling in yellow-bellied marmots: II. Kin selection or parental care?* Animal Behaviour, 53, 173–184.
Cheney, D. L. & Seyfarth, R. M. 1990. *How Monkeys See the World*. Chicago: University of Chicago Press.
Hauser, M. D. 1996. *The Evolution of Communication*. Cambridge, MA: The MIT Press.
Manser, M. B., Bell, M. B. & Fletcher, L. B. 2001. *The information receivers extract from alarm calls in suricates*. Proceedings of the Royal Society of London, Series B, 268, 2485–2491.
Sherman, P. W. 1977. *Nepotism and the evolution of alarm calls*. Science, 197, 1246–1253.

Daniel T. Blumstein

■ Communication—Vocal
Choruses in Frogs

In many species, individuals congregate to produce sounds. Such a group of noisy animals is called a chorus. Choruses are among the most impressive of biological phenomena. In North America, for example, periodical cicadas emerge in the millions and produce painfully loud sounds each day for a week or two. In temperate and tropical areas all over the world, frog choruses form around the borders of large ponds, lakes and rivers and make a continuous din for most nights of the spring and summer. Choruses often contain many different species—a dozen or more is common in frogs and insects. In birds, dawn choruses—named for the time of day when they take place—can also be remarkably diverse. Each species in these complex choruses produces a distinctive sound pattern that differs in pitch, quality, and tempo from all of the other species. This makes it possible to survey biological diversity simply by listening and identifying animals by their sounds.

What are the functions of the sounds produced in choruses? The most general answer is that these sounds advertise the presence of an individual—typically a male—who is ready to mate. In many species, females that are ready to mate are attracted to these sounds and show positive *phonotaxis*; that is, they move toward the source of the sound. In some species of frogs, the female may actually have to touch the male before he detects her presence. In other species, the movement of the female in the male's vicinity is sufficient. He stops calling and begins to court the female or may simply try to grab her.

Advertisement calls also affect other males of the same species. Usually they are repelled—negative phonotaxis—or simply keep their distance. In fact, although the chorus is by definition an aggregation of males, males of the same species usually maintain some minimum spacing. This spacing may be achieved by the perception of the loudness of the sounds of neighbors. In Australian cicadas, for example, males were closer to each other in a thickly vegetated site than in a more open area. The loudness of the sounds at the same distance was less in the first site because of the dampening effects of the vegetation on sound transmission. What happens if calling males get too close to one another? Very often, one or both males switch to a second kind of call—an encounter or aggressive call—which is usually sufficient to cause one of the males to back off. If this doesn't happen, then fighting may occur. Eventually, the males become spaced again, or the loser of the fight may even leave the chorus or stop calling.

Although aggressive calls are more effective in competition between males, they are usually not as attractive to females as advertisement (mating) signals. In some species, males restrict most aggressive signaling to early in the daily or nightly breeding period at a time when they are establishing their calling positions. Females arrive later. In other species, males produce two-part calls: One part attracts females, and the other, tends to repel males. When a rival gets too close, the male first elaborates the aggressive part, but he still continues the female-attracting part in case a female is also nearby. In still other species, males produce a graded series of calls. At one end of the spectrum, calls are highly attractive to females, and at the other end, calls are highly effective at repelling males. Calls with intermediate characteristics are still somewhat attractive to females but less effective at repelling rivals.

Why do males form choruses rather than call on their own where they would not have to compete with other males? One simple reason is that the resources needed by females are usually clumped in space rather than dispersed. For example, most species of frogs and toads need water in which to lay their eggs, and in many insects, individuals of both sexes feed on particular kinds of plants that grow in specific areas, and females may also lay their eggs on these plants. Thus, a male has better chances of attracting a mate in these places than in areas where resources are poor or absent. These benefits offset the costs of competition with other males. Even in species in which males congregate in areas where resources needed by females appear to be absent, such as in grouse and prairie chickens, it has often been shown that females habitually pass through these areas to get to places where food or nesting sites are plentiful.

Two other hypotheses have been proposed to explain why males congregate in order to signal. One is that the sounds of a chorus are louder than those of an individual male and thus can attract females from a greater distance. There is some support for this idea in Australian cicadas, where males aggregating in small bushes attracted more females per male than did males calling in small groups or individually. But in one species of frog, females arrived at the breeding area at the same times and the same numbers even when a researcher removed all of the calling males before the chorus had formed, compared to nights when the chorus was active. The second idea is that particular males produce calls that are especially attractive to females. Other males also recognize these attractive calls and aggregate in the vicinity of this male with the "expectation" that they will be able to mate with excess females that the most attractive male cannot handle.

All three hypotheses could be true to some extent. Once males have congregated in an area where resources are rich, they may benefit from the greater detectability of signals at a distance, and within the chorus, individual males may make their decision about where to call based on their perception of the attractiveness of their neighbors.

The competition among males in a chorus has been classified by researchers according to whether or not males defend the resources needed by females; that is, in some species males defend a food plant or an especially favorable place to lay eggs. In this case, prospective mates may base their decision to mate with the male not just on the quality of his sounds, but also on the quality of the resources he is defending. This form of competition has been termed a *resource-defense polygyny*, which is a mating system in which males that are successful in defending resources can and do mate with more than one female. What about species in which males do not fight for resources but rather defend relatively small areas within the aggregation? Here the female presumably judges some qualities of the male or his signals. Is he in good physical condition? Are his sounds especially attractive? By mating with such a male, the female might avoid disease or parasites, or the male might be able to better fertilize her eggs than other males. Besides these direct benefits, the female's offspring might inherit qualities that make them more likely to survive and reproduce than if they chose a less attractive male. Researchers have termed this form of mating system a *lek polygyny*, and in some species, there is an extreme disparity in mating success. Some males may even attract and mate with most of the females in the population, whereas most other males do not mate at all.

In both resource-defense and lek mating systems, smaller, weaker, or less attractive males may adopt tactics other than calling or displaying. These tactics are less successful, but they offer the male at least some chance of mating. One example is *satellite behavior*. Here, a male does not call but situates himself near a calling male and attempts to intercept females. This tactic can be quite successful, and in green treefrogs nearly half of the satellite males associated with callers intercepted a female that was released by researchers.

Choruses are focal points for competition among males for females or resources needed by females for successful reproduction. Choruses form where such resources are most abundant, and males may defend these resources directly from other males. The sounds produced by males announce the presence of the male to prospective females that are ready to mate and to rival males that might challenge the male for his calling space or for resources he might be defending. Because the males are calling at the same time in the same general area, an added benefit is that the sound of the chorus may attract females from a distance more effectively than the sound of individuals.

See also Communication—Auditory—*Acoustic Communication in Extreme Environments*

Further Resources

Bradbury, J. 1985. *Contrasts between insects and vertebrates in the evolution of male display, female choice, and lek mating.* In: *Experimental Behavioral Ecology and Sociobiology* (Ed. by B. Hoelldobler & M. Lindauer), pp. 271–289. Sunderland, MA: Sinauer Associate.

Emlen, S. T. & Oring, L. W. 1977. *Ecology, sexual selection, and the evolution of mating systems.* Science 197, 185–193.

Gerhardt, H. C. & F. Huber. 2002. *Acoustic Communication in Insects and Anurans: Common Problems and Diverse Solutions.* Chicago: University of Chicago Press.

Carl Gerhardt

■ Communication—Vocal
Communication in Wolves and Dogs

Dogs represent a special kind of domestic animal because many breeds have been intimate social partners with humans for thousands of years, living in a particular kind of relationship with them—some breeds even preferring humans to other dogs. Dogs can be very sensitive to human "body language" as well as to the way humans talk to them. Dogs are even more skillful than great apes at a number of tasks in which they must read human communicative signals. Results of testing suggest that during the process of domestication, dogs have been selected for a set of social–cognitive abilities that enable them to communicate with humans in unique ways. That's where their special talent for human mind reading may have originated. How can we characterize communicative processes in domestic dogs? Which are the changes due to the process of domestication? And how are these questions researched properly?

Closely related canids, such as domestic dogs of various breeds, and their progenitor, the wolf, exhibit (under comparable living conditions), along with conspicuous similarities, a number of dissimilarities in social behavior and its development—with marked intraspecific variability. Pack-living wolves are social canids par excellence. A very high degree of sociality is common to them, a fact that

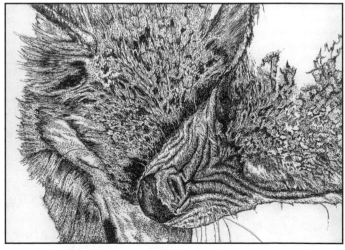

Muzzle-biting assures an individual relationship in wolves.
Courtesy of Dorit Urd Feddersen-Petersen.

may be judged as a preadaptation for domestication. Numerous and highly differentiated communicative skills develop while wolf pups engage in social play. While play-communicating, the pups learn rules to coordinate their lives—codes of social conduct that regulate their behavior. In dogs, play is important to social and cognitive development; however, the predominant channels of communication are variable, and vocalizations generally seem to dominate (for example, using play barkings to prevent play from escalating into aggression).

Comparative studies of wolves and dogs offer excellent opportunities to record constants regarding the development and significance of individual or species-typical behavior particularities—as well as those induced by domestication and breeding. Behavioral studies were first carried out under environmental conditions that closely resembled those of their ancestral species—"semi-natural," variable housing conditions, where to a large extent, animals could live of their own free choice. These conditions allowed dogs to be observed in wide-open-air enclosures living in packs—in comparison to the wolf pack (with equivalent number of members, sex ratio, age structure, and so on).

"Normal behavior" of domestic animals usually is defined as behavior of healthy animals living in this reference system of "semi-natural" environment. Some dog breeds turned out to be unable to cooperate and compete in groups or to establish and maintain a rank order (e.g. toy poodles, golden retrievers). Their interactions were not adaptive, and the members could not cope with challenges from the environment. It is striking that tactical variants of conflict solving (to appease the opponent, to trick him, to animate or to inhibit him), which are common practice with wolves, did not exist in social groups of some dog breeds. These strategies, however, are important for pack unity. Within groups of some dog breeds (toy poodle, pugs, west highland white terrier), trivial conflicts often escalated to damaging fights.

When they did not succeed in removing a threat, the situation became uncontrollable and the stress state remained. Acute stress became chronic. Others (e.g., German shepherds, Alaskan malamutes, several hunting breeds, terriers) were able to evolve coping strategies, making their situation controllable. They were in quite better shape mentally, living under semi-natural conditions.

The groups of dogs differed significantly with respect to 5 of 7 behavioral measures. Those breeds able to adapt to pack living exhibited:

1. much less frequent and less severe aggressive behavior;
2. much more frequent and variable social play;
3. greater social tolerance;
4. more nonagonistic contact behavior, such as social play, more allogrooming (grooming among themselves);
5. many variable barks as a means of subtle communication, occuring within a variety of social situations.

The groups corresponded on two measures: All preferred humans to conspecifics (members of their own species), which led to immediate interruption of interactions with group members when a human being passed the enclosure; and the groups showed a strikingly similar territorial behavior.

What are the reasons for a decreased capability of several dog breeds to form social groups? First, in differing degrees in the various breeds, there is a reduced expressive ability, especially in the optic area, that might cause communication problems or misunderstandings.

Various breeds have lost the highly differentiated capabilities of wolves to show a variety of precise facial displays. For many breeds, lop ears, curly hair, brachycephalic heads, many wrinkles, and so on reduce the possibilities for communication in the facial area, which may lead to misunderstandings.

Many dog breeds are not able to communicate precisely because of an extreme diversity in morphological (structural) characteristics (e.g., American Staffordshire terriers and brachycephalic breeds like pugs). In these and other breeds, there are just fragments of facial display abilities left in the mimic area of the face, which affects such facial possibilities as ear positions, forehead wrinkling, lips (is it possible to retract them, for example, and does this work as a signal or is it covered by hair or skin?). In pugs (or American Staffordshire terriers), for example, the forehead and nose areas are always wrinkled, and teeth baring is not possible because of the prominent flews. This compares with fine details and precise gradation in wolfish diversity of expression.

Thus, several facial regions and many signals are not available for communication in some breeds. Reductions in facial expression (and in the entire expressive behavior) seem to produce problems in social communication and may cause aggression. In an example of another problem, German shepherd dogs seem to look more like wolves, but actually do not seem to be able to communicate as well facially because their facial display abilities have been reduced through breeding. Facial

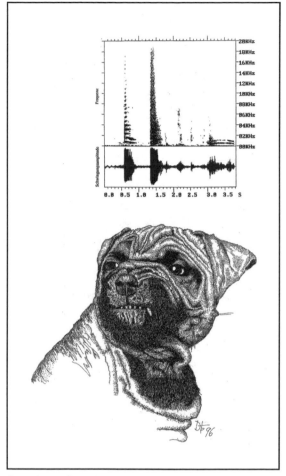

Dogs' vocal patterns.
Courtesy of Dorit Urd Feddersen-Petersen.

reductions are striking, possibly correlated with the enormous extent of skin growth, producing flews and dewlaps. One other channel of communication—vocalization—is partly hypertrophic in most dogs: An abundance of barks seems to exist in highly different social contexts.

As emphasized, domestication and selective breeding of dogs have resulted in a high variability of morphological characteristics such as head size and shape (e.g., the down face in bull terriers), suggesting that there might be large differences in vocal tract length, which could affect interbreed vocal communication because of sound quality differences.

Vocal Communication in Domestic Dogs

Vocalization, especially barking of domestic dogs, still remains a controversial topic. Whereas some authors consider dog barking to be an acoustic means of expression, becoming more and more sophisticated during domestication, others name this sound type "noncommunicative." Barking as a whole seems to be directed to humans or to the circumstances of living with them. Within the pack, barking appeared to cause excitement by means of *allelomimetic*, or contagious, behavior.

Vocal repertoires as studies of individual sound types are rare. There has been almost no work done on low-intensity, close-range vocalizations, yet such types of vocalization are especially important with the more social canids (Schassburger 1993), and hence, with human–dog communication and understanding of dogs. Most of the investigations of this subject published so far are based on auditory sound impressions and lack objectivity. Sonographic methods that graph sounds visually (e.g., amplitude and frequency) facilitate the identification of sounds, and reveal whether subjective classification can be verified by objectively measured parameters. Hence, parameters in bioacoustics differ in classification, a fact which causes subjective differences. Finally, meanings, functions, and emotions have to be examined for all the sounds described and must be discussed in terms of relationships between social structure and signal function, signal emission, and social context as behavioral response, and overlapping channels of communication.

The Structural Characteristics of Major Vocalizations

Canid barks are typically a mixture of regular (harmonic) and irregular (noisy) components. Wolves vocalize by noisy bark components only. Wolves lack *harmonic forms*. These are utterances without a considerable amount of noise, because the energy of the sound is restricted to narrow-frequency bands which are arranged (one above the other) at regular intervals. This is also true for the growl sounds. A comparison among the breeds elucidates that bull terriers, malamutes, and weimaraner hunting dogs show harmonic types of subunits of bark and growl as shown in the table.

According to Schassburger's (1993) notion of the wolf's vocal repertoire, domestic dogs of different breeds evolved neither a stereotyped nor a graded specific sound system; both types of vocal repertoire seem to be of adaptive value to dogs living with humans. In addition, selective breeding supported special sounds in hunting dogs or other working dogs.

But especially close-range vocalizations, concerning the major sound type of most domestic dogs, which is the bark, evolved in highly variable ways. However, the ecological niche of *Hausstand*, a German term referring to the environment in which dogs are raised with humans (Herre & Röhrs 1990), is highly variable, just as the individual differences in the dogs are, which seem to be breed-typical to a great extent.

Thus, complexity within the dog's vocal repertoire, which enhances its communicative value, seems to be achieved by many subunits of bark standing for specific motivations, informations, and expressions. Complexity within the dog's vocal repertoire is extended by the use of mixed sounds in the barking context. Transitions and gradations in communication to a great extent occur via bark sounds: harmonic, intermediate and noisy subunits.

Barking as a Means of Subtle Communication in Dogs: German Shepherd Dogs Compared to European Wolves

The following are some comparative data on the vocal behavior of dogs in specific social contexts. Categories of social contexts are:

1. *Disturbances*: These include disturbances from outside the enclosure (e.g., noises, strange people, or unknown stimuli).

Comparison of Structural Characteristics of the Two Major Vocalizations in Various Dog Breeds and the Wolf

GROWL	Bull Terrier	Alaskan Malamute	KL. Munster- Lander	Standard Poodle	Toy Poodle	Weimaraner Hunting Dog	Wolf
Type of Sound	Noisy	Noisy	Noisy	Noisy	Noisy	Noisy	Noisy
t (ms)	160–2275	93–2800	80–2770	100–3090	131–2487	106–1119	<1000
f dom (Hz)	160–1560	120–1360	80–880	235*	200–600	120–1960	70–580
Type of Sound	Harmonic	Harmonic	Harmonic	Harmonic	Harmonic	Harmonic	Harmon.
t (ms)	280–1937	93–2206	n.a.	n.a.	n.a.	106–1119	n.a.
fo (Hz)	160–1560	120–1360	80–880	235*	200–600	120–1960	70–580
BARK							
Type of Sound	Noisy	Noisy	Noisy	Noisy	Noisy	Noisy	Noisy
t (ms)	160–2275	93–2800	80–2770	100–3090	131–2487	106–1119	<1000
f dom (Hz)	160–1560	120–1360	80–880	235*	200–600	120–1960	70–580
Type of Sound	Harmonic	Harmonic	Harmonic	Harmonic	Harmonic	Harmonic	Harmon.
t (ms)	280–1937	93–2206	n.a.	n.a.	n.a.	106–1119	n.a.
fo (Hz)	160–1560	120–1360	80–880	235*	200–600	120–1960	70–580

n.a. = no analogy, * = average.

2. *Surrounding*: The German shepherds often surrounded one of their social partners and barked at him.

3. *Near food*: Bark sounds were vocalized when protecting resources (meat or bones).

The distribution of the types of barks differed significantly within the above social situations; the sound components (harmonic to noisy) occured in context-specific proportions. In the *near food* category, bark sounds were predominantly noisy or intermediate, with some harmonic and few transitional sounds. In the *surrounding* situations, harmonic barks and transitional sounds dominated. Here the transitional sounds were characterized by abrupt transitions from harmonic to noisy, with harmonic portions occurring throughout the barks, most commonly at the beginning of the sounds. (Parameters of transitional sounds are listed in the graph on p. 390.) The dominant frequencies of the "noisy sound" seem extremely high-pitched and very short in duration.

During *disturbances*, so-called harmonic-stringed bark variants were predominant. According to their auditory impression and to their oscillogram, these variants were characterized by the stringing together of harmonic bark elements (lengthened sounds), in contrast to other harmonic barks which occurred as single sounds. The harmonic string generally was preceded by one or two noisy bark elements. Overall, noisy barks dominated the *disturbance* situation.

Do German shepherds bark in a context-specific way? During *surrounding*—a playful/ agonistic social encounter where an individual is chased and finally gets caught by several animals which surround him and control his behavior—the maximum fundamental

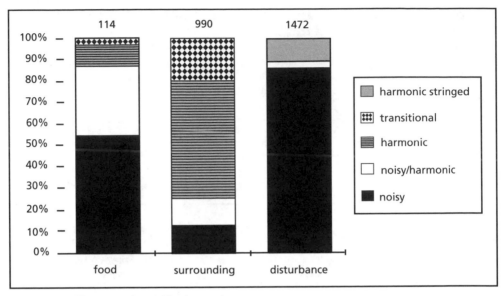

Context-specific proportions of bark sound types.

frequencies of the barking usually exceeded 1000 Hz, whereas after *disturbances* and *near food* situation, the frequencies always fell below this mark.

Regarding proportions, the difference between barking during *surrounding* and after *disturbances* was significant. In *near food* situations, 11 of the 12 (91.7%) harmonic bark elements were vocalized in combination with growls. The difference with respect to all combinations of bark and growl sounds in this context was significant (50 barks combined with growls—41.1%; test of two percentages, p < 0.05).

Typically, *during* disturbance-induced barking, multiple bark elements were strung together. Thus, lengthened sounds which consisted of noisy bark elements resulted. These lengthened sounds were vocalized by one individual at a time, raising the muzzle and opening the mouth slightly to emit each bark element. After *disturbances*, bark sounds frequently

Sound Area	Noisy	Harmonic	
fmax. ampl.<1000 Hz	13	43	
% (N=55)	23.6%	78.2%	
fmax. ampl.	Noisy > Harmonic	Noisy = Harmonic	Noisy < Harmonic
N	40	5	10
% (N=55)	72.7%	9.1%	18.2%
Duration	Range	>0.2 s	<0.2 s
	0.11–0.23 s	1	23
% (N=24)		4.2%	95.8%

Frequency and time measurements of transitional bark sounds. fmax. Ampl. : frequency with maximum amplitude.

overlapped or even masked each other, since normally several dogs barked simultaneously. Agonistic interactions were lacking.

In contrast to this, barking while *surrounding* was obviously directed toward one surrounded individual. Overlaps of barks were seldom heard; the German shepherds barked alternately. The surrounded animal often reacted by baring teeth or wrinkling its muzzle. However, transitions to agonistic behavior did occur, especially when the surrounded animal defended itself vigorously, combining teeth baring, growling, and noisy barks with biting intentions, a combination of combative elements that

"Surrounding"—a playful/agonistic encounter where one individual is surrounded and controlled by others.
Courtesy of Dorit Urd Feddersen-Petersen.

was seldom seen in this context (surrounding). This social encounter was normally initiated with series of transitional bark sounds. These findings suggest that the German shepherds adapt their bark components very precisely to varying social situations.

In the *near food* context just one individual barked. Barking was clearly directed toward a social partner with the following results: no reaction (10 times); hesitation or a pause in mid-behavior (6 times); signals of avoidance tendency or uncertainty (e.g., licking intention, tail tucked, ears flat, head and body low) (7 times); an increase of distance (32 times). The *near food* behavior was successful: A further approach of interactants was never observed. The dog defending its food frequently combined barking with other agonistic elements; however, additional growling did not increase the probability of driving off the intruder. Contrary to this, the German shepherds retreated more often if they were touched by the barking individual. Adding teeth baring and muzzle wrinkling, or raised hackles to the barking occurred quite rarely; therefore it's difficult to draw definite conclusions about the influence of these combative elements on the success of defending the food.

The European wolves reacted to *disturbances* from outside the enclosure (e.g., barking by the German shepherds) with either no alteration of behavior, some orientation responses, or, very seldom, with protective behaviors, but with no sounds emitted.

Surrounding occurred in two variants: a more playful behavior, where the surrounding wolves did not vocalize at all, and an agonistic encounter, during which the surrounding wolves sometimes growled.

In *near food* situations the wolves growled at times, but never barked.

A Summary of Current Knowledge

Some context-specific barks such as shaping of bark sounds could be verified by comparing bark elements in different social situations. Typically, multiple bark elements were strung together. Most often polyphonetic sound sequences were vocalized within a lot of overlaps and even masking of sounds, thus indicating a low degree of coordination of communication within the social group. The dogs' attention was almost exclusively oriented toward the outside of the enclosure, distinguishing this group vocalization from chorus howling in wolves, at least from all types of howling associated with group ceremonies. The disturbance-induced barking of German shepherds probably functions as an information carrier over longer distances, with cues for identification, number of group members and their location. These functions are attributed to wolf howling as well.

The so called harmonic-stringed bark sounds of German shepherds resemble sonagraphically the "bark-howl" of coyotes. However, the "bark-howl" in wolves and dingoes represents sound sequences composed of barks and howls, thus clearly distinct vocalizations. Mech (1970) described a "drawn-out bark" that might be an equivalent of the harmonic-stringed bark: ". . . its [the wolf's] barking had a definite pattern. It consisted of two, occasionally one, sharp barks followed by a more drawn-out bark which ended in a series of softer, lower-pitched barks . . ." Mech (1970) separated this sound as a "threatening or challenging bark" from the "alarm bark." Most probably, the drawn-out bark is seldom used by wolves—this described sound was vocalized by a wolf that approached the howling observer. During disturbance-induced barking, the German shepherds regularly vocalized harmonic-stringed barks and raised their muzzles, thus, the position of the head was similar in wolves and dogs, but wolves never howled simultaneously.

Surrounding was always characterized by the polyphonetic sound sequences used by two individual German shepherd dogs, obviously a cooperating duo, since rarely were there overlaps of bark elements. Fine tuning was already achieved by the initiation of *surrounding*: Usually one dog just made eye contact with the other, and they started surrounding. Their behavior indicated a playful context—even though there was no role reversal, and transitions to agonistic behavior did occur. Considering play "as a concept analogous to intelligence, in that it is a continuous quality with various dimensions, and no discrete boundaries" (Heinrich & Smolker 1998), this "play" classification of surrounding seems plausible, even through the behavior of the surrounding animals might fall "on the extreme end where "pleasure/fun" is the proximate motivation . . ." (Heinrich & Smolker 1998). This suggestion is in accordance with the structure of the vocalized bark sounds. Barks during surrounding were predominantly harmonic, but transitional sounds were frequently vocalized also, in sharp contrast to the other social situations. According to the motivation-structural rules, listed by Morton (1977), the more purely tonelike (i.e., harmonic) sounds are an expression of submission, appeasement, fearfulness, or friendliness. The higher the frequency and the more purely "tonal" the sound, the more fearful or friendly the sender. The transitional barks, with their abrupt transition from harmonic to noisy, might be an expression of ambivalence. These sounds reflected the behavior of the *surrounding* German shepherds, which was characterized by rapid changes between approach and attack, and an increase of distance or withdrawal. The participating German shepherds combined some signals of uncertainty or aggressiveness with otherwise relaxed facial expression and gestures. Such combinations seem very characteristic of playful situations in wolves. According to Schassburger (1993), the "intermediate" sounds (moan, whine-moan, growl-moan) that may occur in wolves as harmonic, noisy or intermediate in structure, express ambivalence on the one hand, and playful motivation on the other hand. In wolves, the intermediate sounds as expression of comfort and discontent seem to correspond with being in control or being controlled during play (for a discussion of play signals, metacommunication and intentionality see Bekoff & Allen 1997, 1998).

The interactions in the *near food* context are classified as agonistic. *Near food* the barks were evidently directed toward a social partner, and they functioned as releasers (stimuli that elicit specific behaviors). These barks must be classified as distance-keeping or distance-increasing sound signals within a social group. The same goes for the growl sounds, which were vocalized by the wolves. However, adult wolves almost never combine harmonic and noisy sounds in any given context. If wolves combine these sound types at all, it is via intermediate sounds (no barks!); consequently, according to Schassburger (1993), the concept of ambivalence seems to be of no importance for the sound combinations in wolves. In German shepherds, expressions of ambivalence by means of sound combinations occurs as a rule, for example, near food. The wolves sometimes growled in this situation, but they never barked. Schassburger (1993) found that bark sounds contributed only 2.3% to the overall vocalizations in timber wolves, and even these rare barks occurred in different contexts, that is, while warning and in defense, threatening or attacking. Mech (1970) described a "battle barking" which accompanied the "protest snapping" of middle-ranked wolves. Contrary to this, in German shepherds, high-ranked animals also barked. Schassburger (1993) observed the following interaction between wolves: "At the same time, while lying on the ground, he [the inferior wolf] yelped with each muzzle poke of Chippewa's [the superior wolf] . . ." The harmonic "distress calls" were vocalized simultaneously—a sound duel so to speak—thus indicating the agonistic context and, once again, representing coordinated vocalizations that were well suited to the respective behavioral context.

Consequently, the barks of German shepherds were not only context-specific, but they were also communicative and elicited specific responses from the social partners. Barks conveyed definite messages about the motivation and intention of the vocalizing individual. This communicative significance of the German shepherds' bark sounds is even more heightened based on their greater variability in comparison with wolves. For example, low-ranked German shepherds that were threatened by a superior social partner barked harmonically. In comparable situations, wolves vocalized different harmonic sounds (e.g., yelp or whine sounds), and high-ranking wolves always used noisy barks, characterized as protest sounds (see Schassburger 1993). Following Schassburger (1993) one might characterize the German shepherds' barking as a nearly complete sound system on its own, for the bark sounds cover the whole range from harmonic to intermediate to noisy, including several variations within these overall sound structures. These data indicate fundamental differences between the aggressive communication of wolves and German shepherd dogs caused by the process of domestication and selective breeding. This investi-

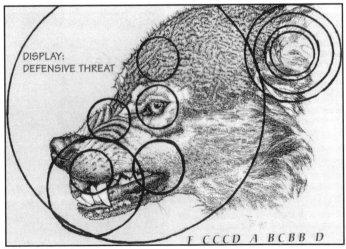

Facial indications of a defensive threat.
Courtesy of Dorit Urd Feddersen-Petersen.

gation not only furthers our knowledge about domestication, but it is also important for a better understanding of the aggressive behavior of domestic dogs, and this understanding facilitates a better, more sophisticated communication between man and dog, and can reduce potential dangers in dealing with dogs.

Wolves never emitted harmonic barks, nor harmonic growls, whereas dogs evolved a differentiated sound system of barks and growls (harmonic, intermediate and noisy sub-units). Categories of functions regarding interactions with barking/growling included, for example, play soliciting, social play, caregiving, and social contact, such as affiliative acts like grooming. The vocalizations of dogs seemed to develop into an increasingly communicative component of social interaction, especially via bark differentiations. Vocalizations changed in adaptation to living with humans. Thus, the hypothesis of domestication-induced hypertrophy (or increased development) of barking, and its increasing sophistication, as a result of living with predominantly verbalizing humans, cannot be mistaken. Living in packs, as wolves do, a variety of finely-graded facial displays, that is, visual expressions as a whole, seem to be advantageous over barking. It is also seen that barking is loud while competing or solving conflicts, causes excitement, and induces mood changes in group members. The fine-tuned mimics and gestures of wolves serve them better in establishing and maintaining a very differentiated social hierarchy.

Dogs are highly variable creatures, "formed" by the power of human imagination. Thus, selective breeding produced "extremes" with respect to morphology and behaviors. Finely-graded mimic communication is impossible in pugs or, for example, a Dogue de Bordeaux. However, dogs who look "more wolfish" show just as many mimical reductions. Some reasons may be the social life with humans and the lack of selective pressure to evolve sophisticated mimical expressions. As a rule, domesticated wolves prefer the vocal channel of communication.

See also Behavioral Physiology—*Vision, Skull Shape, and*
Behavior in Dogs
Communication—Auditory—*Acoustic Communication*
in Extreme Environments
Communication—Vocal—*Social Communication in*
Dogs: The Subtleties of Silent Language

Further Resources

Bekoff, M. 2001. *Social play behaviour.* Journal of Consciousness Studies, 8, 81–90.

Bekoff, M. & Allen, C. 1997. *Species of Mind: The Philosophy and Biology of Cognitive Ethology.* Cambridge, London: MIT Press.

Bekoff, M. & Allen, C. 1998. *Intentional communication and social play: How and why animals negotiate and agree to play.* In: *Animal Play: Evolutionary, Comparative, and Ecological Perspectives* (Ed. by M. Bekoff & J. A. Byers), pp. 97–114. Cambridge, New York, Melbourne: Cambridge University Press.

Hare, B., Brown, M., Williamson, C. & Tomasello, M. 2002. *The Domestication of Social Cognition in Dogs.* Science, 298, 1634–1636.

Herre, W. & Röhrs, M. 1990. *Haustiere—zoologisch gesehen.* Stuttgart, New York: Gustav Fischer.

Mech, L. D. 1970. *The Wolf.* New York: Natural History Press.

Morton, E. S. 1977. *On the occurrence and significance of motivation-structural rules in some bird and mammal sounds.* American Naturalist, 111, 855–869.

Schassburger, R. M. 1993. *Vocal Communication in the Timber Wolf,* Canis lupus, Linnaeus. *Advances in Ethology 30.* Berlin, Hamburg: Paul Parey.

Tembrock, G. 1976. *Canid vocalizations.* Behavioural Processes, 1, 57–75.

Dorit Urd Feddersen-Petersen

■ Communication—Vocal
Jump-Yips of Black-Tailed Prairie Dogs

A black-tailed prairie dog flings the front half of its body into the air with a loud "AH" sound, and returns to the ground with a softer "aaahh." Immediately several others do the same; a chain reaction may result, with large numbers of individuals joining in. The highly visible movement, two-part "AH-aaahh" call, and often contagious nature of the behavior makes the *jump-yip* one of the most recognizable displays on the prairie. But what is this signal? Why would a prairie dog make itself so conspicuous—both visibly and auditorially—when predators lurk all around?

Work by Carl Koford, John King, W. John Smith, John L. Hoogland, and others has helped us begin to understand black-tailed prairie dogs and the role communication signals such as the jump-yip play in their everyday lives. Black-tails (*Cynomys ludovicianus*) are one of 4–5 recognized prairie dog species. All prairie dogs are rodents, not dogs, and are part of the rodent subgroup *Sciuridae*, that includes woodchucks, marmots or "rock chucks," tree squirrels, chipmunks, and ground squirrels. All prairie dog species are smallish—about 30–45 cm (12–18 in) long, excluding tail, and about 500–1500 gm (1–3 lb) in weight, depending on season—light brown ground-dwellers who live in burrows and forage on various grasses and occasional insects.

There are two major groupings of prairie dogs, black-tails and white-tails. They differ in where and how they live as well as in tail characteristics. Black-tails have a relatively long tail (about 60–100 cm; 24–39 in) with black fur covering the last 4–5 cm (1.5–2 in), making the tail very visible and giving them their common name. They live in lower grassland prairies where there is little cover, and they are extremely visible to predators. They are also extremely social, and have a number of important vocal and visual signals they use as they interact.

Black-tail "towns" are large, often covering hundreds of acres. (Historically, they were even larger, but expansion of ranching and agriculture, housing development, other habitat loss and hunting by humans has greatly decreased the size of towns and the number of individuals.) Each town is divided into small, contiguous territories, or *coteries*. Several individuals live together in a coterie and are "friendly" to one another. Coterie members defend their space from adjacent neighbors in ritual "challenges" that define the boundaries between territories. Theoretically, this system of having many contiguous territories and being "friendly within, unfriendly between" enables the black-tails to have lots of eyes and ears available to watch for predators and give warning, while making it easy to know what to do when encountering the many other prairie dogs that are present.

Predation is a real and constant threat to black-tails. Hawks, badgers, coyotes, black-footed ferrets, and wolves, before they were hunted to extinction in the plains, all prey on them. With no vegetation to hide behind, black-tails need time to run to their burrows to escape. Loud sharp barks serve as warnings that can be heard across even large colonies. They are often followed by repeated short barks that serve as continuous reminders of possible danger. Visible signals accompany these calls, including stopping all activity and looking alert, standing up on back legs alert, or running to burrows with tails flicking and pausing there, alert, ready to run underground. Warnings can last a minute or two, or half an hour or longer, and other activities—eating, defending territories, grooming—are put on hold or frequently interrupted during this time, as individuals continue to monitor the warnings being given around them.

How is a black-tail to know that the warning is over and regular activities can resume once again? The jump-yip seems tailor-made to serve as an "all clear" signal. Loud and visible, it can be seen and heard across the colony. Black-tails who give the call and those individuals around them usually immediately return to their pre-warning activities after this signal is given.

The call also seems to serve a similar purpose after territorial boundary disputes. After jump-yipping, individuals who were tense, alert, and ready to challenge one another, return to other non-challenge behaviors, such as eating, greeting coterie members, and sunbathing.

An unfortunate consequence of jump-yipping is the danger it can bring to the signaler. Being loud and visible can make an individual an obvious target for a predator that has not in fact left the area. Being wrong about an "all-clear" can be fatal. There is a natural selection, then, to jump-yip only at the right times, and most jump-yips are probably reliable because of this. Another unfortunate consequence is not so easily dealt with. Humans "hidden" in vehicles seem to find it fun to shoot prairie dogs when they are in the middle of a jump-yip, and many are lost in this way.

See also Communication—Vocal—*Referential Communication in Prairie Dogs*

Further Resources

Graves, R. A. 2001. *The Prairie Dog, Sentinel of the Plains.* Lubbock: Texas Tech University Press.
Hoogland, J. L. 1995. *The Black-Tailed Prairie Dog: Social Life of a Burrowing Mammal.* Chicago & London: The University of Chicago Press.
King, J. A. 1959. *The social behavior of prairie dogs.* Scientific American, 201, 128–140.
Koford, C. B. 1958. *Prairie dogs, whitefaces, and blue grama.* Wildlife Monographs, 3, 1–78.
Smith, W. J. 1977. *The Behavior of Communicating.* Cambridge, MA: Harvard University Press.

Penny Bernstein

■ Communication—Vocal
Referential Communication in Prairie Dogs

Communication is the transfer of information from one animal to another. *Referential communication* is the transfer of information about events or objects in the external environment of the animal, such as the presence of predators or food resources (Bradbury & Vehrencamp 1998). One form of referential communication can be the alarm calls that animals give in response to a predator. When animals give alarm calls, the calls may express only the animal's emotional state upon seeing the predator, such as fear. In that case, the call would not be considered referential, because it provides no information to another animal about the predator, but is simply an expression that could be elicited by anything that might frighten the animal. Calls that are expressions of the animal's internal state or emotions are known as *motivational*. For example, imagine that you went around the corner of a building and bumped into a hungry tiger that had just escaped from the local zoo. You might yell, "aaahh!" as an expression of your fear at encountering a tiger. There is no information in your yell about tigers, and all a listener could conclude is that you were badly frightened by something. On the other hand, the alarm call could contain specific information about the predator—the species of predator, or its size and shape, or its level of danger to you and your friends. Then the alarm call would be considered referential. Suppose now that when you bumped into the tiger, you yelled to your friends, "big hungry tiger!" As you can probably guess, referential and motivational signals need not be mutually exclusive, because you can yell "tiger" in a tone of voice that would convey to your friends that you are scared.

Gunnison's prairie dogs (*Cynomys gunnisoni*) communicate referentially in their alarm calls. These animals are ground squirrels that live in grasslands in northern Arizona, northern New Mexico, and southern Colorado. They live in colonies called prairie dog towns, and are eaten by a number of predators, such as coyotes, red-tailed hawks, and domestic dogs. Historically, they have been hunted as food by the Native Americans inhabiting the Arizona-New Mexico-Colorado region of the United States, and presently they are frequently shot by humans for sport. Probably because so many different predators come to the prairie dog towns to hunt, the prairie dogs have evolved a sophisticated system of communication that is able to convey information to other prairie dogs about the species of predator that is hunting, and also about the individual description of the predator, such as its color, size, shape, and speed of travel.

There are at least four acoustically-different alarm calls. One call is elicited by humans, another is elicited by coyotes, a third is elicited by domestic dogs, and the fourth is elicited by red-tailed hawks (Placer and Slobodchikoff 2000). An important component of testing

Prairie dogs have been shown to have quite sophisticated vocal communication skills.
© *Anthony F. Chiffolo.*

whether referential communication occurs is playing back the alarm call to the animals when no predator is present and assessing the response of the animals. If the animals respond to the playback in the same way as they respond to the sight of the predator, then we can conclude that the call contained some information about the predator. This is particularly the case when there are different alarm calls. If the animals respond in the same way to each different call—for example, if they always run to their burrows every time they hear an alarm call—then we do not know if the alarm call conveyed any information beyond the fear that the caller felt upon seeing the predator. On the other hand, if the escape responses differ among the different predators, then playback experiments can show us that the animals understand the specific information that might be encoded in each alarm call. For example, vervet monkeys (*Cercopithecus aethiops*) have three different alarm calls: one for leopards, another for eagles, and a third for snakes. The escape responses differ for the different calls. Playbacks of the different calls elicit the same escape responses as when the call was made by a monkey in the presence of a predator (Cheney & Seyfarth 1990).

Prairie dogs have different escape responses for the different predators (see the chart following). When a human appears, the animals run to their burrow in a colony-wide response and jump inside. When a red-tailed hawk dives down into the colony, only the animals in the immediate flight path of the diving hawk run into their burrows and jump inside, whereas the animals outside the flight path of the hawk stand on their hind legs and

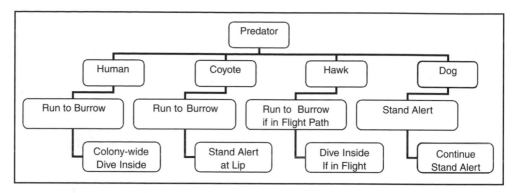

Escape responses of prairie dogs to different predators.

watch what is happening. When a coyote appears, all the animals that have been foraging run to the lip of their burrows and stand on their hind legs, watching the progress of the coyote until the coyote comes close to them, and then run inside. When a domestic dog approaches the colony, all the animals stand on their hind legs wherever they were foraging and watch the progress of the dog. If the dog comes too close, then they run to their burrow and disappear. When either dogs or coyotes approach the colony, animals that were inside the burrows come up to the surface, so that nearly the entire population of the town is standing up on their hind legs above ground, watching the predator. Playbacks of the different calls when no predator is present produce the same escape responses as when the predator is hunting in the colony, suggesting that the information encoded in the alarm calls is understood by the animals that hear the alarm call (Kiriazis 1991). As with the vervet monkeys, prairie dogs have qualitatively different escape responses to different predators, and can use the information encoded in the alarm call to determine which escape response to choose.

Prairie dogs can also encode additional information, in the form of descriptions of the color, size, and shape of the predator. Experiments using humans wearing different color shirts have shown that prairie dogs incorporate information into their calls about different colors, as well as the general size and shape of the humans (Slobodchikoff et al. 1991). Prairie dogs have a dichromatic form of color vision, which means that they can see blues, greens, and yellows, but not oranges and reds (Jacobs & Pulliam 1973). In one experiment, two humans first walked through a colony, one at a time, one wearing a yellow shirt and the other wearing a gray shirt. Then they traded shirts, and each separately walked through the colony again. All of the acoustic components of the human-elicited call remained the same for each of the humans, except that the component that coded for color changed from yellow to gray to correspond to the different shirt colors. In another experiment, four humans that differed in size and shape each wore the same white laboratory coat, which tended to obscure their physical description. Under these circumstances, the prairie dogs encoded much more similar information about each of the humans than when the humans were wearing differently colored shirts and the size and shape differences were more obvious (Slobodchikoff et al. 1991).

In addition to describing the species of predator and the color, size, and shape of the predator individual, prairie dogs can also describe objects that they have never seen before. In one experiment, the prairie dogs were shown three silhouettes—a coyote, a skunk, and an oval. Each silhouette was painted black and was run out in the colony on a pulley system, with appearances of the different silhouettes determined randomly so that the prairie dogs could not predict either when or what kind of silhouette was going to appear. The coyote silhouette and the oval silhouette were the same size, so they projected the same size of image on a prairie dog's retina. The skunk silhouette was smaller, mirroring the relative size differences between skunks and coyotes. The prairie dogs saw two of these objects—skunks and coyotes— frequently. The prairie dogs had never seen before the other object—the oval. All of the prairie dogs had one type of call for the skunk silhouette, and a different call for the coyote silhouette, a call that was very similar to the calls elicited by live coyotes. However, all of the prairie dogs had yet another call, different from any other call, for the oval silhouette. Each prairie dog called the same way for the oval silhouette, suggesting that they encoded information about the physical properties of the oval (Ackers & Slobodchikoff 1999). Perhaps prairie dogs have a lexicon of descriptive elements in their neural system, and choose elements out of that lexicon to describe objects in their environment, whether or not they have seen those objects before.

See also Communication—Auditory—*Acoustic Communication*
        in Extreme Environments
    Communication—Vocal—*Alarm Calls*
    Communication—Vocal—*Choruses in Frogs*
    Communication—Vocal—*Vocal Communication*
        in Wolves and Dogs
    Communication—*Jump-yips of Black-tailed Prairie Dogs*

Further Resources

Ackers, S. H. & C. N. Slobodchikoff. 1999. *Communication of stimulus size and shape in alarm calls of Gunnison's prairie dogs.* Ethology, 105, 149–162.

Bradbury, J. W. & S. L. Vehrencamp. 1998. *Principles of Animal Communication.* Sunderland, MA: Sinauer Associates.

Cheney, D. L. & R. M. Seyfarth. 1990. *How Monkeys See the World.* Chicago: University of Chicago Press.

Jacobs, G. H. & K. A. Pulliam 1973. *Vision in the prairie dog: spectral sensitivity and color vision.* Journal of Comparative and Physiological Psychology, 84, 240–245.

Kiriazis, J. 1991. *Communication and Sociality in Gunnison's Prairie Dogs.* Ph.D. Dissertation, Flagstaff: Northern Arizona University.

Placer, J. & Slobodchikoff, C. N. 2000. *A fuzzy-neural system for identification of species-specific alarm calls of Gunnison's prairie dogs.* Behavioural Processes, 52, 1–9.

Slobodchikoff, C. N., Kiriazis, J., Fischer, C., & Creef, E. 1991. *Semantic information distinguishing individual predators in the alarm calls of Gunnison's prairie dogs.* Animal Behaviour, 42, 713–719.

Con Slobodchikoff

■ Communication—Vocal
Singing Birds: From Communication to Speciation

Some birds utter only a simple song, while others sing a stunning variety of complex notes, variable in pitch and duration. Some species, such as bittern or ostrich, use sound as low as a foghorn, while hummingbirds and cowbirds often sing so high-pitched that it is hardly audible to the human ear. Some songs sound melodious, others have something peculiar to

Portraits of Sound

Hans Slabbekoorn

The investigation of acoustic variation in bird songs has gone through an enormous revolution the last few decades with the development of modern sonographic analyses techniques. With accessible software one can now easily digitize large numbers of song recordings and visualize them on the screen, after which detailed measurements can be taken with high repeatability. Three types of graphics are common tools for acoustic investigation. The *amplitude wave* form depicts sound intensity fluctuations over time: The *x*-axis reflects the passing of time whereas the height of the spikes above and below this axis indicates the sound volume. The *power spectrogram* displays pitch versus amplitude, summated for a segment of sound, or an entire song. It shows the distribution of sound power through the spectrum depicting the sound energy present at each frequency. The *sonogram* includes information on both pitch and amplitude of the sound through time. The *x*-axis again represents time, the *y*-axis pitch, and information on the amplitude range, from silence to the loudest sound, is depicted by the color scale from black and blue to yellow and red. Acoustic measurements that are often included in studies of acoustic variation are: Fmax = maximum frequency, Fmin = minimum frequency, Fpeak = frequency of peak amplitude, Dnote = note duration, Dphrase = duration of a repeated phrase, and Dsong = duration of the whole song. See the sonogram of a song of a great tit in the color photo section.

them: resembling a bouncing ping-pong ball such as in the wren tit of California or human laughter such as in the Australian kookaburra. Singing behavior is extremely variable among species, but why do birds sing? And, do all species, singing in so many different ways, accomplish the same thing? One thing is certain: Birds do not sing just for fun. All the variety in songs found in different bird species, whether simple and short or long and complex, is likely to serve the singer well in achieving higher reproductive success. For each species, the singers of today are the offspring of previous generations that were successful breeders singing a similar melody. Those that sang not or differently must have been less successful and hence failed in transmitting their genes to the current generation. Variable success rates of individual singers may depend on the ecological settings of the breeding environment and affect evolutionary change of whole populations. Furthermore, song differences often play a role in keeping species apart, and song variation within species potentially plays a role in the process of speciation. Hence, the study of the singing behavior of birds is interesting from ethological, ecological, and evolutionary perspectives, especially because of the close link with reproductive success. Therefore, the first crucial question is: How can singing a song lead to more offspring?

Functions of Birdsong

The two main functions of singing in birds have a clear impact on the reproductive output of the singer and are associated with competition among males and attraction of females. In many species, male birds use song in competition with other males for access to territories, food, and females. They acoustically signal their presence and motivation to defend their local surroundings and to prevent intrusions by competing birds.

The role of song in this context has been tested experimentally in different ways. In one experiment, territorial males of Scott's seaside sparrows were muted temporarily, which made them silent warriors, because they had many more conspecific intruders to fight off than the experimental controls that were caught and released without being muted. In another setup, pairs of great tits were removed completely from their natural territories and replaced with speakers that played back great tit song or a control sound. Territories from which song was broadcast remained empty much longer than territories that were acoustically vacant to great tit ears. Evidently, singing helps in keeping competitors out and prevents the singer from spending energy and time on chases and fights that may also be risky in the sense that predators may have an easier catch during such turbulent events. If some song variants are more effective tools in regulating conflicts than others, male responsiveness may direct evolutionary change (for example, toward songs with high stereotypy or ones that are difficult to produce).

The second signaling function in the context of sexual selection is the attraction of potential mates and the subsequent seduction to mate. Experiments with pied flycatchers showed that male dummies attracted more females when a speaker was broadcasting the male song. It is likely that the acoustic stimulus of a song guides females of many species toward their eventual mates, after which visual stimuli may also play a role in the actual decision to mate. Besides attracting females to close quarters, singing may stimulate the courtship behavior of the female. For example, even in the absence of the visual stimulus of a male cowbird, receptive females of this species will exhibit a copulation–solicitation display after hearing a playback of male song. This display is crucial for subsequent mounting and fertilization of the eggs and clearly indicates a positive evaluation of the male singer as a suitable mate. For some dove species, it is known that the male cooing stimulates ovarian development, although it is crucial that the female herself is also able to coo in return to actually accelerate egg production. Even after being paired up a female bird may still be able to hear several neighbors singing in addition to her own mate, and attractive songs may persuade her to switch mates or select one of the neighbors to father her offspring through extrapair copulations. If some song variants are preferred over others, female mate choice may direct evolutionary change, for example, toward more complex or longer songs.

Singing may serve other functions beyond sexual selection, but these are often specific to particular species. Such functions are generally related to a direct impact on survival of the singing bird himself or his offspring and therefore subject to natural selection. In the house wren, for example, females that are incubating on the nest will usually wait to leave the nest until her mate is singing nearby. The benefit of this behavior may be that the singing indicates to the female that no predators are present, or just that the male himself is present to guard the nest in her absence. The common yellowthroat sings in flight especially when uninvited visitors such as humans enter its territory. The flight direction of the male bird is usually away from his nest, which is hidden in the vegetation and contains his precious offspring. This behavior will likely distract the potentially dangerous intruder and lure it away from the nest. Another example of a different function is singing as a pursuit-deterring signal in the skylark. Completely outside the breeding season, male skylarks may start to sing in response to merlins in hot pursuit. Fleeing males that are capable of singing complete songs instead of a few stutters are more successful in escaping the chasing bird of prey. Here, it is assumed that singing signals to the predator that the prey is in good condition and will not be easy to catch, which makes the merlin give up sooner.

Message and Meaning

Song has to convey some information about the singer to the audience, because just general sounds will not help male birds keep competing males out of their territory or attract females to mate with them. Both the singer and the audience (the sender and the receiver) can benefit from information transfer. For example, if a territorial male could make clear in advance that he is a strong and determined male that will be difficult to fight, many competitors will prefer a detour. The male will also benefit if he attracts only females of his own species, which he can convince of his good health and parenting quality over a distance. Similarly, if song can convey information about the sender, receivers may also benefit from first hearing out a potential opponent or mate. Rivals would only approach males that they can beat, and females would only inspect males at close quarters that they consider to be good fathers to their offspring. So, how is such detailed information encoded in the song? And how can we be sure that birds themselves are able to decode?

An acoustic signal may carry a message through a correlation between singer characteristics and acoustic characteristics. The nature and detail of the message encoded in song varies and can comprise many aspects of the singer. First of all, every bird species has its own species-typical song, and therefore song carries a message about species identity. Species recognition by voice may be very important when many different species are acoustically broadcasting their presence at the same time in low-visibility forest, or when closely related species are very similar in plumage and body size. Second, males and females of the same species may sound different. We focus here on male song because males are usually the most vocal sex, although in many species females also sing, especially in the tropics. Third, strong males may sound different from weak males. This is especially useful if strong males can make sounds that are impossible to make if one is weak; only then can females be sure that a certain signal belongs to a strong partner and not to one that is bluffing. Fourth, males may be highly motivated to defend their territory, or they may be less interested in it because local food sources are few. Such motivational variation can be reflected in acoustic details of the songs sung and be available to males in the neighborhood that are deciding on whether or not to challenge the resident male. The fifth and final example of an acoustic message: Birds of one species usually do not sound exactly the same throughout their distribution. Songbirds often have dialects, with individuals at one location all singing songs that are more similar to each other than to the songs of individuals at another location. A female benefiting from finding a male that is adapted to the local environment may choose among males singing the local dialect and avoid mating with males with a foreign dialect. These examples illustrate that the message potentially communicated through song variation may concern species identity, sex, strength, motivation, and dialectal information, but there are of course more possible correlates between singer and song.

Testing Receivers

If different song variants trigger different behaviors, one can conclude that the birds are able to detect the acoustic differences, and that they perceive them as meaningful. This can be confirmed by different responses to playback of the songs without the physical presence of the singer. Playbacks simulate an intruder that is announcing its presence vocally. One can test acoustic variation within and beyond the natural range by using both original recordings and experimentally manipulated songs. An artificial change in one acoustic parameter, for example the song frequency or trill rate, could lead to a significant decrease in the response of territorial males. Such behavioral variation is interpreted as a lack of recognition; the song seems to be classified as not being of the right species. Similarly, playback of longer songs or more complete songs leads to stronger responses by territorial males in several species. Presumably, long and complete songs indicate a competitor that is highly motivated to stay and needs an energetic reply to be convinced that it is better to leave the territory. This playback setup is a valuable low impact method that can answer many basic questions, although first removing birds from their territory and then testing the potency of song variants in keeping new colonizers out may better reflect the natural situation in which song is used most often.

Female responses to song variants are usually tested in captivity, because unpaired females are hard to locate in the wild and, if already paired, the male is likely to be the most responsive. Results from tests in artificial aviary conditions can be extrapolated to the natural situation with caution. Females in captivity either respond spontaneously to song exposure, for example by approaching the speaker or by copulation solicitation displays, or they are trained to peck keys to get a song as a reward. In the latter situation, females may hear one song variant when pecking one key and another when pecking a second key. Approach bias to one of two speakers, variation in display rate, and relative key pecking rates are all behavioral measurements that may indicate a preference to hear one song over the other, and thus suggesting a preference for one singer over the other. Female white-crowned sparrows, for example, produce more solicitation displays in response to their natal dialect than to foreign dialect songs, although this is only true when they have not heard foreign dialect songs in their first spring as an adult. Female zebra finches peck keys at a higher rate when rewarded with playback of a repertoire of different song types instead of a repetition of the same song. Interestingly, male zebra finches only sing one song type and lack the repertoire that females prefer.

Playing back sound to birds is used not only to test the perception of information carried by the sound in field playbacks and preference tests, but also to test auditory sensitivity. In contrast to the first two tests, meaning should not play a role when testing the capacity to detect sound of a certain frequency or recognize sounds of a particular type. Therefore, such laboratory studies use either a neurophysiological approach measuring firing activity directly in the neurons of the sensory system, or operant conditioning where birds are trained to detect, discriminate or categorize sounds, with food as reward. It is possible that birds are able to detect acoustic variation that bears no meaning, and therefore one can distinguish between the smallest acoustic change that can be detected by the sensory system, the "Just-Noticeable-Difference" and the smallest acoustic change that leads to a different behavioral response in the wild or under seminatural conditions, the "Just-Meaningful-Difference."

Songs Fitting the Bill and Matching the Ear

Singing plays the same roles in many different species, but they each use a different song. Being different allows species recognition based on sound, but how do these song differences arise? Song characteristics are determined by several factors that set limits as to what can be produced and what is more likely to evolve.

First, song features strongly depend on body size and shape. Not all birds are equally capable of generating sound variation. One of the most obvious correlations between body and sound is that larger-bodied species usually have lower-frequency songs. Frequencies emitted by a bird depend directly on what sounds are produced at the syrinx and how these sounds are altered on their way out via the vocal tract, consisting of thrachea, throat

Dove Coo Code Decoded

Hans Slabbekoorn

The coos of male collared doves are different from female collared doves coos. Both sexes produce series of three-element coos, but the male sound is lower in pitch and contains fewer overtones. Male and female doves start to produce the first coo-like sounds at 12 weeks of age; the males become distinct, only gradually, and after about 60 weeks they produce a fully-developed male coo structure that remains more or less the same across years. The gradual change in coo characteristics from juvenile to adult provides acoustically encoded information about age to individuals that hear the cooing. Furthermore, no two adult males have exactly the same coo characteristics. One aspect that varies among them is whether the coo elements show only a gentle rise and fall in frequency or instead show a discrete jump to a higher frequency. The number of elements with such frequency jumps in a series of coos varies with body weight: Heavier males produce more jumps which suggests that they may not be easy to generate. The jumps emerge through a physical phenomenon related to size and shape of the vocal tract and speed of air flow during coo production. Doves that are able to produce many coo elements with discrete frequency jumps may be in better shape, have stronger muscles or have a greater lung capacity—all of which are factors interesting to communicate to potential mates and rivals (is the sound of a "weak" or "strong" male?).

Acoustic variation across cooing individuals may broadcast information about sex, age and strength, but the message can only become available to the doves themselves if they are able to detect and recognize the variation. Operant conditioning techniques in the laboratory are helpful in showing the auditory sensitivity for certain frequency changes and to show whether doves are able to classify consistently, for example, coos as being from a male or female. However, playback experiments in the field are necessary to show not only whether they are able to detect and recognize the variation, but also that they perceive it as meaningful in a natural context. Male collared doves do not respond differently to playback of male and female coos in their territory. The lack of a difference in response in the field does not exclude the possibility that they are able to detect and recognize the acoustic variation. It may well be that the presence of a male or female intruder does not trigger a different response from an already-paired male. Male collared doves do respond more strongly to playback of coos that contain discrete frequency jumps than to coos without, which indicates that in this case the acoustic variation conveys some relevant information about the simulated intruder. The stronger response suggests that the cooing is evaluated as being of a more dangerous competitor which requires a high state of arousal to put off. See the sonograms of collared dove coos in the color photo section.

and beak. Larger birds usually have a larger syrinx and thus produce a lower sound to start with. Larger birds also have a larger and longer vocal tract, which results in resonance features that favor lower over higher frequencies. This generates a kind of filter mechanism: The source sound produced inside the bird is not the same when heard on the outside; part of it may be attenuated or completely filtered out, whereas other parts sound loud and clear.

The bill is at the very end of the vocal tract, but still may affect the sound that comes out: Opening it shortens the vocal tract and raises resonance frequencies. As a consequence, bill opening and closing during song is generally correlated to producing respectively high and low song components. Similarly, birds with long bills are likely to sing with relatively low frequencies. For example, species with shorter bills sing higher songs among neotropical woodcreepers. Interestingly, in a few bird species, internal morphology has evolved in such a way that the outside appearance is not a good indicator for the sounds that are produced. In several owl species, females are larger than males, but they hoot at a higher pitch because they have a smaller syrinx. In the trumpeter swan and the whooping crane, among other species, the trachea is of disproportional length, coiled up within the space available in their necks. The elongated vocal tract length has a dramatic impact on the resonance characteristics, making their sounds much lower than in similar-sized species.

A second factor that affects song characteristics is what birds are able to hear. The frequency range used for song is obviously tuned to the sensory capacities of the ear. It may be important to hear sounds beyond what is sung by members of ones own species, but if song is to have a communicative role within the species, it should in the first place be audible. Therefore, sound production and perception usually evolve in concert and often mirror each other in acoustic characteristics. On average, the frequency range in which birds hear best is very similar to the auditory sensitivity of humans: They are usually most sensitive between 0.5 and 5.0 kHz, and detection of sounds becomes more and more difficult for frequencies below 0.1 and above 10.0 kHz. Songbirds tend to use higher frequencies within this range and are also relatively more sensitive to higher frequencies. Non-songbirds tend to use lower frequencies and concordantly have especially good hearing for the lower frequencies.

Birds of prey are often more sensitive to sounds compared to other birds, especially nocturnal hunters that depend on hearing for localizing prey. An example of a comparative study on auditory sensitivity illustrates well the interplay between hearing and producing sound. Sparrowhawks hunt great tits and other small songbirds, and when the prey detects the predator in time, it will use alarm calls that may increase the chances of escape. These alarm calls signaling the presence of an aerial hunter are long drawn notes of high pitch at 8.0 kHz. Sparrowhawks outcompete great tits in auditory sensitivity almost throughout the frequency range, except for a peak difference at 8.0 kHz, where the prey outcompetes the predator and consequently finds a safe channel to communicate secretively.

The third factor that has an impact on how birdsong sounds, besides the production and perception, is the evolutionary history of a species. The capacity to produce and perceive sounds are directly related to morphology, physiology and neurobiology, and species differences in these body characteristics will be reflected in acoustic characteristics. Body characteristics only change gradually over evolutionary time, and therefore close relatives, say two species of doves, will be more similar than a dove and a parrot. Likewise, a new sparrow species will arise with a more or less sparrow-type beak and sparrow-type syrinx, and therefore will also have a song typical for sparrow-like birds. Consequently, closely related species usually sound alike.

Singing Conditions in the Field

In addition to the intrinsic factors mentioned above determining the limits to acoustic variation, signaling conditions in the bird's environment will influence which songs work best. Sound that leaves the beak at one location will not remain unchanged until it is received by another bird. Sound transmission through the environment will lead to several changes such as amplitude attenuation (sound becomes less loud while traveling through the air and vegetation); spectral degradation (high-frequency sounds attenuate faster than low-frequency sounds, and therefore the relative amplitude of different song parts will change with distance); and temporal degradation (echoes will arrive later at a receiver than sound that travels in a straight line, and will fill silent pauses and cause overlap with subsequent notes). Song attenuation and degradation will depend on the amount and type of vegetation between sender and receiver, but also on climatic conditions like wind exposure, temperature and humidity. As different habitats vary with respect to these conditions, the optimal sound characteristics for a song will also vary with the habitat birds are singing in.

Habitat-dependent sound transmission has been put forward as the selection pressure responsible for habitat-typical song characteristics across species. Low-frequency sound transmits better through dense vegetation because the longer wavelengths are less affected by reflection and absorption. In more open habitat, air turbidity leads to irregular amplitude fluctuations which affects longer song elements more than short and repetitive elements. Song comparisons of bird communities in closed forest habitat and more open woodland habitat reveal patterns in line with the optimal acoustic design for each habitat. In the forest, birds tend to use low-pitched songs with relatively long and slow notes. In the open, birds more often sing higher-pitched songs with rapid sequences of syllables and trills. This pattern indicates that different species may change their songs over time in the same direction when exposed to the same selection pressures, and consequently converge toward similar acoustic characteristics determined by the habitat.

Another environmental factor that has an impact on whether information transfer via song works well is ambient noise. A favorable signal-to-noise ratio is crucial for detection and recognition of the acoustic details in song. Sounds from other birds that have songs of similar frequency and syllable shape may lead to confusion, whereas more continuous, loud, and nearby sound sources may mask songs completely. Ambient noise will also vary among habitats because it depends on the presence and abundance of noise sources that can be strongly linked to climatic regions or vegetation types. Low-frequency noise from wind and running water are common in open and riparian habitat, while high-frequency noise from other birds, frogs and insects are typical for forest and marshland habitat, especially in tropical and Mediterranean climates. Noise interference may cause a considerable barrier for acoustic communication, and if noise is not equally distributed throughout the spectrum, the use of some song frequencies may be favorable over others.

If noise spectra are consistent for a certain habitat, birdsong may be shaped by the variable interference levels, because song elements of frequencies that experience low competition by ambient noise are more likely to be effective than those with high levels of competing noise. Therefore, both auditory sensitivity and sound production may mirror the spectral profile of ambient noise, especially when noise levels are high relative to song amplitude. Red-winged blackbirds, for example, sing a song for which the part that travels long distance falls within the range of 2.5 to 4.0 kHz. This frequency band coincides with a relatively quiet noise window in their marshland breeding habitat; just above low-frequency noise caused by wind turbulence and just below high-frequency noise caused by calling insects. Not only natural noise sources will affect bird signaling conditions. In most urban areas, cars, trains, and trucks generate a continuous noise curtain of low-frequency sound. Great tits are originally forest birds, but are now also successful city breeders. One of the reasons for that may be their capacity to adjust their song to local noise conditions. Males in noisy territories limit the use of their frequency range to higher song notes

Echoes in the Forest

Hans Slabbekoorn

The green hylia is a common songbird of dense tropical rainforests in Africa. This small-sized bird uses a simple relatively low-pitched song: two long-drawn notes that remain at the same frequency of about 4.0 kHz. Such an acoustic design is typical for many other unrelated bird species in this habitat, presumably the result of song convergence through shared environmental selection pressures. Besides the advantage of low frequencies that transmit well through vegetation and the long-drawn nature of the notes which would be distorted in open windy habitat, the acoustic design may even benefit from more sound transmission characteristics related to singing in dense forest. Echoes from the many twigs and leaves will fill the one pause in this song and thereby may be obscuring some information; but otherwise echoes will only increase the amplitude and duration of the song, because echoes always remain of the same frequency. This phenomenon can be experienced when whistling at one frequency into an echo tube: Keep on producing sound and you will not hear a discrete echo but only a loud and piercing accumulation of sound that forms a continuous signal. As a consequence of this physical phenomenon of the echoing forest, green hylia sing a longer and louder song than they would have produced in the open with the same amount of energy. The fact that green hylia males respond more strongly to playback of songs with artificially elongated notes suggests that they may actually benefit from the echoes and in a way exploit the transmission conditions of the forest. See the sonogram of a green hylia song in the color photo section.

compared to nearby birds that sing in a relatively quiet territory. Apparently they do not need their full spectral capacity to communicate all necessary detail. Both examples of red-winged blackbirds and great tits suggest that songs can be changed over time to adjust to noisy conditions, thereby optimizing signaling conditions.

Song Changes in Time

Acoustic changes in song can be classified into three different time scales. First, birds may change their song characteristics in response to a certain stimulus or situation. Examples of such plasticity can be found in each of the acoustic domains: frequency, time, and amplitude. European blackbird's song consists of relatively low and slow whistles followed by a series of short, higher notes with strong frequency modulations. A male singing at low intensity, at a central perch in his territory with no competitors around, will sing mostly low whistles and few high-pitched notes. In contrast, when in conflict with a competitor, song is strongly biased toward notes of the high-frequency type. This so-called "strangled song" signals high motivation to fight for the territory. Yellow warblers sing two song types with distinct song endings: an accented version which is longer, addressing females, and an unaccented version addressing males. The presence of male or female yellow warblers nearby may therefore elicit a different type with a different duration. In response to high noise levels, birds raise the volume of their song. Such an amplitude change is called the *Lombard effect* and also occurs in human speech, for example, in a conversation at a noisy party. All three of these examples of immediate changes concern song plasticity within an individual's current repertoire.

Secondly, birds may also change their song over a longer time period of weeks, months, or years. This acoustic change is very common in the process of song development from the start of song practicing at juvenile age to fully crystallized adult song. These changes may occur independent of learning (for example, in doves or in Galliformes like chicken and quail), or they may be largely determined by experience through learning from song tutors. Song learning is a remarkable phenomenon that has evolved independently in three taxonomic groups: the songbirds, parrots and hummingbirds. Parallels with language learning in humans have drawn the attention of many researchers to singing birds and especially to the neuronal processes underlying learning, memorizing and imitating song features. Birds that learn their song are flexible in the acoustic details that will make up their adult song, but they develop abnormal song if they do not hear song of their own species at the right time. For many species, song copies can only be acquired during a sensitive phase early in life, after which they loose the ability. The zebra finch is an example of such a close-ended learner with a sensitive period from 30 to 65 days of age. Other species remain capable of copying song from others, and they refresh their repertoire every spring (e.g., the canary), or they extend their song repertoire throughout life (e.g., the European starling). These acoustic changes accrue over a longer time period and pre- and post-change variants may not occur simultaneously in an individual's repertoire, but they still fall within the life-time plasticity of birds.

The third level of acoustic change concerns shifts over an evolutionary time period. Because sounds do not fossilize, we can only deduce acoustic change through evolutionary time by comparing current song characteristics of different species. Closely related species may sound more alike than distantly related species, but they are different. Presumably the ancestral species of two new bird species sang a song from which both new species evolved their own distinctive songs. One or both may have diverged from that ancestral song, and they may have evolved into different directions. Acoustic changes may also have been in the same direction, but in that case one diverged more than the other which rendered them distinct. Independent of other species or whether or not one species split up into two, the typical song features of one species may fluctuate through evolutionary time. Note that the acoustic change at this third time scale covers different generations instead of a change in the song features of one individual.

Comparative studies not only yield insight into whether species-typical songs must have evolved, but they may also shed light on the role played by the factors mentioned above related to production, perception, and environmental conditions. For example, closely related dove species may still have very similar coos with respect to frequency use because syrinx features and body size may not have changed much and allow only limited variation. On the other hand, the temporal pattern of coo elements may show a strongly diverged rhythm, as gating a different sound–silence pattern may be an easier change. Different closely related reed warbler species do vary considerably in song pitch, even when body size has not diverged appreciably. Here changes in species-typical song features may have been directed by sound transmission properties of their respective habitats and associated singing posts. Species that usually sing within dense reedbeds use lower frequencies than species that prefer more open habitat or song posts above the vegetation. Some may even sing in flight and thereby exploit the favorable sound transmission layer in their habitat which allows communication with higher-pitched songs.

Song Changes in Space

Songs of different males of the same species can vary not only in time but also in space. There may be gradual acoustic changes with geographic distance, or more discrete changes between populations on different islands, on different sides of mountain chains, or even abrupt differences between clusters of birds that are not separated by any physical barrier. Examples are easy to find across bird taxonomy. Zebra doves in Southeast Asia have geographic coo variants that differ in the number of coo elements; songs of willow flycatchers in North America vary in duration and frequency with latitude and elevation; contact calls of Amazon parrots sound different at different roosts; violet-ear hummingbirds have neighborhoods of birds that sing the same song type, and white-crowned sparrows on Alcatraz Island have a distinct terminal syllable not found a few kilometers away on Angel Island, both in the San Francisco Bay.

Most famous examples of geographic variation in song concerns bird species that learn the fine details of their song, such as amazon parrots, violet-ear hummingbirds, and especially songbirds like the white-crowned sparrows. Many songbird species exhibit dialectal variation—a form of geographic variation with discrete steps in the acoustic change, such that songs of a group of males are similar to each other, but distinct from those of a neighboring group. The occurrence of dialects in songbirds is a mere consequence of the fact that males copy song elements from each other. Moreover, variation among species in the learning behavior of individual birds has population-level consequences for the geographic pattern of dialects. For example, different sensitive periods will have a different impact on dialect formation. Species in which males learn very early in life will have their father or one of his neighbors as a tutor. Those that learn later, after they dispersed to the area where they will breed themselves, will have their own territory neighbors as tutors. The early learners will carry acoustic features along when dispersing, which will have a homogenizing effect on geographic variation. The late learners will converge after dispersal to local singers and promote locally distinct acoustic clusters. Other factors that will influence the dialectal patterns are dispersal distance, song repertoire size, and copying accuracy.

The cultural transmission of song through learning from each other may also cause song changes over multiple years for populations and may even cause changes in species-typical features over evolutionary time. The latter means that no evolutionary change in morphology, physiology, or neurobiology may be needed to evolve a new species-typical song. In contrast, variation in time and space in the song of bird species that do not learn, such as zebra doves and willow flycatchers, must reflect variation in at least one of these body characteristics.

Song Divergence and Speciation

The role of song learning in the formation of dialects has been relatively undisputed, in contrast to the question of whether dialects have an impact on the exchange of breeding birds between populations. Different species sound different, and song plays a role in keeping them apart reproductively. Could it be that different populations of the same species that are singing a different dialect are precursors to shortly emerging new species? If males settle more easily among same-dialect birds and females preferentially mate with males of their father's dialect, population differences may increase and reproductive divergence may become more and more complete until breeding exchange approaches zero and one biological species has turned into two.

The evidence gathered so far has not led to unequivocal results, and an obvious complication is that we just live too short a time to observe the speciation process which takes place over an evolutionary time scale. We will have to use contemporary indications to infer the process. In several species, females do prefer song dialects that represent their natal population, but there are also studies that failed to find such a preference. Furthermore, female preferences seem to be flexible: If they have heard multiple dialects or have experience with a foreign dialect, they do not necessarily choose a males singing the dialect of their natal population. Male responses in playback studies show slightly more consistent results. Local dialects usually trigger a stronger response than foreign dialects, and this may indicate that local males have a more effective song. In some species it has been shown that males sharing many song characteristics with neighbors have more success in territory establishment and maintenance than those with acoustically deviating songs.

If the behavioral response by male and female birds in relation to dialectal variation has an impact on the exchange rate of breeding birds between populations, genetic differences should accrue. Several studies on only a few species have explored whether geographic patterns of song variation match with genetic variation. Some studies find such a spatial concordance, for example, for the white-crowned and rufous-collared sparrow. However, many other studies, including the same species, do not. This implies that populations with different dialects do not necessarily differ in their neutral genetic make-up.

The ambivalent evidence for a promoting role for dialects in the process of genetic and reproductive divergence may be partly due to the plastic nature of song learning and dialect formation. First, dialects may occur independent of habitats, which reduces the chance that birds with a different dialect are also better adapted to a different habitat. Females are less likely to be very selective to song dialects if males from different dialect areas are equally good mates. In the little greenbul in Cameroon, populations in different forest types sing a different song, but also have different morphology. In this case, environmental selection probably drives divergence in both body and song characteristics and, as a result, female preferences for males with own-habitat songs may yield an increased reproductive success.

A second consequence of song learning is that individual plasticity may allow males to switch unnoticed between different dialect areas. Especially for species in which males are able to adjust their songs after dispersal, females may not be certain to get a locally adapted male if they select one that sings a local song. On the other hand, song learning may also accelerate song divergence between populations. For example, urban great tits sing within a frequency range that correlates with the ambient noise conditions in their territory. The most likely explanation for this is that they learn selectively, and given that individual great tits are able to adjust to their individual territories, population-wide adjustment to habitat conditions could cause habitat-dependent song divergence. Anyway, it is clear that song divergence has a high potential to play a critical role in speciation, but that role may vary with the species depending on their learning behavior, and depending on fitness consequences of song-based mate choice related to the ecological context. We certainly only scratched the surface of this intriguing area of research.

Conclusions

Song plays an important role in acoustic communication of birds. Effective acoustic signals may benefit singers and audience through a regulatory impact on territorial competition and mate attraction. A message may be encoded in song through correlations between acoustic variation and sender characteristics, which can become meaningful when decoded in the natural context. The benefits for the birds, translated into reproductive success, are the driving force behind evolutionary changes in song directed by variable success rates for acoustic variants. The capacity to produce and hear sounds and the evolutionary history of a species determine the possibilities and constraints to these changes, whereas environmental conditions like sound transmission and interference add a set of shaping factors. As a consequence songs vary in time and space. Species that have the ability to learn are especially variable, and their evolutionary changes depend on genetic and cultural transmission. The role of song in communication, sexual selection, and speciation, together with its esthetic value, will guarantee future studies that may lead to further scientific progress and exciting new insights.

See also Behavioral Phylogeny—*The Evolutionary Origins of Behavior*
     Communication—Vocal—*Variation in Bird Song*

Further Resources

Baptista, L. F., & Kroodsma, D. E. 2001. *Avian bioacoustics, A tribute to Luis Felipe Baptista.* In: *Handbook of Birds of the World. Volume 6* (Ed. by J. del Hoyo & A. Elliott), pp. 11–52. Barcelona, Spain: Lynx Edicions.

Catchpole, C. K. & Slater, P. J. B. 1995. *Bird Song: Biological Themes and Variations.* New York: Cambridge University Press.

Marler, P. & Slabbekoorn, H. 2004. *Nature's Music: The Science of Birdsong.* San Diego, CA: Academic Press.

Slabbekoorn, H. & Smith, T. B. 2002. *Bird song, ecology, and speciation.* Philosophical Transactions of the Royal Society, London, B. 357, 493–503.

Ten Cate, C., Slabbekoorn, H. & Ballintijn, M. R. 2002. *Bird song and male-male competition: Causes and consequences of vocal variability in the collared dove* (Streptopelia decaocto). Advances in the Study of Behavior, 31, 31–75.

Hans Slabbekoorn

■| Communication—Vocal
Social Communication in Dogs: The Subtleties of Silent Language

Domestic dogs and their wild relatives—wolves, coyotes, jackals, and foxes—communicate simple and complex social messages in many different ways. There are theoretical and practical reasons for developing an understanding of how dogs and other canids (members of the dog family) communicate. This essay concerns the hows and whys of social communication in dogs and how this information can help us live in greater harmony with our "best friends."

The pioneering research performed by Michael W. Fox clearly showed that all canids use their bodies to communicate a wide variety of messages, some extremely subtle and some not so subtle. They also use many different vocalizations, gaits, and odors to communicate with other individuals. In the drawing on p. 412, a schema of a dog's body language during social displays, stemming from the research of Michael W. Fox, shows how body postures, ear positions, facial expressions (including changes in the size of the pupils and whether an individual is staring at another individual or avoiding eye contact), and tail positions (upwards of 12 different positions) can vary along a scale when dogs and their relatives want to send a specific message to another individual. Individuals can choose to signal different moods when there are variations in their motivational state. However, some components of a complex social display are involuntary, resulting from the action of the autonomic or involuntary nervous system.

It is obvious that there is a flow between the different positions in this diagram that indicate remarkable subtleties of mood and intention, all of which communicate information that is important to convey to other individuals. When various vocalizations, gaits, and

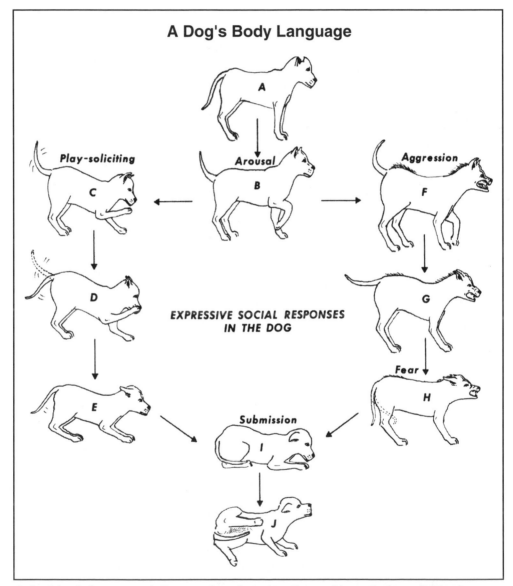

A dog's body language. A–B, neutral to alert attentive positions. C, play-soliciting bow. D–E, active and passive submissive greeting, note tail wag, shift in ear position, and the distribution of weight on fore and hind limbs. I, passive submission with J, rolling over and presentation of inguinal-genital region. F–H, gradual shift from aggressive display to ambivalent fear-defensive-aggressive posture.
Courtesy of Michael W. Fox.

odors are combined with different aspects of body language, the communication patterns used during social interactions by members of the dog family are rather complex. Social displays can also be extremely subtle and often fleeting. Members of the dog family are highly vocal, and there are also subtle but important variations in the numerous sounds that they produce. For example, coyotes are considered to be the most vocal of North American wild mammals, and the scientific name of coyotes, *Canis latrans*, means "barking dog."

Dogs, however, also have a silent language which includes their body language and odors. As social animals, it is important for dogs to be able to communicate subtleties and different degrees of moods, desires, and intentions to other individuals, often very rapidly and on the run. Dogs have evolved a complex system of social communication that does just this. They are excellent communicators, sending clear and unambiguous messages when they have to do so.

In the diagram in position "A," the dog (called Maddy) is alert. When she becomes aroused ("B") she elevates her tail and perks up her ears. Depending on what she notices, she may become aggressive or playful. Moving counterclockwise to position "C," Maddy solicits play from another individual by lowering her body and extending a paw. She might also slide into a "play bow" by lowering her body and crouching on her forelimbs. While bowing Maddy might also elevate and wag her tail and display a "play face" in which her mouth is slightly open as if to prepare to bite a playmate. Postures "C" and "D" are also called "active submission" and are gestures a dog might use not only to solicit play with another dog but also to greet you when you come home. If Maddy feels threatened her "active submission" might change to "passive submission," in which she lowers her body ("E"). She might eventually lie down with her tail tucked and head lowered ("I"). If she wants to signal complete submission, she will roll over ("J"), a position that indicates, "I am truly submissive." Maddy might also pull her lips back horizontally into what is called a submissive "grin," and she might urinate as a sign of complete deference.

Moving clockwise in the diagram, after arousal Maddy might threaten another dog or display aggressive intentions by raising her tail, exposing her teeth by lifting her lips, and standing tall ("F"). If Maddy is highly *aggressive*, her hackles (the fur on her back) might stand up. This is called *piloerection*, an involuntary response due to the activity of the autonomic nervous system that also is responsible for upright feathers in birds and changes in pupil size. During piloerection Maddy's body size is enhanced, especially when she arches her neck. Maddy may then threaten another dog and bare her teeth by retracting her lips vertically, displaying a small mouth or "aggressive pucker," and snarling or growling. She might then walk stiffly as part of her threatening body language. If she attacks, her ears will be folded back to protect them. *Threat* signals can be combined in various intensities depending on how aroused Maddy is. If she loses some confidence, she might lower her body slightly, drop her tail, and pull back her ears ("G"). If she is frightened, she will continue to lower her body, may tuck her tail, and pull her ears way back against the side of her head ("H"). Finally, she may submit ("I" and "J").

Although this diagram indicates the relationship between subtle differences in body posture that communicate different messages, it does not necessarily indicate a smooth continuum. For example, if Maddy were to meet a strange dog in a park, she could progress from either a solicitation of play to submission or, if the other dog is aggressive, she might immediately submit, letting the other dog know that she is not a threat. Likewise, if Maddy's play solicitation is accepted, she will not progress to passive submission, but will likely head off for a romp with her new friend.

Using their body language, dogs communicate very complicated messages—for example, a dog can be both confident and aggressive ("F"), or a dog can be both fearful and aggressive ("H"). When a dog exhibits both fear and aggression, ethologists refer to this as a *conflict situation* because the animal simultaneously has a desire to flee based on fear and to attack based on aggression. For example, fear-biting dogs simultaneously show elements of fear and submission and aggression by raising their hackles (a signal of aggression), tucking their tail between their legs (a signal of fear and submission), and pulling back their ears into the flattened submissive position. People sometimes get bitten by "fear biters" because they think the dog is being submissive when the dog is actually saying, "I'm frightened and if you get any closer, I will bite." People who get bitten by fear biters often make the mistake of concentrating on a dog's face instead of reading his or her body language as a whole.

In order to understand the messages that dogs are trying to communicate, it is essential to read the whole dog and not concentrate on only one element in a complex social signal that is made up of many different parts. For example, when a dogs wags his tail, it is often a friendly signal, but it can also be a sign of agitation: The only way to know for sure is to notice the dog's complete body language, including his ear position and facial expression. To complicate matters, dogs also use a range of odors and vocalizations, beyond our sensory abilities, to communicate with each other. Thus, when dogs communicate with each other, they are aware of a variety of signals including body postures, smells, and vocalizations, which, when combined, constitute a social display.

It is easy to study the social communication of dogs first hand at dog parks, at home, or just about anywhere where there are dogs. As a first step, it is useful to describe the posture, taking note of the position of various parts of the body during different social contexts (greeting, playing, threatening, submitting, or fighting, for example). Keeping track of vocalizations is important, as is noting whether an individual *scent marks* (urinates) while performing a social display. Naming different postures and vocalizations also is useful as you develop an *ethogram*. Taking videos of the animals and then watching them in slow motion or one frame at a time is one method for learning about the subtleties of social communication and appreciating how rapidly messages can be sent even when dogs are running here and there. I have watched some video footage of dogs at play over 100 times, and each time I watch it I see something new. Two of my favorites dogs, Sasha and Woody, were quite playful, and for years I studied the details of how they played with one another. One day, while viewing a video that I had watched at least 50 times, I saw Sasha stumble and hurt her front leg while playing. I was astounded that neither my students nor I had noticed this before.

The observational skills acquired by studying dogs can also be used to study other animals. My early experiences with dogs helped me to understand the communication patterns used by other canids and other carnivores, including hyenas, domestic and wild cats, and bears. If you take the time to watch dogs, you will immediately discover not only how complex their social communication can be, but also how much fun it is to try to figure out what our dog friends are "saying" to one another and to us.

The social dynamics among dogs living in dog *packs* are as complex as the social dynamics that have been observed in wolves and other carnivores. Understanding basic dog behavior and body language is not only scientifically important, but on a practical level, it is very useful for deciding how to interact with and "read" the companion dogs who share your home and the dogs you meet elsewhere. People who deliver mail or make house calls know only too well how important it is to be able to reliably deduce what a dog is likely to do when they meet. Children can benefit from this knowledge because often youngsters meet dogs at eye level and can unintentionally make them uncomfortable and experience conflict because they are staring directly at the dogs' eyes. More than 60% of dog bites occur within family situations, so it is important to understand your own dog's body language. When in doubt take caution. You, your friends, and the dogs will benefit from showing some humility about your knowledge of the complexities of social communication in dogs. I once approached Mishka, a rather large malamute with whom I had lived for 4 years, too rapidly on my hands and knees, and she told me in no uncertain way that I had transgressed into her space in a way that was unacceptable. Usually a mild-mannered dog even when I played vigorously with her, Mishka snarled at me and growled softly telling me to back off. Sufficiently humbled, I did just that and a few minutes later we romped around as if nothing had happened.

Now, all of these cautionary statements do not mean that dogs are unsafe. We all know that most dogs are friendly and predictable companions. And learning more about the ways in which dogs communicate socially will add to the comfort with which we interact with our canine friends. What is so exciting about studying the social communication of dogs and other animals is not only the challenge and frustration of trying to answer the question, "What is it like to be dog (or another animal)?" but the possibility of solving some of the mysteries of animal behavior. Just when we think we know it all—and we do know quite a lot about dog behavior—we learn that there is much more research to be done. Curious minds can discover fascinating bits of information about the hidden and not-so-hidden lives of other animals.

See also Applied Animal Behavior—*Social Dynamics*
        and Aggression in Dogs
        Communication—Vocal—*Communication*
        in Wolves and Dogs
        Emotions—*Laughter in Animals*
        Methods—*Ethograms*
        Play—*Dog Minds and Dog Play*
        Social Organization—*Social Order and Communication*
        in Dogs

Further Resources

Dodman, N. H. 1996. *The Dog Who Loved Too Much: Tales, Treatments, and the Psychology of Dogs.* New York: Bantam Books.
Dodman, N. H. 2002. *If Only They Could Speak: Stories about Pets and their People.* New York: Norton.
Fox, M. W. 1971. *The Behaviour of Wolves, Dogs and Related Canids.* London: Jonathan Cape.
Fox, M. W. 1972. *Understanding your Dog: An Original Study of the Behavior Patterns of Dogs.* New York: Coward, McCann & Geoghegan.

Marc Bekoff

■ Communication—Vocal
Social System and Acoustic Communication in Spectacled Parrotlets

The major concern of the study of animal behavior is the question of how the behavior of an individual maximizes the number of its own genes in the next generations (*genetic fitness*). In social species the group members may have a big influence on the individual's ability to reproduce or to maximize its fitness in other ways. Thus it is important to study the underlying mechanisms of social behavior in order to find out how social systems were established and maintained.

Experiments have shown that spectacled parrotlets discriminate between the contact calls of different social categories and individuals.
Courtesy of Ralf Wanker.

A good model organism for the study of social systems and social relationships is the spectacled parrotlet (*Forpus conspicillatus*). Spectacled parrotlets live in groups in eastern Panama, most parts of Colombia, and western Venezuela, are about 12 cm (5 in) long, and are sexually dimorphic so that males and females can be distinguished very well. The male has a bluish-green plumage with a ring of blue feathers around the eyes, a blue rump, and blue primary and secondary wing coverts. The female shows a more yellow–green plumage lacking all blue markings. They are very common in semi-open cultivated areas in the subtropical regions. Because they are very common birds for European aviculturists, there is detailed information about feeding and breeding available.

There were three main questions of interest in the study of the social system and the acoustic communication of the spectacled parrotlets: How is the social system of spectacled parrotlets organized, are the parrotlets able to recognize individuals or social classes by acoustical cues, and what cues do they use? Individual and social class recognition is generally assumed to be a prerequisite for establishing and maintaining social systems.

From a combination of studies both of spectacled parrotlets in captivity in Hamburg and in the in natural habitat in Colombia, a model of the social organization was drawn. The birds were held in three indoor aviaries, the largest measuring 20 m² (215 ft²), in groups of up to 24 individuals. The group compositions resembled those found in nature, with long-paired adult pairs, recently-paired adults, unpaired mature birds, and unpaired subadults and juveniles. In a series of playback experiments, we tested if spectacled parrotlets use contact calls for vocal recognition.

The results of the studies showed that spectacled parrotlets live in a complex system of social relationships. Soon after fledging, the young establish close sibling relationships which are important for successful socialization, pairing, and reproduction. They grow up in a kindergarten with other juveniles of the same age. Results from playback experiments showed that spectacled parrotlets discriminate between the contact calls of different social categories and individuals. Adult birds preferred to respond to the contact calls of their mates. Subadult individuals recognized the contact calls of their siblings. During the period of pair bond formation, the affiliative contacts to the siblings decrease, but the parrotlets continue to respond to the calls of their sibling, which might facilitate the re-establishment of close sibling relationships in case of mate loss.

Further Resources

Bradbury, J. W. 2003. *Vocal communication in wild parrots*. In: *Animal Social Complexity: Intelligence, Culture and Individualised Societies* (Ed. by F. B. M. de Waal & P. L. Tyack). Cambridge, MA: Harvard University Press.

Hauser, M. D. 1998. *The Evolution of Communication*. Cambridge, MA: The MIT Press.

Wanker, R., Apcin, J., Jennerjahn, B. & Waibel, B. 1998. *Discrimination of different social companions in spectacled parrotlets* (Forpus conspicillatus): *Evidence for individual vocal recognition*. Behavioral Ecology and Sociobiology, 43, 197–202.

Wanker, R. & Fischer, J. 2001. *Intra- and interindividual variation in the contact calls of spectacled parrotlets* (Forpus conspicillatus). Behaviour, 138, 709–726.

Wells, R. S. 2003. *Dolphin social complexity: Lessons from long-term study and life history*. In: *Animal Social Complexity: Intelligence, Culture and Individualised Societies* (Ed. by F. B. M. de Waal & P. L. Tyack). Cambridge, MA: Harvard University Press.

Wilson, E. O. 1975. *Sociobiology*. Cambridge, MA: The Belknap Press of Harvard University Press.

Ralf Wanker

■ Communication—Vocal
Variation in Bird Song

Many people enjoy hearing birds sing with their amazing diversity of songs and singing styles. With just a bit of careful attention you can take enjoyment of bird song to the next level and start to appreciate the diversity of song and singing behavior that is the subject of much current research. Below are examples of types of variation and suggestions for how you might experience and document that variation.

Song Sequencing

Among birds that sing, repertoires range from one to thousands of song types per bird. Species with multiple song types have a variety of patterns of song delivery. In some species individuals sing from several to many renditions of one song type before switching to another—a style called *eventual variety singing*. Singers in other species switch rapidly among song types, rarely repeating a song twice in a row—*immediate variety singing*. These singing behaviors can be heard by listening carefully to some common species. For example, in North America, song sparrows, tufted titmouses, northern cardinals, and red-winged blackbirds are eventual variety singers, whereas red-eyed vireos and many other vireos (but not the white-eyed vireo), wood thrushes and many other thrushes, and gray catbirds sing with immediate variety.

Song Type

A careful listener can make a record of the different songs of some species. The technique championed by Aretas Saunders has proven useful for many people. As you listen to a bird song, record what you hear as a series of marks and squiggles on paper. Your system need make sense only to you. Most people find that they can record more detail, faster and more reliably, with a free-form system than with transliterations or musical notation. Species that sing with eventual variety allow you to hear several renditions of the same song type and thus to refine your representation.

Song Repertoire Size and Interaction with Song Repertoires

With a system of recording song types, such as suggested above, you can attempt to record the entire song repertoire of an individual bird. You can then look at individual variation in song repertoire size, song sharing between neighboring singers, and differences between species in song repertoire size and song sharing. Species that sing with a moderate degree of eventual variety, such as song sparrow, tufted titmouse, and red-winged blackbird, are probably the easiest to study in this way; a reasonable estimate of song repertoire size may take several hours of listening and note taking. The longer you listen to a bird the more confident you can be that you have heard and recorded most of an individual's song types. In some species, singers in certain social situations match song types with neighbors, and you may be able to detect this *matched countersinging* by keeping track of the songs used by two neighbors singing at the same time.

Some species have very large song repertoires and sing with relatively little or no repetition of song types, and it is difficult to keep track of all their different songs. Nonetheless, you can estimate song repertoire size for individuals of such species. Make a descriptive note of a distinctive, easy to recognize song type in the singing of a northern mockingbird; then count the number of transitions between song types that the bird makes before you hear that distinctive song again. You now have a crude estimate of that individual's song repertoire size; by repeating this exercise several times with the same bird and averaging the counts, you can refine the estimate.

The significance of song repertoires and of the variation in repertoire size among individuals of a species and between species remains a subject of active debate and research. Singers of some species use matched countersinging, as described above, in territorial interactions, and several researchers have suggested that females might prefer large repertoire sizes in some species. There is no consensus yet on the function(s) of song repertoires.

Annual and Diel (24-hour) Variation in Singing Behavior

Bird species have varied patterns of singing over the course of a day and a year; careful listening may allow you to note interesting patterns in quantity of song and in singing behaviors used at different times. For those who like to get up early, possible projects include keeping track of the time that different species start singing. Especially in the breeding season, and depending on latitude, some species start singing an hour or more before sunrise; a few species sing throughout the night. Keeping track of the time that song starts and the amount of singing throughout a year can give a feel for the annual cycle of a species. Such patterns in song use can suggest song function. For example, an increase in song around the time of pair formation suggests that song functions in mate attraction, but one must remember that such a correlation in time is not definitive evidence for song function. Changes in territorial defense may occur at the same time as pair formation, and an increase in song could function in either, both, or neither of these phenomena.

As you listen to the singing of a species over the course of a day, you might note not just changes in the amount of singing, but also changes in song types and singing behaviors. Many species sing with greater variety at dawn than later in the day. For example, a yellow warbler's dawn bout, lasting roughly from an hour to a half hour before sunrise in late May and early June, might contain a dozen different song types sung with little repetition, whereas the same bird in full daylight might sing one song type repetitively. With the yellow warbler you might not be able to tell by ear that different song types are used in daytime and dawn singing, but with some species the song types are readily distinguished. Wallace Craig's classic study of the dawn singing of the eastern wood-pewee was done by ear by volunteers who could easily distinguish the three song types of this species and record their sequence of delivery; only two of those song types are commonly used in daytime singing. The function(s) of the dawn bout and of the difference in styles of singing between dawn and day are not understood. It is intriguing that in many cases the songs and singing behaviors used largely at dawn are the ones used in territorial interactions during the day. There are at least a dozen different hypotheses for the function of dawn song and relatively few data for choosing among these hypotheses. Your data, if systematically collected, could help to address this puzzle.

See also Communication—Vocal—*Singing Birds: From*
Communication to Speciation

Further Resources

A field guide to the birds of your area will help you to determine which species occur where you live and to identify them. There are cassettes and CDs available with samples of the songs of bird species.

The Birds of North America series has good literature reviews on the songs of North American birds species. Handbook series, such as the *Handbook of the Birds of the Western Palearctic*, provide briefer reviews for the birds of other continents.

Burt, J. M., Bard, S. C., Campbell, S. E., Beecher, M. D. 2002. *Alternative forms of song matching in song sparrows*. Animal Behaviour, 63(6), 1143–1151.
(*An example of how birds use song repertoires in social interactions.*)
Craig, W. 1943. *The song of the wood pewee*, Myiochanes virens *Linnaeus: A study of bird music.* N. Y. State Museum Bulletin, 334, 1–186.
(*A classic study of singing behavior done by ear.*)
Nice, M. M. 1943. Studies in the life history of the Song Sparrow. II. The behavior of the Song Sparrow and other passerines. Transactions of the Linnaen Society of New York, 6, 1–328.
(*An example of the degree to which one can study song seriously without recording and analysis equipment.*)
Saunders, A. A. 1935. *A Guide to Bird Songs*. New York: Appleton-Century.
(*Out of print, but widely available on the used book market. It presents a useful graphic method for representing bird songs.*)
Staicer, C. A., Spector, D. A. & Horn, A. G. 1996. *The dawn chorus and other diel patterns in acoustic signaling.* In: *Ecology and Evolution of Acoustic Communication in Birds* (Ed. by D. E. Kroodsma & E. H. Miller), pp. 426–453. Ithaca, New York: Cornell University Press.
(*A review of the literature on diel variation in bird singing behavior.*)

David A. Spector

■ Communication—Vocal
Vocalizations of Northern Grasshopper Mice

Mention mouse vocalizations and people typically think of squeaks. But our favorite rodent, the grasshopper mouse, is decidedly nontypical—both in natural history and in vocal ability. The three species of this genus are fierce, albeit diminutive, predators of the western United States that make their living eating scorpions and other invertebrates. And like the top canid predator in the same habitat, they howl.

The call of the grasshopper mouse was first described by naturalists Vernon Bailey and C. C. Sperry in 1929 and has since attracted attention from researchers and wildlife

A northern grasshopper mouse.
Courtesy of Robert Sikes.

documentary crews alike. We became interested in identifying the purpose of the howl but in the course of our work came to realize that this signature vocalization is only the tiniest fraction of the grasshopper mouse repertoire. As it turns out, the auditory world of the grasshopper mouse is an extremely noisy one, with rival males busily proclaiming their ownership of adjoining territories, and familiar companions keeping up an almost nonstop chatter. This world lies just beyond our perception without the aid of high-end audio equipment. However, using our technological advantage we can catch a glimpse of what the mouse's ear hears.

To place the vocal abilities of grasshopper mice into perspective, it is important to have a bit of background on their unique natural history. There are three species of grasshopper mice in the genus *Onychomys*, which is found exclusively in the western United States. All species share similar biology, so descriptions of a single species, the northern grasshopper mouse (*Onychomys leucogaster*), will suffice for all. Grasshopper mice are fierce predators and are highly territorial. Their territories are much larger than those of similar-sized rodents probably owing to their predatory lifestyle, and they actively defend their space against conspecifics (members of the same species). In the wild, grasshopper mice are relatively scarce and widely distributed. Grasshopper mice are stocky rodents, half again as heavy as the ubiquitous house mouse, with short tails and long claws that add dexterity for handling prey. In fact, the name *Onychomys* means "clawed mouse." These inhabitants of the short-grass prairies and arid semidesert are preyed upon by larger carnivores such as foxes, coyotes, and owls. However, studies show that *Onychomys* remains are rather scarce in the nests or stomachs of predators, which may be attributed to the scarcity of grasshopper mice in a given area, the repulsion of predators by the musky odor of grasshopper mice, or the ability of grasshopper mice to avoid detection.

The howl of this species, or more precisely the loud–long call, has a *fundamental frequency* (the lowest natural frequency of a harmonic series) of around 13,800 Hz and lasts for just over a second. This frequency is at the upper end of the audible range for adult humans (humans can detect sounds from 20 to 20,000 Hz, but the range usually decreases as we age) and hence is easily missed or misinterpreted because sounds much like a high-pitched squeaky hinge or wind noise. From recordings we know that the call has considerable power; the call probably would exceed the volume of a lion's roar or a timber wolf's howl if the mouse were of similar size. This call is a territorial claim that allows other grasshopper mice to know that, yes, this is my territory and I stand willing to defend it against all comers.

Both the physical nature of the call and the way it is emitted are consistent with long-distance communication. Sonograms of the loud–long call show an almost pure sine wave. Sounds with this characteristic may stand out in an otherwise chaotic sound environment. Although the simple wave may not have the capacity to encode a great deal of information, the location, sex, identity, and body condition of the calling animal can be advertised to surrounding conspecifics. The loud–long call is emitted with maximal power to extend as far as possible into the environment. The cool desert night air, present at the time when these nocturnal mice are most active, establishes a temperature gradient that allows the sound waves to propagate along a "transmission channel" near the ground that extends the travel distance of the call. Grasshopper mice further maximize distance by climbing on rocks or mounds to vocalize and even then point their noses up to howl rather than projecting their voice into surrounding vegetation and structure. The loud–long call also contains multiple harmonics (integer multiples of the fundamental frequency extending into the ultrasonic region). Because higher frequencies are degraded more than lower frequencies by distance, the higher-frequency harmonics of these calls may provide important distance information to the receiver so that the call can function as a spacing cue as well as individual recognition.

Loud–long calls are most often made by males placed in proximity with unfamiliar males and seldom by males exposed to females. Rather than loud howls, in these latter close-contact situations, especially between males and females, grasshopper mice often emit a call that is similar to the loud–long call in sonograms, but at a markedly lower intensity. Upon closer inspection, we discovered that these soft–long calls were rich in amplitude modulations producing a much more complex waveform than the loud–long call. We also identified a whine call during close-contact encounters between males when a dominant–submissive situation was evident. In these situations, the submissive male emitted a whine along with submissive behaviors supposedly to abate the aggression of the dominant male. Also during close-contact interactions we discovered a third type of call that was the most common form of communication during these situations and lay entirely in the ultrasonic region of the frequency band (and hence is completely inaudible to humans). As with many discoveries, serendipity played a role in our identification of these ultrasonic vocalizations. Our recording equipment was detecting what we thought was broad-band interference well above the frequency of the calls in which we were interested, but as we magnified this section of the sonogram in attempts to isolate the suspected electronic interference, we realized that the frequency sweeps were too irregular to be equipment related and were interspersed perfectly with the soft–long calls. The "interference" we were seeing was almost continuous vocal communications ranging in frequency from 20 to 80 kHz.

The level of complexity in the soft–long calls (amplitude modulations) and the ultrasonic calls (amplitude and frequency modulations) has the capacity to encode much more information and hence be an effective form of communication. Further, both calls are softer than the loud–long call, and the ultrasonic calls are of higher frequency. These characteristics limit transmission of information beyond the individual's immediate vicinity and minimize the probability of attracting predators.

In addition to the calls described above, we have identified another four adult and five pup vocalizations, many of which have not previously been described. Despite this progress, we feel that we are just beginning to document and understand the vocal capability of grasshopper mice. The treasure trove of information we have found with just this single species makes one wonder what other auditory worlds we are missing because we simply haven't bothered to look, or because we expect communication to be in a form with which we are instantly familiar.

See also Communication—Auditory—*Acoustic*
       Communication in Extreme Environments
    Communication—Auditory—*Ultrasound in Small*
      Rodents

Further Resources

Bailey, V., & Sperry, C. C. 1929. *Life history and habits of grasshopper mice, genus* Onychomys. U. S. Department of Agriculture Technical Bulletin, 145, 1–19.

Hafner, M. S., & Hafner, D. J. 1979. *Vocalizations of grasshopper mice (genus* Onychomys). Journal of Mammalogy, 60, 85–94.

McCarty, R. 1978. *Onychomys leucogaster*. Mammalian Species, 87, 1–6.

Riddle, B. R. 1999. *Northern grasshopper mouse /* Onychomys leucogaster. In: *The Smithsonian Book of North American Mammals* (Ed. by D. E. Wilson & S. Ruff), pp. 588–590. Washington and London: Smithsonian Institution Press.

Robert S. Sikes & Tommy G. Finley

■ |Comparative Psychology

Defining Comparative Psychology

Comparative psychologists are interested in the adaptive significance of behavior, the evolution of brain-behavior relationships, the individual development of behavior in a comparative perspective, and the learning and cognitive capacities of animals. An attempt to accomplish these goals requires not only knowledge of psychological theories and techniques, but also an understanding of comparative neuroscience, ecology, developmental biology, evolutionary theory, and social psychology. Thus, comparative psychology is an interdisciplinary field that can provide an integrative perspective that bridges the social and biological sciences. It is hoped that such integration will contribute to a better understanding of the general principles governing behavior, as well as to the development of new ideas that will provide tools for behavioral interventions in all fields of application.

Comparative psychology in relation to other fields of study in the social and biological sciences.
Courtesy of Mauricio R. Papini.

Historical Roots

Interest in animal behavior begins in prehistoric times. This is hardly surprising for a species that has relied on group hunting, domestication of animals, and agriculture for prosperity, as is the case with modern humans. The domestication of dogs and other animals was based on artificial selection for a behavioral character that could be characterized as "tamability." To ancient thinkers we also owe the first intellectual efforts to explain animal behavior in terms of natural processes. Aristotle (fourth century BCE) wrote accurate descriptions of brood parasitism in European cuckoos and of the parental behavior of catfish, both indicating meticulous observations given the peculiar difficulties of recording these two types of behavior. In his treatise on animal behavior, Aristotle concluded that behavior can be explained in terms of two independent causes: a mechanistic cause that refers to physiological processes and a functional cause that refers to the purpose of behavior. Interest in understanding animal behavior can be documented from ancient Greek and Roman thinkers, to the Arab scientists of the Middle Ages, and to the European Renaissance. Modern comparative psychology emerged in the second half of the nineteenth century. The impetus was provided by the confluence of Darwin's theories of evolution by natural selection and mental continuity across species, with the techniques of experimental psychology.

Adaptive Significance of Behavior

Complex characteristics (e.g., color vision) that evolved because they conferred greater reproductive success than alternative versions of the same trait (black/white vision) are called *adaptations*. A demonstration of the adaptive significance of behavior requires evidence that a behavior promotes reproductive success (number of offspring produced by individuals during their life). For example, lions that live in large groups accrue a greater reproductive success than lions that live in small groups or in isolation. As a result, lions tend to live in stable social groups. Some traits, such as the size of the finch's beak, vary in accordance with ecological changes. After a drought season, large seeds are more abundant than small seeds and the surviving finches exhibit larger beaks (which are more effective at opening large seeds). When implemented in the laboratory, this is called artificial selection. Suppose fruit flies displaying *positive geotaxis* (tendency to walk toward the ground) are paired with each other, and those displaying *negative geotaxis* (tendency to walk upward) are also inbred. After approximately 600 generations of such selection, mating occurs preferentially among flies with the same geotaxis tendency. Reproductive isolation among animals whose ancestors were capable of interbreeding is the hallmark of the evolution of new species.

In many species, males are characterized by a large body size, colorful body features, and more intense aggressive behavior than females. Such sexual dimorphism demonstrates that different selective pressures operate on males and females. Competition among males and the preference of females for certain traits drive the evolution of sexual dimorphism. For example, female peacocks prefer to mate with males displaying exuberant tails because they tend to be older. The female's choice ensures that her offspring will inherit genes that promote a long life and thus would tend to maximize her reproductive success. Similarly, men display sexual attraction toward women whose waist is about 70% the size of their hips. Such proportion indicates a relatively high fitness (low incidence of various diseases and high fertility).

Reproductive success also depends on the animal's ability to prepare for incoming events. In territorial species, males defend a territory from other

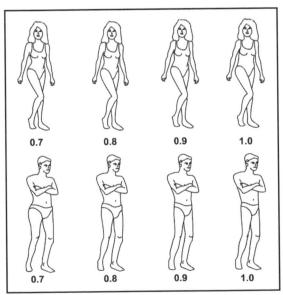

Waist-to-hip ratio and attraction. Women are rated as increasingly more attractive as the ratio approaches a value of 0.7. Interestingly, a ratio closer to 0.7 feminizes the figure of men.
Courtesy of the American Psychological Association.

males and use it for courtship and mating with females. Experiments in which an incoming animal is signaled by a stimulus (e.g., a light) demonstrate that resident males profit by engaging in courtship earlier, producing a larger volume of semen, and siring a greater number of offspring than males lacking a signal. Studies demonstrating that signal learning (called *conditioning*) promotes reproductive success provide compelling evidence for the adaptive significance of learning.

Brain and Behavior in Evolutionary Perspective

As the organ responsible for producing behavior, the study of brain evolution is of paramount importance for comparative psychologists. Vertebrates (fishes, amphibians, reptiles, birds, and mammals) are characterized by possessing a dorsal central nervous system (CNS) with subdivisions that can be easily recognized in all species. The main sections of the CNS (spinal cord, rhombencephalon, mesencephalon, diencephalon, and telencephalon) are recognizable in all vertebrates, as are the main neurotransmitter systems (cholinergic, dopaminergic, and serotoninergic systems). The telencephalon (brain hemispheres) is the part that exhibits the greatest species diversity in organization. Relatively complex behaviors, including courtship, aggression, parental care, learning and memory, and complex perceptual and motor capacities are related to telencephalic activity. There is evidence that some telencephalic functions are common to many species. For example, the hippocampus is known to contribute to spatial learning in mammals, birds, and fish.

When a brain function is particularly important for a species, then the CNS nucleus that deals with that function is enlarged. This is called the principle of proper mass. Catfish, for example, exhibit enlarged vagal nuclei (processing gustatory and facial information) associated

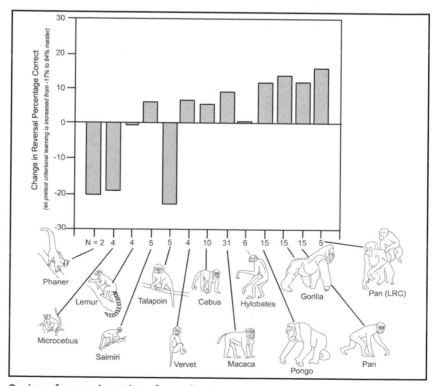

Brains of several species of vertebrates.
From Curtis, H. & Barnes, N. S. 1989. Biology, 5th edn. New York: Worth; figure 43-3, p. 883.

with feeding habits consisting of filtering food particles from the substratum. Birds that rely on spatial learning to locate food stored days or weeks earlier have a relatively large hippocampus.

Studies of brain size in vertebrates demonstrate two major points. First, the overall size of the brain (and other organs) is a direct function of body size. Larger animals also have larger brains (and larger hearts, stomachs, etc.). This suggests that the evolution of a large brain is passively driven by selection for large body size. Second, some species have larger brains than others, even when body size is constant. Generally speaking, the brain of a reptile is about 10 times smaller than that of a bird or mammal of equal body size. This suggests that natural selection may, in some cases, promote *encephalization*, that is, a trend toward the evolution of increased brain size. Encephalization has occurred in several animal lineages, including hominids, but comparative psychologists have yet to fully understand its behavioral significance. Research with mammals suggests that relative brain size may correlate with the ability to learn new problems and to benefit from extensive training. Brain size appears to be connected to behavioral flexibility.

Comparative Developmental Psychology

A simple-minded approach to behavioral development would posit that a behavior is either innate or acquired, giving rise to nativist and environmentalist views, respectively. Classic ethologists and radical behaviorists championed these views, and they reached maximum tension during the 1950–60s, when the debate became known as the *nature–nurture controversy*. Upon reflection, however, it makes just as little sense to argue that genes determine behavior (because genes can only produce proteins), as it does to imagine that experience can change the CNS without any genetic involvement (because gene transcription is needed for many such changes). Today, most comparative psychologists adhere to the *epigenetic view* according to which development is the result of complex interactions across levels of organization from genes, to cells, networks, and organism–environment interactions.

Epigenetic interactions are illustrated by studies of the development of primate social behavior. Mothers and infants can form attachments characterized by individual recognition, preference for the attached object, and separation distress. After separation, infant monkeys exhibit increased levels of stress hormones, impaired immune function, and vulnerability to infections. Tactile stimulation plays a fundamental role in the development of such attachments. Infants form strong attachments to mother models covered with cloth, even if they do not provide milk. Such attachments regulate the infant's emotional response to novelty, its ability to interact with peers, and its adult social behavior. For example, rhesus monkeys raised in isolation display abnormal and ineffective copulatory behavior as adults. Socially deprived females that become mothers provide deficient maternal care for their infants, often being aggressive toward them. Whereas primates are genetically predisposed to develop social attachments, normal adult behavior requires a specific kind of early social experience.

Comparative Learning and Cognition

The study of learning and cognition is at the core of comparative psychology. Associative learning (also called conditioning) refers to the ability to link events that occur contiguously either in time or in space. Examples of associative learning from daily experience include predicting rain when we see cloudy skies, salivating when we think about pizza, expecting a good grade after studying hard for an exam, or becoming restless right before an exciting sports game. In these cases, a prediction is made because previous experience has resulted in an association between two events, A → B, such that the presence of A evokes B from memory.

Conditioning is a mechanism that detects event sequences. Because all environments tend to be structured in terms of sequences of events, it is not surprising that conditioning phenomena are common among animals.

There are about 35 animal phyla, and conditioning has been studied in only a handful of them. There is no evidence of conditioning in sponges (which have no nervous system) and in jellyfish and other cnidarians (which possess a diffuse neural net), and only controversial evidence in planarians (simplest bilateral animals). The presence of conditioning in species that have simple nervous systems (the nematode *C. elegans* has only 302 neurons, compared to the billions of neurons in the human brain), suggests that conditioning does not require a complex neural architecture, as it was previously thought. The learning skills of insects in general, and honeybees in particular, are astonishingly similar to those of mammals.

Can nonhuman animals learn concepts, acquire abstract ideas, perform mathematical operations, and understand language? The word *cognition* is used to refer to such skills that are generally thought to be based on mechanisms other than conditioning. Experiments show that rats, pigeons, dolphins, moneys, and apes, among other animals, can be trained to use abstract concepts (causality), to estimate quantities, to count, and to understand complex linguistic commands. Whether such cognitive capacities are based on the same processes as those operating in humans remains to be determined. For the moment, however, comparative psychology has expanded our understanding of behavior and enriched our appreciation of nature.

Further Resources

Boakes, R. 1984. *From Darwin to Behaviorism. Psychology and the Minds of Animals.* Cambridge, UK: Cambridge University Press

Domjan, M. 1998. *The Principles of Learning and Behavior.* 4th edn. Pacific Grove, CA: Brooks/Cole.

Gottlieb, G. 1992. *Individual Development and Evolution: The Genesis of Novel Behavior.* New York: Oxford University Press.

Greenberg, G., & Haraway, M. M. (Eds.) 1998. *Comparative Psychology: A Handbook.* New York: Garland Press.

Harlow, H. F. 1971. *Learning to Love.* San Francisco, CA: Albion Press.

Macphail, E. M. 1982. *Brain and Intelligence in Vertebrates.* Oxford: Oxford University Press.

Papini, M. R. 2002. *Comparative Psychology. Evolution and Development of Behavior.* Upper Saddle River, N.J.: Prentice Hall.

Shettleworth, S. J. 1998. *Cognition, Evolution, and Behavior.* Oxford: Oxford University Press.

Mauricio R. Papini

■ Conservation and Behavior
Species Reintroduction

Conservationists increasingly use behavioral studies in programs to conserve and recover threatened and endangered species. Species declines and extinction rates are increasing dramatically, mainly due to human activities. Humans are polluting, destroying, transforming, and fragmenting natural environments at unprecedented rates. They are also exploiting species until many no longer exist as viable populations. These activities inevitably affect the behavior and ecology of species, including interactions between species. For example, hunting and commercial exploitation of species (e.g., killing of monkeys and apes for the bushmeat trade in Africa and international fishing and whaling) may negatively affect social behavior and reproduction as the social structure of a species is completely disrupted.

Human impacts on natural environments affect how animals perceive their environment and how they communicate with each other. For example, sonic investigations of the sea floor (*sound pollution*) appear to interfere with whale movements, social behavior, and reproduction since whales communicate by sound across hundreds of miles of ocean. City lights (*light pollution*) disrupt turtle reproduction on centuries-old breeding beaches because turtles use changes in moonlight cycles to orient their ascent to the beaches, dig nest holes, and deposit eggs. Chemical pollution may affect many mammal species that communicate by scent marking as human deposited odors overwhelm their *pheromones* (chemicals used for communication).

People also introduce non-native species that compete with or prey upon native species. For example, introductions of dingos, cats, red foxes, and other predators decimated Australia's native marsupial fauna, which lack appropriate antipredator behaviors. Similarly, on the Galapagos Islands, many of the bird and reptile species that contributed to the evolution of Darwin's theory of evolution suffer from interactions with introduced goats and cats to which they are not behaviorally adapted.

Because of the inter-relatedness and mutual dependence of species in the food chain, an impact on one species may affect an entire food chain. Sea otters, an endangered species from Alaska's Aleutian Islands, were slowly recovering from low numbers after a ban on hunting several decades ago. Recently, however, sea otter numbers again began declining; they are being taken by killer whales, or orcas, whose original diet of great whales and pinnipeds was reduced dramatically by hunting. The orcas adopted new foraging strategies to cope with the loss of their original prey, thus creating additional problems for sea otter conservation.

Species require a large number of behaviors to survive and reproduce (see the following figure). Conservation programs should strive to understand, as completely as possible, the behavior and ecology of threatened and endangered species before developing a set of objectives for a preservation program. For example, if selective logging removes the major tree species that a bird (e.g., red-cockaded woodpecker) or mammal (e.g., golden lion tamarin monkey) uses for nesting, providing sufficient nest boxes spaced at appropriate distances may be all that is required to recover that species.

Giant pandas feed on bamboo almost exclusively and formerly migrated seasonally up and down mountainsides in southwestern China to feed on the choicest new bamboo shoots. But human activities now restrict movements, and many giant pandas have died of starvation following massive die-offs of bamboo that occur naturally every 50–100 years. Without the ability to move to different areas, giant pandas cannot survive unless they have at least two species of bamboo within their habitat.

Understanding animal behavior is often crucial to successful conservation. This point is illustrated in the following discussion on the relevance of behavior to captive breeding and reintroduction, which also stresses that behavioral considerations are important to many other conservation strategies as well.

Captive Breeding Programs

Increasingly, zoos and other captive breeding facilities maintain endangered species with hope of eventually reintroducing them to the wild. Understanding an animal's behavioral adaptations is key to successful management in captivity. Recreating the appropriate social organization is one key to success. Golden lion tamarin numbers were declining in captivity until research determined that zoos should maintain males and

Behavioral Variables to Consider When Considering Species Conservation Needs or When Planning a Reintroduction.

Habitat Selection
  Foraging (see below)
  Water & Minerals
  Reproduction (see below)
  Refugia (i.e. Escape Predation)
  Hibernation/Aestivation

Reproduction
  Mating
    Finding Mates
    Selection of Courting/Breeding Site
    Courtship
    Mate Selection
    Acquiring Mates
    Copulation
  Producing & Raising Young
    Nest/Den Site/Birthing Ground Selection
    Imprinting
    Nesting/Denning Behavior
    Parental and Helper Care
      Nursing
      Food Provisioning
      Predator Avoidance
      Predator Defense

Other Intraspecific Behavior
  Social Structure
    Group/Solitary/Colonial
    Dominance/Submission
    Grooming
    Resource Sharing

Cooperative Group Defense
Home Range Use
Territoriality
Dispersal
Communication
Competition
  Resource Defense
  Territoriality

Interspecific Behavior
  Predators
    Identification
    Avoidance
    Mutual Predator Defense
  Predation (see Foraging)
  Competition
    Resource Defense
  Methods of Ecological Separation

Foraging
  Locating Food
  Obtaining Food
  Food Processing
  Consuming Food
  Caching Food

Locomotion
  Circadian, Lunar, Annual Activity Cycles
    Movement Type (swimming,
      brachiating, etc.)
  Daily Movement Patterns
  Seasonal Migrations

Adapted and expanded from Reading, R., Miller, B. & Stanley-Price, M. 1998. Reintroducing animals into the wild: Lessons for giant pandas. In: Proceedings of Giant Panda Reintroduction Workshop *(Ed. by S. Mainka, Beijing), Ministry of Forestry China Protecting Giant Panda Project Office and WWF–China Programme.*

females as mated pairs with their offspring, or adults would not reproduce successfully. Many zoos incorrectly kept these small monkeys in large social groups typical of many other small primates, like rhesus monkeys. Additionally, subadult tamarins learn how to provide parental care while living in their family group. Removing subadults too early often resulted in poor parenting behavior, including the abuse of young.

In some species, like cheetahs, females must have the opportunity to choose their mate, thus they require exposure to multiple males when in heat. For other species, population density is important for breeding. For example, flamingoes in zoos must be kept in groups of more than 20 animals for reproduction to occur.

Tree shrew mothers only nurse their young once every 48 hours and do not sleep or interact with them at other times. Tree shrews therefore require multiple nesting areas in their enclosures to raise young successfully. In many ungulate species, like deer, keepers separate young and mothers during the day except during nursing times, and enclosures must include hiding places for the young since in nature mothers and young remain separated most of the time, and the young hide in vegetation to avoid predators.

Reintroduction

Reintroductions using captive-bred animals or wild-born individuals (often termed *translocations*) are complex, risky, and multidimensional. Thus, the World Conservation Union's (IUCN) Reintroduction Specialist Group, the American Zoo and Aquarium Association's Reintroduction Advisory Group, and other conservation organizations developed guidelines and recommendations to improve chances for success. Most reintroduction specialists recommend developing holistic, interdisciplinary programs to address the wide variety of factors influencing reintroduction success, and behavioral considerations are paramount.

The IUCN defines a *reintroduction* as "an attempt to establish a species in an area which was once part of its historical range, but from which it has been extirpated or become extinct." Other related terms are *translocation*, the "deliberate and mediated movement of wild individuals to an existing population of conspecifics" (individuals of the same species); *reinforcement* or *supplementation*, the "addition of individuals to an existing population of conspecifics"; and *conservation* or *benign introduction*, "an attempt to establish a species, for the purpose of conservation, outside its recorded distribution but within an appropriate habitat and eco-geographical area." For simplicity, in this essay, the term *reintroduction* will be used to refer to all of these actions.

In general, reintroductions should never be conducted without first performing a detailed pre-release assessment. The most important questions to ask are: "What was the original cause of the species' decline?" and "Has that cause been successful addressed (or can it be)?" If not, it is irresponsible to conduct a reintroduction that may well cause more harm than good (for example, by disrupting the social structure of a local population). If a species still exists in an area, it should be allowed to expand naturally if at all possible, and reintroduction should only rarely be used, for instance, to increase genetic diversity. To be truly successful, the reintroduced animals must not only survive, but also reproduce and raise offspring that reproduce. This requires careful attention to several variables, many of which are behavioral.

The Importance of Behavior to Reintroduction Success

A large and complex array of biological factors influences the success of reintroductions, including genetic, demographic, behavioral, physiological, and habitat/release site considerations. Although all of these variables are important to reintroduction success, people often underappreciate behavioral considerations. Reintroduced animals must be able to perform a wide range of behaviors efficiently, including behaviors associated with

- finding, obtaining, consuming, and caching food;
- identifying and avoiding predators;
- finding, courting, acquiring, and copulating with mates;
- producing, rearing, and protecting young;
- communicating and interacting with conspecifics, such as defending resources and territories;

Preble's Meadow Jumping Mouse and Culverts

Carron A. Meaney

Preble's meadow jumping mouse is an unusual mouse. It has kangaroo hind legs, swims, hops, jumps, can go a mile in short order, and hibernates for seven months each winter. A threatened subspecies of the meadow jumping mouse, it occupies *riparian systems* (the banks of creeks and rivers) only in Colorado and Wyoming. Populations of this mouse have been greatly reduced in the past 20 to 50 years due to development, gravel mining, overgrazing, and other human activities. There are efforts currently underway to protect the mouse and its habitat through the Endangered Species Act. One activity that comes with development is road building and widening. When culverts are installed along a stream occupied by this mouse, for the creation or widening of a road or highway, there is the possibility that a single population will become fragmented into two small populations. Small populations are more vulnerable to extirpation than larger ones because there are fewer individuals to breed and/or to withstand a calamity such as a flood or other natural disaster.

With the use of motion-detecting infrared cameras set up inside culverts, scientists hope to see how the Preble's meadow jumping mouse will react to these culverts in various circumstances. Courtesy of Carron Meaney.

Some small mammals, such as deer mice, will use ledges on the side of a culvert, but others, such as voles, don't. The Colorado Department of Transportation has set aside funding to study this question in relation to Preble's meadow jumping mice: Will these mice use ledges in culverts to pass under roadways, and/or will more animals pass through a culvert with a ledge installed compared to no ledge? This question will be addressed with the use of motion-detecting infrared cameras set up inside the culverts. In this manner, we can determine whether the mice, and other small mammals, use the ledges at all. A more detailed study will be conducted to determine if the ledges cause an increase in use of the culverts. Because we have found that the number of mice passing through a culvert without a ledge appears to be related to the population density of the mice that particular year, we will incorporate animal density as a factor in comparing the usage rates.

This particular project is concerned with a rare mouse, but many species of wildlife are impacted by roads. Many animals are killed directly by cars, and others are hampered in their ability to move through the landscape in order to find food, mates, and resting places. If we can develop applied solutions that work, we can do a lot to help wild animals maintain their population numbers sufficiently to avoid extinction in the face of ongoing development. The International Conference on Wildlife Ecology and Transportation convenes every couple of years to share information and research on wildlife and transportation issues. North Carolina State University has a Center for Transportation and the Environment, and the Federal Highways Administration and many state transportation departments are starting to help fund this type of research.

Further Resources

Meaney, C. A., Ruggles, A. K., Clippinger, N. W., & Lubow, B. C. 2003. *The impact of recreational trails and grazing on small mammals in the Colorado piedmont.* The Prairie Naturalist, 34 (3/4), 115–136.

http://itre.ncsu.edu/cte/wildlife.htm—A searchable database hosted by the Center for Transportation and the Environment

http://itre.ncsu.edu/cte/ICOET/ICOET2001.html—Proceedings of the 2001 International Conference on Ecology and Transportation

- interacting with conspecifics within and outside their group;
- moving efficiently in complex three-dimensional environments in trees, on land, underground, in the air or in aquatic environments;
- orienting themselves during daily movements throughout their home range;
- migrating seasonally; and
- choosing the appropriate habitat for foraging, escaping predation, hibernating, estivating, resting, nesting, and denning.

In addition, these behaviors often vary with the gender and age of an animal and the context. Young animals often display greater behavioral plasticity than adults and so may respond better to a new habitat.

Foraging requires being able to find appropriate food, capture or otherwise acquire that food, and handle and consume it. For example, captive-bred black-footed ferrets must know how to capture and kill prairie dogs. Golden lion tamarins must learn how to search for insects under the bark of trees or in bromeliads and how to differentiate frogs from toads (which may be toxic). Other species must learn to extract nuts from shells or meat from a porcupine. For some species, caching or storing food for later consumption is essential; for example, squirrels bury and must know where to retrieve nuts. Leopards often drag carcasses into trees. Many species such as elephants migrate seasonally to find food, and older animals remember the proper route to take—something hard to re-create in a captive environment. In other cases, different preferred foods, for example, a fruit tree, may only occur seasonally within the home range, and members of the group must remember its location.

Foraging skills vary widely among species. Some need little training or experience. For example, most young shorebirds are able to begin foraging within hours of hatching and apparently know what represents food and how to acquire it with little or no training; so do most *herbivores* (plant eaters) like antelopes. Other species, especially *omnivores* (plant and meat eaters) and predators, often require substantial learning and skill development before they can forage adequately. Indeed, many large carnivores remain with adults (often just their mothers) for years while they develop their hunting skills. Therefore, reintroducing antelopes and tigers requires very different approaches.

Reintroduced animals often also represent prey for other species. As such, they must be able to identify potential predators and avoid them. To reduce wasting energy, animals must learn to avoid responding to animals that pose no threat, either because they are not predators, they are too small, or they are not actively hunting. In some monkey species, young respond to a wide range of potential "predators," but learn to fine-tune their responses by observing older group members. Proper avoidance skills include flight (e.g., many antelopes), freezing (e.g., many frogs), or seeking refuge (e.g., prairie dogs into burrows) or escape terrain, like steep cliffs (e.g., mountain sheep). Different predators require different responses. For example, golden lion tamarins produce alarm calls and drop down from the tree canopy when raptors fly overhead, but produce a mobbing call and mob any snake that crosses their path. Conservationists often attempt to teach captive-born young to respond appropriately to predators before they are released. However, such training is labor intensive and may not confer a long-term survival advantage after release. The advantage in pre-release conditioning for predators may come just after release. At that time, a newly released animal spends a great deal of time exploring a new environment, and the less naïve it is, the better chance it has of establishing a home range.

Humans often represent the most important species for a reintroduced animal to avoid. Teaching reintroduced animals to avoid humans is especially difficult for captive-bred animals that often become habituated to people. For example, in reintroductions of large carnivores, animals raised in captivity experience higher human-caused mortality than do releases of wild-caught animals. Captive-raised Florida panthers wandered through farms more frequently than wild-caught and translocated pumas. Releasing some species of captive-raised large carnivores can put both the animal and nearby humans at risk. Keepers raise captive-bred condors with condor "puppets" to avoid habituating them to people. Keiko, the killer whale, prefers to remain in human company and has been unable to join a wild group.

Survival alone will not ensure reintroduction success. Animals must also reproduce and rear their young. This includes finding potential mates, courting and gaining access to that mate, knowing how to copulate, and successfully producing and raising young. Often simply finding a mate is difficult, especially for solitary animals that live in low densities. And low densities are common in reintroduction programs, especially in the early stages when only a few animals have been released. Finding a mate may depend on the ability to identify chemical signals in the urine of receptive females.

U.S. Fish and Wildlife staff with two red wolf pups bred in captivity. The red wolf is an endangered species that is currently found in the wild only as experimental populations in Tennessee and North Carolina.
Courtesy of U.S. Fish and Wildlife.

But simply locating a potential mate is not enough. In some species, males (or occasionally females) must perform elaborate courtship and mating rituals, and those that perform with greater skill usually gain access to more mates. The performance of these behaviors often improves over time as an animal gains experience and exposure to competitors. Interestingly, some species show little interest in mating if no competitors are around; that this may negatively affect breeding success in captivity was shown with cheetahs. In giant pandas, a relatively solitary species, early social experience helps with interactions during mating.

After becoming impregnated, a female mammal must successfully carry her fetus to term and give birth. Birds usually must build a nest and successfully brood their clutch. Other egg-laying species, like most reptiles, amphibians, fishes, and invertebrates, may also use nests, but often must simply find a suitable place to deposit their eggs. Finally, many species, especially most mammals and birds, must successfully rear their offspring. This includes varying amounts of parental care, such as providing food, carrying young, and protecting them from predators, all of which may require training. Especially for zoo-born animals like golden lion tamarins, lack of experience in helping raise younger siblings may lead to inappropriate parental behavior as an adult.

Many important behaviors are associated with interactions between members of the same species (i.e., intraspecific interactions). This includes vocal and nonvocal sound communication, chemical communication (e.g., depositing scent, urine, and feces), facial expressions, body postures, and physical displays. Wolves, for example, communicate within and between packs using a complex array of vocalizations (e.g., howls), chemical signals (often mixed with urine or feces), facial expressions (e.g., baring the teeth and pinning back the ears), and tail and body postures (e.g., raising the hair on the nape of the neck). We are just beginning to understand many of these communication behaviors. Signals may help define territories or defend potential mates, food resources, or other resources, like nest sites. For example, golden lion tamarins have a "long-call" to communicate with neighboring groups or when a group member has wandered out of sight. Captive-bred and reintroduced tamarins respond differently than wild-born tamarins to experimental playbacks of the long-call until they gain experience in the wild.

Social species (i.e., those that live in groups) tend to require more attention to intraspecific behaviors than do more solitary species. Thus, to improve chances for success, reintroductions involving captive-bred animals often first establish social groups comparable to what occurs in the wild (such as wolf packs) and then release the animals together to allow animals to develop proper interaction skills. But, even solitary species like giant pandas have some social interactions, and proper management in captivity and reintroduction require providing social experience. The timing of a reintroduction may affect success. For swift foxes, researchers found that releasing litters in the autumn, about the time young normally disperse, resulted in greater survival than releasing similar groups the following spring during the mating season.

Reintroduced animals must be able to choose appropriate habitat for foraging, escaping predation, denning or nesting, resting, hibernating or *estivating* (short-term hibernation), and more. For some species, this requires seasonal migrations, often over long distances. For example, conservationists teach reintroduced whooping cranes migration routes between Canada and the United States by actually leading animals in flight using ultralight planes. Other species must learn shorter movement patterns, often on a daily basis and covering only a few miles—experiences not typically available in captivity.

Females of many species cue in on the best available habitat, whereas males often cue in on areas that support resident females. Thus, it may make sense to try to establish females prior to releasing males on a site. For example, translocated male pumas moved over large areas until they located females. Results from research on these projects suggest that puma releases should start with dispersal-aged females (12 to 27 months of age), followed by young males.

Species raised in captivity may require training in locomotion skills that they cannot acquire in small, captive environments, such as moving through three-dimensional arboreal habitats or flying or swimming long distances. Imagine being raised in a house and never going outdoors until you reach 20 years of age. How would you find the mall and then return home? Newly reintroduced golden lion tamarins often get lost as they wander through their complex tropical forest habitat in Brazil. In addition, site fidelity and homing behavior often affect success, because animals often try to return from where they originated, especially for animals translocated from wild populations. Habituating animals to a release site, by holding them in large pens for example, appears to help reduce homing behavior for many species. Habituation also permits animals to acclimate to a site and to hone other behaviors, such as locomotion, social skills, and foraging.

Captive vs. Wild Source Animals

Captive breeding and reintroduction is an important conservation strategy for some species, especially those extirpated from the wild such as California condors, Guam rails, and black-footed ferrets. In addition, captive animals can be used for education and research even if individuals are never reintroduced into the wild. Reintroductions of wild-born animals are preferable to those using captive-born animals because the latter may lose behavioral skills after generations in captivity. And the erosion of behavioral skills usually results in lower survivorship after release. In addition, captive breeding is usually expensive in time, space, and money. Captive-bred animals should be released only if there are no other alternatives.

Using animals in a reintroduction program after generations in captivity raises a number of important behavioral issues. The captive environment imposes artificial selection pressures (e.g., domestication) on animals, eroding the genetic basis for important morphological, physiological, and behavioral traits. Learned behavioral traits can be lost much more rapidly than genetically-based ones, and different species respond variably to captive conditions; but more generations in captivity will likely increase the loss of survival skills. So, while captive animals may still exhibit the correct behaviors in given situations, they may not perform them efficiently enough for survival in the wild. As a result, reintroduction programs that use wild animals tend to lower mortality rates and succeed better than those that use animals from captivity.

Using captive-reared animals in reintroduction programs often requires substantial pre-release and post-release training of animals, acclimatization periods, and other intensive assistance to increase chances for survival. Such protocols are referred to as "soft releases," as opposed to "hard releases" in which animals are simply turned loose without training, acclimatization, or post-release support. Pre-release preparation and post-release training often help restore survival traits, but usually not to full efficiency. This is especially true for behaviors that are environmentally cued or culturally transferred, such as caring for and rearing young, following migration routes, hunting, or finding food resources. Effective development of adaptive behaviors requires the correct environment for learning (often including a skilled parent) or, in the case of critical periods, the correct stimulus at the proper time during development. These contexts are difficult to replicate in captivity. Furthermore, selection for tameness and other adaptations to the captive environment will likely become increasingly serious as the number of generations in captivity increases. Beach mice exhibit much greater variability in response to a model of predator after multiple generations in captivity. Presumably, these animals would not survive if released into a natural environment.

"Soft" releases are generally considered more effective in reintroduction programs involving captive-bred animals, but are often precluded due to the far greater costs involved. Adding animals from captivity into an existing wild population also raises important disease, genetic, and demographic considerations. For example, animals in captivity are often exposed to far more diseases than wild animals because of their proximity to individuals of the same and other related species that might be carrying pathogens.

Research & Evaluation

Research and evaluation are crucial to a well-performing reintroduction program and should include behavioral, as well as survivorship and reproductive, goals. Without active research, it is difficult to know which methods are most efficient and effective. This requires maintaining accurate and detailed records. Research and monitoring should include evaluating how well goals were achieved and knowledge gained. In the early stages of a reintroduction, learning may be the most important goal, as it will ultimately lead to more rapid success. Survivorship and reproduction rates are ultimately the best indicators of reintroduction success; however, during these early stages, there may be insufficient data on these variables, and mortality can be high and quite variable. As such, evaluating behavioral traits can provide valuable information to guide and improve the program in its early stages. The best behavioral traits to monitor are those that cost the least to study, can be reliably recorded, and reflect how well the reintroduction is progressing. For example, radio telemetry often allows researchers to examine how far reintroduced animals move, what times of the day they are active, and what habitats they use. These biologists can then compare these data with similar information from studies of natural populations to assess how "natural" the reintroduced animals are acting. Better reintroduction programs test using two different release strategies (e.g., soft vs. hard releases), and examine how well animals released under each strategy perform relative to animals in natural populations.

Further Resources

Caro, T. (Ed.). 1998. *Behavioral Ecology and Conservation Biology.* New York: Oxford University Press.

Clemmons, J. R. & Buchholz, R. (Eds.). 1997. *Behavioral Approaches to Conservation in the Wild.* New York: Cambridge University Press.

Gosling, L. M. & Sutherland, W. J. (Eds.). 2000. *Behaviour and Conservation.* New York: Cambridge University Press.

IUCN. 1998. *IUCN Guidelines for Re-introductions.* Gland, Switzerland: IUCN.

Kleiman, D. 1989. *Reintroduction of captive mammals for conservation: Guidelines for reintroducing endangered species into the wild*, Bioscience, 39, 152–161.

Miller, B., Ralls, K., Reading, R., Scott, J. M., & Estes, J. 1999. *Biological and technical considerations of carnivore translocation: A review.* Animal Conservation, 2(1), 59–68.

Miller, B., Reading, R., & Forrest, S. 1996. *Prairie Night: Black-Footed Ferrets and the Recovery of Endangered Species.* Washington, D.C.: Smithsonian Institution Press.

Olney, P., Mace, G., & Feistner, A. (Eds.). 1993. *Creative Conservation: Interactive Management of Wild and Captive Animals.* London: Chapman and Hall.

Reading, R. & Clark, T. 1996. *An interdisciplinary examination of carnivore reintroductions.* In: *Carnivore Behavior, Ecology, and Evolution*, Vol. II (Ed. by J. Gittleman), pp. 296–336. Ithaca, New York: Cornell University Press.

Stanley-Price, M. 1989. *Animal Re-introductions: The Arabian Oryx in Oman.* New York: Cambridge University Press.

Richard P. Reading, Devra G. Kleiman, & Brian J. Miller

■ Conservation and Behavior
Wildlife Behavior as a Management Tool in United States National Parks

Often the thought of national parks conjures up images of stunning rock formations, awesome trees, beautiful vistas and magnificent wildlife. The United States National Park Service (NPS) was established by Congress through the Organic Act of 1916 to preserve and protect the natural and cultural resources for future generations. National parks managers are charged with protecting a variety of natural and cultural resources, for example, sites associated with threatened and endangered species, unique ecosystems, American Indians, enslaved people from Africa, battlefields, and geologic formations.

It is important to understand that the National Park Service mission is to protect these resources as well as to provide opportunities for people to enjoy and experience resources without compromising their conservation. This "dual" mission presents a challenging task for resource managers. All wildlife, plants, geologic formations, ecological processes and artifacts are protected under NPS policy. Under certain circumstances, native pests are allowed to function unimpeded. Native wildlife, an important facet of the diverse natural resources that the National Park Service protects, may at times present difficult management decisions for park managers. Each situation must be reviewed on a case-by-case basis in accordance with policy, cultural, and environmental regulations in attempt to accomplish this important mission.

A most critical tool is knowledge, understanding and appreciation for the behavior of the species of wildlife in parks. Wildlife includes all animals, hence refers to all living creatures in our national parks. Large charismatic animals we often associate with national parks include bison, bear, elk, deer, otters, prairie dogs, wolves and skunk, to mention only a few. In addition, birds are animals of great popularity and concern, such as the bald eagle, peregrine falcon, and a tremendous number of smaller birds such as pinion jays and painted buntings. However, the majority of animals, the all-important species that constitute the base of the food chain, often go unnoticed or are forgotten in the realm conservation. Animals such as fish, reptiles (including venomous snakes), amphibians and the phylum *Arthropoda* which includes insects, arachnids (spiders), crustaceans (crabs, lobsters), centipedes, and millipedes, are key to the survival of all higher animals. There are many land and water based animals including a host of microscopic animals as well, most of which are yet undiscovered.

Some animals are considered pests and threaten the survival of other species such as endangered plants, or threaten cultural resources such as historic structures, landscapes and features, and developed zones. Sometimes animals may present a public health threat to the safety of employees and visitors. Knowledge of an invasive species' foraging and activity patterns, behavioral sex differences, sensory and learning abilities, and other behaviors of invasive wildlife may lead to more effective management of these destructive invasive animals. For example, in Great Smoky Mountains National Park, a population of wild hogs exist within the park. This is a species not native to the North American continent and was introduced in the area of Great Smoky Mountains National Park in the early 1900s (prior to the establishment of the park). This invasive animal is a threat to many native species of plants and wildlife in the park. The hogs eat native plants, snails, bird eggs, salamanders, and even small mammals; and if that is not destructive enough, their behavior of rooting while foraging for food is destructive to habitat as evidenced by what has been described as rototiller-like action upon the landscape. Understanding behavior patterns of this destructive invasive wild hog, such as the solitary nature of males, breeding behaviors, seasonal movements, and other behaviors, have assisted park management in attempts to reduce their numbers in the park, thereby fulfilling the mission of protecting native plants and wildlife.

Balancing all the many valuable resources that national parks attempt to preserve is challenging. In addition, National Park Service policy states that national parks and the resources contained within them, including animals, should be observed, enjoyed, and appreciated by visitors to such places. In fact, all resources are important tools for park managers to interpret for educational and conservation purposes. Through interpretation and education, park managers provide opportunities for visitors to enjoy and understand the biological dependence of all living things, thus fostering personal connections with those resources. It is through personal intellectual and emotional connections with resources that most people begin to develop caring attitudes and provoke positive conservation actions. Communicating to visitors how and why animals behave the way they do is one of the most effective ways to accomplish the National Park Service mission goals.

Knowledge about how animals behave is an essential component necessary for park managers to assess and determine the best management practices for protecting wildlife and natural systems. For example, animals do not recognize park boundaries and often move beyond jurisdictional or human-designated boundaries in order to fulfill their biological needs, such as seeking mates or foraging for food. Some wildlife, such as wolves, skunks, and raccoons, may extend their *home range* (living space that the animal shares with other animals and does not defend) and even their *territory* (the living space that an animal will defend against intruders) which can result in a failure of park management attempts to protect a particular animal species and create difficulties with park neighbors.

Some animals move from place to place within a region seasonally to search for food or potential mates that can also take them out of the protective boundary of a national park. American bison or buffalo herds such as those in Yellowstone National Park have presented management challenges because of their natural behavior. One ongoing challenge is managing the behavior of bears and other mammals that are *omnivorous* (will eat both plant and animal prey). Wild young black bears, for example, learn from their mothers that there is security in darkness and to avoid the scent of humans. However, when a young black bear accidentally comes across the remains of human food, it begins to associate an easy meal with the scent of humans, often causing the animal to alter its behavior and thus become more day active, approach human areas, and perhaps even make contact with humans. If purposely fed by a human, this compounds their nuisance behaviors, making them more susceptible to poaching, starvation, and even aggressive encounters with humans that can lead to the eventual death of the "problem" bear. Park managers must understand and carefully assess an individual animal's behavior before making any management decisions such as relocating an animal.

In order to address wildlife management issues properly, it is imperative that the behavior and the biology of the species are understood. How will the animal interact with other species that reside in the park? Is the species a predator? How susceptible is the species to predation by others? There are a host of questions that need to be known about a species' behavior that can make or break an attempt at reintroduction. This is especially true of animal species that may be reintroduced to the wild for the first time. What happens to animals that were raised in captivity in the absence of predators, and that then must struggle to survive without the benefit of learning predator avoidance behaviors?

Questions with regard to how other management decisions will affect a species' behavior must also be assessed. For example, how will vegetation changes as a result of exotic plant removal or prescribed fire programs effect behavior patterns of some species? Such questions become extremely complicated when considering lesser known small mammals, birds, reptiles, and insects.

Understanding animal behaviors assists park managers in protecting cultural resources in national parks. For example, in Mesa Verde National Park, native digger bees were burrowing inside the inner walls of an historic kiva build by the ancestral Puebloan people. The burrowing action was destroying the original fabric (mud). After consultation with *hymenopteran* or bee experts, it was determined that these were solitary bees seeking nesting sites. This particular animal digs a tunnel, deposits one egg in each tunnel and provisions it with food. The adult hunts for food and places it inside each tunnel for the developing bee larvae to feed upon. The larval bee will consume the food and pupate in the nesting chamber, emerging as an adult later in the season. Experts recommended that park management respond by placing a sand snake (cloth tube filled with sand) on top of the kiva to hold a piece of dark cloth in place, concealing the opening at the top of the kiva from the bees' sight. This was done during the 2–3 week nest-seeking season with minimal visual impact on the resource for visitors to see. In addition, this spring event provides opportunities for park rangers to educate visitors both about the challenges of cultural resource protection and about animal behavior.

Specific knowledge of wildlife behavior assists park managers to determine how best to protect resources and how to provide positive experiences for visitors with minimal impact upon the lives of animals. For example, knowing the behavioral needs and patterns of whales at Glacier Bay National Park and Preserve is critical when designing tour boat activities. Behavioral information about bald eagles and other birds helps park managers design tours to avoid interference with breeding and nesting behavior. Understanding seasonal behavioral patterns of reptiles and amphibians helps park managers make decisions about road closures or building animal "causeways" so that they may cross a busy road without being preyed upon by a human automobile!

National parks protect populations of wild animal species and their habitats thereby protecting the local biological diversity of ecosystems that is important for their continued existence. National parks often are the places where animals that are on the brink of extinction are protected in such a way that these animals continue to exist in as close to their natural habitat as possible.

It is important to realize that while protecting and preserving animal genetic and biological diversity, it is also essential to understand wildlife behavior. Wildlife behavior is a key component in accomplishing the broader goals of conserving genetic diversity, species diversity and ecosystem integrity. Understanding the way wild animals typically behave in their own native environment (sometimes called species-typical behavior) is an important component for future wildlife conservation. Wildlife behavior and how animals adapt to changes in their habitat affect the fitness of the population—finding mates and breeding that allows them to pass on their genes (referred to as *inclusive fitness*), acquiring food, avoiding predators, finding shelter and nesting sites, parenting and other aspects of daily survival including opportunities for young animals to learn and develop survival skills. To what degree are wild animals able to cope with added stressors provided by human activities in national parks? How much of an impact do stressors such as noise, proximity, and human interactions, to mention only a few, have on behavior? Will these factors change species-typical behavior to a degree that begins to impact significantly entire populations of animals over time?

Wildlife behavioral studies in national parks provide opportunities for broad understanding of animal behavior that can have implications for managing entire species and habitats. Behavioral research in national parks can help scientists and managers to better understand the relationships and effects of altered habitats and impacts, such as absence of predators, presence of exotic species, noise pollution, and a plethora of other variables. Behavioral studies of animals in relatively undisturbed, compared with disturbed, habitats may help to determine why some species can adapt and persist while others do not. Information gathered from such studies in conjunction with traditional biological studies and with activities adjacent to a park area, will have important management implications. National parks can serve as optimal sites for such behavioral study. It is important to understand animal behavior regarding animal response to reintroductions, translocations and population augmentations.

Finally, conservation frequently involves resolving conflicts between wild animals and humans. Conservation programs are often challenged by having to justify the protection of species that threaten human lives or damage agricultural products. What factors elicit animal aggression or promote habituation to humans? Habitat alteration can bring previously separate species into contact, with the result that one dominant species drives competitive species or hosts that lack defenses to extinction. What impact do noises associated with human activities in parks have upon the behaviors of various animals, for example, amphibians during mating season? Understanding people and animal behavioral interaction provides clues as to how best to manage both people and animals in national parks as well as providing insight into the complexities of human and animal minds. The understanding and appreciation of wildlife behavior is most certainly an important tool for better management and protection of the various resources in the national parks.

See also Wildlife Management and Behavior

Further Resources

Caro, T. (Ed.) 1998. *Behavioral Ecology and Conservation Biology*. New York: Oxford University Press.
Clemmons, J. R. & Buchholz, R. (Eds.). 1997. *Behavioural Approaches to Conservation in the Wild*. New York: Cambridge University Press
Gosling, L. M. & Sutherland, W. J. (Eds.) 2000. *Behaviour and Conservation*. New York: Cambridge University Press.
Helfman, G. S. 1999. *Behavior and fish conservation: introduction, motivation, and overview*. Environmental Biology of Fishes, 55, 7–12.
Knight, J. 2001. *If they could talk to the animals*. Nature, 414, 246–247.
Sarrazom, F. & Barbault, R. 1966. *Reintroduction: challenges and lesson for basic ecology*. Trends in Ecology and Evolution, 11, 474–478.
Sutherland, W. J. 1998. *The importance of behavioral studies in conservation biology*. Animal Behaviour, 56, 801–809.
Visit www.nps.gov for information on all of the national parks.

Daniel R. Tardona

■|Cooperation

From 1963 to the present, four paths to the evolution and maintenance of cooperation in animals have developed—namely kin-selected cooperation, reciprocity, by-product mutualism, and group selection. Before we examine these different paths to cooperation, it is important to begin with a discussion of definitions. This is particularly important when tackling terms like cooperation, which has both an everyday usage, as well as a more technical definition within the sciences.

Cooperation is an outcome, not an action. It is an outcome that despite *potential* costs to an individual, is in some manner good for members of a group. Under this definition, achieving cooperation requires some sort of collective action. A related definition is now necessary. The phrase "to cooperate" has two common usages. It can mean either: 1) to achieve cooperation—something the group does; or 2) to behave cooperatively; that is, to behave in such a manner that makes cooperation possible (something the individual does) even though the cooperation will not actually be realized unless other group members also behave cooperatively. (Here, when I employ the phrase "to cooperate," I mean "to behave cooperatively.")

While philosophers, psychologists and economists have all contributed to the study of cooperative behavior, they have generally failed to address a fundamental problem with respect to cooperative behavior: How can cooperation persist *over long periods of time* when there seem to be so many different ways that individuals who don't cooperate (cheaters) can circumvent the system? Why don't we see cheaters slowly increase in frequency, replacing their cooperating peers? If individuals benefit from others cooperating, why should they ever cooperate, since they obtain the resources that cooperators obtain, but do not pay the costs associated with cooperation?

To fully address the above issues, the science of evolutionary biology must be considered. Here, using natural selection as our guidepost, we can examine the various means by which cooperative strategies can evolve over vast stretches of time, as well as touch on the reasons cooperation often fails to manifest itself in certain conditions.

With some background now behind us, we can examine each of the paths to cooperation in more detail. We begin with kin selection, and subsequently examine reciprocity, by-product mutualism, and group selection.

Kin Selection

Imagine a field in which a group of ground squirrels live. Now, seemingly out of the blue, a hawk, or some other avian predator, begins a deadly dive from the air, targeting the squirrels in the field for its next meal. Then a piercing shriek is emitted; an alarm call given by one squirrel has alerted the others of the impending danger. Squirrels dash around madly, doing whatever they can to reach their burrow, or at least attempt some other retreat.

From an evolutionary perspective, a riddle emerges here, and only kin selection can help us solve it. Why should any individual squirrel be the one to emit an alarm call? If nothing else, alarm callers are likely the single most obvious creature in the entire field. Why should a squirrel attract the hawk in its

This ground squirrel alerts his family by emitting an alarm call.
© *Richard R. Hanson / Photo Researchers, Inc.*

direction? Why not let someone else take the risks? To answer these questions we need to recognize that the group of squirrels being warned by an alarm caller is not a random assortment of animals. Instead, ground squirrels live in societies full of relatives, and this makes a huge difference in terms of evolution and behavior.

Under the evolutionary definition, blood relatives are more likely to carry the same genes than are a handful of unrelated individuals. In the nomenclature of evolution, these are called "identical by descent" genes because the likelihood of sharing them is related to descent from some common ancestral relative. For example, the two most common ancestors that sisters share are their mother and father, whereas for cousins the common ancestors would be a grandmother and grandfather. What this means is that when a squirrel risks its own life to save the lives of its relatives, it is not a completely altruistic act because its relatives most likely carry the same genes that it does. In essence then, since sisters, for example, share 50% of their genes that are identical by descent, saving two sisters and losing your life is equivalent to an even genetic swap, whereas saving three sisters and losing your life is a net positive for all parties involved. And if you are only *risking*, rather than necessarily *giving up* your life, you might do it even if fewer relatives were around.

Half of the riddle has now been solved—the half surrounding why *anyone* should give alarm calls. But, *who* gives a call? One important component in our story of the alarm calling ground squirrels who risk their lives to save relatives is that gender seems to play a role in establishing who our cooperative sentinel is. In ground squirrels, it is generally the females, not males, who put their necks on the line. Why? The answer once again is kinship, this time mediated by decisions about where to live.

In many species of animals, males and females do not make the same decisions about where to set up residence. When individuals are mature enough to survive on their own, they are faced with a difficult decision: Should they live near where they grew up or emigrate elsewhere? Remaining in the old neighborhood clearly has both costs (e.g., competition from local rivals) and benefits (e.g., security), as does moving to a new location. Furthermore, the costs and benefits can be very different depending upon whether one is male or female. Indeed in the animal kingdom, we see every combination possible: Males stay/females leave, males leave/females stay, both sexes leave when reaching maturity, and both sexes remain in the natal home area. In squirrels, it is the males that leave home to set up shop elsewhere and the females that stay in their old stomping grounds.

Once males leave and females stay put, an interesting imbalance arises. Females are surrounded by relatives, whereas males are in areas with complete strangers. This asymmetry in relatedness favors dangerous alarm calling in females, since they will be assisting kin; however, the same can't be said for males. Blood kinship, then, explains both *why* anyone would put themselves in harm's way and call out when a predator is sighted, and *who* would be most likely to do so. If this is correct, then an interesting prediction can be made. Under some very unusual circumstances, females are forced to emigrate to new groups. In such cases, these newcomers, despite being female, should be among the least likely individuals to give alarm calls when they are needed. Sure enough, that is precisely what the scant data available on this question seems to suggest.

Reciprocity

Instead of imagining a field of ground squirrels, now envision a walk alongside a stream in the Northern Mountains of Trinidad, West Indies. In many streams in Trinidad, the water is crystal clear and the behavior of guppies can be seen from the bank. If you are patient, you will probably observe a pair of guppies breaking away from their group and approaching a dangerous predator that has been sighted. This alone is bizarre: The fish could just as easily have headed for cover, rather than toward this menace. But not only do some fish take the risks inherent in approaching a predator, they actually seem to return to their group and somehow pass on the information they just obtained.

The risk-taking guppies are trapped in a dilemma. The temptation to cheat is always there—the best thing that could happen for fish 1 is that fish 2 takes the risks and passes the information on to the other guppy for free. But if both individuals opt to wait for the other to go out and do the dirty work, they may be worse off than if they had just done it as a team. Many other examples of this dilemma pervade everyday animal life. The following question then arises: Is there an escape from this dilemma? Can we identify any sort of co-operation in this example? The answer is yes, providing a *pair* of individuals finds themselves in a similar predicament *many times*.

When partners are paired up many times, evolutionary theory predicts that individuals should use what is called the "tit-for-tat" strategy; that is, start out cooperating with your partner, but if he cheats on you, cheat right back in the same manner he did. Despite having a brain not much bigger than a pinhead, guppies appear to use the tit-for-tat rule on their expeditions out toward potentially dangerous predators. Each fish keeps track of what the other is doing when both go out to examine the predator. Should one fish lag behind, the other fish slows down so that the distance between it and its partner does not become too great. To top it off, guppies genuinely prefer to spend their time associating with other guppies who cooperated with them during their danger-filled sorties, presumably to be in their vicinity again should the situation arise once more.

By-Product Mutualism

Let us once again switch hypothetical venues and place ourselves in Africa, watching lion behavior (from a safe distance). Watching lionesses hunt a water buffalo on the plains of Africa is both a savage and beautiful event. In other instances, lions don't just chase after a gazelle in a random manner; instead a lion hunt is a coordinated action aimed with one goal—a gazelle meal. In water buffalo hunts, often one lioness will flush the water buffalo, and one or more of the others will chase it; or in gazelle hunts each hunter will come at the prey from an angle that cuts maneuvering down to a minimum for the gazelles. Every lion stands something to gain from a hunt, but they need others to accomplish that goal.

If they don't cooperate, every lion suffers a cost—in that many prey items, particularly large items such as a water buffalo, cannot be killed by a lone lion hunter. Cooperation is

Lions tend to hunt using a form of teamwork called by-product mutualism.

Beverly Joubert/National Geographic Image Collection.

required for such prey types. This type of cooperation, one in which joint action is predicated only on the self interest of all parties involved, may be the most common type of animal cooperation and is labeled *by-product mutualism*. The situation is dramatically different when it comes to prey that can be taken by a single lion hunter. In this case cooperation is not beneficial, but harmful, because a small prey that could have provided a meal for a single lion instead needs to be shared among a number of hunters.

Group Selected Cooperation

The Sonoran Desert is home to a fascinating ant species called *Acromyrmex versicolor*. In this species, underground nests are initiated by queens that are unrelated (thus ruling out any kinship element to the story). Despite not being relatives, queens in this species are friendly—no one attacks anyone else in the nest (rare in insects in general), all food is shared equally by everyone (equally rare in insects), and all the queens have about the same number of offspring. Yet in the midst of all this cooperation, a riddle emerges in that only a single queen in the whole nest goes out and gets food for everyone. This is odd because the underground nest is a relatively safe place to be, but going out of the entrance into the desert is dangerous because many predators await any individual who does so. On top of all that, *which* queen ultimately becomes the group's food provider seems to be determined by chance—no one is coerced into taking the job.

How could such dramatic, "for the good of others," cooperative behavior ever come to be? If these queens are not related, why should any one of them accept the role of food gatherer with all of its risks, given that the food that comes into a nest is divided between all group members equally and ultimately leads to all nestmates producing similar numbers of offspring? The answer to these questions lies in changing perspectives. Rather than viewing this strictly in terms of the costs and benefits to the *individual* who is the forager, one must expand the notion of costs and benefits to the group. This perspective is called *group selection*.

Group selection requires one to examine the effect the behavior has on the individual undertaking it *and* on those around it. If the behavior is beneficial to all involved, no obstacles exist to its evolution. If the behavior has negative effects on all parties involved, then such behavior disappears very quickly. But what about the queens in our example above? Here we have a case where the behavior (food gathering) has a negative impact on the forager, but a positive impact on the group. When should we expect this sort of cooperation to persist? The answer: When the group-level positive effects outweigh the individual-level negative effects.

Despite costs to the queen, who takes all the risks associated with sorties from the safety of the nest, the group-level benefits of this action are so great that the behavior persists. In *Acromyrmex versicolor*, once queens raise the offspring in the nest, everyone comes up to the surface and "warfare" occurs—that is, intense fighting between individuals in different nests. Only one group will survive. The odds of surviving such between-group conflict is directly related to the number of fighters each group has, and that in turn is related to how much food queens had and how many offspring they could produce.

Imagine nests in which no one would go out and forage and everyone would try to produce just whatever progeny they could from their stored body fat. Such a nest would surely perish in the nastiness to come, and hence the group-level benefits of having a specialized food gatherer outweigh the costs to the individual undertaking the action.

Implications of Animal Cooperation

In addition to animal cooperation being an inherently interesting subject, it has what might be thought of as a more practical application. Animal cooperation shows us what to expect when the complex web of human social networks, as well as the laws and norms found in all human societies, are absent, and so these studies act as a sort of baseline from which to operate. Animals show us a stripped down version of what behavior in a given circumstance would look like without moral will and freedom. Only with this understanding of what a particular behavior looks like outside the context of some moral code, can we use human morality to focus in on and foster cooperation in our species.

Studies on cooperation in animals can be used to help us better understand and promote human cooperation in at least two ways. Both begin with a thorough search for common factors that have been found in many cases of animal cooperation. Once such factors are uncovered, we can:

1. Identify areas in which we see failures in cooperation in our own species, and then use our knowledge of animal cooperation to add critical factors to the human scenario with which we are concerned. For example, if we have found that cooperation in animals is common when individuals frequently interact, we have identified a factor promoting animal cooperation. We can then examine how we might add this factor to the human scenario with which we are concerned.

2. Identify human situation that might contain some of the factors that we know from animal studies tend to favor cooperation, and use various techniques to amplify these factors and hence enhance the probability of cooperation.

Animal cooperation may not be either the rule or the exception in nature, but it occurs often enough to make it an irresistible subject. It is also important to note, that though natural selection is the driving force shaping cooperation in animals, "natural selection thinking" should not *necessarily* be the guidepost humans use in shaping our own behavior. And by no means should humans fall into the trap of the "naturalistic fallacy"—that because something "is" in nature, then it "ought" to be that way. Whether natural selection favors something in nonhumans, or even whether natural selection might favor a behavior in humans if we stripped away culture, is not the primary issue with respect to shaping human cooperative tendencies. Rather, the critical point is that humans can use animal examples as a means to focus our moral compasses on the right elements, from the right perspective, in such a manner as to make human cooperation more probable. We may very well fail, but at least we will have used all the scientific tools we have at our disposal in trying.

See also Cognition—*Social Cognition in Primates and Other Animals*
Emotions—*Emotions and Affective Experiences*
Empathy
Friendship in Animals
Play—*Dog Minds and Dog Play*
Play—*Social Play Behavior and Social Morality*
Social Evolution—*Optimization and Evolutionary Game Theory*

Further Resources

Axelrod, R. 1984. *The Evolution of Cooperation*. New York: Basic Books.
Dugatkin, L. A. 1997. *Cooperation among Animals: An Evolutionary Perspective*. New York: Oxford University Press.
Gadagkar, R. 1997. *Survival Strategies : Cooperation and Conflict in Animal Societies*. Cambridge, MA: Harvard University Press.
Hamilton, W. D. 1963. *The evolution of altruistic behavior*. American Naturalist, 97, 354–356.
Hardin, G. 1968. *The tragedy of the commons*. Science, 162, 1243–1248.
Ridley, M. 1996. *The Origins of Virtue*. New York: Viking Press.
Sober, E., & Wilson, D. S. 1998. *Unto Others*. Cambridge, MA: Harvard University Press.
Trivers, R. L. 1971. *The evolution of reciprocal altruism*. Quarterly Review of Biology, 46, 189–226.

Lee Alan Dugatkin

■ Corvids
The Crow Family

The "corvids" or crow family (Corvidae) is almost cosmopolitan in distribution and includes about 103 species, including not only crows, but also ravens, magpies, jays, nutcrackers, and choughs. Corvids are distantly related to birds of paradise and bowerbirds, and they still share characteristics with them, including bright coloration (crows and ravens seem black from a distance but reveal metallic sheens of green, blue, and purple from up close) and spectacular behavior. Bowerbirds lack bright coloration but share with corvids an attraction to and collection of bright and novel objects, and the males of the former build and maintain elaborate structures where they display these objects that, instead of body/feather ornaments, entice females for mating.

Corvids are widely, and probably correctly, perceived as the brains of the bird world, and a number of species are of great interest because of their sociality, phenomenal memory, food hoarding, elaborate play, ingenuity, and tool use.

Most corvids are either social or highly gregarious for at least part of their lives. Some crows, such as the European rook, *Corvus frugilegus*, stay together through life, and they breed in dense colonies. Other crows, including the American crow, *C. brachyrhynchos*, breed solitarily. However, the young and unmated birds aggregate into flocks, and birds of all ages may sleep at night in communal roosts of thousands or even hundreds of thousands of birds in the wintertime. In the spring the birds disperse to breed solitarily in their territories. Some stay on site year round. In some species of crows and jays (but not necessarily in all populations of the same species), several individuals beside the parents may help at the nest.

In North America the blue jay, *Cyanocitta cristata*, breeds solitarily, but in early spring the birds gather in noisy groups, possibly to find mates. Other jays stay in permanent small family flocks, and the young of one generation help the parents rear young of the next generation.

In common with many other corvids, the common raven, *Corvus corax*, are semisocial when they are either young or unmated, and may form temporary alliances, traveling in bands with other nonbreeders. As in other corvids, only the females incubate, and mates may form life-long attachments. Ravens defend territories against other ravens, including against nonbreeders who may in turn team up to overpower the defenses of territory-holding adults. Magpies, *Pica pica*, have roughly overlapping behavior to that of crows, jays, and ravens; at some times the birds band together, but they usually breed apart.

When encountering a temporary surplus of food, possibly all corvids hide and store some of that food for future use. The nutcrackers, *Nucifraga* (both the European and North American species) rely on cached seeds/nuts collected in the fall to sustain them through the winter and into spring. They make their food caches on generally windswept slopes, and can recover over 80% of thousands of different caches made several months earlier. Gray jays, *Perisoreus canadensis*, who live in northern areas where there are deep winter snows, also rely on food stored during summer and early fall for their winter survival, but they attach (with a specialized saliva that hardens and forms a glue) this food onto tree limbs where it does not become buried under deep snow.

When in the company of a horde of competitors at a carcass, ravens cache meat largely to secure as much as possible for themselves in the few hours before the others remove it all. Ravens take care not to be seen by others where they hide food, because they are able to remember the precise locations not only of their own caches, but also of those they see others make. Recent research suggests that ravens anticipate each other's actions, as if knowing their intentions.

A wolf tries to keep ravens from its wapiti kill. Ravens have been known to make caches of meat when competitors are around.

Tom Murphy/National Geographic Image Collection.

Some crows have learned to open nuts and clams by dropping them from the air onto rocks or placing them in front of cars at intersections. An Australian crow, *Corvus moneduloides*, routinely makes tools, fashioning two types of picks out of pandanus leaves, and then expertly uses these picks to extract grubs from crevices. These behaviors could be due to large components of either genetic programming or trial-and-error learning. One of the more impressive demonstrations of a corvid's previously unprogrammed behavior is the ability of some naive, untutored ravens to pull up food suspended on a long string. Their behavior involves not only the use of the string as a tool, but also performing a long series of behavioral steps of using this tool correctly to get the food. Other experiments suggest that they have a mental representation of the tool-using act so as to anticipate physical consequences of the string–food connection.

Corvids have for centuries impressed humans with their presumed cleverness. Biologists had generally scoffed at such presumptions. However, in the last several decades they have become ever more popular animals in the study of both social behavior and cognition. The results of these studies indicate them to be more sophisticated animals than was imagined.

See also Cognition—*Caching Behavior*
     Tools— *Tool Use and Manufacture by Birds*

Further Resources

Angell, T. 1978. *Ravens, Crows, Magpies, and Jays*. Seattle: University of Washington Press.
Goodwin, D. 1986. *Crows of the World* 2nd. edn. UK: St. Edmundsbury Press, Ltd.

Heinrich, B. 1999. *Mind of the Raven: Investigations and Adventures with Wolf-Birds.* New York: HarperCollins.

Kilham, L. 1989 *The American Crow and the Common Raven.* College Station: Texas A&M University Press.

Bernd Heinrich

■ Coyotes
Clever Tricksters, Protean Predators

"Old man coyote" is an amazing and adaptable being. Coyotes are true survivors, chiefly because of their adaptable ways. William Bright, in his collection of stories titled *A Coyote Reader*, notes: "Coyote is the trickster par excellence for the largest number of American Indian cultures." Native peoples have portrayed coyotes as a sly tricksters, thieves, gluttons, outlaws, and spoilers because of their uncanny ability to survive and to reproduce successfully in a wide variety of habitats and under harsh conditions. The cartoon character, "Wile E. Coyote," exemplifies these features. Coyotes generally survive not only their encounters with other nonhuman predators (though they are losing out to gray wolves in some parts of Yellowstone National Park, but not in others), but also with humans who attempt to control them using brutal methods, and who also hold well-organized community hunts in which the person who kills the most coyotes wins a prize.

Coyotes belong to the dog family (Canidae). They are closely related to wolves, domestic dogs, jackals, dingos, dholes, and foxes. Coyotes originally inhabited open plains and grasslands in North America. Due to intense control and management and the absence of such competitors as wolves, coyotes now thrive in habitats ranging from dry, warm deserts, to wet grasslands and plains, to forests, to colder climates at high elevations (up to about 10,000 feet or 3,200 meters). Coyotes can be found between northern Alaska and Costa Rica and throughout mainland United States and Canada. They also live in large urban cities such as Los Angeles, California, and New York City. A coyote once tried to enter an elevator in an office building in Seattle, Washington!

Coyotes are genuine masters of behavioral flexibility. I have studied coyotes for more than 25 years and my research along with that of my colleagues has shown that talking about "the" coyote is

A coyote sits in an urban parking lot.
© *Glenn Oliver / Visuals Unlimited.*

misleading. The moment one begins making rampant generalizations, they are proven wrong. Coyotes show great variation in their social organization. In some areas coyotes live like typical gray wolves, in resident packs of 4–8 individuals that are closely-knit extended families consisting of overlapping generations of parents, young-of-the-year, and adult helpers who do not reproduce but could if they left the group. In other habitats they live either as mated pairs or as roaming single individuals, showing little or no attachment to a particular site. During our 8-year study of coyotes living around Blacktail Butte in the Grand Teton National Park just outside of Jackson, Wyoming, we discovered that coyotes living as few as 1.5 km (1 m) apart showed variations in social organization.

Packs typically exist when there is enough food to allow some young-of-the-year to remain with their parents and older siblings, rather than to go out or to disperse on their own. Pack formation in coyotes appears to be an adaptation for the defense of food and territory rather than for the acquisition of live prey. Coyotes, unlike wolves, only rarely actively try to capture large prey such as deer or elk. Furthermore, coyotes are sometimes territorial and sometimes they do not defend territorial borders.

Male and female coyotes form dominance hierarchies. Sometimes dominance relationships among littermates persist for more than a year. Coyotes are usually monogamous. Pair bonds between the same male and female can last upwards of 4 years. In packs, usually only the dominant male and dominant female mate each year. Coyotes breed once a year, in early-to mid-winter depending on locale. Courtship lasts 2 to 3 months. The gestation period averages 63 days and average litter size is 6 pups with an even sex ratio (same number of females and males) at birth. Pups are born blind and helpless, usually in an excavated den, and emerge from the den at about 2–3 weeks of age. Pups are weaned at about 5–7 weeks of age, at which time they begin to eat meat provided by parents and other adults. Parents often receive help from other group members in rearing young and in defending food and territory borders both within and outside of the breeding season. Some helpers go on to inherit the territory in which they were born (called the *natal area*), breed, and receive help from individuals to whom they previously provided care, whereas others strike out on their own and attempt to find a mate. During our study of coyotes living in the Grand Teton National Park, my students and I discovered that it took more than two coyotes to successfully defend a territory from intruding coyotes, and that single coyotes were significantly less successful in doing so. Pack members working together were successful in driving off intruders 75% of the time, whereas single individuals had a 33% success rate. Thus, there are benefits to living in a group that is larger than just the parents and their young offspring who do not help in territorial defense. Coyotes can live as long as 18 years in captivity, but in the wild few live longer than about 6–10 years. Coyotes can produce fertile hybrids when they mate with gray wolves, red wolves, and domestic dogs.

In addition to showing great behavioral flexibility, coyotes also enjoy a highly varied diet including plant and animal matter and inanimate objects such as boots and gloves. They have been observed to fish and to climb trees in pursuit of food. Coyotes are also successful scavengers. Diet varies seasonally and in different regions. Coyotes rely primarily on vision while hunting. They typically hunt small rodents (mice and squirrels) alone and do not regularly group hunt large ungulates such as deer and elk.

Coyotes also show variations in home range and territory size. Individual coyote ranges can be as large as 50–70 km² (19–27 mi²). Home range area is affected by the presence of other coyotes and competitors and the amount of population control to which individuals are subjected. The distribution of food also influences territory and home range size. When food is abundant in small areas, coyotes tend to spend most of their time in that place.

Coyotes communicate using many different olfactory (odor), vocal (11 different sounds), and visual (different facial expressions, gaits, postures, and tail positions) signals. Males often deposit urine to scent mark territorial boundaries. Female urine attracts males when they are ready to mate. Coyotes are considered to be the most vocal of North American wild mammals. The scientific name of coyotes, *Canis latrans*, means "barking dog." Coyotes growl, huff, bark, bark–howl, whine, yelp, and howl. Group yip–howling occurs following a reunion or during greeting. Group howling and lone howling are used as contact calls to announce location when individuals are separated from group members. Coyotes bark when they are alarmed or when they threaten others.

In a nutshell, coyotes are quintessential and resourceful opportunists who defy profiling as individuals who predictably behave this way or that. Because of their slyness and behavioral flexibility, they are fascinating and challenging to study.

Coyotes are also a very important part of the ecological web in various communities because they help to regulate species at different trophic levels. The biologists Kevin Crooks and Michael Soulé studied the complex interrelationships among coyotes, other medium-sized predators (*mesopredators*) such as domestic cats, opossum, and raccoons, and scrub birds including California quail, wren tits, Bewick's wrens, greater roadrunners, and cactus wrens living near San Diego, California. Their research is an excellent example of the importance of long-term projects that investigate complex webs of nature that are not obvious at first glance. Crooks and Soulé discovered that scrub bird diversity, the number of different species present, was higher in areas where coy-

Female coyote and her young howling.
© *Charlie Heidecker / Visuals Unlimited.*

otes were either present or more abundant. Domestic cats, opossum, and raccoons avoided areas where coyotes were most active. Coyotes often kill domestic cats where they cohabit.

Unlike wild predators, including coyotes, domestic cats are recreational hunters; they continue to kill birds even when bird populations are low. Crooks and Soulé found that 84% of outdoor cats brought back kills to their homes. Cat owners reported that each outdoor cat who hunted returned on average 24 rodents, 15 birds, and 17 lizards to the residence each year, a large number of victims. The level of bird predation was unsustainable, and least 75 local extinctions have occurred in these areas over the past century.

Because of their incredible adaptability, coyotes have been hunted for more than a century because some people consider them to be pests. Wildlife Services (formerly called Animal Damage Control), a branch of the United States Department of Agriculture (U.S.D.A.), slaughters tens of thousands coyotes each year (about 86,000 in 1999, 10% more than in the previous year despite claims that Wildlife Services is switching to nonlethal techniques) because coyotes supposedly are rampant predators on livestock. Livestock protection programs cost taxpayers about $10–11 million annually. In Colorado, more than 90% of Wildlife Service's money ($1.1 million) is spent on lethal control of native wildlife. Federal extermination efforts have been conducted since 1885, and during the past 50 years about 3.5 million coyotes have been killed. Killing methods—trapping (28% of control efforts), poisoning (21%), shooting from airplanes (aerial gunning, 33%), and snaring and other procedures (18%)—are considered by many to be inhumane and indiscriminate, and other predators, domestic dogs, and endangered species also fall victim. In Colorado alone, during the 1999–2000 harvest season, about 26,000 coyotes were killed by private hunters.

Coyotes are often controlled using aircraft. Aerial gunners killed almost 31,000 coyotes in 1999 (along with 17 ravens, 180 red foxes and 390 bobcats). According to the Boulder-based

A coyote (Canis latrans) curled up on the ground.
Tom Murphy/National Geographic Image Collection.

conservation organization Sinapu, there have been 18 airplane crashes since 1989 resulting in 7 deaths and 21 injuries. The cost of aerial gunning to taxpayers ranges from $180 to $800 per animal. This comes to about $5.7 million spent on aerial gunning annually. Often tens of thousands of dollars are used to capture a single coyote who might be responsible for a few hundred dollars of damage or not blamable at all. A study conducted at Utah State University showed that gunning down coyotes from helicopters is actually ineffective in that there is not a significant decrease in predation on livestock. In another study done at Utah State, coyotes, some of whom were seriously injured, were kept in leghold traps for long periods of time to determine the effects of tranquilizers to keep them calm when they were in pain.

Coyotes are a wonderful example of a sly animal for whom it is essential to know about their versatile behavior and their ability "to wear many different hats." Wanton killing does not work because little attention is paid to the versatile behavior of these adaptable predators. And disease and unsanitary conditions frequently cause more livestock death than do coyotes or other predators. Only rarely is "the problem" coyote caught or killed, and when coyotes are killed others take their place. There is also some preliminary evidence that, in areas where coyotes are killed, their birth rates and litter size increase, the result of which is the maintenance or increase in coyote numbers. If the positive correlation between level of exploitation and coyote reproductive productivity turns out to be true, this would be another reason not to remove coyotes from an area in order to control their numbers. If we knew more about coyote behavior and also spent more time trying to understand why they are such adaptable beings, we could develop more effective and humane ways to coexist with them.

Loved or hated and feared by many, coyotes have defied virtually all attempts to control their clever ways. Coyotes are amazing animals. They offer valuable lessons in survival. Their incredible behavioral flexibility, their ability to adapt to innumerable habitats and ecological niches, and their intimate role in numerous webs of nature are good reasons for studying them in more detail rather than needlessly slaughtering them. Though coyotes try our patience, they are a model animal for learning about adaptability and success by non-human individuals striving to make it in a human dominated world. Coyotes, like Proteus the Greek who could change his form at will and avoid capture, are truly "protean predators." They are a success story, perhaps hapless victims of their own success.

See also Wolf Behavior—*Learning to Live in Life or Death Situations*

Further Resources

Ballard, W. B., Carbyn, L. N. & Smith, D. W. 2003. *Wolf interactions with non-prey*. In: *Wolves: Behavior, Ecology, and Conservation*. (Ed. by L. David Mech & L. Boitani), pp. 259–271. Chicago: University of Chicago Press.

Bekoff, M. (Ed.) 1978/2001. *Coyotes: Biology, Behavior, and Management.* New York: Academic Press. Reprinted in 2001 by Blackburn Press, West Caldwell, New Jersey.

Bekoff, M. 2001. *Cunning coyotes: Tireless tricksters, protean predators.* In: *Model Systems in Behavioral Ecology* (Ed. by L. Dugatkin), pp. 381–407. Princeton: Princeton University Press.

Bekoff, M. & Gese, E. 2003. *Coyote,* Canis latrans. In: *Wild mammals of North America: Biology, Management and Conservation* (Ed. by J. Chapman & G. Feldhamer), pp. 467–481. Baltimore, MD: John Hopkins University Press.

Bekoff, M. & Wells, M. C. 1986. *Social behavior and ecology of coyotes.* Advances in the Study of Behavior, 16, 251–338.

Crooks, K. R. & Soulé, M. E. 1999. *Mesopredator release and avifaunal extinctions in a fragmented system.* Nature, 400, 563–566.

Two videos (*Song Dog* and *Wolf Pack*) that contain much information on the behavior of coyotes and their interactions with other predators are available from the wildlife photographer, Bob Landis who lives in Gardiner, Montana). See his entry, Careers—*Wildlife Filmmaking.*

Marc Bekoff

■ Craig, Wallace
(1876–1954)

Wallace Craig (1876–1954) was an American zoologist and psychologist who made important contributions to the study of animal instinct. His ideas proved of special importance to Konrad Lorenz when the latter was developing his own theory of instinct, and thus the conceptual foundations of the new science of ethology.

Craig studied under Charles Otis Whitman at the University of Chicago. Whitman taught him the importance of studying animal behavior comparatively. Working with Whitman's collection of pigeons, Craig focused in particular on the behavior of the ring dove, but he also made observations on other pigeon species, including the passenger pigeon, a species which was about to become extinct.

Craig earned his Ph.D. from the University of Chicago in 1908 and then went to teach psychology at the University of Maine. In 1918 he published the single most significant paper of his career, entitled "Appetites and Aversions as Constituents of Instincts." There he explained that instincts were something more than innate reflex actions, contrary to what many people in his day supposed. He maintained that appetites arising with the organism, not reflexive responses to stimuli coming from outside the organism, were what set an animal's cycles of instinctive behavior into motion. He also observed that the longer it had been since an animal performed a certain instinctive action, the easier it became for that action to be "released" by stimuli that were weaker or less appropriate than the kind of stimuli that normally released it.

Among Craig's other interesting publications was a paper of 1921 entitled: "Why Do Animals Fight?" He maintained in this paper that aggression is not an instinct. Pigeons, he found, have no special appetite for fighting. A pigeon does not seek to fight other birds, nor does it seek to prolong a fight when in one. When it did fight, it did so simply to rid itself of a stimulus that irritated it.

After Konrad Lorenz read Craig's paper of 1918, he began to emancipate himself from the chain-reflex image of animal instincts. On the other hand, he paid no attention to Craig's analysis of why animals fight and to Craig's denial that aggression in animals is an instinct. Lorenz's own theory of aggression, made popular in the 1960s in his book,

On Aggression, portrayed aggression as an instinct that in people needed to be channeled into productive directions in order for it not to have disastrous consequences.

Craig had to give up his teaching in the early 1920s when he became too hard of hearing to function effectively in the classroom. He died in 1954 in poverty and relative obscurity. At least some of his ideas lived on, however, in the new science of ethology.

See also History—*A History of Animal Behavior Studies*
Lorenz, Konrad Z. *(1903–1989)*
Tinbergen, Nikolaas *(1907–1988)*
Whitman, Charles Otis *(1842–1910)*

Richard W. Burkhardt, Jr.

■|Culture

Cultural diversity is a central fact of human life. Humans use elaborate systems of communication to coordinate behavior. Societies vary greatly in language, diet, dress, and countless other traditions. Children eagerly learn and imitate these behaviors, giving cultures continuity over generations. In recent years scientists have found that the basic processes underlying culture—the passing on of behavioral traditions—are not unique to humans. Multigenerational cultural learning has now been documented in several other animal species.

To find out if other animals transmit cultures, scientists must first define what they mean by cultural behavior, and then show how it is different from other behavior. Put simply, the *culture* label applies to any species in which a community can be clearly and consistently distinguished from other communities by its unique behaviors that are not determined by genetic or environmental factors.

Research on nonhuman cultural behavior is a relatively new field of scientific inquiry. The two major ways to approach the study of cultural behavior are controlled laboratory experiments on social learning and language acquisition, and field research describing behavioral variation in natural habitats. The experimental method investigates cognition, memory and communication: Is the species capable of imitation and/or teaching? The second method, sometimes called the ethnographic approach, explores the results of those processes in natural communities: What are the animals' actual behavior patterns or traditions?

The earliest field studies to provide insights into nonhuman cultures began in the 1960s when Jane Goodall began observing chimpanzees in Gombe National Park in east Africa. Over the decades, Goodall patiently observed chimps fashioning sticks to forage for ants, practicing traditional grooming habits, and even forming political alliances and fighting protracted battles. Studies of other chimpanzee communities also described an array of behaviors that were unique to specific groups.

In 1992, William McGrew reviewed chimpanzee tool use in the wild and found that the wide range of group-specific behaviors that emerged from the studies could only be explained by cultural learning. In 1999, scientists compared seven long-term studies combining over 151 years of chimpanzee observation. They found patterns of variation far more extensive than had previously been documented for any species except humans. Specific behavior patterns, including tool use, grooming and courtship behaviors, were found to be customary or habitual among some communities, while completely absent in others. Chimps in some groups learned to use leaves to scrape termites out of nests, for example, while other chimp communities employed sticks. In each group, young chimps watched their mates closely, absorbing the lessons of each of their actions. The scientists concluded that each chimpanzee community behaves in ways that are culturally distinct from others, which was previously unrecognized in nonhuman species.

Other primates have also been found to exhibit cultural behavior. Groups of orang-utans have been found to share distinct "tricks of the trade" for feeding, nesting and communicating, and scientists say these behaviors represent humanlike culture. In the early 1950s, scientists studied Japanese macaque monkeys and found that one young monkey named Imo, when given potatoes, would wash them before eating. Potato washing soon spread to older kin, siblings, and playmates, and other members of the troop, and yet this is the only group of monkeys in the world to do it.

In another example, certain Japanese macaques pick up and handle stones. Initially spread only among individuals of the same age, this behavior was later passed down from older to younger individuals in successive generations. This example shows the influence of social networks in the transmission phase of novel behavior: Social networks determine how the behaviors are initially transmitted. Various other factors may also influence cultural transmission: environment, gender, age, and other social and biological variables. For example, unlike potato washing, stone handling declines when individuals mature.

Cultural behavior appears to have also evolved in other animals. Elephants are now believed to share information culturally. Elephants can produce and perceive rumbles below the range of human hearing. These "infrasounds" pass easily over long distances and allow elephants to broadcast information by booming out thunderous calls that only they can hear. The resulting dialogue allows elephants to wander widely and unpredictably while still keeping in touch. Researchers are just beginning to unravel the range of information elephants can share with these infrasonic sounds. In 2001 researchers found that matriarchs retain social knowledge in African elephants. The oldest individuals in a group inform the group as a whole and thus pass their knowledge down over generations.

Scientists have also examined cultural traditions in whales and dolphins, known as cetaceans. In at least four (of about 80) species—the bottlenose dolphin, orca, sperm whale and humpback whale—a diverse array of cultural behaviors have been found. Experiments

A herd matriarch clashes with an intruder near the Chobe River.

Chris Johns/National Geographic Image Collection.

by Louis Herman on captive dolphins suggest that they understand syntax and word order, and can use abstract representations of objects, actions and concepts to guide their behavior, suggesting that they typically use symbols or language systems. The substantial brain

Springer's Homecoming

Howard E. Garrett

In the fall of 2001, an orphaned orca calf less than 2 years old became separated from her pod and wandered several hundred miles into Puget Sound, Washington. Orca pods travel day and night and may cover 100 miles each 24 hours. Young calves keep up with the pod by riding alongside their mothers' flanks, where the waves are parted and resistance is reduced. Without that help, the calf may not have had the strength and endurance to keep up. By mid-January, 2002, the calf had arrived at one of the best fishing holes for steelhead salmon in Puget Sound. Most scientists would not have believed that an orca calf so young could have caught fish and taken care of herself. Eventually identified as "Springer," from the A pod of the "Northern Resident" orca community, the calf found companionship in her chosen location within view of the Seattle skyline. Family bonds for both males and females among orcas of Springer's clan last a lifetime, unlike any other mammal known. Springer had visitors in boats and ferries full of fascinated commuters moving by at regular intervals.

On July 13, in an unprecedented and widely televised journey, the little calf was lifted aboard a high-speed catamaran in a container half-filled with water and ice for the 400-mile, 12-hour journey to her home waters off Telegraph Cove, British Columbia. Just days before Springer's arrival, her own family pod had returned from Alaska to their summer feeding grounds off the north tip of Vancouver Island. Springer's grandmother (A24) and great aunt (A11) were there.

As soon as she was lowered into her home waters, Springer vocalized loudly. At 1:30 a.m. Springer's family came near her pen. She became very excited and began to breach and call out loudly while nudging the net in the direction of her family's calls. Springer stayed very excited all night after the encounter.

Early the next afternoon, the whales returned to the bay. Immediately Springer made an "obvious vocal connection" with the pod, led by A11, Springer's great aunt. They milled and mingled for a few minutes, then lined up side by side facing Springer. They gradually approached within a few hundred yards of her net pen. Springer again showed that she wanted to go by vocalizing and pressing on the net, and the team agreed that the time was right. Springer grabbed a fish in her mouth, but did not eat it. The net was dropped. There was not a dry eye around the pen as Springer swam off, holding the salmon in her mouth, seemingly as an offering or a sign of trust.

She went charging off toward the pod, but then, about 100 yards away from them, she stopped and played with some kelp. Apparently rejoining your orca pod after a long absence is not automatic, even when you know the right calls. The pod moved slowly to the east while Springer turned gradually westward. Later Springer turned and followed a few hundred yards behind her pod.

Four days after her release she was seen sandwiched between a 16-year-old orphaned female named A51 and her younger brother. She stayed very close to them for over a week. Twice the female intervened as Springer headed off toward boats. Apparently the rule is: Real whales don't play with boats. In August and September Springer was consistently in the company of her grandmother's sister and other members of the A4 pod. Springer's reunification with her family seems to be a complete success. She is finally back among those who really know her.

See also Culture—*Orangutan Culture*
Culture—*Whale Culture and Conservation*
Dolphins—*Dolphin Behavior and Communication*

anatomy found in most cetaceans contributes to the implication that whales and dolphins live as members of cultural communities.

Ethnographic studies show that there are at least two different forms of orcas, or killer whales, inhabiting the waters around Vancouver Island, Canada, and the inland waters of Washington State in the United States. The two forms, known as "residents" and "tran-sients," use the same habitat but differ in vocalizations, diet, appearance, behavior, social structure, and genetic makeup. *Residents* eat only fish and squid and live in stable matrilineal *pods* averaging 12 animals—all offspring remain with their mothers for life and there is no known case of individuals changing pods in over 30 years of study. Unlike residents, *transient* orcas eat only marine mammals and live in smaller pods, averaging just three animals per pod, which allows more efficient predation. Transient orcas occasionally leave their natal pods and travel with other transient groups. Even then, they maintain their family's vocal dialect, indicating they may resume contact.

A5 pod of the Northern Resident orca community in a resting pattern.
© Joe Alicia.

In 2001 Luke Rendell and Hal Whitehead suggested: ". . . there is good evidence for cultural transmission in several cetacean species. There is observational evidence for imitation and teaching in killer whales. . . . The complex and stable vocal and behavioural cultures of sympatric groups of killer whales (*Orcinus orca*) appear to have no parallel outside humans and represent an independent evolution of cultural faculties." (309)

The distinctive vocal traditions found in each orca community provide the most compelling support for this conclusion. Each pod of resident orcas uses a characteristic set of 7–17 discrete calls, which are maintained despite extensive associations between pods. Some pods share up to 10 calls, and pods which share calls can be grouped together in acoustic *clans*, suggesting another level of population structure. Canadian researcher John Ford found four distinct clans within two resident communities, suggesting that the observed pattern of call variation is a result of dialects being passed down the generations through vocal learning while being modified over time. Thus, given the lack of dispersal,

Springer, the orphaned orca is welcomed back to her family. Left to right: A61, A51, A73 (Springer), A43, A60, A69— September 22, 2002.
© *Ellen Hartlmeier.*

acoustic clans appear to reflect common matrilineal ancestry, and the number of calls any two pods share reflects their relatedness, similar to the way human dialects and accents reflect relatedness.

In addition to the orcas' varying vocal traditions, they demonstrate vastly different food choices, variations in group size and different behavioral repertoires. Southern resident orcas sometimes perform a ritual when greeting one another, in which any combination of two or three pods line up at the surface facing the other, then slowly swim toward each other. Usually this is followed by "intermingling" in which members of different pods form tight clusters comprised of 10 to 15 whales, each cluster including members of all pods present, sliding and rolling around each other. Such ceremonies have not been observed in any other community of orcas. Members of one orca community in Canada often take turns rubbing their bellies and sides along a shallow shelf containing small round pebbles.

Some examples of cultural behavior in cetaceans are obvious, while other indications appear only as the result of long-term research. The combination of direct observation and long-term study has yielded some fascinating results. Some bottlenose dolphins in Shark Bay, Australia, carry sponges on their rostra (snouts), possibly to forage for fish buried in the sand. All members of the population experience the same mixed habitat, so ecological explanations for this behavioral variation can be discounted. Other members of the population are seemingly aware of the technique, as evidenced by the occasional observations of sponging in other individuals, but they do not fully adopt it; thus variation is unlikely due to any genetic ability. However, the calf of one of the regular spongers itself uses sponges, suggesting cultural transmission from mother to offspring.

Cultural learning has survival value and is now understood to occur among many species in the animal kingdom. In a few species there are well-defined behavioral differences between groups that occupy the same habitat which indicates the transmission of culture throughout the group and across generations. Humans have always been biologically similar to other animals—now we know there are cultural similarities as well.

Further Resources

De Waal, F. 2001. *The Ape and the Sushi Master: Cultural Reflections of a Primatologist.* New York: Basic Books.

Ford, J. K. B., Ellis, G. M., Balcomb, III, K. C. 1994. *Killer Whales: The Natural History and Genealogy of Orcinus Orca in British Columbia and Washington State.* Seattle, WA: University of Washington Press.

Goodall, J. 1990. *Through a Window: My Thirty Years with the Chimpanzees of Gombe.* Boston, MA: Houghton Mifflin.

McGrew, W. C. 1992. *Chimpanzee material culture: Implications for Human Evolution.* New York: Cambridge University Press.

Payne, K. 1998. *Silent Thunder, In the Presence of Elephants.* New York: Simon & Schuster.

Rendell, L. & Whitehead, H. 2001. *Culture in whales and dolphins.* Behavioral and Brain Sciences 24, 309–382.

Howard E. Garrett

■ Culture
Orangutan Culture

Distribution and Taxa

The orangutan, together with all of the great apes, is one of our closest relatives. Although this relationship is a not a simple linear ancestry, *Homo sapiens* and the orangutan had a common ancestor living about 10–12 million years ago. The ancestral orangutans were larger than extant ones, and they are thought to have lived a less arboreal existence, but over many generations evolutionary pressures, such as predation, might have forced them back into the trees. Their ground predators include the clouded leopard and the Sumatran tiger but, in general, orangutans have had relatively few predators, apart from humans.

Orangutans once ranged over much of China (as far as Beijing) and Southeast Asia, but over the last 10,000 years their terrain gradually shrunk, and now orangutans survive on only two islands: Sumatra and Borneo. Borneo is the third largest island on earth (746,951 km^2), a landmass bigger than the entire United Kingdom and France, but orangutans typically have only ever been found in less than half of this landmass, partly because they prefer lowlands and partly because some areas have too few fruiting trees. Increasing human population has also effectively reduced their living space and has done so usually in the very spots that orangutans prefer (peat swamp/lowlands).

Sumatra orangutans (*Pongo pygmaeus abelii*) live in and around the Gunung Leuser National Park, where there is an estimated number of 5,000–9,000 individuals. The other subspecies, the Bornean orangutan (*Pongo pygmaeus bornensii*), is found in the east-Malaysian states of Sabah and Sarawak, and on the Indonesian side of the island of Borneo, called Kalimantan. Wild orangutans declined from 80,000 in the 1980s to a quarter of this in the 1990s. By 1996 it was 23,000, and just a year later estimates were down by over a quarter of that number again to about 15,000. The actual number of protected wild orangutans (protected by virtue of living in one of the national parks) at the turn of the 21st century was barely 4,000. These figures are estimates because it is difficult to assess their numbers reliably. However, agreement has been reached that number of orangutans has been declining rapidly. Most research reports of the last decade also stress that even populations in protected areas may not be sustainable in the long run.

There is much variability among the two subspecies. Orangutans in Sarawak differ from those in Sabah to a greater extent than the Sumatran orangutans differ from the Bornean orangutans as a whole. They may even differ not only in behavior but also in their anatomy and physiology.

Orangutans are so well-adapted to their environment, they are often considered "umbrella" species, a barometer for all wildlife and the general health of the forest. Courtesy of Gisela Kaplan.

Culture in Our Relatives, The Orangutans

Anne E. Russon

Orangutans, the great red apes of Borneo and Sumatra, are about the last species one would expect to have cultures. It is not strange to find non-human species with cultures, if cultures are basically systems that allow members of a community to share behaviors that they learn. Bird song dialects are an example: Birds of some species learn different songs depending on the community in which they are raised. But one would hardly expect cultures in a semisolitary species, like the orangutan, because cultures make most sense in species that live in groups. The discovery that orangutans have cultures has therefore been important news to science.

The main clue that a species has cultures is that different communities learn different behaviors, for no obvious reason. Orangutans everywhere, for example, make a kiss–squeak vocalization when they are irritated, but only orangutans in Gunung Palung, Borneo hold leaves to their mouths while kiss–squeaking. This may make the sound louder but if so, it is hard to understand why orangutans elsewhere do not do it, too. Orangutans everywhere also build nests in which to sleep. In two communities, they make "raspberry" sounds while nest building—but orangutans in Suaq Balimbing, Sumatra "raspberry" just before finishing nest building, whereas those at Kinabatangan, Borneo "raspberry" just before starting. Nobody knows why. Orangutans rescued from captivity and rehabilitated to life in the wild also show hints of culture, in fads that run through groups of friends. One year several youngsters who regularly foraged together began banging termite nests against hard objects, apparently to break them open. This is odd because they already knew how to crack nests apart by hand and banging rarely, if ever, works.

A young orangutan breaks open a termite nest by banging it on a hard object.
Courtesy of Anne Russon.

The most important example concerns obtaining *Neesia* fruit seeds. These seeds are embedded in a mass of stinging hairs inside an extremely hard husk. Orangutans simply break *Neesia* husks open by force at Gunung Palung, but at Suaq Balimbing they make a stick probe, insert it through splits in the husk that open as the fruit ripens, scrape away the stinging hairs, and then pry the seeds out.

The *Neesia* tool technique is a complex skill, and it qualifies orangutan cultures as more complex than most nonhuman cultures. Beyond sharing simple signals and knowledge about habitat features, such as identifying foods or predators, orangutans, because they are exceptionally intelligent, can learn and share more complex skills and behaviors. And they live in especially flexible and tolerant societies so they have access to a richer mix-and-match of knowledgeable companions. This is not to say that sharing complex skills comes easily. Although highly intelligent, orangutans still need years to learn their most complex skills and extensive help to do it. Infants and juveniles get 7–8 years' help with the basics from their mothers—the longest of any primates. They often beg or steal half-prepared food, for instance (their mothers let them do it and may even tutor). Finishing an already started job is easier than starting from scratch, so begging and stealing nets them help with learning as well as food. Adolescents and near adults are highly sociable (adulthood is what turns them into lovers of solitude). They may travel together for days on end, making opportunities for sharing complex skills across the whole community. Orangutans probably share complex skills in bits and pieces, over long periods of time, with many different members of their community.

Orangutans share, with chimpanzees, cultures that are intermediate in complexity between those of other nonhuman species and humans. Their cultures suggest stepping stones that may have paved the way for the evolution of our more complex human cultures. Cultures of this intermediate complexity were thought to have evolved about 4–6 years ago, with the common ancestor of chimpanzees and humans. The orangutan discovery means they probably evolved in the common ancestor of chimpanzees, humans, and orangutans, about 14 million years ago. Our human qualities then have a much longer evolutionary history than we have believed, and more of them than we thought possible are shared with our close relatives, the great apes.

See also Culture

Further Resources

Fox, E. A., Sitompul, A. F., & van Schaik, C. P. 1999. *Intelligent tool use in wild Sumatran orangutans.* In: *The Mentalities of Gorillas and Orangutans: Comparative Perspectives* (Ed. by S. T. Parker, R. W. Mitchell & H. L. Miles), pp. 99–116. Cambridge, UK: Cambridge University Press.

Peters, H. 2001. *Tool use to modify vocalisations by wild orang-utans.* Folia Primatologica, 72(4), 242–244.

Russon, A. E. 2003. *Developmental perspectives on great ape traditions.* In: *Towards a Biology of Traditions: Models and Evidence* (Ed. by D. Fragaszy & S. Perry), pp. 329–364. Cambridge, UK: Cambridge University Press.

Van Schaik, C. P., Ancrenaz, M., Borgen, G., Galdikas, B., Knott, C. D., Singleton, I., Suzuki, A., Utami, S. & Merrill, M. 2003. *Orangutan cultures and the evolution of material culture.* Science, 299, 102–105.

Van Schaik, C. P., Deaner, R. O. & Merrill, M. Y. 1999. *The conditions for tool use in primates: Implications for the evolution of material culture.* Journal of Human Evolution, 36, 719–741.

Van Schaik, C. P. & Knott, C. 2001. *Geographic variation in tool use in* Neesia *fruits in orangutans.* American Journal of Physical Anthropology, 114, 331–342.

Variability in Behavior

Orangutans show a remarkable ability to adapt to different environments, and their cognitive abilities and sociality vary accordingly. Captive and ex-captive orangutans engage in social play and a wide range of other social behavior that has rarely been observed in wild populations. In the laboratory, orangutans display well-developed abilities to manipulate objects as tools and to solve problems. Orangutans undergoing rehabilitation in reserves in Indonesia and Malaysia display more tool using than do wild orangutans, who have been observed to use tools in their natural environment far less frequently than wild chimpanzees. Hence one cannot speak of "the orangutan." Knowing one orangutan may not help to understand all others, and the difference between a captive and a wild population of orangutans may be profound.

Problem Solving

Common to all orangutans, no matter what the environment, is their extraordinary problem-solving ability. Unbelievably, orangutans were once thought to be the least intelligent of all great apes, but we now know that orangutans can solve the same range of problems as chimpanzees. In certain fine manipulation tasks they may even exceed other apes. And this is perhaps not surprising because fine object manipulation is needed constantly in their natural habitat.

Tool Using

Orangutans show remarkable abilities to use ropes as tools. Tool manufacture has now been observed in wild orangutans on Sumatra involving fashioning of sticks to serve specific food extraction processes (such as enticing ants out of crevices or using a bamboo to soak up water from an otherwise inaccessible tree hole water supply) or using ready-made tools for the same purpose on Borneo. Other instances are the use of leaves. Rehabilitating orangutans have been observed choosing a large leaf and placing regurgitated food on it which they then take to quiet areas for feeding. They have also been observed using leaves to wipe their infants' feces from their coats. Wild orangutans may use some of their "tool using capacity" in social contexts: There is a form of "social tool use" in which gestures are used to obtain something or to direct the behavior of another. For example, an infant may take hold of the mother's hand and pull in order to obtain food being held by the mother. In this case the mother's hand is being used as a tool to obtain the food. Orangutans as young as 7–10 months of age use this method of sampling the food that the mother is eating. On the whole, we have far fewer examples of tool using of wild orangutans than of chimpanzees. This could mean either that this is because we have had fewer opportunities to observe orangutans in the wild (which is difficult to do) or that the rare sightings of tool using may indeed be a significant difference from other apes. Researchers believe that arboreality tends not to encourage tool use.

Many of the tool-using behaviors have been learned by imitating human behavior. At Tanjung Putting (West Kalimantan), for instance, the orangutans learned to imitate many human activities, including cooking, making fire, cutting wood, washing, and so on. On seeing workmen bridging the river with logs they soon adopted the behavior themselves.

Moving in Trees

Orangutans are the heaviest of any species living in trees. In fact, they are a canopy species. To live and move at dangerous heights carrying a substantial body weight requires constant alertness. One wrong step and they can fall to their deaths or break limbs. Locomotion by *clambering* (the way of moving and scrambling) is seen as one reason why orangutans have to have high cognitive ability. Orangutans move through the trees by carefully calculating which tree and branch can carry them, on which to swing and on which one to climb. Bridging gaps is often achieved by swinging back and forth on a flexible straight tree with increasing excursions until they manage to reach across the gap to the next tree. This behavior is not used for independent locomotion by infants until they are over 4 years old. Infants are actually afraid to climb. In rehabilitation, much time has to be spent giving them the confidence to do so. In the wild, they have to hang on to the mother and watch and learn, and even then it will take the best part of 10 years before they are strong and skilled enough to carry out such feats of bridging gaps.

Communication

We still know relatively little about the communication system of orangutans in comparison to those of the other apes. One of the first systematic attempts to describe vocalizations of orangutans was undertaken by J. MacKinnon. To date we know of about 18 different calls, and these are mostly distress or warning calls. There are two call types emitted only by adult males. One is the "fast call" heard after conflict or other contact. The other is the so-called "long call," a most distinctive and even spine-chilling sound. It is a high intensity bellow, produced by first filling the cheek pouches with air and then expelling this air over the vocal chords for a prolonged period. The long call can be heard over large distances, and it is commonly considered to be a territorial call either to keep other males away or to attract sexually receptive females into the male's territory. Infants produce a range of vocalizations of their own, including high-frequency distress calls as well as low-intensity purring, contentment calls. Communication also occurs via body gestures and sometimes by using tools. The presence of human observers often leads wild orangutans to throw sticks at researchers, gesticulating or jumping up and down on a branch. A direct stare is also used at times, and adult males, especially in Sumatra, at times engage in branch breaking and leaf thrashing in a similar manner as has been observed in alpha male chimpanzees.

The face is another important site for communication. Orangutans appear to display many of the same facial expressions as other primates, including humans. They show the relaxed open mouth display used in play and the bared teeth display of submission. There are threat displays with wide open mouths. According to MacKinnon, fear is expressed by drawing back the sides of the mouth, exposing the teeth and making a grimace. Mild worry is expressed by pouting of lips, and a threat is accompanied by a "trumpet" mouth (shaping the lips into a trumpet) and deep grunts followed by gulps.

Another important form of communication is eye gazing. Juveniles may beg for food from the mother by shifting their gaze back and forth between the mother's eyes and the food item, as also occurs in chimpanzees. In human contact, orangutans largely avert their gaze during interaction with a trainer. Kaplan and Rogers (2000) found that eye gazing occurs in age classes, and that various parts of the body are focused upon. Adults may communicate with juveniles in complex ways that include limbs, the body and to a significant extent the face. Infants look at their mother's faces, but not exclusively, whereas juveniles

In spite of the variability among the two subspecies of orangutans, they share a number of cultural similarities with homo sapiens.
Courtesy of Gisela Kaplan.

observe the faces of other juveniles and may be gauging emotional states and readiness to play or fight. Each age group of orangutans observed attended to different signals. The research also established that Bornean orangutans tend to use sideway gazing far more often than direct looking and this may enhance the potential of eye-gaze following by conspecifics (members of the same species) because it exposes a large area of eye white. At the same time sideways looking and avoidance of direct gazing for prolonged periods is a characteristic of social communication in Bornean orangutans and one that apparently differentiates them from the other great apes.

There have also been some attempts (five by the mid-1990s) to teach American sign language to orangutans. One orangutan named Chantek learned 140 signs, and he invented extra signs. His vocabulary increased over the period of the study, and each day he regularly used from one- to two-thirds of his existing vocabulary. The rate at which he acquired new signs was marginally better than that found previously in the language-trained chimpanzees, such as Nim. Chantek's comprehension abilities were also tested on the standard Bayley Scale for Infant (mental) Development, and other tests at 5 years old showed his abilities equivalent to a 4-year old child. Of course, such tests are partly done to measure apes against humans. One could not put a 4-year-old child into the forest on its own because it would die, but a 5-year-old orangutan, if reared in the wild, has some chance of survival on its own. In other words, the cognitive abilities we measure as humans are not the entire skill and knowledge base that orangutans need to have in order to survive to adulthood.

Development

The main social unit of orangutans is the mother with her offspring who rears her offspring, one at a time, and she does so completely on her own. During the first 2 years the infant is carried by the mother, with increasing individual explorations by the infant of nearby branches. The infants get weaned at around 2 years of age, but juveniles stay with the mother until they are about 6 or even 7 years of age. Orangutans enter a subadult stage in which they may occasionally meet other youngsters, and they usually do not reproduce until well in their teens—about 12 years of age for females and 15 or older in males. Subadulthood in males is often defined as the period between the ages of 8 to 15 years of age, after which time they may rapidly grow the pronounced cheek pads so distinctive in the Borneon orangutan (*Pongo pygmaeus pygmaeus*) as well as the full laryngeal sac which enables the male orangutan to make calls that are audible for miles. This means that orangutans have extremely long childhoods followed by subadulthood, and all this time they are learning about their environment and perfecting their skills before they start reproducing.

Play

In the orangutan, we find plenty of social play in captive and rehabilitating individuals but there is relatively little evidence of play in the natural environment. Its occurrence is rudimentary compared to chimpanzees and gorillas. Free-ranging infants in their natural habitat often have just one companion, their mother, who will play only rarely. Play behavior, to which we tend to ascribe important social and survival functions, is dependent on group interaction. This is true of nonhuman primates and humans. In play, infants are said to develop the skills that they will later need for survival. Although orangutan juveniles show the ability to play extensively in unnatural crowding and rehabilitation situations, it would be difficult to maintain that their survival depends on play learning. When orangutans gather on fruiting trees, the infants play with each other.

Mating and Reproduction

Adult males and females form *consortships* (brief, but intense associations) that last for about 2 weeks, and such relationships are often initiated by the female. During this time, the pair mates frequently, and the orangutans often show great tenderness towards each other. Thereafter, the male goes his own way and will have no further part in the birth or in the raising of the infant. Sexual intercourse in orangutans lasts longer than in any of the great apes and is always executed in the frontal facing position, unlike mating practices in other great apes. For females, onset of sexual maturity is around 7–8 years of age (although in captivity often much earlier) and menopause occurs in the mid- to late-40s, as in many human females. The menstrual cycle itself is identical to the human menstrual cycle (22–30 days) and, unlike other primates, there are no visible external signs when female orangutans are in estrus. Both orangutan and human females have the longest pregnancies of any of the great apes. In both, the gestation period is about 9 months. Pregnancy generally follows the same patterns as in humans. Birth intervals of all great apes are relatively long. For mountain gorillas these are about 4–5 years, and for chimpanzees 6 years, whereas for orangutans it is about 8 years. Orangutan females usually raise no more than 4 offspring during a lifetime.

Lifestyle

Orangutans differ substantially from other primates, and especially from the other great apes, in their social organization. Orangutans are less well described by referring to troops, domination hierarchies, or communities. They tend to move solitarily or in small *natal* groups (i.e., mother with offspring). It is difficult to impose upon the orangutan the kind of vocabulary with which we tend to describe human society, as a basically gregarious, group-living organization. Words like *competition* or *status*, for instance, can do relatively little to confer knowledge and meaning on orangutan behavior. However, there is a strong sex and age difference in sociality among orangutans. Subadult males will usually choose another female for company rather than a male. In general, males seem to be increasingly intolerant of each other with growing age. Bashing of leaves and snagging can be seen when one adult male hears another adult male call, and hostilities are greatest between adult males. Adolescent females are the most gregarious and either spend their time in their natal group (family of origin, including mother and possibly another sibling) or alone, and almost half their time with others, either with other adolescent females or with subadult males. Orangutan males are largely solitary. Adults do not spend more than around 10% of their time in social encounters with each other, and most of this contact is between consorting males and females. Adult females, by contrast, spend very little time alone because they have years of rearing of a single offspring. In addition, adult females may spend time with others outside the mother-offspring dyad or triad—largely with a consorting adult male and occasionally with subadults or other adult females.

Orangutans have loosely defined ranges rather than territories. The common assumption of territorial claims of male orangutans (i.e., defending territory and females), has only rarely been observed in Bornean orangutans. This is an assumption taken from chimpanzee research where increased intolerance and territorial behavior of males has been shown. Rather, there is evidence as Galdikas suggests that adult males defend their own *personal* space rather than a large area that may be defined as a territorial space. Orangutans like best to live in lowland areas that have plenty of fruiting trees but, depending on geography and opportunity, can be seasonally seminomadic. Nomadic behavior is regulated by the fruiting times of the trees on which they feed. Larger congregations of orangutans occur when a favorite tree (usually a fig or durian tree) ripens. Hence, we usually distinguish among three patterns of lifestyle:

1. a sedentary lifestyle consisting of resident orangutans who move across their very large terrains, but never leave it because the terrain holds all their needs;

2. the commuters who live and sleep largely in one terrain but commute to other areas when fruits are ripening elsewhere (and not present in their area); and

3. the nomads (or wanderers) who need to keep moving in search of food. This is the fate of many orangutans in areas degraded by logging, fire and human expansion.

Population Status

The difference between decline of orangutans in the past and the present is that today it is very dramatic. Before the beginning of the twentieth century there were probably over 200,000 in Borneo and Sumatra. In the first modern wave of destruction, orangutans were shot by big game hunters, adventurers, plantation owners and by scientists who shipped their carcasses back to European museums. They never recovered from this attack, and their populations became more and more fragmented as humans pushed into their territory and logging assumed large-scale proportions. The rainforests of Borneo and Sumatra, thought to be the oldest remaining on earth, older even than those of the Congo and the Amazon, have been under siege from the timber industry for a long time. Less than 15% of undisturbed low and highland rainforests remain at the turn of the twenty-first century, and that percentage is getting lower all the time. Even in national parks, chain saws can be heard day in and day out. Logging and agriculture are changing the whole region, and reforestation is increasingly an impossible dream.

How would this affect orangutans? Orangutans are flexible, resourceful, and not fussy about what they eat. They are master generalists who can exploit any food source to advantage (over 400 different edible items have been identified in their diet). They know where to find food anywhere from the ground to the top of the canopy. They extract roots, they eat berries growing on the ground or on stems, they find proteins such as ants and other small insects, reptiles and eggs. They eat leaves, fruits, nuts and even bark. They are highly intelligent in problem solving. They are resourceful and will move if they need to do so, becoming

---

Did You Know?

- Orangutans are the largest tree-living animal and the only great ape that lives in Asia.

- Males (317 lb/144 kg) are more than twice the size of females (143 lb/65 kg).

- In the wild they live for up to 40 years or more,

- Orangutans sleep in large nests. They build a fresh nest high in the trees each night. Nest-building practice commences at about 14 months of age, and is perfected by the end of juvenile development (at about 5 to 6 years). Infants and juveniles sleep with their mother, older juveniles make a "bed" for themselves. A nest is also constructed when an orangutan feels ill or a female is in the later stages of pregnancy and wishes to rest.

- Sometimes, orangutans build nest-like structures to provide them with shelter from heavy rain.

- More than half an orangutan's feeding time is spent eating fruit—wild figs and durians are their favorite food.

---

seminomadic or sedentary as circumstances require. Furthermore, they usually succeed in raising their offspring (unless these get poached). Indeed, so adapted are they to their environment that they are generally regarded as an *umbrella* species, a barometer for all wildlife and the general health of the forest. If orangutans are not doing well, no other species is. The fact that their numbers continue to decline sharply indicates that the forest and all its inhabitants are in trouble. Their plight is not just that of their own species but it represents the dramatic decline of the entire rainforest system in Southeast Asia.

Some wrongly think that rainforest indicates fertility of the soil and a fast rate of growth. Nothing could be further from the fact. The soil on Borneo and most tropical islands in the region is extremely poor, and rates of growth are therefore extremely slow. Some trees take 60 years to reach sexual maturity and 200 years to reach full height while most logging cycles are 25-year cycles, not allowing full growth or regeneration.

The forest contributes about $8 billion annually to the Indonesian economy alone and provides 80% of the plywood used in the United States home building industry. The international timber industry, led by western countries, is a keen customer. Illegal logging (and poaching of orangutans) has also increased over the past decades despite some intentions of the Indonesian and Malaysian governments to protect conservation sites. Roads also mean easier access for poachers. Without international demand for cheap timber items, there would be no need to cut down the oldest rainforest in the world. Often the same countries that buy the timber also try to help save the orangutan. We cannot blame the orangutans that they need a lot of space and many fruiting trees. These needs can only be accommodated in the wild.

Further Resources

Call, J. 1999. *Levels of imitation and cognitive mechanisms in orangutans*. In: *The Mentalities of Gorillas and Orangutans* (Ed. by S. T. Parker, R. W. Mitchell, & H. L. Miles), pp. 316–341. Cambridge: Cambridge University Press.

Custance, D. M., Whiten, A., Sambrook, T., & Galdikas, B. M. F. 2001. *Testing for social learning in the "artificial fruit" processing of wildborn orangutans* (Pongo pygmaeus), *Tanjung Putting, Indonesia*. Animal Cognition, 4, 305–313.

Galdikas, B. M. F. & Insley, S. J. 1988. *The fast call of the adult male orangutan.* Journal of Mammalogy, 69, 371–375.

Kaplan, G. & Rogers, L. J. 2000. *The Orang-utans: Their Evolution, Behavior, and Future.* Cambridge, MA: Perseus Publishing.

Kaplan, G. & Rogers, L. J. 2002. *Patterns of eye gazing in orangutans,* International Journal of Primatology, 23, 501–526.

MacKinnon, J. 1974. *The behaviour and ecology of wild orang-utans* (Pongo pymaeus). Animal Behaviour, 22, 3–74.

MacKinnon, J. 1974. *In Search of the Red Ape.* London: William Collins.

Parker, S. T. & K. R. Gibson (Eds.) 1990. *"Language" and Intelligence in Monkeys and Apes: Comparative Developmental Perspectives,* Cambridge: Cambridge University Press.
     See especially pp. 356–378 and pp. 511–539.

Rijksen, H. D. & Meijaard, E. 1999. *Our Vanishing Relative. The Status of Wild Orang-utans at the Close of the Twentieth Century.* Dordrecht, Netherlands: Kluwer Academic Publishers.

Rogers, L. J. & Kaplan, G. 1996. *Hand preferences and other lateral biases in rehabilitant orang-utans* (Pongo pygmaeus pygmaeus), Animal Behaviour, 51, 13–25.

Rogers, L. J. & Kaplan, G. (Eds.). 2003. *Comparative Vertebrate Cognition: Are Primates Superior to Non-primates.* Kluwer Primatology Series: *Developments in Primatology: Progress and Prospect.* New York: Kluwer Academic/Plenum Publishers.

Russon, A. E. 1999. *Orangutans' imitation of tool use: A cognitive interpretation.* In: *Mentalities of Gorillas and Orangutans* (Ed. by S. T. Parker, R. W. Mitchell, & H. L. Miles), pp. 119–145. Cambridge: Cambridge University Press.

Small, M. (Ed.). 1984. *Female Primates: Studies by Women Primatologists.* New York: Alan R. Liss. (especially pp. 217–235.)

Smith, E. O. (Ed.) 1978. *Social Play in Primates.* New York: Academic Press (pp. 113–142).

Snowdon, C., Brown, C. H. & M. M. Petersen, (Eds.) 1982. *Primate Communication.* Cambridge: Cambridge University Press.

teBoekhorst, I. J. A. 1989. *The social organization of the orangutan pongo-pygmaeus in comparison with other apes.* Lutra, 32, 74–77.

Van Schaik, C. P., Fox, E. A., & Sitompul, A. F. 1996. *Manufacture and use of tools in wild Sumatran orangutans.* Naturwissenschaften, 83, 186–188.

Van Schaik, C. P., & Knott, C. 2001. *Geographic variation in tool use in* Neesia *fruits in orangutans.* American Journal of Physical Anthropology, 114, 331–342.

Gisela Kaplan

■ Culture
Whale Culture and Conservation

"Culture" and "conservation" both rate high in the consciousness of modern humans. But they are usually juxtaposed in only one context, the conservation of valued elements of human culture, such as art or rare languages. I believe that *culture*—behavior shared by groups of individuals which is transmitted by social learning—is not restricted to humans, and that the biodiversity that conservationists strive to conserve includes this culture as well as genes.

One of the groups of animals in which culture is most recognized is Cetacea, the whales and dolphins. Whale culture possesses some special features which enhance its relationship with conservation. One of these is sheer scale. For instance, the songs of the humpback whale are ever-changing, but within an entire ocean basin the preferred song changes in lockstep. An even more important feature is *cultural sympatry*. Culturally-distinct groups of members of the same whale species are frequently found using the same areas, creating multicultural societies and conservation dilemmas.

The most controversial case is that of the orcas (or killer whales) which live off British Columbia and Washington state, crossing the United States–Canada border. This orca population is structured at several levels. There are mammal-eating and fish-eating "types" which differ morphologically, behaviorally, genetically, and acoustically. There are "communities" of fish-eaters which have distinctive ways of life and rarely interact, and within communities there are acoustically-distinct "clans," "pods," and "matrilineal groups." At all of these levels the differences between sets of orcas seem to be primarily cultural in origin and nature.

All the orcas are threatened, but the nature and severity of the threat varies. The mammal eaters have accumulated alarming levels of chemical pollutants, while the fish eaters have seen their food resources, primarily salmon, decimated. The "southern resident" community has been declining over the past 10 years. The reasons for this are not clear, but might include extraordinary levels of attention by whale watchers during the summer months.

The Canadian Committee on the Status of Endangered Wildlife in Canada implicitly recognized the importance of these cultural differences, dividing these orcas into three "nationally significant populations," and classifying the southern resident community as "endangered," the northern resident community as "threatened," and the mammal eaters as "threatened." The United States authorities took the opposite approach, dismissing the cultural differences, lumping the communities and types into a "healthy" North Pacific orca population, and so substantially diminishing the legal protection that can be given to the southern residents when they are on the United States side of the border. The case is now in court.

As a second example, consider the sperm whale. Females of this huge nomadic species are always found in groups of about 20 animals. When the animals socialize they communicate using patterns of clicks. These patterns are characteristic of acoustic clans, each of which may contain tens of thousands of animals and cover thousands of miles. Off the Galápagos Islands groups come from two principal clans: the "regular" clan which make regularly spaced patterns (e.g., "click-click-click-click"); and the "plus-one" clan who insert a pause before the final click (e.g., "click-click-click-click-pause-click"). Groups of the two clans behave distinctively, the regular groups making frequent turns and coming in close to the islands whereas the plus-one groups stay further from shore and move in straighter lines. These distinctions seem to be cultural. However, they lead to important differences in how the animals interact with their environment. The regular groups' behavior seems better suited to normal Galápagos conditions, and they have substantially higher feeding success. However, when El Niño strikes, the waters warm and become unproductive. Both clans suffer, but now the plus-one groups have the advantage. With global warming it is likely that the frequency of El Niños will increase, and median conditions will become more like those of an El Niño. Preserving the cultural diversity, and thus behavioral diversity, of sperm whales will help them survive the coming changes.

Despite the paucity of behavioral studies of cetaceans, these are just two examples among many. The behavior, survival and identity of an individual whale depends strongly on what she and her social partners have learned from each other—their culture. If we want to conserve whales as whales, we must conserve whale culture.

See also Culture
      Dolphins—*Dolphin Behavior and Communication*

Further Resources

Boran, J. R., & S. L. Heimlich. 1999. *Social learning in cetaceans: hunting, hearing and hierarchies.* Symposium of the Zoological Society, London 73, 282–307.

Ford, J. K. B., G. M. Ellis, & K. C. Balcomb. 2000. *Killer Whales,* 2nd edn. Vancouver, B.C.: UBC Press.

Rendell, L., & H. Whitehead. 2001. *Culture in whales and dolphins.* Behavioral and Brain Sciences, 24, 309–324.

Rendell, L., & H. Whitehead. 2003. *Vocal clans in sperm whales* (Physeter macrocephalus). Proceeding of the Royal Society of London B 270, 225–231.

Whitehead, H., & L. Rendell. 2004. *Movements, habitat use and feeding success of cultural clans of South Pacific sperm whales.* Journal of Animal Ecology, 73, 190–196.

Yurk, H. 2003. *Do killer whales have culture?* In: *Animal Social Complexity: Intelligence, Culture, and Individualized Societies* (Ed. by F. B. M. de Waal & P. L. Tyack), pp. 465–67. Cambridge, MA: Harvard University Press.

Hal Whitehead

■ |Curiosity

> *All animals feel Wonder, and many exhibit Curiosity . . . I then placed a live snake in a paper bag, with the mouth loosely closed, in one of the larger compartments [of the monkey-house at the London Zoo] . . . monkey after monkey, with head raised high and turned on one side, could not resist taking a momentary peep at the dreadful object lying quietly at the bottom.*
>
> CHARLES DARWIN, THE DESCENT OF MAN (1873), P. 72.

Curiosity, along with the related term *exploration*, refers to a persistent tendency to seek out parts of the environment that are new or not completely understood, and to try to find out more about them. Scientists often say that curiosity is the reason they became scientists, and it's probably part of the reason you picked up this book.

Humans have recognized that curiosity exists for a long time. For example, the explorer Meriwether Lewis (of the Lewis and Clark expedition that explored the Western United States in 1804–1806), even wrote about animal curiosity in his journal:

> The buffalo Elk and Antelope are so gentle that we pass near them while feeding, without apearing [sic] to excite any alarm among them, and when we attract their attention, they frequently approach us more nearly to discover what we are, and in some instances pursue us a considerable distance . . . —April 25, 1805

Many different types of animals have demonstrated curious behavior, including examples from every group of *vertebrates* (animals with backbones).

Curiosity is one of the biggest unsolved mysteries about behavior. Many theories—explanations for behavior—that say we only act to solve a problem or fix a deficiency (like being hungry or cold) run aground because they can't explain why animals seek out novelty. What's more, laboratory experiments have shown that animals will work hard for a chance to explore a new place. The idea captured in the famous bit of folk wisdom that says "curiosity killed the cat" adds to the puzzle. Why should an animal leave its home, where it is relatively safe, and venture out into the world out of curiosity, where it runs the risk of injury or being eaten by a predator? Finally, when an animals spends time and energy on any activity, there is an "opportunity cost," meaning that they can't do other things with that time and energy. For example, if a human spends an hour watching television, that hour is no longer available for eating, dating, or sleeping. This applies to curiosity: An animal that spends time investigating its world is choosing exploration and investigation instead of other ways it might spend its time, and these other things it could have been doing might be things that would have led to something that was useful right away.

Laboratory experiments have shown that animals that are exploring aren't simply looking for food. Animals, even hungry ones, continue to investigate novel objects after they learn that these objects aren't edible. Animals will also work for a long time at a dull task for the chance to look at something interesting, and will pass through uncomfortable conditions when the only reward is the opportunity to investigate a new place. There must be something good that happens when an animal acts this way, something to offset the costs. What might that be?

The most likely explanation is that curious animals learn things that are useful when they need to solve life's problems. If an animal isn't busy solving an immediate problem such as finding food, shelter, water, or a mate, it might be beneficial for them to use their time learning more about their environment. The information gained this way may or may not be important, but there's at least a chance that it will come in handy later. Animals that use their time this way could be better off than those that don't, as long as the cost of being curious isn't too high.

This trade-off between costs and benefits is important, and plays an important role in many decisions about how to act. For example, rats that explored more in one experiment were more likely to be eaten by a predator than rats who were more timid and stayed in sheltered locations. On the other hand, in a different laboratory experiment, being curious had real benefits. Rats that explored more found an escape route from an experimental environment, and rats that explored less didn't find this route. Later, when the rats (who were tested individually) were chased by a radio-controlled simulated predator, the rats that had found the escape route were able to escape, while the other rats did not escape. This experiment shows that the information animals gather while they are exploring can be important to them in facing challenges that occur later. We don't know exactly how animals balance costs and benefits in a specific situation; it's a problem to be solved by future studies.

The idea that exploration might be a way to create chances for learning gets support from experiments showing that the way animals investigate their environments is affected by their previous experience, and is not simply an expression of some preprogrammed behavior. Other studies have shown that things animals learn in one situation actually change the way those animals investigate the next new object they encounter. Still other experiments discovered that the specific component behaviors animals use when investigating an object (such as sniffing an object or placing one paw on it) are put together into sequences. Animals put these sequences together in ways that can be described by the same sorts of rules used to describe grammar in language, and these "grammars" are changed by learning experiences. All of this points to an important connection between exploration and learning.

Research has also shown us that there are two types of activity that are often lumped together under the heading of curiosity. One type is sometimes called *stimulus seeking* (searching for stimulation). This might be similar to leaving your house so that you can walk around the neighborhood and look for something to do. An animal that engages in stimulus seeking is creating the possiblity that it might encounter a situation where learning could occur, but it is also increasing the chance that it could encounter danger. The other type of behavior happens in response to an event or detecting some kind of sensory stimulation; the animal's attention is drawn and it changes its behavior so that it can interact with the source of this stimulus. This is known as *investigation*.

One important feature in the environment that leads to curiosity is the element of novelty or unfamiliarity. Animals *habituate* (stop paying attention to) objects and events that are completely familiar, because they don't provide any new information. Novelty, on the other hand, causes arousal and perhaps sometimes even stress. When an object or event is novel, an animal has no basis for knowing whether it is potentially useful or dangerous. Curiosity is the response of an animal to novelty, and this can lead to investigation if the new thing doesn't seem like a threat (and withdrawal if it does).

Animal professionals have become more systematic about using animal curiosity to design healthier and more interesting environments for captive animals, such as those in zoos. Displaying curiosity may be one of the signs that an animal is getting an appropriate level of stimluation in its environment; on the other hand, animals that are not investigating their environment might be bored. This could be a sign of inadequate housing conditions. Recent studies with many different species lend support to this idea.

A better understanding of curiosity and the role it plays in how animals live their lives would make a major contribution to a complete understanding of animal behavior.

See also Exploratory Behavior—*Inquisitiveness in Animals*

Further Resources

Archer, J. E. & Birke, L. I. 1983. *Exploration in Animals and Humans*. New York: Van Nostrand Reinhold.
Berlyne, D. E. 1960. *Conflict, Arousal, and Curiosity*. New York: McGraw-Hill.
Glickman, S. E. 1971. *Curiosity has killed more mice than cats*. Psychology Today, 5(5), 55–56, 86
Renner, M. J. (1990). *Neglected aspects of exploratory and investigatory behavior*. Psychobiology, 18(1), 16–22.

Michael J. Renner